The Science of Disasters

Springer
Berlin
Heidelberg
New York
Barcelona
Hong Kong
London
Milan
Paris
Tokyo

Physics and Astronomy

ONLINE LIBRARY

http://www.springer.de/phys/

Armin Bunde   Jürgen Kropp
Hans Joachim Schellnhuber

The Science
of Disasters

Climate Disruptions, Heart Attacks,
and Market Crashes

With 170 Figures Including 57 Color Figures

Springer

Professor Armin Bunde
Justus-Liebig-Universität Giessen
Institut für Theoretische Physik
Heinrich-Buff-Ring 16
35392 Giessen, Germany
E-mail: bunde@physik.uni-giessen.de

Dr. Jürgen Kropp
Potsdam-Institut für Klimafolgenforschung
Postfach 601203
14412 Potsdam, Germany
E-mail: kropp@pik-potsdam.de

Professor Hans Joachim Schellnhuber
Potsdam-Institut für Klimafolgenforschung
Postfach 601203
14412 Potsdam, Germany
E-mail: john@pik-potsdam.de

Library of Congress Cataloging-in-Publication Data applied for.

Die Deutsche Bibliothek - CIP-Einheitsaufnahme

The science of disasters: climate disruptions, heart attacks, and market crashes/
Armin Bunde; Jürgen Kropp; Hans Joachim Schellnhuber. –
Berlin; Heidelberg; New York; Barcelona; Hong Kong; London; Milan; Paris; Tokyo:
Springer, 2002 (Physics and astronomy online library) ISBN 3-540-41324-3

ISBN 3-540-41324-3 Springer-Verlag Berlin Heidelberg New York

Springer-Verlag Berlin Heidelberg New York
a member of BertelsmannSpringer Science+Business Media GmbH

http://www.springer.de

© Springer-Verlag Berlin Heidelberg 2002
Printed in Germany

Typesetting: Camera ready copy by the editors
Cover design: *de'blik, Berlin*

Printed on acid-free paper   SPIN 10785694   57/3141/tr   5 4 3 2 1 0

Preface

With this work authors, editors, and the publisher offer a very stimulating and modern book to a broad audience. It is stimulating and modern for several reasons. First it contributes to the culture of monographs as it has developed today. Second, it dwells on interdisciplinary cooperation. Third, it reflects recent progress in nonlinear dynamics of complex systems, one of the most modern fields in science. Fourth, it immediately captures the reader's attention: when she/he starts to turn over the leaves, the reader will be momentarily attracted by the rich variety of topical issues and fields of problems, will feel stimulated by the most inspiring introductory presentations as well as by the many details of the exciting problems, which he finds in 14 chapters. He will be seized by all the ideas, results, methods - and cannot easily stop reading.

We scientists are realizing more and more that we have to put more emphasis on monographs. Their part is a summarizing overview on original work, a unifying presentation of a larger field, is eliminating the sometimes meandering trails of research, short-cutting its unnecessary detours. It is to open up new associations and to provide a convenient learning for the newcomer. To share the enormous effort of writing a monograph, authors or teams of authors today each write a single, coherent chapter. Together all chapters add up to an even richer presentation. This volume is another successful example of this type of modern monographs.

There is common agreement that the frontiers of research are particularly exciting if interdisciplinary problems are addressed. Many interesting questions need the knowledge of experts from various different fields to find a satisfactory answer. Interdisciplinary work may and often does even start a new discipline. But it always starts by proper combination of existing excellence in the fields that contribute. Also, the different views of different specialists on the same subject lead to new insight, new conceptions, which no single discipline would have developed from its own. The present work exemplifies all this very clearly. Physicists, meteorologists, biologists, economists, etc. describe their answers in four main parts of the volume: Basic methods and notions, features of climate predictability, biodynamics, and econophysics. Inter- or better multidisciplinarity in its purest!

Science has developed recently an impressive set of characteristic notions and approaches to deal with complex systems. These all live under the headline of nonlinear dynamics and/or deterministic chaos. Nonlinearity here is not just a deformation of a basically linear behavior, characterized by the conceptions of superposition and mode dynamics, with some quantitative effects. It lends, in contrast, qualitatively new and specific signatures to the phenomena observed and to their understanding by theoretical models. Concepts as scaling, fractality (mono- or multi-), geometrical symmetry, etc., are features of strong nonlinearity, of a new quality. In this 'additive monograph' very different but most fascinating questions are explored with the nonlinear, complex systems' approaches and their methods: climate, physiology and heart phenomena, sensory biology, behavioral science, financial markets. And they can be explored by the same methods because of the universality, which we have found in nonlinear dynamics. Universality is one of the big surprises and lies at the very heart of nonlinear complexity. Successful methods are time series analysis, wavelet expansion, entropy, complexity measures, etc. These allow to derive statistical description for unpredictable macroscopic phenomena like the weather or the climate, to sharpen the idea of prediction itself, prediction despite of chaos and sensitive dependence on the starting conditions. They give rise to a nonequilibrium thermodynamics despite of the lack of time scale separation as we know it from micro versus macro systems. Universality seems to enable us to treat clouds, DNA sequences, or stock markets with the same apparatus of methods. Detrending analysis opens the possibility to separate general trends from the sometimes huge fluctuations in the time development. The extremely successful conception of linear response to external forcing can be transferred from condensed matter physics to nonlinear complex systems, etc., etc. A most inspiring, beautiful ballpark of nonlinear science methods together with applications in several fields of urgent interest, which use it, is offered in this volume. The reader will also notice surprising feedback from the various applications to the development of the basic concepts.

This finishes my rush through the arguments, why this is a stimulating and modern book and what the reader is about to learn from it. I am sure that it will find a large audience, that it will induce more, new multidisciplinary cooperation, will be a stimulating introduction for newcomers and also be a handbook for the expert researchers.

$$*\qquad*\qquad*$$

Marburg

Siegfried Grossmann
May 2002

Introductory Remarks

Everyday life is pervaded by critical phenomena that unfold on specific spatiotemporal scales. Typical examples are stock markets crashes, panic dynamics in overcrowded locations, the outbreak of cancer, or the occurrence of sudden cardiac death. The potential breakdown of the Atlantic thermohaline circulation, the abrupt shift of climate regimes, or the irreversible decline of natural resources due to overexploitation, on the other hand, are some of the precarious developments associated with 'Global Change', i.e. the present transformation of the planetary environment by humanity. All these singular phenomena might have, whenever they occur, tremendous negative effects on our lives, and they often result from super-complex dynamics involving the interaction of innumerable systems parts.

In non-scientific terms, such events are commonly referred to as crises, catastrophes or disasters. Their scientific investigation was restricted - until very recently - to disciplinary communities accumulating utterly specific knowledge about the specific phenomenon in question but ignoring the structural generalities this phenomenon may share with critical dynamics observed in many other fields. Thus heart attacks 'belonged' to physiologists as much as stock market crashes 'belonged' to investment analysts, and there seemed to exist no bridge to connect so fundamentally distinct camps. Driven and motivated by stunning progress in nonlinear complex systems science, however, scholars from various disciplines have now set out to realize pertinent similarities rather than differences, and to unravel what may ultimately be called the *universal anatomy of disaster*.

In fact, the so-emerging transdisciplinary scientific enterprise will be interested in all aspects of catastrophic systems failure: the primary causes as well as the generic repercussions; the potential precursor signals as well as the typical development of the phenomenon and its relaxation to a non-critical state; the formal analysis as well as all types of applications which may save thousands of human lives in the not-too-distant future. This enterprise will benefit tremendously from existing methodologies developed over recent decades in branches like structural stability theory or synergetics, but it will also have to generate novel schemes and techniques that are particularly suited to studying - and possibly predicting - singular behavior. One urgent reason for improving our understanding of disaster dynamics is certainly anthropogenic climatic change, the impacts of which are likely to be dominated by extreme events and abrupt changes in geophysical, ecological, and socioeconomic systems.

Considerations of this sort provided the motivation for the editors to organize, with the generous support of the Heraeus Foundation, a topical workshop called 'Facets of Universality in Climate, Biodynamics and Stock Markets' at Giessen in 1999. This meeting brought together researchers from rather separate, if not alienated communities, and nevertheless was a splendid success. Therefore, it was decided to preserve and even enrich the main insights and messages from that workshop by putting together a monograph that is sufficiently tutorial to introduce students to the general issue of disaster science and at the same time sufficiently elaborate to satisfy the curiosity of active investigators. In order to achieve this, mini-teams of workshop lecturers and participants were composed by the editors and invited to prepare a suite of review articles on the most relevant topics discussed. Thus a number of scholars became intellectually 'married' to others with whom they had never directly collaborated.

The emerging *science of disaster* will have to cover and plough a huge cognitive field, so the first steps have to explore the envisaged domain in a rather eclectic way. In fact, the Giessen Workshop focused on a few issues only, notably advanced analysis and interpretation schemes for empirical datasets (like continuous time series) recording transient and singular systems dynamics. In this context, it is crucially important to find out whether the available data exhibit pure or mixed scaling behavior, or whether they do not scale at all. On the other hand, the 'countdown to disaster' in many relevant systems may be characterized by log-periodic fluctuations that can even serve as a yardstick for measuring the temporal distance from the incipient crisis. The latter tool would actually constitute something like the philosopher's stone for disaster prevention. Quite generally, most sensitive and discriminative techniques have to be employed in order to identify typical 'fingerprints' in the given datasets. Novel analysis strategies are able, for instance, to distinguish extrinsic trends from intrinsic variability - something that is absolutely essential for the assessment of global warming records from the last decades and the attribution of climate change to anthropogenic interference. The adoption of other modern methodologies from statistical physics such as wavelet approaches, Hurst analysis, entropy concepts, and multifractal measures will be instrumental for the advancement of the new science.

When it comes to explaining the 'drivers of disasters', however, there is growing awareness that even mild and regular nonlinearities may cause disruptive systems responses. Thus there is certainly 'sensitive dependence' of catastrophe dynamics on structural details of the underlying causal web (re-placing the causal chain of the linear heuristic philosophy). Here we enter plain *terra incognita*.

After these words of both hope and caution, let us now briefly discuss how this book is organized. Corresponding to the topics of the Giessen Workshop, the book consists of four parts. The first part, written by Ebeling et al. (Chap. 1) and Arneodo et al. (Chap. 2) introduces important notions and analysis methods, i.a. the multifractal formalism that is particularly relevant when analyzing nonlinear complex systems.

The second part is devoted to the climate system, its predictability and the possible occurrence of disasters. First, in Chap. 3, Fraedrich and Schönwiese show how the conventional tools of meteorology can be used to obtain a comprehensive view of the European climate. In Chap. 4, the reader is introduced by Hasselmann to the principles behind state-of-the-art climate modeling and obtains an insight into the difficulties one is faced with when attempting to predict future climate development. In Chap. 5, Bunde et al. discuss the occurrence of long-range correlations in the atmosphere and show how they can be used as an (uncomfortable) test-bed for complex climate models. Finally, Chap. 6 by Kropp et al., describes possible disaster scenarios which might affect the entire Earth, and how management or prevention strategies can be derived if only uncertain process knowledge is on hand.

The third part of the book focuses on the wide field of biodynamics. Ivanov et al. (Chap. 7) and Liebovitch et al. (Chap. 8) use similar techniques as described in Chaps. 2 and 5 to analyze heartbeat records. They point out the existence of long-range correlations in heart rhythm and show how modern time series analysis can be used to distinguish between healthy and pathologic situations. In Chap. 9, Bunk et al. describe new methods of investigation that were originally developed for the analysis of astronomical images and time series, and are very helpful in the detection of 'biological disasters' such as skin cancer and other tumors. Chapter 10, by Moss and Braun, addresses the important problem of stochastic synchronization in sensory biology. Finally, in Chap. 11, Helbing et al. show that computer simulations are powerful tools to study the occurrence of panics in crowds, and give hints on how to avoid these situations.

The final part of the book directs the reader to the peculiarities of stock markets, including crashes. First, in Chap. 12, Mantegna and Stanley discuss key economic concepts such as arbitrage or the efficient market hypothesis and relate them to 'universal aspects' observed in the empirical analysis of stock price dynamics in financial markets. In Chap. 13, Lux and Ausloos review scaling phenomena in financial records and discuss the implications of scaling for the theoretical modeling of the price formation process. Finally, in the last chapter of this book, Sornette et al. present a set of models for market price fluctuations, and discuss in detail the onset of critical crashes and their precursor phenomena.

We wish to thank first and foremost the authors for their cooperation and many of our colleagues, in particular Dr. S. Brenner for many useful discussions. We also wish to thank Dr. C. Ascheron, P. Treiber and A. Duhm from Springer-Verlag for their continuous help during the preparation of this book. We hope that this monograph can be used as a textbook for graduate students, for teachers at universities preparing courses or seminars, and for researchers in a variety of disciplines who are about to encounter the phenomena described here in their own work.

*   *   *

Giessen, Potsdam
May 2002

Armin Bunde
Jürgen Kropp
Hans Joachim Schellnhuber

Contents

Part I. General

1. Entropy, Complexity, Predictability, and Data Analysis of Time Series and Letter Sequences 3

By W. Ebeling, L. Molgedey, J. Kurths, and U. Schwarz (With 9 Figures)

1.1	Introduction	3
1.2	Conditional Entropies and Predictability	4
1.3	Concepts of Complexity	6
1.4	Applications to Biosequences and Other Information Carriers ..	10
1.5	Applications of Entropy Concepts to Data Analysis	12
1.6	Applications of Complexity Concepts	16
1.7	Conclusion	21
References		23

2. Wavelet Based Multifractal Formalism: Applications to DNA Sequences, Satellite Images of the Cloud Structure, and Stock Market Data 27

By A. Arneodo, B. Audit, N. Decoster, J.-F. Muzy, and C. Vaillant (With 22 Figures)

2.1	Introduction	28
2.2	The Wavelet Transform Modulus Maxima Method for the Multifractal Analysis of 1D signals	32
2.3	Wavelet Based Fractal Analysis of DNA Sequences	46
2.4	The 2D Wavelet Transform Modulus Maxima Method for the Multifractal Analysis of Rough Surfaces...............	59

2.5   Application of the 2D WTMM Method
       to High-Resolution Satellite Images of Cloud Structure 75
2.6   Beyond Multifractal Analysis
       with Wavelet-Based Space-Scale Correlation Functions:
       Revealing a Causal Information Cascade
       in Stock Market Data 84
2.7   Conclusion 94
References 95

Part II. Climate Systems

---

3. Space-Time Variability of the European Climate ... 105

By K. Fraedrich and C.-D. Schönwiese (With 16 Figures)

3.1   Introduction 105
3.2   Time and Space Scales: Peaks, Gaps, and Scaling 106
3.3   Europe's Climate: Storm Tracks, Grosswetterlagen,
       and Climate Zones 113
3.4   Climate Trends: Europe at the End
       of the Twentieth Century 129
3.5   Conclusion 137
References 138

4. Is Climate Predictable? 141

By K. Hasselmann (With 17 Figures)

4.1   Introduction 141
4.2   Weather and Climate 142
4.3   Climate Prediction of the First Kind: ENSO 144
4.4   Stochastic Climate Models 147
4.5   Climate Predictions of the Second Kind:
       Global Warming 153
4.6   Linear Response Relations 156
4.7   Detection and Attribution of Climate Change 159
4.8   Nonlinear Signatures in Linear Response 163
4.9   Conclusion 165
References 167

5. **Atmospheric Persistence Analysis:**
   Novel Approaches and Applications 171

By A. Bunde, S. Havlin, E. Koscielny-Bunde, and H.J. Schellnhuber
(With 7 Figures)

5.1   Introduction 171
5.2   Analysis of Meteorological Methods 173
5.3   The Modeling Approach 175
5.4   Record Analysis: Detrending Techniques 176
5.5   Analysis of Temperature Records 181
5.6   Analysis of Simulated Temperature Records 184
5.7   Conclusion 187
References 189

6. **Assessment and Management of Critical Events:**
   The Breakdown of Marine Fisheries
   and The North Atlantic Thermohaline Circulation. 193

By J. Kropp, K. Zickfeld, and K. Eisenack (With 12 Figures)

6.1   Introduction 193
6.2   The Role of Market Mechanisms
      in Marine Resource Exploitation 195
6.3   Could Europe's Heating System
      be Threatened by Human Interference? 204
6.4   Conclusion 213
References 214

Part III. Biodynamics

7. **Fractal and Multifractal Approaches**
   in Physiology 219

By P.Ch. Ivanov, A.L. Goldberger, and H.E. Stanley (With 22 Figures)

7.1   Introduction 219
7.2   Limitations of Traditional Techniques 222
7.3   Monofractal Analysis 227
7.4   Multifractal Analysis 240
7.5   Conclusion 251
References 254

8. Physiological Relevance of Scaling of Heart Phenomena 259

By L.S. Liebovitch, T. Penzel, and J.W. Kantelhardt (With 11 Figures)

8.1   Introduction 259
8.2   Methods of Scaling Analysis 261
8.3   Heart Rate During Sleep 267
8.4   Timing Between Arrhythmic Events 277
8.5   Conclusion 279
References 280

9. Local Scaling Properties for Diagnostic Purposes .. 283

By W. Bunk, F. Jamitzky, R. Pompl, C. Räth, and G. Morfill
(With 16 Figures)

9.1   Introduction 283
9.2   Reductionism 284
9.3   Scaling Index Method 286
9.4   Applications 288
9.5   Conclusion 308
References 309

10. Unstable Periodic Orbits and Stochastic Synchronization in Sensory Biology 311

By F. Moss and H. Braun (With 10 Figures)

10.1   Introduction 311
10.2   Unstable Periodic Orbits in Physical and Biological Systems ... 319
10.3   Synchronization of Stable Periodic Orbits
        in the Paddlefish Electroreceptor
        with an External Periodic Stimulus 324
10.4   Conclusion 326
References 327

11. Crowd Disasters and Simulation of Panic Situations 331

By D. Helbing, I.J. Farkas, and T. Vicsek (With 9 Figures)

11.1   Introduction 331
11.2   Observations 335

11.3 Generalized Force Model of Pedestrian Motion 336
11.4 Simulation Results 338
11.5 Conclusions 347
References 349

Part IV. Nonlinear Economics

---

12. Investigations of Financial Markets Using Statistical Physics Methods 353

By R.N. Mantegna and H.E. Stanley (With 6 Figures)

12.1 Introduction 353
12.2 Econophysics 355
12.3 An Historical Note 357
12.4 Key Concepts...361
12.5 Idealized Systems in Physics and Finance..................363
12.6 Empirical Analysis 363
12.7 Collective Dynamics 367
12.8 Conclusion 368
References 369

13. Market Fluctuations I: Scaling, Multiscaling, and Their Possible Origins 373

By F. Lux and M. Ausloos (With 5 Figures)

13.1 Introduction 373
13.2 Scaling in the Probability Distribution of Returns 374
13.3 Temporal Dependence 384
13.4 Multiscaling, Multifractality, and Turbulence
      in Financial Markets......................................393
13.5 Explanations of Financial Scaling Laws 397
13.6 Conclusion 403
References 406

14. Market Fluctuations II: Multiplicative and Percolation Models, Size Effects, and Predictions 411

By D. Sornette, D. Stauffer, and H. Takayasu (With 7 Figures)

14.1 Stylized Facts of Financial Time Series 411
14.2 Fluctuations of Demand and Supply in Open Markets.........415

14.3 Percolation Models 421
14.4 Critical Crashes 427
14.5 Conclusion 432
References 433

Glossary 437

Subject Index 441

List of Contributors

Alain Arneodo
Centre de Recherche Paul Pascal
Université de Bordeaux
33600 Pessac, France
arneodo@crpp.u-bordeaux.fr

Benjamin Audit
Centre de Recherche Paul Pascal
Université de Bordeaux
33600 Pessac, France
audit@crpp.u-bordeaux.fr

Marcel C. Ausloos
Institut de Physique
Université de Liège
4000 Liège, Belgium
marcel.ausloos@ulg.ac.be

Hans A. Braun
Institut für Physiologie
Universität Marburg
36033 Marburg, Germany
braun@mailer.uni-marburg.de

Armin Bunde
Institut für Theoretische Physik
Justus-Liebig-Universität
35392 Giessen, Germany
bunde@physik.uni-giessen.de

Wolfram Bunk
Max-Planck-Institut für
extraterrestrische Physik
85741 Garching, Germany
whb@mpe.mpg.de

Werner Ebeling
Institut für Physik
Humboldt-Universität zu Berlin
10115 Berlin, Germany
ebeling@physik.hu-berlin.de

Klaus Eisenack
Potsdam Institut für
Klimafolgenforschung
14412 Potsdam, Germany
eisenack@pik-potsdam.de

Nicolas Decoster
Centre de Recherche Paul Pascal
Université de Bordeaux
33600 Pessac, France
decoster@crpp.u-bordeaux.fr

Illés J. Farkas
Department of Biological Physics
Eötvös University
1117 Budapest, Hungary
fij@angel.elte.hu

Klaus Fraedrich
Meteorologisches Institut
Universität Hamburg
20146 Hamburg, Germany
`fraedrich@dkrz.de`

Ary Goldberger
Cardiovascular Division
Havard Medical School
Massachusetts 02215, USA
`agoldber@bidmc.havard.edu`

Klaus Hasselmann
Max-Planck-Institut
für Meteorologie
20146 Hamburg, Germany
`klaus.hasselmann@dkrz.de`

Shlomo Havlin
Department of Physics
Bar-Ilan University
Ramat-Gan 52100, Israel
`havlin@ophir.ph.biu.ac.il`

Dirk Helbing
Institut für Wirtschaft
und Verkehr
Technische Universität Dresden
01062 Dresden, Germany
`helbing@trafficforum.org`

Plamen C. Ivanov
Center of Polymer Studies
and Department of Physics
Boston University
Boston, MA 02215, USA
`Plamen@argento.bu.edu`

Ferdinand Jamitzky
Max-Planck-Institut für
extraterrestrische Physik
85741 Garching, Germany
`fxj@mpe.mpg.de`

Jan W. Kantelhardt
Institut für Theoretische Physik
Justus-Liebig-Universität
35392 Giessen, Germany
`kantelhardt`
`@physik.uni-giessen.de`

Eva Koscielny-Bunde
Institut für Theoretische Physik
Justus-Liebig-Universität
35392 Giessen, Germany
`Eva.Koscielny-Bunde`
`@uni-giessen.de`

Jürgen Kropp
Potsdam Institut für
Klimafolgenforschung
14412 Potsdam, Germany
`kropp@pik-potsdam.de`

Jürgen Kurths
Institut für Physik
Universität Potsdam
14469 Potsdam, Germany
`juergen@agnld.uni-potsdam.de`

Larry S. Liebovitch
Center for Complex Systems
and Brain Sciences
Florida Atlantic University
Boca Raton, USA
`liebovitch@walt.ccs.fau.edu`

Thomas Lux
Fachbereich Ökonomie
Universität Kiel
24118 Kiel, Germany
`lux@bwl.uni-kiel.de`

Rosario N. Mantegna
Dipartimento di Energetica
e Tecnologie Relative
Universita' di Palermo
90128 Palermo, Italy
`mantegna@unipa.it`

Lutz Molgedey
Institut für Physik
Humboldt-Universität zu Berlin
10115 Berlin, Germany
molgedey@physik.hu-berlin.de

Gregor E. Morfill
Max-Planck-Institut für
extraterrestrische Physik
85741 Garching, Germany
gem@mpe.mpg.de

Frank Moss
Center for Neurodynamics
University of Missouri
St. Louis, MO 63121, USA
mossf@umsl.edu

Jean-Francois Muzy
Centre de Recherche Paul Pascal
Université de Bordeaux
33600 Pessac, France
muzy@crpp.u-bordeaux.fr

Thomas Penzel
Fachbereich Humanmedizin
Universität Marburg
35043 Marburg, Germany
penzel@mailer.uni-marburg.de

René Pompl
Max-Planck-Institut für
extraterrestrische Physik
85741 Garching, Germany
pompl@mpe.mpg.de

Christoph Räth
Max-Planck-Institut für
extraterrestrische Physik
85741 Garching, Germany
cwr@mpe.mpg.de

Hans Joachim Schellnhuber
Potsdam Institut für
Klimafolgenforschung

14412 Potsdam, Germany
john@pik-potsdam.de

Christian-D. Schönwiese
Institut für Meteorologie
und Geophysik
Goethe-Universität
60054 Frankfurt, Germany
schoenwiese
@meteor.uni-frankfurt.de

Udo Schwarz
Institut für Physik
Universität Potsdam
14469 Potsdam, Germany
USchwarz@agnld.uni-potsdam.de

Didier Sornette
Laboratoire de Physique
de la Matière Condensée
Université de Nice
06108 Nice Cedex, France
sornette@naxos.unice.fr

H. Eugene Stanley
Center of Polymer Studies
and Department of Physics
Boston University
Boston, MA 02215, USA
HES@BU.EDU

Dietrich Stauffer
Institut für Theoretische Physik
Universität Köln
50937 Köln, Germany
stauffer@thp.uni-koeln.de

Hideki Takayasu
Sony Computer Science
Laboratories Inc.
Tokyo 141-0022, Japan
takayasu@csl.sony.co.jp

XX List of Contributors

Tamás Vicsek
Department of Biological Physics
Eötvös University
1117 Budapest, Hungary
vicsek@angel.elte.hu

Cedric Vaillant
Centre de Recherche Paul Pascal
Université de Bordeaux

33600 Pessac, France
vaillant@crpp.u-bordeaux.fr

Kirsten Zickfeld
Potsdam Institut für
Klimafolgenforschung
14412 Potsdam, Germany
zickfeld@pik-potsdam.de

Part I

General

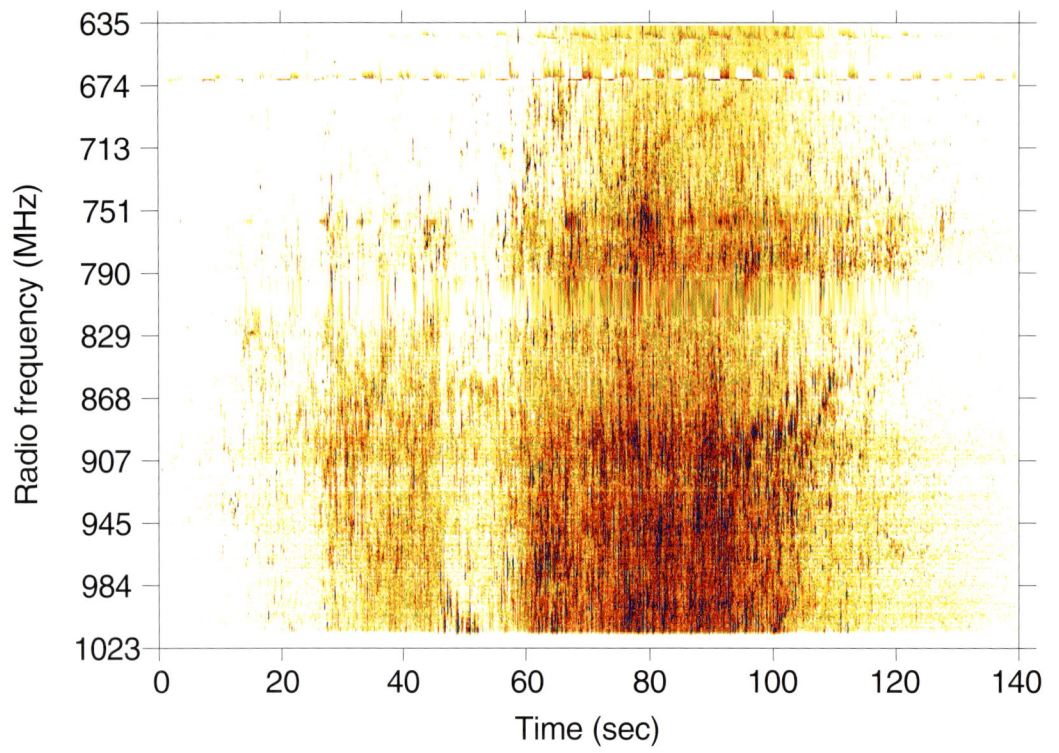

1. Entropy, Complexity, Predictability, and Data Analysis of Time Series and Letter Sequences

Werner Ebeling, Lutz Molgedey, Jürgen Kurths, and Udo Schwarz

The structure of time series and letter sequences is investigated using the concepts of entropy and complexity. First, conditional entropy and mutual information are introduced and several generalizations are discussed. Further, several measures of complexity are introduced and discussed. The capability of these concepts to describe the structure of time series and letter sequences generated by nonlinear maps, data series from meteorology, astrophysics, cardiology, cognitive psychology, and finance is investigated. The relation between the complexity and the predictability of information strings is discussed. The relation between local order and the predictability of time series is investigated.

1.1 Introduction

The category of entropy was introduced in 1864 by Rudolf Clausius into physics and in a different context in 1949 by Claude Shannon into information theory. Applications to the structure of sequences have been given by Shannon [1.71], who published in 1951 the seminal paper on 'Predictions and Entropy of Printed English'. Later Shannon's approach was also applied to other languages [1.20, 1.88], to biosequences, and to many other information carriers [1.3, 1.13, 1.14, 1.17, 1.18, 1.20, 1.27, 1.30, 1.32–1.35, 1.40, 1.52, 1.89]. The extension of Shannon's concept to the investigation of dynamic processes is due to Kolmogorov and Sinai [1.44, 1.75].

The concept of the Kolmogorov-Sinai entropy belongs to the key concepts of the modern theory of dynamical systems [1.69].

◄ **Fig. 1.0.** Dynamic spectrograms of millisecond spikes. The data were recorded by the frequency-agile solar radio spectrometer in ETH Zürich on 6 June 1983. Color coding of the flux intensity: low flux - bright, high flux - dark. The resolution in frequency is 1 MHz and in time 0.2 s

A few years later Kolmogorov developed a concept for the characterization of the complexity of sequences [1.45]. Several related concepts were developed later [1.1, 1.23, 1.24, 1.31, 1.42, 1.46, 1.47, 1.51, 1.70, 1.84].

Entropy and complexity concepts provided new tools for the investigation of irregular time series which play a great role in many branches of science [1.41].

Our investigation of time series is restricted to the concepts of conditional entropies, mutual information, and complexity, as well as to certain generalizations.

Our working hypothesis is that many time series and most information carriers as texts, pieces of music, and biosequences are not first order Markov processes but have higher order correlations. Further, we expect some structural analogies to strings generated by nonlinear processes.

In some cases we expect the existence of long-range correlations. This hypothesis was checked by the analysis of the behavior of the dynamic entropies, the mutual informations, and other correlation measures [1.12, 1.18, 1.33, 1.90]. Here, this line of investigations will be continued. In comparison to our earlier work special attention is paid here to the relation between local order and the predictability of time series.

1.2 Conditional Entropies and Predictability

In physics the entropy concept is connected with the names of Boltzmann, Gibbs, Einstein, Onsager, Prigogine, and others. The relation between physical and information-theoretical concepts has been discussed by Maxwell, Szilard, Brillouin, and other researchers [1.3, 1.20].

Here, we are mainly concerned with the applications of the entropy concept to time series and to information carriers. In order to proceed let us assume that the processes or structures to be studied are modeled by trajectories on discrete state spaces having the total length L. Let λ be the length of the alphabet. Further let $A_1, A_2, \ldots, A_n$ be the letters of a given subtrajectory of length $n \leq L$. Let further $p^{(n)}(A_1, \ldots, A_n)$ be the probability to find in the total trajectory a block (subtrajectory) with the letters $A_1, \ldots, A_n$. Then we may introduce the entropy per block of length n:

$$H_n = -\sum p^{(n)}(A_1, \ldots, A_n) \log p^{(n)}(A_1, \ldots, A_n) \,. \qquad (1.1)$$

From the block entropies we derive the conditional (dynamic) entropies by the definition

$$h_n = H_{n+1} - H_n \,. \qquad (1.2)$$

Further we define $r_n = 1 - h_n$ as the average predictability of the state following immediately after a measured n-trajectory.

These quantities are called by Shannon n-gram entropies. The limit of the conditional entropy for large n is the entropy of the source (Kolmogorov-Sinai entropy). We have seen that the predictability of processes is closely connected with the conditional (dynamic) entropies. Let us consider now a certain section of length n of the trajectory, a time series or another sequence of symbols $A_1, \ldots, A_n$, which often is denoted as a subcylinder. We are interested in the uncertainty of the predictions of the state following after this subtrajectory of length n. We define now the expression

$$h_n^{(1)}(A_1, \ldots, A_n) = \sum p(A_{n+1}|A_1, \ldots, A_n) \log p(A_{n+1}|A_1, \ldots, A_n)^{-1} \quad (1.3)$$

as the uncertainty of the next state (one step into the future) of the state following behind the measured trajectory $A_1, \ldots, A_n$ ($A_i \in alphabet$). Here, and in the following all logs are measured in λ-units. We note that in these units the inequality holds:

$$0 \leq h_n^{(1)}(A_1, \ldots, A_n) \leq 1 . \quad (1.4)$$

Further we define

$$r_n^{(1)}(A_1, \ldots, A_n) = 1 - h_n^{(1)}(A_1, \ldots, A_n) \quad (1.5)$$

as the predictability of the next state following after a measured subtrajectory, which is a quantity between zero and one [1.21, 1.22, 1.68].

We note that the average of the local uncertainty

$$h_n = h_n^{(1)} = \langle h_n^{(1)}(A_1, \ldots, A_n) \rangle \quad (1.6)$$
$$= \sum p(A_1, \ldots, A_n) h_n^{(1)}(A_1, \ldots, A_n)$$

leads us back to Shannon's n-gram conditional entropy. Let us consider other possible generalizations. If we want to predict the state which follows not immediately after the observed n-string, but only after k steps into the future we may define the quantity

$$h_n^{(k)}(A_1, \ldots, A_n) = \sum p(A_{n+k}|A_1, \ldots, A_n) \log p(A_{n+k}|A_1, \ldots, A_n)^{-1} . \quad (1.7)$$

This is the uncertainty of the state which occurs k steps into the future after the observation of an n-block. Further we may define accordingly the local predictabilities [1.21]

$$r_n^{(k)}(A_1, \ldots, A_n) = 1 - h_n^{(k)}(A_1, \ldots, A_n) . \quad (1.8)$$

For $n = 1$ the average predictability is closely related to the mutual information (transinformation) [1.36, 1.37] and the information flow [1.68]. The mutual information can be expressed in terms of our predictabilities by

$$I(k) = r_1^{(k)} - r_0 \,,$$

where $r_0 = 1 - H_1$ is the predictability of a letter, if no previous knowledge is available. For systems with long memory it makes sense to study a whole series of predictabilities with increasing n-values (where n is denoted by the lower index)

$$r_1^{(k)}, r_2^{(k)}, r_3^{(k)}, \dots, r_m^{(k)} \,,$$

where m is an estimate for the length of the memory. Since

$$r_{n+1}^{(k)} \geq r_n^{(k)}$$

the average predictability may be improved by taking into account longer blocks. In other words, one can gain advantage for prediction by basing the predictions not only on actual states but on whole trajectory blocks which represent the actual state and its history. Let us mention that the conditional entropies may be exactly calculated for several model systems [1.14, 1.19, 1.30]. In our empirical investigations described below we considered only the mutual information, also called transinformation, which is a special case of our concepts. The mutual information is explicitly defined as

$$I(n) = \sum_{A_i A_j} p^{(n)}(A_i, A_j) \log \left[\frac{p^{(n)}(A_i, A_j)}{p^{(1)}(A_i) p^{(1)}(A_j)} \right] \,. \tag{1.9}$$

For completeness let us further define $I(0) = H_1$ and $I(-n) = I(n)$. The mutual information is a special measure of correlations [1.32, 1.33, 1.36, 1.37, 1.40, 1.63] which is closely related to the autocorrelation function [1.61, 1.77].

For our analysis the following relations between the entropies and predictabilities defined above and the mutual information are of special importance

$$h_1 = H_1 - I(1) \,, \tag{1.10}$$

$$r_1^{(n)} = I(n) + r_0 \,. \tag{1.11}$$

In other words, the predictability of a letter n steps ahead is the sum of the mutual information and the overall predictability of letters $r_0 = 1 - H_1$. As shown by several authors [1.27, 1.36, 1.40, 1.82, 1.83], the mutual information is also a reliable measure for the correlations of letters in the distance n. Every peak at n corresponds to a strong positive correlation [1.36, 1.37].

1.3 Concepts of Complexity

Observational data from astrophysical, geophysical, or physiological experiments are typically quite different from those obtained in laboratories. Often we have rather short, noisy, irregularly sampled time series. More important, however, is that nonstationary and very complicated behavior in time is usually observed. In such cases, well-known global characteristics of the underlying processes, such as periodicities or fractal dimension, do not provide a sufficient description. With respect to modeling the question often arises whether the data have any structure at all, for example correlations, and of what kind the structure is. The concept of complexity is an appropriate approach to analyze such data.

During the past decade numerous definitions of complexity have been proposed, e.g. [1.2, 1.4, 1.8, 1.26, 1.30, 1.45, 1.54, 1.55, 1.58, 1.73, 1.74, 1.84], and successfully used in various fields, ranging from information processing, e.g. [1.7, 1.45], and theory of dynamical systems, e.g. [1.30, 1.84, 1.87]), to thermodynamics, e.g. [1.8–1.10, 1.73, 1.74], astrophysics [1.31, 1.70], geophysics [1.86], evolution theory, e.g. [1.49, 1.50, 1.72], and medical diagnostics, e.g. [1.47, 1.65]. Recently, some generalizations of this approach to analyze two-dimensional objects have been proposed [1.28, 1.29, 1.65]. Many of these definitions rely on the intuitive impression that complexity should reflect some hidden order of a phenomenon, which nevertheless possesses a certain degree of randomness. Neither well-ordered nor completely disordered objects are seemingly complex; thus complexity appears somewhere at the borderline between disorder and order. Formally, this implies that complexity is a convex function of the disorder, provided the latter is appropriately defined (see [1.73, 1.74] for a discussion).

Quite commonly, the definition of disorder is based on the comparison of the Boltzmann-Gibbs-Shannon entropy H_1 (1.1) with the maximal possible entropy of the system H_1^{max}. The value of the maximal entropy H^{max} depends on the nature of the system, but for the simplest case when N states are available, the maximal entropy is achieved for the equiprobable distribution:

$$H^{\mathrm{max}} = \log_2 N. \tag{1.12}$$

It is important to note that Shannon entropy is a measure of randomness, i.e. it assigns highest complexity for white noise-like behavior, where past and future are uncorrelated. Other popular measures, especially the algorithmic complexity [1.42] or the approximative entropy [1.62] have the same property; we call this class traditional measures of complexity.

Such a characterization is, however, not sufficient for many systems, especially in nonlinear dynamics. We, therefore, present another kind: nontraditional or alternative measures of complexity which relate highest complexity at phase transitions, e.g. the onset of chaos.

A straightforward notion of such an alternative measure has been recently proposed by combining disorder and order. The disorder is defined as H/H^{max} and correspondingly the disorder-based complexity $\Gamma_{\alpha,\beta}$ [1.73, 1.74]:

$$\Gamma_{\alpha,\beta} = (1 - H/H^{\text{max}})^{\alpha}(H/H^{\text{max}})^{\beta} \,. \tag{1.13}$$

For $\alpha > 0$ and $\beta > 0$, $\Gamma_{\alpha,\beta}$ is a convex function of the disorder. Other values of these parameters may correspond to alternative definitions of complexity [1.73, 1.74].

The relevance of the complexity measure $\Gamma_{\alpha,\beta}$ as introduced by (1.13) has been demonstrated in application to the logistic map and to one-dimensional spin-systems [1.73, 1.74]. We, however, wish to stress two features of the definition in (1.13): (i) it exploits a concept of 'maximal possible entropy', which for some systems may not be easily computed and even unambiguously defined, and (ii) it lacks the accounting for inherent correlations in the system, which are certainly an important component of order, and thus of complexity. By construction this measure relates zero complexity to most random behaviors e.g. the equiprobable distribution, as well as to simple ordered states.

Grassberger [1.30] introduced another approach to complexity which is based on the differences of block entropies h_n (1.2). This effective measure complexity (EMC) is defined as

$$\text{EMC} = \sum_{n=1}^{\infty} n \left(h_{n-1} - h_n \right) . \tag{1.14}$$

EMC describes the behavior of the local difference h_n as it converges towards the dynamical entropy of the dynamical system. It can also be written as an average Kullback information, for instance in terms of conditional probabilities, as demonstrated in [1.53].

It is easy to see that EMC vanishes both for most random cases such as white noise, and for constant symbolic strings. It goes to infinity in period doubling sequences, i.e. it goes to infinity along this typical route to chaos.

One of the most interesting complexity measures is the renormalized entropy, originally introduced by Klimontovich [1.43] in thermodynamics. It takes into account that the energy of an open system changes with its control parameter, which makes a direct comparison of Shannon entropies impossible. The main idea is that the Shannon entropy for different system states is normalized to a fixed value of mean effective energy. This approach, loosely speaking, renormalizes the entropy obtained from a time series $x(t)$ of a certain system state in such a manner that the mean effective energy coincides with that of a reference state $x_r(t)$.

Starting from these two time series, we can easily estimate the corresponding probability distributions $f(x)$ and $f_r(x)$. By using formal arguments from thermodynamics the effective energy is defined as

$$h_{\text{eff}}(x) = - \log f_r(x)\,. \tag{1.15}$$

The renormalization of f_r into $\tilde{f}_r$ is constructed such that the mean effective energies $\langle h_{\text{eff}} \rangle$ of f and $\tilde{f}_r$ are equal. To make this idea operational, we first represent the distribution in terms of the canonical Gibbs distribution

$$\tilde{f}_r(x) = \exp\left(\frac{\Phi(T_{\text{eff}}) - h_{\text{eff}}(x)}{T_{\text{eff}}} \right), \tag{1.16}$$

which can be rewritten as

$$\tilde{f}_r(x) = C(T_{\text{eff}}) \exp\left(-\frac{h_{\text{eff}}(x)}{T_{\text{eff}}} \right), \tag{1.17}$$

where T_{eff} and $\Phi(T_{\text{eff}})$ are the effective temperature and the free effective energy respectively. Because h_{eff} can be calculated from (1.15), there are two unknowns in (1.17): $C(T_{\text{eff}})$ and T_{eff}. They are determined from the following two conditions:
(a) Normalization:

$$\int \tilde{f}_r(x)\, dx = 1\,, \tag{1.18}$$

(b) Equality of mean effective energy:

$$\int h_{\text{eff}}(x) \tilde{f}_r(x)\, dx = \int h_{\text{eff}}(x) f(x)\, dx. \tag{1.19}$$

Hence, $\tilde{f}_r$ fulfills the properties wanted. Consequently, we can compare the Shannon entropies of f and $\tilde{f}_r$:

$$H = - \int f(x) \log f(x)\, dx \quad \text{and} \quad \tilde{H}_r = - \int \tilde{f}_r(x) \log \tilde{f}_r(x)\, dx\,. \tag{1.20}$$

For that the renormalized entropy difference

$$\Delta\tilde{H} = H - \tilde{H}_r \tag{1.21}$$

is introduced. It is important to note that $\Delta\tilde{H}$ is a relative measure that depends on the reference state chosen.

If this reference state is chosen suitably, then $\Delta\tilde{H}$ relates smaller values to periodic and random behavior than to chaotic dynamics [1.64].

There are some other alternative measures of complexity, such as the epsilon-complexity [1.8] or the fluctuation complexity. We compared the properties of these measures in detail for the logistic map [1.84] and found that there is till now no outstanding alternative measure of complexity; each of them is sensitive to certain structural changes. The proper choice of these measures is context-dependent. Therefore we recommend to compare the proper choice of these measures in each special application. We will demonstrate in Sect. 1.6 how efficiently these measures can be used in several applications.

1.4 Applications to Biosequences and Other Information Carriers

In genetic data banks one can find nowadays a very large number of DNA-sequences and there is an urgent need for the development of tools which formalize their analysis (see, for example, also Chap. 2). Formally, DNA-sequences are linear strings written on an alphabet consisting on four letters

$$X = A, C, G, T\,. \qquad (1.22)$$

A genome contains about 1–100 billion nucleotides and corresponds therefore to a very long string, which, however, consists in general of several pieces. The strings which are available for a statistical analysis comprise in general $10^4 - 10^6$ letters. Since Gatlin's pioneering work '*Information Theory and the Living System*' [1.27], the calculation of entropic measures for genetic sequences as conditional entropies and mutual informations has found many fruitful applications [1.15, 1.32, 1.33, 1.40]. In particular, we mention the development of criteria to differ between coding and noncoding regions [1.36, 1.37]. For genetic sequences several authors have pointed out the existence of long-range correlations [1.12, 1.33, 1.40, 1.61, 1.77, 1.82, 1.83].

However, an analysis of the average uncertainties (the dynamic entropies) yields rather high values. Measured in bits the limit uncertainty is in most cases larger than 1.8 bits, i.e. larger than 0.9 in λ units [1.27, 1.32]. For this reason the average dynamic entropies do not seem to be the appropriate instrument to analyze DNA-strings. However, as we can show, local investigations of the entropy and the mutual information might be very powerful for the analysis of long-range correlations. In [1.16] we presented the lowest uncertainties of predictions for the interval 9 800–10 300 of the DNA of the virus HIV2BEN. In spite of the fact that the average uncertainty is rather high ($h_3 = 0.94$), we may find special positions where the uncertainty is lower than 0.83 [1.16]. For example, if we observe the triple GTC, then with large probability an A or a G is expected to follow. If however the triple AAC is observed then the most probable continuation is either A or C. These simple rules locally increase the predictability.

Since for DNA a huge volume of material on the mutual information is available [1.32, 1.40] we will not go here into further detail. Let us only mention that DNA-strings show some (formal) analogies to texts.

Let us discuss now in brief several results which are available for texts. We have studied for example Melville's Moby Dick ($L \approx 1\,170\,200$) and Grimm's Tales ($L \approx 1\,435\,800$). Our methods for the analysis of the entropy of sequences were explained in detail elsewhere [1.16]. We have shown that at least in a reasonable approximation the scaling of the entropy against the word length is at large n given by a root law.

For example a reasonable fit of the data obtained for texts on the 32-alphabet (measured in $\log(32)$ units) reads

$$h_n \approx (0.25/\sqrt{n}) + 0.07 \ . \tag{1.23}$$

The dominating term is given by a root law corresponding to a rather long memory tail. We mention that a scaling law of the root type was first found by Hilberg who made a new fit for Shannon's original data [1.38]. We used our own data for $n = 1, \dots, 26$, but also included Shannon's result for $n = 100$. The slow decay of the conditional entropies may be interpreted by the existence of long correlations in texts.

Let us now briefly summarize results obtained from using other measures of correlations [1.12]. At first we have calculated the algorithmic entropy according to Lempel and Ziv which is introduced as the relation of the length of the compressed sequence (with respect to a Lempel-Ziv compression algorithm) to the original length. The results obtained for the Lempel-Ziv complexities (entropies) of several DNA sequences and for texts were compared with the exponents of the mean square fluctuation of the composition and with diffusion exponents [1.12]. Further we studied also the power spectrum which is defined as the Fourier transform of the correlation function. The results of spectra calculations for the original file of the Bible, for Moby Dick and for the same files shuffled on the word level or on the letter level correspondingly were presented in a foregoing work [1.12].

We have shown that the spectra of the original texts have a characteristic shape with a well-expressed low frequency part. This indicates again the existence of long-range correlations in texts [1.12]. Similar results were obtained for DNA sequences [1.61,1.77]. We see from this analysis that DNA sequences also show some type of long correlations. However, as shown in [1.12] the long-range correlations are only due to slow changes in the composition. As a matter of fact, the composition (the letter content) of DNA (and also of texts) fluctuates slowly with a wavelength of $10^2 - 10^3$. This fluctuation of the composition is evidently the main reason for the observed characteristic exponents. Shuffling destroys these properties [1.12].

Let us go now to the investigation of protein sequences [1.16]. As is well known protein structures play a fundamental role in all living processes [1.16]. The building blocks of the proteins are the 20 amino acids which we denote by the letters of the alphabet

$$X = A, C, E, G, \dots , Y, W \,. \tag{1.24}$$

In this way the primary structure of a protein can be mapped to a linear string on an alphabet with 20 letters. Typically the protein strings have a length between 10^2 and 10^4. In other words, protein strings are much shorter than DNA strings. Further, protein strings show a very high degree of randomness [1.23] but nevertheless they contain also much intriguing information connected with their function and their history [1.5, 1.6]. All this together leads to serious difficulties in the statistical analysis of the primary structure of proteins. A possible way to reduce these difficulties is to use reduced alphabets [1.39].

Here, we shall use nevertheless the full alphabet; this restricts our investigation to a rather small statistical significance. As a prototype we considered in an earlier work [1.16] the sequences PACVSPEN and PAPHUMAN which have a length $L \simeq 5\,000$ and for those we have calculated the mutual information. The predictability was obtained by adding $r_0 = 0.05$. We obtained several well expressed peaks which show structural regularities where the predictability is a little bit better than in other places.

Looking at the length and at the size of the alphabet, protein sequences show at least a formal similarity to musical strings. Therefore we have made a comparison of the mutual information of both these types of information carriers [1.16]. In the mentioned work [1.16] we calculated the mutual information and the predictabilities for the Beethoven Sonatas 10 no. 2 and no. 3, for the Beethoven Sonatas 28 and 48, and for Mozart's pieces KV 311 and 330. The peaks show that there exist strong correlations between two notes at certain distances. In this way analogies between pieces of music and protein sequences have been shown.

1.5 Applications of Entropy Concepts to Data Analysis

Prediction of strong noisy data using classical linear methods usually fails to give an accurate and reliable confidence level of the prediction. Moreover, the linear methods are dominated by the most frequent events. However, predictability may not be constant in time and may be even higher for rare events. The concept of entropy and local predictability in combination with classical methods is a good candidate to give reliable results. Applications of these concepts to meteorological strings were given in [1.59, 1.85] and to nerve signals in [1.15].

In the following our concept will be demonstrated on daily stock index data S_t: Dow Jones 1900–1999 (27 044 trading days) (regarding the analysis of economic records see also Chaps. 12–14). Since the stock index itself has an exponentially growing trend one uses daily logarithmic price changes

$$x_t = \ln(S_t) - \ln(S_{t-1}). \tag{1.25}$$

A direct application of the entropy concept requires a partitioning of the real value data x_t into symbols A_t of an alphabet having the length λ. To find an optimal partition and alphabet is a process of maximizing the Kolmogorov-Sinai entropy. However, for strong noisy signals with short memory an equal frequency of the letters is near to optimal. To be concrete $\lambda = 3$ and $A_t = 0; x_t < -0.0025$ (strong decrease in the stock value), $A_t = 2; x_t > 0.0034$ (strong increase), $A_t = 2$ (intermediate) were chosen.

The result of the calculation of the local uncertainty $h_n(A_1, \dots, A_n)$ for the next trading day following behind an observation of n trading days $A_1, \dots, A_n$ according (1.3) for $n = 5$ is plotted in Fig. 1.1 and Fig. 1.4. The local uncer-

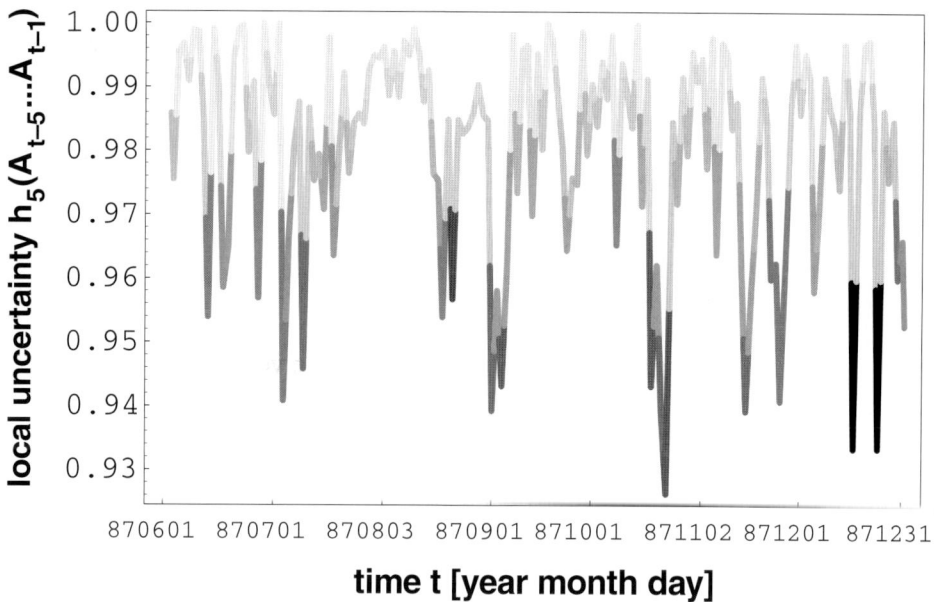

Fig. 1.1. Local uncertainty of the sixth symbol when five symbols are seen for the second half of 1987. The gray value codes the level of significance calculated from a surrogate with memory of two. Dark represents a large deviation from the noise level (good significance). Note the higher predictability following the October crash

tainty is almost near one, i.e. the local predictability is very small. However, behind certain patterns of stock movements $A_1, \dots, A_n$ the local predictability reaches 8% - a notable value for the stock market, which is usually pure random. The mean predictability over the full dataset is less then 2% (see Fig. 1.2). The question of the significance of the prediction is treated by calculating a distribution of local uncertainty $h_n^S(A_1, \dots, A_n)$ by help of surrogates. The surrogate sequences have the same two point probabilities $p^{(2)}(A_2|A_1)$ as the original sequence [1.11, 1.60, 1.66, 1.67, 1.79]. The level of significance K is calculated as

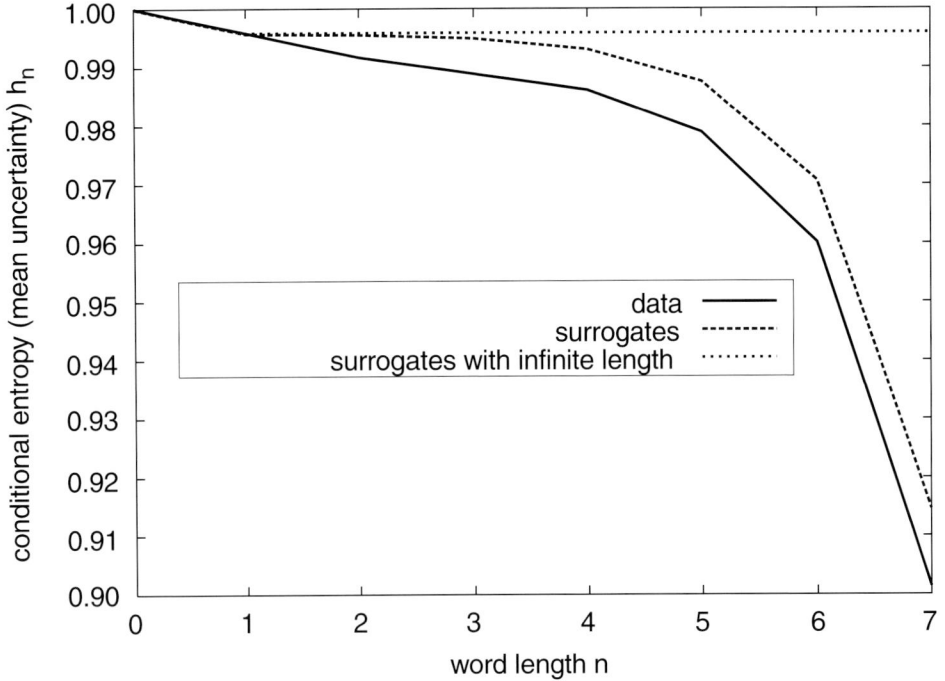

Fig. 1.2. Conditional entropy $h_n = H_{n+1} - H_n$ as a function of word length n

$$K_n(A_1, \dots, A_n) = \frac{h_n(A_1, \dots, A_n) - \langle h_n^S(A_1, \dots, A_n)\rangle}{\sigma} \quad , \qquad (1.26)$$

where $\langle h_n^S(A_1, \dots, A_n)\rangle$ is the mean and σ is the standard deviation of the local uncertainty distribution for the word $A_1, \dots, A_n$.

Assuming Gaussian statistics $|K| \leq 2$ represents a confidence level greater than 95%. However, the local uncertainty distribution (Fig. 1.3) is more exponential distribution like. Therefore, larger K values are required to guarantee significance. For the analyzed dataset a word length up to six seems to give reliable results.

Fortunately higher local predictabilities coincides with larger levels of significance as seen in Fig. 1.1 and from Table 1.1. Since we used a time series over a very long period we have to address the problem of nonstationary by dividing the original time series into smaller pieces. Furthermore instead of producing surrogates on the level of symbols one can discuss surrogates obtained by models of stock markets like ARCH/GARCH models [1.56]. This has been done in [1.57].

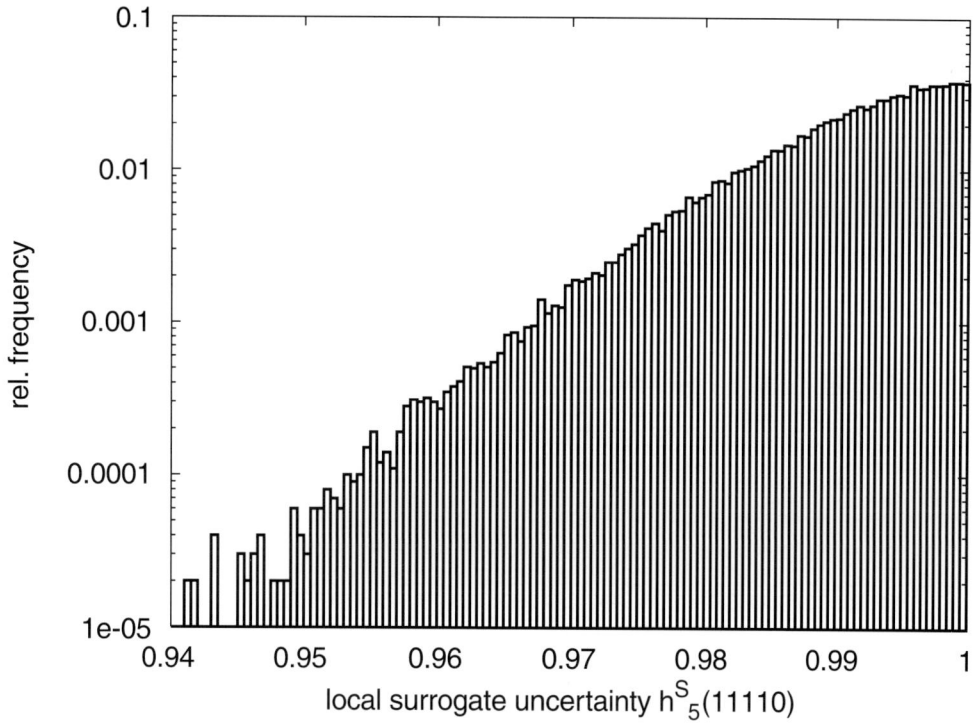

Fig. 1.3. Local uncertainty distribution of the surrogate sequence for the word 11110

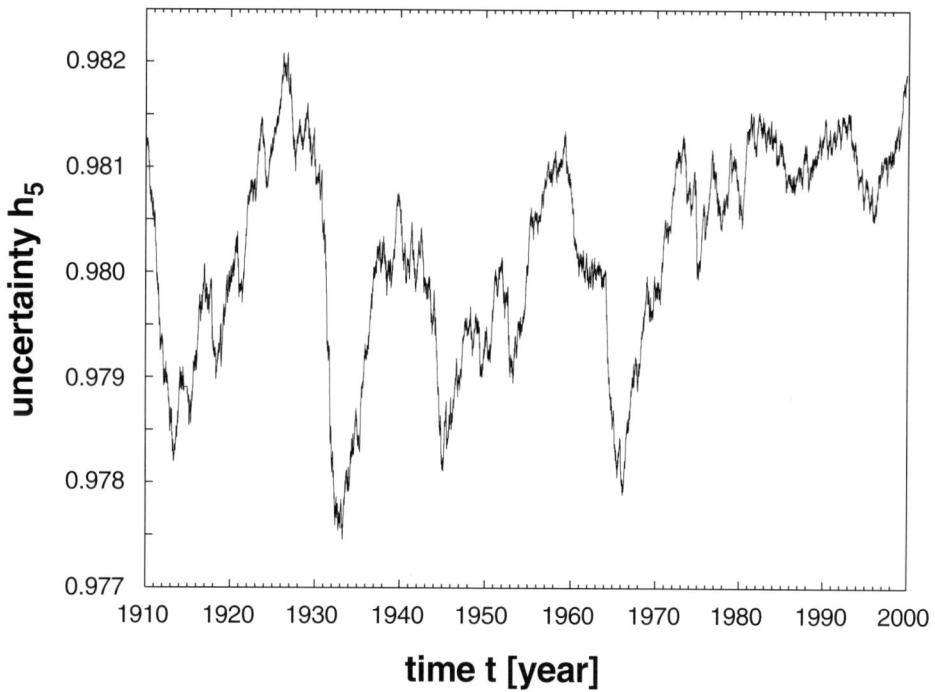

Fig. 1.4. Moving exponential average of the local uncertainty h_5 with an half-life period of 5 yr for the full dataset

Table 1.1. Words with the highest predictability

Word	Uncert.	K	Word	Uncert.	K	Word	Uncert.	K
020	0.971	−27.9	1112	0.954	−14.3	11110	0.919	−9.5
112	0.971	−30.5	0000	0.957	−12.0	11120	0.926	−9.1
110	0.977	−23.4	1110	0.958	−15.0	20000	0.926	−6.9
120	0.981	−29.4	0110	0.960	−14.9	11112	0.931	−8.7
000	0.982	−18.5	0020	0.961	−12.5	10120	0.933	−5.2
212	0.983	−19.5	1102	0.962	−12.4	22202	0.933	−9.4
202	0.984	−22.9	2020	0.966	−10.6	00000	0.934	−6.4
111	0.985	−20.1	0200	0.968	−10.6	11011	0.937	−9.7
121	0.985	−12.4	0202	0.969	−14.6	02000	0.939	−5.7
012	0.987	−14.2	0120	0.971	−12.3	02020	0.941	−5.1
102	0.988	−10.5	2112	0.971	−9.0	00020	0.943	−6.4

1.6 Applications of Complexity Concepts

We have applied the concept of measures of complexity to experimental data where other tools of linear as well as nonlinear data analysis fail. The areas of application range from astro- and geophysics, via physiology to cognitive psychology [1.25, 1.28, 1.29, 1.31, 1.46, 1.47, 1.64, 1.65, 1.70, 1.86].

Such experimental data consists usually of real numbers. Therefore, the first necessary step is to transform them into a symbolic string, i.e. the data are transformed into a series of the same length but the elements are only a few symbols. In doing so one loses a certain amount of detailed information, but some of the invariant, robust properties of the dynamics are retained. In the best case, such a transformation generates a Markov partition. However, in most examples of natural systems we know neither the existence of such a partition nor their construction. Therefore, more pragmatic transformations have to be used which may be not Markovian ones [1.81, 1.84].

We distinguish static transformations, where the transformation is based on a few thresholds (see application to cardiology) and dynamic ones, where we consider the step-to-step difference of data points adjacent in time (see application to cognitive complexity). The choice of the kind of transformation is of particular importance. In the case of rather small data records, as typically occurs in applications, a transformation into only a few symbols is to recommended. In this manner, the transformation is context-dependent. If possible, such a coarse-graining should, hence, be based on some physical motivation. Otherwise, we highly recommend comparing different transformations into symbols.

1.6.1 Cardiology: Detecting High-Risk-Patients for the Sudden Cardiac Death

Every year several hundred thousand persons die due to sudden cardiac death that is caused by cardiac fibrillation. Without any warning it even can occur in subjects who are up to this time apparently healthy or medically inconspicuous. By use of conventional methods only 30% of all affected persons are diagnosed as high-risk-patients. This is an important challenge to nonlinear dynamics.

The aim of our investigation is to get by means of nonlinear dynamics a clear improvement of the detection rate of persons with a high risk for sudden cardiac death. An essential point is the finding of new parameters, which describe the complex processes and their interactions for detecting those high-risk-patients, who could not recognized by traditional - mostly linear methods.

The basis of our analysis is the heart rate variability (HRV) which is yielded from 30 min and 24 hr ECG-measurements, i.e. from noninvasive methods (For other investigations regarding heartbeat dynamics see Chaps. 7 and 8.). It is important to emphasize even in case of healthy volunteers, the variability shows a broad variety of structures. This is essentially caused by the fact that the system which generates the HRV has to be considered as an open one whose energy changes temporally as well as from person to person.

The following application of the concept of complexity leads to an improved risk stratification: From various tests it emerges that for our purpose at least four different symbols are necessary. The most appropriate transformation is a static one:

$$s_i = \begin{cases} 0 & \text{if} \quad t_i > (1+a)\mu \\ 1 & \text{if} \quad \mu < t_i \le (1+a)\mu \\ 2 & \text{if} \quad (1-a)\mu < t_i \le \mu \\ 3 & \text{if} \quad t_i \le (1-a)\mu \,, \end{cases} \tag{1.27}$$

where t_i are the RR-intervals (Fig. 1.5), μ is their mean value, and $a = 0.1$. The RR-interval is the time distance between two successive heartbeats.

To analyze the symbolic strings, Shannon and Renyi entropies of length-3 words are calculated. As expected, the Shannon entropy is not so useful as the generalized Renyi entropies. We use in particular the $H_k^{(q)}$ for $q = 0.25$ and for $q = 4$ to describe the complexity. It is interesting to note that already the distribution of length-3 words yields a criterion for a distinction of both groups: for persons with high cardiac risk, this distribution is mainly concentrated on about 10 words (of 64 possible ones), whereas healthy persons are characterized by a more uniform distribution.

The renormalized entropy is especially related to compare different states of one system. An important problem of its application is to choose a suitable reference state, i.e. a special person in our case. We choose that healthy person as a reference person who has the largest renormalized entropy. Note that this choice does not sensitively influence the results. From this, we indeed get an

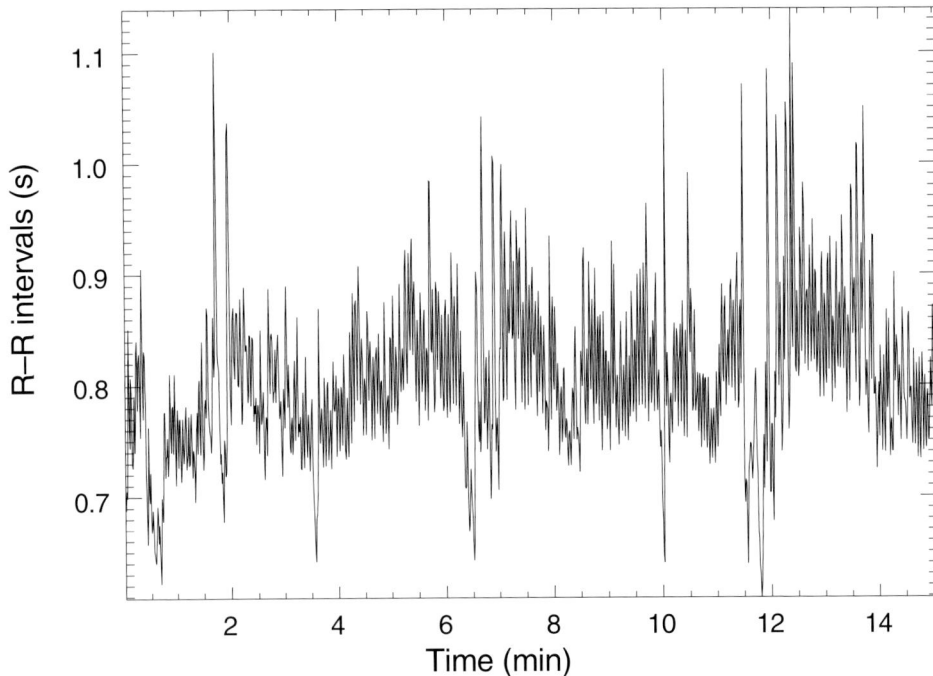

Fig. 1.5. Tachogram of a healthy person

indication for high cardiac risk in two directions. If $\Delta\tilde{H}$ (1.21) is very low, a strongly reduced variability is expressed and, on the other hand, if $\Delta\tilde{H}$ is rather large, an exceptional variability is indicated.

It is important to note that none of these measures of complexity alone is sufficient to describe the risk. Therefore, we combine several measures of complexity with traditional methods from the time and frequency domain. Applying this combination method to ECG measurements from 572 survivors of acute myocardial infarction, we get a significantly better prediction of high cardiac arrhythmia risk than the standard measurement of global heart rate variability [1.80].

1.6.2 Cognitive Psychology: Synchronization and Coordination of Movements

The abilities to perform precise movements, coordinate movements between different limbs, or adjust them to external performance constraints, constitute general, but highly complex human capacities. Biologists and psychologists have for a long time been interested in these capacities in order to gain insights into the functionality of the central nervous system. More recently, the dynamic systems perspective has been applied to a number of phenomena related to motor control [1.78] and human development of cognition and action [1.76].

Krampe et al. [1.48] investigated the production of bimanual rhythms in a large number of subjects differing in musical (pianist) skills, and also age. One task in these experiments required polyrhythmic performance, that is, the combination of different rhythms in the two hands (see Fig. 1.6).

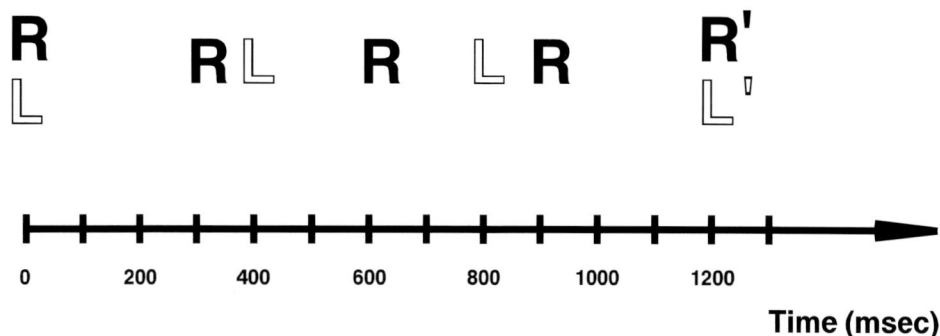

Fig. 1.6. Bimanual rhythm production: the figure illustrates the sequence of finger movements and their ideal timing for a polyrhythm (three beats per cycle in the left hand vs. four beats per cycle in the right hand) at a cycle duration of 1 200 ms. R refers to the right, L to the left index finger. A cycle starts with simultaneous strokes of the two fingers

The task was performed on two keys of an electronic piano connected to a computer which measured the time intervals between successive keystrokes. The external control parameter, that is the prescribed tempo, was experimentally varied across trials for each subject between 800 ms and 8 200 ms cycle duration; for highly skilled subjects, performance was assessed at speeds as fast as 500 ms per cycle. After a short synchronization phase during which subjects could play along with the rhythm generated by the computer and adjust to the prescribed tempo, participants had to continue their performance without external support for another 12 cycles. A complete continuation trial consists of 36 intervals between left, and 48 intervals between right hand keystrokes, which should ideally be of equal duration for a given hand.

Methods from symbolic dynamics permit us to investigate whether the empirically observed variation in the duration of these intervals can be described as an orderly sequence of violations of the prescribed duration within trials, and whether these systematic patterns emerge or dissolve as a function of the external control parameter performance tempo. We have chosen one particular coding scheme to get symbolic strings (Fig. 1.7). The coding scheme used here transforms continuous measures of interval duration to a dichotomous variable based on local comparisons between successive keystrokes performed with the same hand. A given interval terminated by a left hand keystroke (only these data are considered here for illustration) receives a value of '1', if its duration exceeds the duration of the preceding interval between left-hand keystrokes, otherwise the value is '0'. Values of '1' are indicated by black pixels in Fig. 1.7. Already this simple transformation clearly suggests a qualitative shift in strate-

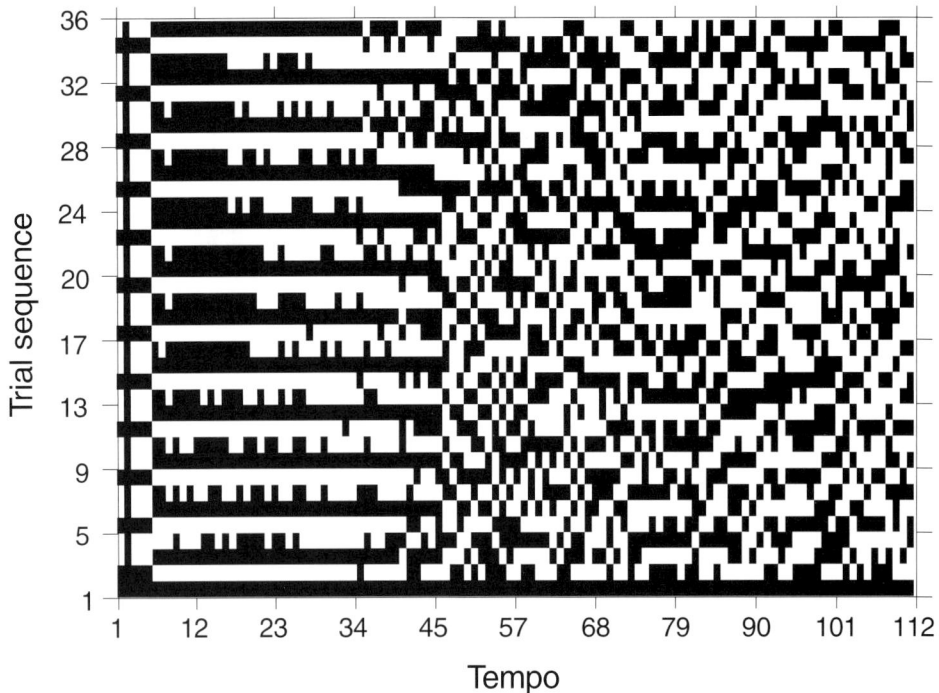

Fig. 1.7. Phase shift in the performance of a three against four polyrhythm as a function of tempo in one subject. Trials have been sorted by tempo for this illustration. The trial number is provided by the x-axis and corresponds to the external control parameter (tempo); trial number 1 refers to a cycle duration of 800 ms, trial number 112 to a tempo of 8 200 ms per cycle. The position of a single interval within a given trial is provided by the y-axis. Only intervals from the left hand (1–36) are shown. Black pixels indicate intervals which are longer than their immediate predecessor, otherwise a white pixel is set. A change of the pattern can be observed in a region around trial number 47, which corresponds to a performance tempo of 1 400 ms per cycle

gies for the realization of the polyrhythmic task as a function of performance tempo. The calculation of different measures of complexity exhibits that this transition is significant and that it is typical for young and old amateurs but also for concert pianists [1.25]. This qualitative transition has been modeled using a delayed feedback control. We conclude that the complexity of coordinated bimanual movements results from interactions between nonlinear control mechanisms with delayed feedback and stochastic timing components.

1.6.3 Astrophysics: Organization of Solar Spikes

The variation of solar electromagnetic radiation and the particle emission is mainly caused by solar activity. Flares are the most violent manifestation of solar activity. They are caused by a rapid release of energy stored in the coronal magnetic field. Understanding the flare phenomenon requires us to identify and model a large variety of physical processes involved. From observations of solar radio emission we know that the impulsive phase of this primary energy release in flares is fragmented into a multitude of substructures, called spikes (Fig. 1.0) which are triggered almost simultaneously. Therefore, properties of spikes give some detailed insight into the nature of this impulsive phase. Depending on the assumed energy release and emission processes, two types of fragmentation are now under discussion: a scenario of global organization (spikes are emitted in a succession of similar events by the same system) or a scenario of local organization (many systems are triggered by an initial event).

We have searched for interrelations of spikes emitted simultaneously at different frequencies during the impulsive phase of a flare event [1.70]. To characterize such complex spatio-temporal patterns, such as dynamic spectra (Fig. 1.0) measured in solar radio astronomy, we use quantities of symbolic dynamics, such as Shannon information and algorithmic complexity. This approach is appropriate to characterize these patterns, whereas the popular estimate of fractal dimensions and related techniques fail here.

In the case of the analyzed dynamic spectra the length of the symbolic strings is about 400. To improve the statistics, we concatenate up to ten symbol strings of successive scans of the dynamic spectrum. We observe in all cases that the only effect of such concatenation is a smoothing in Shannon information and algorithmic complexity. The Shannon information and algorithmic complexity of the symbol sequence are used for comparing the considered observation with others, or with models (surrogates) and for characterizing the observation. These measures concern the whole frequency region, i.e. the global source region. This way, we find out that global organization is also apparent in quasi-periodic changes of these Shannon information and algorithmic complexity in the range of 2–8 s (Fig. 1.8). Our analysis of spike events suggests that the structure in frequency is not stochastic but a process in which spikes at nearby locations are simultaneously triggered by a common exciter.

1.7 Conclusion

Let us summarize the main results obtained here.

- The dynamic entropy (uncertainty of next state) is connected with predictability.
- The Shannon entropy and algorithmic complexity are measures of randomness, but alternative measures, such as EMC or renormalized entropy, relate highest complexity to phase transitions and are, therefore, more appropriate to describe complex systems.
- Typical time series and information carrying sequences (DNA, texts, proteins, music) show correlations on many scales (including those of very long range).

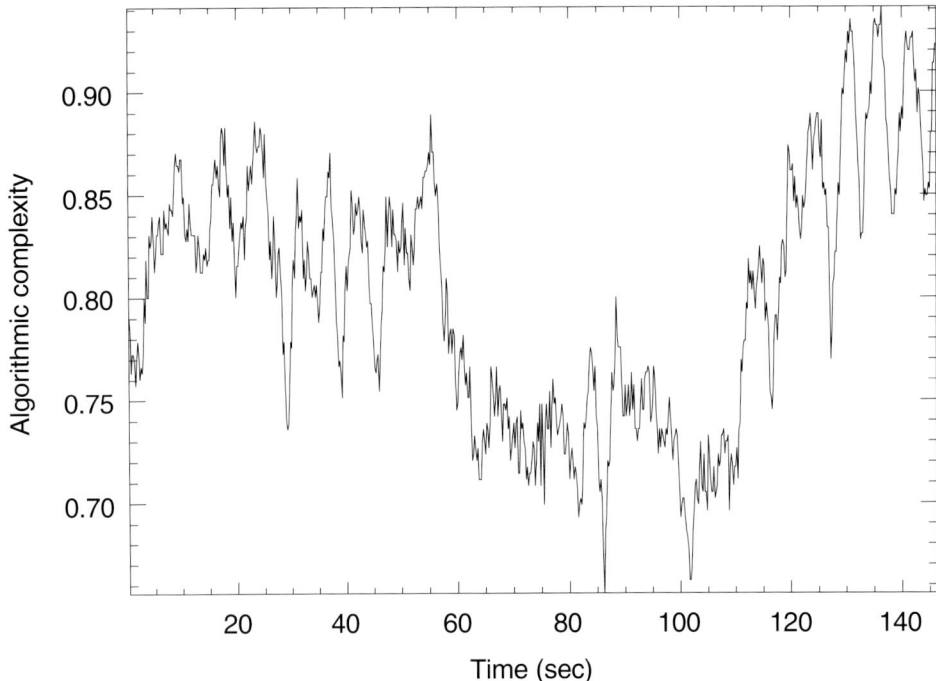

Fig. 1.8. Algorithmic complexity calculated from the irregular spike pattern presented in Fig. 1.0

- Measures of complexity are useful tools in various fields (e.g. physiology, astro- and geophysics) where other techniques of linear and nonlinear data analysis fail.

Our results show that the dynamic entropies and other complexity measures are an appropriate measure for studying the predictability of evolutionary processes. Of particular interest are local studies of the predictabilities of certain local histories. Long correlations are of specific interest since they improve the predictability. This means that one can in principle improve the predictions by basing the predictions at longer observations. Further we can conclude that there are specific substrings, which are relatively seldom, where the local uncertainty is much smaller than the average, i.e. the predictability is much better than on average. In other words, there are specific situations where the predictability is much better than the average predictability. It may be of practical importance to find out all substrings which belong to this particular class.

There was no space here to discuss the relation to other measures of long-range relations based on methods of statistical physics, as e.g. algorithmic entropy, correlation functions, mean square deviations [1.12], $1/f^\delta$ noise [1.12, 1.82, 1.83], and scaling exponents [1.12, 1.61, 1.77].

In conclusion we would like to express the hope that the analysis of entropies, predictabilities, and other complexity measures could be developed to useful instruments for studies of the large-scale structure of a rather broad class of time series from various fields and information-carrying sequences.

Acknowledgements. The authors thank A.O. Benz, R. Engbert, F.-W. Gerstengarbe, H. Herzel, M.A. Jimenez-Montano, R. Kliegl, R. Krampe, F. Moss, A. Neiman, C. Nicolis, G. Nicolis, T. Pohl, P. Saparin, R. Steuer, A. Voss, P.C. Werner, and A. Witt for many fruitful discussions and a collaboration on special topics of the problems discussed here.

References

1.1 H. Atmanspacher, J. Kurths, H. Scheingraber, R. Wackerbauer, and A. Witt, Open Syst. Info. Dyn. **1**, 269 (1992).

1.2 H. Atmanspacher, C. Rath, and G. Wiedemann, Physica A **234**, 819 (1997).

1.3 R. Badii and A. Politi, *Complexity: Hierarchical Structures and Scaling in Physics* (Cambridge University Press, Cambridge, 1997).

1.4 C.H. Bennett, in *Complexity, Entropy and the Physics of Information*, edited by W.H. Zurek (Addison-Wesley, Reading, 1990).

1.5 A. Berman et al., Proc. Natl. Acad. Sci. USA **91**, 4044 (1994).

1.6 E. Kolker and E.N. Trifonov, Proc. Natl. Acad. Sci. USA **92**, 557 (1995).

1.7 G.J. Chaitin, J. ACM **13**, 547 (1996).

1.8 J.P. Crutchfield and K. Young, Phys. Rev. Lett. **63**, 105 (1989).

1.9 J.P. Crutchfield and D.P. Feldman, Phys. Rev. E **55**, R1239 (1997).

1.10 S. Lloyd and H. Pagels, Ann. Phys. **188**, 186 (1988).

1.11 K. Dolan, A. Witt, M.L. Spano, A. Neiman, and F. Moss, Phys. Rev. E **59**, 5235 (1999).

1.12 W. Ebeling, A. Neiman, and T. Pöschel, in *Coherent Approach to Fluctuations* (Proc. Hayashibara Forum 1995) (World Scientific, Singapore, 1995).

1.13 W. Ebeling and G. Nicolis, Europhys. Lett. **14**, 191 (1991).

1.14 W. Ebeling and G. Nicolis, Chaos, Solitons Fractals **2**, 635 (1992).

1.15 W. Ebeling, M.A. Jimenez-Montano, and T. Pohl, in *Festschrift devoted to Jagat N. Kapur*, edited by Karmeshu (New Delhi, 1999).

1.16 W. Ebeling and C. Frömmel, Biosystems **46**, 47 (1998).

1.17 W. Ebeling and T. Pöschel, Europhys. Lett. **26**, 241 (1994).

1.18 W. Ebeling, T. Pöschel, and K.F. Albrecht, Int. J. Bifurcat. Chaos **5**, 51 (1995).

1.19 W. Ebeling, J. Freund, and K. Rateitschak, J. Bifurcat. Chaos **6**, 611 (1996).

1.20 W. Ebeling, J. Freund, and F. Schweitzer, *Entropie, Struktur, Komplexität* (Teubner, Leipzig, 1998).

1.21 W. Ebeling, in *Nonlinear Dynamics, Chaotic, and Complex Systems*, edited by E. Infeld, R. Zelazny, and A. Galkowski (Cambridge University Press, Cambridge, 1997).

1.22 W. Ebeling, Physica D **109**, 42 (1997).

1.23 W. Ebeling and M.A. Jimenez-Montano, Math. Biosci. **52**, 53 (1980).

1.24 W. Ebeling and K. Rateitschak, Discrete Dyn. Nat. Soc. **2**, 187 (1998).

1.25 R. Engbert, C. Scheffczyk, R.T. Krampe, M. Rosenblum, J. Kurths, and R. Kliegl, Phys. Rev. E **56**, 5823 (1997).

1.26 D.P. Feldman and J.P. Crutchfield, Phys. Lett. A **238**, 244 (1998).

1.27 L. Gatlin, *Information Theory and the Living System* (Columbia University Press, New York, 1972).

1.28 W. Gowin, P.I. Saparin, J. Kurths, and D. Felsenberg, Radiology **205**, 428 (1997).

1.29 W. Gowin, P.I. Saparin J. Kurths, and D. Felsenberg, Technol. Health Care **6**, 373 (1998).

1.30 P. Grassberger, Int. J. Theor. Phys. **25**, 907 (1986); Physica A **A 140**, 319 (1986).

1.31 A. Hempelmann and J. Kurths, Astron. Astrophys. **232**, 356 (1990).

1.32 H. Herzel, A.O. Schmitt, and W. Ebeling, Phys. Rev. E **50**, 5061 (1994).

1.33 H. Herzel, W. Ebeling, and A. Schmitt, Chaos, Solitons Fractals **4**, 97 (1994).

1.34 H. Herzel, W. Ebeling, A.O. Schmitt, and M.A. Jimenez-Montano, in *From Simplicity to Complexity in Chemistry*, edited by A. Müller et al. (Vieweg, Braunschweig, 1996).

1.35 H. Herzel, W. Ebeling, and I. Grosse, *Proc. Conf. Bioinf.* (GBF Monogr. 18, Braunschweig 1995).

1.36 H. Herzel and I. Grosse, Physica A **216**, 518 (1995).

1.37 H. Herzel and I. Grosse, Phys. Rev. E **55**, 1 (1997).

1.38 W. Hilberg, Frequenz **44**, 243 (1990).

1.39 M.A. Jimenez Montano, Bull. Math. Biol. **46**, 641 (1984).

1.40 W. Li and K. Kaneko, Europhys. Lett. **17**, 655 (1992).

1.41 H. Kantz and T. Schreiber, *Nonlinear Time Series Analysis* (Cambridge University Press, Cambridge, 1997).

1.42 F. Kaspar and H.G. Schuster, Phys. Rev. A **36**, 842 (1987).

1.43 Yu.L. Klimontovich, *Turbulent Motion and the Structure of Chaos* (Kluwer, Dordrecht, 1991).

1.44 A.N. Kolmogorov, Dokl. Akad. Nauk USSR **124**, 754 (1959).

1.45 A.N. Kolmogorov, Probl. Inf. Theory **1**, 3 (1965).

1.46 J. Kurths and U. Schwarz, Space Sci. Rev. **68**, 171 (1994).

1.47 J. Kurths, A. Voß, P. Saparin, A. Witt, H.J. Kleiner, and N. Wessel, Chaos **5**, 88 (1995).

1.48 R. Krampe, R. Kliegl, and U. Mayr, *The Fast and the Slow of Bimanual Movement Timing*, Res. Rep. (Max-Planck-Institute for Human Development, Berlin, Germany, 1993).

1.49 P.T. Landsberg, Phys. Lett. A **102**, 107 (1984).

1.50 P.T. Landsberg, in *On Self-Organization*, edited by R.K. Misra, D. Maas, and E. Zwierlein (Springer, Berlin, 1994).

1.51 A. Lempel and J. Ziv, IEEE Trans. Inf. Theory **IT-22**, 75 (1976).

1.52 L. Levitin and Z. Reingold, Chaos, Solitons Fractals **4**, 709 (1994).

1.53 K. Lindgren and M. Nordahl, Complex Syst. **2**, 409 (1988).

1.54 R. Lopez-Ruiz, H.L. Mancini, and X. Calbet, Phys. Lett. A **209**, 321 (1995).

1.55 D.W. McShea, Biol. Physiol. **6**, 303 (1991).

1.56 L. Molgedey, Int. J. Theo. App. Finance **3**, 417 (2000).

1.57 L. Molgedey and W. Ebeling, Eur. Phys. J. B **15**, 733 (2000).

1.58 A. Neiman, B. Shulgin, V. Anishchenko, W. Ebeling, L. Schimansky-Geier, and J. Freund, Phys. Rev. Lett. **76**, 4299 (1996).

1.59 C. Nicolis, W. Ebeling, and C. Baraldi, Tellus **49A**, 10 (1997).

1.60 X. Pel and F. Moss, Nature **379**, 618 (1996).

1.61 C.K. Peng et al., Phys. Rev. E **49**, 1685 (1994).

1.62 S.M. Pincus, Proc. Natl. Acad. Sci. USA **88**, 2297 (1991).

1.63 B. Pompe, J. Stat. Phys. **73**, 587 (1993).

1.64 P. Saparin, A. Witt, J. Kurths, and V. Anishenko, Chaos, Solitons Fractals **4**, 1907 (1994).

1.65 P.I. Saparin, W. Gowin, J. Kurths, and D. Felsenberg, Phys. Rev. E **58**, 6449 (1998).

1.66 A.O. Schmitt, H. Herzel, and W. Ebeling, Europhys. Lett. **23**, 303 (1993).

1.67 A.O. Schmitt, W. Ebeling, and H. Herzel, Biosystems **37**, 199 (1996).

1.68 C. Schittenkopf and G. Deco, Physica D **94**, 57 (1996).

1.69 H.G. Schuster, *Deterministic Chaos: An Introduction* (VCH, Weinheim, 1988).

1.70 U. Schwarz, J. Kurths, A. Witt, and A.O. Benz, Astron. Astrophys. **277**, 215 (1993).

1.71 C. Shannon, Bell Systems Tech. **30**, 50 (1951).

1.72 J.S. Shiner, in *Self-Organization of Complex Structures: From Individual to Collective Dynamics*, edited by F. Schweitzer (Gordon and Breach, London, 1996).

1.73 J.S. Shiner, M. Davison, and P.T. Landsberg, Phys. Rev. E **59**, 1459 (1999).

1.74 P.T. Landsberg and J.S. Shiner, Phys. Lett. A **245**, 228 (1998).

1.75 Ya. B. Sinai, Dokl. Akad. Nauk USSR **124**, 768 (1959); **125**, 1200 (1959).

1.76 L.B. Smith and E. Thelen (eds.), *A Dynamic Systems Approach to the Development of Cognition and Action* (MIT, Cambridge, MA, 1994).

1.77 H.E. Stanley et al., Physica A **205**, 214 (1994).

1.78 G.E. Stelmach and J. Requin (eds.), *Tutorials in Motor Behavior II* (North Holland, Amsterdam, 1992).

1.79 E.N. Trifonov and V. Brendel, *Gnomic – A Dictionary of Genetic Codes* (VCH, Weinheim, 1987).

1.80 A. Voss, K. Hnatkova, N. Wessel, J. Kurths, A. Sander, A. Schirdewan, A.J. Camm, and M. Malik, Pace **21**, 186 (1998).

1.81 H. Voss and J. Kurths, Phys. Rev. E **58**, 1155 (1998).

1.82 R.F. Voss, Phys. Rev. Lett. **68**, 3805 (1992).

1.83 R.F. Voss, Fractals **2**, 1 (1994).

1.84 R. Wackerbauer, A. Witt, H. Atmanspacher, J. Kurths, and H. Scheingraber, Chaos, Solitons Fractals **4**, 133 (1994).

1.85 P.C. Werner, F.-W. Gerstengarbe, and W. Ebeling, Theor. Appl. Climatol. **62**, 125 (1999).

1.86 A. Witt, J. Kurths, F. Krause, and K. Fischer, Geophys. Astrophys. Fluid Dyn. **77**, 79 (1994).

1.87 A. Witt, A. Neiman, and J. Kurths, Phys. Rev. E **55**, 5050 (1997).

1.88 A.M. Yaglom and I.M. Yaglom, *Probability and Information* (Kluwer, Dordrecht, 1983).

1.89 H.P. Yockey, *Information Theory and Molecular Biology* (Cambridge University Press, Cambridge, 1992).

1.90 M. Zaks, A. Pikovsky, and J. Kurths, Physica D **117**, 77 (1998).

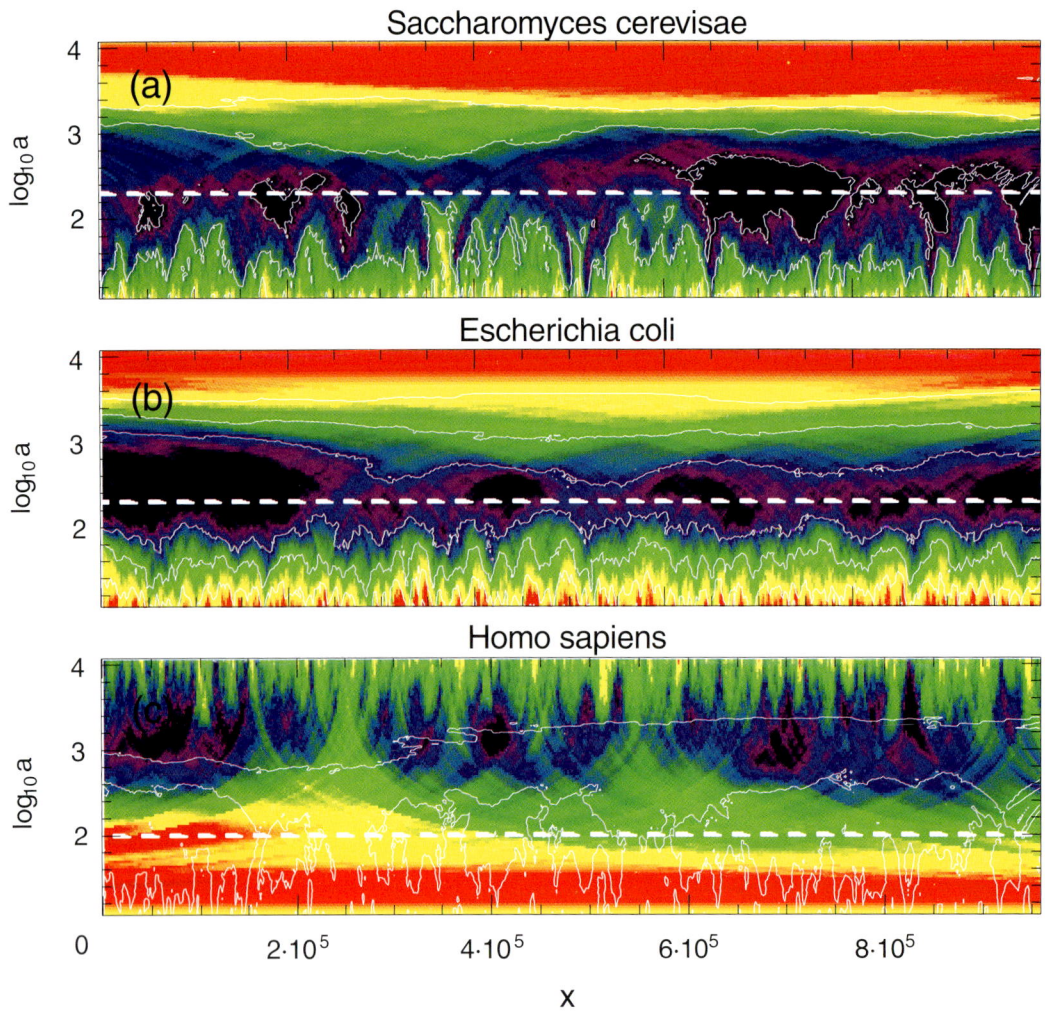

Saccharomyces cerevisae

Escherichia coli

Homo sapiens

2. Wavelet Based Multifractal Formalism: Applications to DNA Sequences, Satellite Images of the Cloud Structure, and Stock Market Data

Alain Arneodo, Benjamin Audit, Nicolas Decoster, Jean-Francois Muzy, and Cedric Vaillant

We elaborate on a unified thermodynamic description of multifractal distributions including measures and functions. This new approach relies on the computation of partition functions from the wavelet transform skeleton defined by the wavelet transform modulus maxima (WTMM). This skeleton provides an adaptive space-scale partition of the fractal distribution under study, from which one can extract the $D(h)$ singularity spectrum as the equivalent of a thermodynamic function. With some appropriate choice of the analyzing wavelet, we show that the WTMM method provides a natural generalization of the classical box-counting and structure function techniques. We then extend this method to multifractal image analysis, with the specific goal to characterize statistically the roughness fluctuations of fractal surfaces. As a very promising perspective, we demonstrate that one can go even deeper in the multifractal analysis by studying correlation functions in both space and scales. Actually, in the arborescent structure of the WT skeleton is somehow uncoded the multiplicative cascade process that underlies the multifractal properties of the considered deterministic or random function. To illustrate our purpose, we report on the most significant results obtained when applying our concepts and methodology to three experimental situations, namely the statistical analysis of DNA sequences, of high resolution satellite images of the cloud structure, and of stock market data.

◀ **Fig. 2.0.** Local estimate of the r.m.s. $\sigma(a, x)$ of the WT coefficients of the A DNA walk, computed with the Mexican hat analyzing wavelet $g^{(2)}$. $\sigma(a)$ is computed over a window of width $l = 2\,000$, sliding along the first 10^6 bp of the *yeast* chromosome IV (**a**), *Escherichia coli* (**b**), and a human contig (**c**). $\log_{10} \sigma(a) - 2/3 \log_{10} a$ is coded using 128 colors from black (min) to red (max). In this space-scale wavelet like representation, x and a are expressed in nucleotide units. The horizontal white dashed lines mark the scale a^* where some minimum is observed consistently along the entire genomes: $a^* = 200$ bp for *Saccharomyces cerevisiae*, $a^* = 200$ bp for *Escherichia coli*, and $a^* = 100$ bp for the human contig

2.1 Introduction

In the real world, it is often the case that a wide range of scales is needed to characterize physical properties. Actually, multiscale phenomena seem to be ubiquitous in nature. A paradigmatic illustration of such situation are *fractals* which are complex mathematical objects that have no minimal natural length scale. The relevance of fractals to physics and many other fields was pointed out by Mandelbrot [2.1] who demonstrated the richness of fractal geometry and stimulated many theoretical, numerical, and experimental studies. Actually, in many situations in physics as well as in some applied sciences, one is faced with the problem of characterizing very irregular functions [2.1–2.20]. The examples range from plots of various kinds of random walks, e.g. Brownian signals [2.21, 2.22], to financial time series [2.19, 2.20, 2.23] (cf. also Chap. 13), to geological shapes [2.1, 2.16], to medical time series [2.9, 2.13, 2.14] (cf. also Chaps. 7–8), to interface developing in far from equilibrium growth processes [2.6, 2.10, 2.17], to turbulent velocity signals [2.15], and to DNA 'walks' coding nucleotide sequences [2.24–2.26]. These functions can be qualified as *fractal functions* [2.1, 2.22, 2.27, 2.28] whenever their graphs are fractal sets in $\mathbb{R}^2$ (in this introduction we will only consider functions from $\mathbb{R}$ to $\mathbb{R}$). They are commonly called *self-affine* functions since their graphs are self-affine sets which are similar to themselves when transformed by anisotropic dilations, i.e. when shrinking along the x-axis by a factor λ followed by a re-scaling of the increments of the function by a different factor λ^{-H}. This can be stated mathematically in the following way: if $f(x)$ is a self-affine function then, $\forall x_0 \in \mathbb{R}$, $\exists H \in \mathbb{R}$ such that for any $\lambda > 0$, one has

$$f(x_0 + \lambda x) - f(x_0) \simeq \lambda^H \big(f(x_0 + x) - f(x_0) \big) . \tag{2.1}$$

If f is a stochastic process, this identity holds in law for fixed λ and x_0. The exponent H is called the *roughness* or Hurst exponent [2.1, 2.5, 2.10]. Let us note that if $H < 1$, then f is not differentiable and the smaller the exponent H, the more singular f. Thus the Hurst exponent provides indication of how globally irregular the function f is. Indeed H is supposed to be related to the fractal dimension $D_{\mathrm{F}} = 2 - H$ of the graph of f [2.1, 2.5, 2.10, 2.22].

In various contexts, several methods have been used to estimate the Hurst exponent. For instance, for the growth of rough surfaces [2.5, 2.6, 2.10, 2.17, 2.18], the average height difference between points separated by a distance l has been considered to scale as:

$$\langle |f(x + l) - f(x)| \rangle \sim l^H . \tag{2.2}$$

Alternatively, the root-mean square (r.m.s.) of the height fluctuations over a distance l is a quantitative measure of the width or thickness of the rough interface [2.5, 2.6, 2.10, 2.17, 2.18]:

$$w(l) = \left[\langle f^2 \rangle_l - \langle f \rangle_l^2\right]^{1/2} \sim l^H \; . \tag{2.3}$$

Note that in the pioneering analysis of DNA sequences [2.24, 2.25], the r.m.s. fluctuations about the average displacement of the DNA walk have been mainly used to detect the presence of long-range correlations ($H \neq 1/2$). A more classical method consists in investigating the scaling behavior of the power spectrum as a function of the wavevector k [2.1–2.20]:

$$S(k) \sim k^{-(2H+1)} \; . \tag{2.4}$$

But some care is required when using these methods, since they may lead to conflicting estimates of the Hurst exponent [2.29, 2.30]. Limited resolution as well as finite-size effects are well known to introduce biases in the estimate of H. Moreover, on a more fundamental ground, these methods are not adapted when the fractal function under consideration is not an homogeneous fractal function with a constant roughness associated to a unique exponent H [2.31, 2.32].

Fractal functions can possess multi-affine properties in the sense that their roughness (or regularity) may fluctuate from point to point [2.31–2.36]. To describe these *multifractal* functions, one thus needs to change slightly the definition of the Hurst regularity of f so that it becomes a local quantity [2.31, 2.32]:

$$\left| f(x+l) - f(x) \right| \sim l^{h(x)} \; . \tag{2.5}$$

This 'local Hurst exponent' $h(x)$ is generally called the *Hölder exponent* of f at the point x. A more rigorous definition of the Hölder exponent, as the strength of the singularity of a function f at the point x_0, is given by the largest exponent such that there exist a polynomial $P_n(x - x_0)$ of order $n < h(x_0)$ and a constant $C > 0$, so that for any point x in the neighborhood of x_0, one has [2.37–2.42]:

$$\left| f(x) - P_n(x - x_0) \right| \leq C \left| x - x_0 \right|^h \; . \tag{2.6}$$

If f is n times continuously differentiable at the point x_0, then one can use for the polynomial $P_n(x - x_0)$, the order-n Taylor series of f at x_0 and thus prove that $h(x_0) > n$. Thus $h(x_0)$ measures how irregular the function f is at the point x_0. The higher the exponent $h(x_0)$, the more regular the function f.

The aim of a quantitative theory of multi-affine functions is to provide mathematical concepts and numerical tools for the description of the fluctuations of regularity of these objects based on some limited amount of information. In the early 1980s, a phenomenological approach to the characterization of fractal objects was proposed and advanced: the *multifractal formalism* [2.43–2.51]. In its original form, this approach was essentially adapted to describe statistically the scaling properties of *singular measures* [2.45, 2.46, 2.51]. Notable examples include the invariant probability distribution on a strange attractor [2.45, 2.46, 2.51], the distribution of voltage drops across a random resistor network [2.2, 2.5, 2.11], the distribution of growth probabilities on the boundary of a diffusion-limited aggregate [2.5, 2.6, 2.52], and the spatial distribution of dissipative regions in a turbulent flow [2.15, 2.48, 2.53, 2.54]. This formalism relies upon the determination of the so-called $f(\alpha)$ *singularity spectrum* [2.45], which characterizes the relative contribution of each singularity of the measure: let S_α be the subset of points x where the measure of an ϵ-box $B_x(\epsilon)$, centered at x, scales like $\mu(B_x(\epsilon)) \sim \epsilon^\alpha$ in the limit $\epsilon \to 0^+$, then, by definition, $f(\alpha)$ is the Hausdorff dimension of S_α: $f(\alpha) = \dim_{\mathrm{H}}(S_\alpha)$. Actually, there exists a deep analogy that links the multifractal formalism with that of statistical thermodynamics [2.55–2.57]. This analogy provides a natural connection between the $f(\alpha)$ spectrum and a directly observable spectrum $\tau(q)$ defined from the power-law behavior, in the limit $\epsilon \to 0^+$, of the partition function [2.45]:

$$Z_q(\epsilon) = \sum_i \mu\Big(B_i(\epsilon)\Big)^q \sim \epsilon^{\tau(q)}, \qquad (2.7)$$

where the sum is taken over a partition of the support of the singular measure into boxes of size ϵ. The variables q and $\tau(q)$ play the same role as the inverse of temperature and the free energy in thermodynamics, while the Legendre transform

$$f(\alpha) = \min_q \Big(q\alpha - \tau(q)\Big) \qquad (2.8)$$

indicates that instead of energy and entropy, we have α and $f(\alpha)$ as the thermodynamic variables conjugate to q and $\tau(q)$ [2.36, 2.45–2.47, 2.50]. Let us mention that the so-called generalized fractal dimensions D_q [2.49, 2.58] are nothing else than $D_q = \tau(q)/(q-1)$. Most of the rigorous mathematical results concerning the multifractal formalism have been obtained in the context of dynamical system theory [2.46, 2.51] and of modeling of random cascades in fully developed turbulence [2.53, 2.59, 2.60]. It has been developed into a powerful technique (e.g. box-counting algorithms, fixed-size and fixed-mass correlation algorithms) [2.49, 2.61–2.66] accessible also to experimentalists. Successful applications have been reported in various fields [2.8, 2.11, 2.13, 2.15] and the pertinence of the multifractal approach seems, nowadays, to be well admitted in the scientific community at large.

There have been several attempts to extend the concept of multifractals to *singular functions* [2.33, 2.34]. In the context of fully developed turbulence [2.15], the intermittent character of turbulent velocity signals was investigated by calculating the moments of the probability density function (PDF) of (longitudinal) velocity increments $\delta v_l(x) = v(x+l) - v(x)$, over inertial separation [2.15, 2.33]:

$$S_p(l) = \left\langle \left| \delta v_l \right|^p \right\rangle \sim l^{\zeta_p}. \tag{2.9}$$

By Legendre transforming the scaling exponents ζ_p of these *structure functions* (SF) of order p [2.67], one aims to estimate the Hausdorff dimension $D(h)$ of the subset of $\mathbb{R}$ where velocity increments behave as $\delta v_l \sim l^h$ [2.33]:

$$D(h) = \min_p \left(ph - \zeta_p + 1 \right). \tag{2.10}$$

In a more general context, $D(h)$ can be defined as the spectrum of Hölder exponents for the singular signal under study, in full analogy with the $f(\alpha)$ singularity spectrum (2.8) for singular measures. However, as discussed in the next section, there are some fundamental limitations to the structure function approach which intrinsically fails to fully characterize the $D(h)$ singularity spectrum [2.68].

Our purpose here is to report on an alternative strategy that we have proposed and which is likely to provide a unified thermodynamic description of multifractal distributions including measures and functions [2.31, 2.32, 2.35, 2.36, 2.42]. This approach relies on the use of a mathematical tool introduced in signal analysis: the *wavelet transform* [2.69–2.81]. The wavelet transform has been proved to be very efficient to detect singularities [2.37–2.42, 2.77–2.80]. In that respect, it is a well adapted technique to study fractal objects [2.77, 2.82–2.86]. Since a wavelet can be seen as an oscillating variant of a box (i.e. a 'square' function), we will show, as a first step, that one can generalize the multifractal formalism by defining new partition functions in terms of wavelet coefficients [2.31, 2.32, 2.35, 2.36, 2.42]. In particular, by choosing a wavelet which is orthogonal to polynomial behavior up to some order N, one can make the wavelet transform blind to regular behavior, remedying in this way for one of the main failures of the classical approaches (e.g. the box-counting method in the case of measures and the structure function method in the case of functions). The other fundamental advantage of using wavelets is that the skeleton defined by the *wavelet transform modulus maxima* (WTMM) [2.40, 2.41] provides an adaptive space-scale partition of the fractal distribution under study, from which one can extract the $D(h)$ singularity spectrum [2.31, 2.32, 2.35, 2.36, 2.42]. As a second step, we will demonstrate that one can go even deeper in the multifractal analysis by studying correlation functions in both space and scales [2.87, 2.88]. Actually, in the arborescent structure of the wavelet transform skeleton, is somehow uncoded the multiplicative cascade process that underlies the multifractal properties of the considered function. To illustrate our purpose, we will report on the most significant results obtained when applying our concepts and methodology to three different experimental situations, namely the statistical analysis of DNA sequences, of high resolution satellite images of the cloud structure, and of stock market data.

2.2 The Wavelet Transform Modulus Maxima Method for the Multifractal Analysis of 1D signals

The continuous wavelet transform (WT) is a mathematical technique introduced in signal analysis in the early 1980s [2.89–2.91]. Since then, it has been the subject of considerable theoretical developments and practical applications in a wide variety of fields [2.69–2.81]. The WT has been early recognized as a mathematical microscope that is well adapted to reveal the hierarchy that governs the spatial distribution of the singularities of multifractal measures [2.84–2.86]. What makes the WT of fundamental use in the present study is that its singularity scanning ability equally applies to singular functions than to singular measures [2.31, 2.32, 2.35–2.42, 2.77–2.80, 2.82, 2.83].

2.2.1 The Continuous Wavelet Transform

The WT is a space-scale analysis which consists in expanding signals in terms of *wavelets* which are constructed from a single function, the *analyzing wavelet* ψ, by means of translations and dilations. The WT of a real-valued function f is defined as [2.89–2.91]:

$$T_\psi[f](x_0, a) = \frac{1}{a} \int_{-\infty}^{+\infty} f(x)\psi\left(\frac{x - x_0}{a}\right) dx , \qquad (2.11)$$

where x_0 is the space parameter and a (> 0) the scale parameter. The analyzing wavelet ψ is generally chosen to be well localized in both space and frequency. Usually ψ is required to be of zero mean for the WT to be invertible. But for the particular purpose of singularity tracking that is of interest here, we will further require ψ to be orthogonal to low-order polynomials [2.31, 2.32, 2.35–2.42]:

$$\int_{-\infty}^{+\infty} x^m \psi(x)dx , \quad \forall m , \quad 0 \le m < n_\psi . \qquad (2.12)$$

As originally pointed out by Mallat and collaborators [2.40, 2.41], for the specific purpose of analyzing the regularity of a function, one can get rid of the redundancy of the WT by concentrating on the WT skeleton defined by its modulus maxima only. These maxima are defined, at each scale a, as the local maxima of $|T_\psi[f](x, a)|$ considered as a function of x (Fig. 2.1c). As illustrated in Fig. 2.1d, these WTMM are disposed on connected curves in the space-scale (or time scale) half-plane, called *maxima lines*. Let us define $\mathcal{L}(a_0)$ as the set of all the maxima lines that exist at the scale a_0 and which contain maxima at any scale $a \le a_0$. An important feature of these maxima lines, when analyzing singular functions, is that there is at least one maxima line pointing towards each singularity (Fig. 2.1d) [2.32, 2.40–2.42].

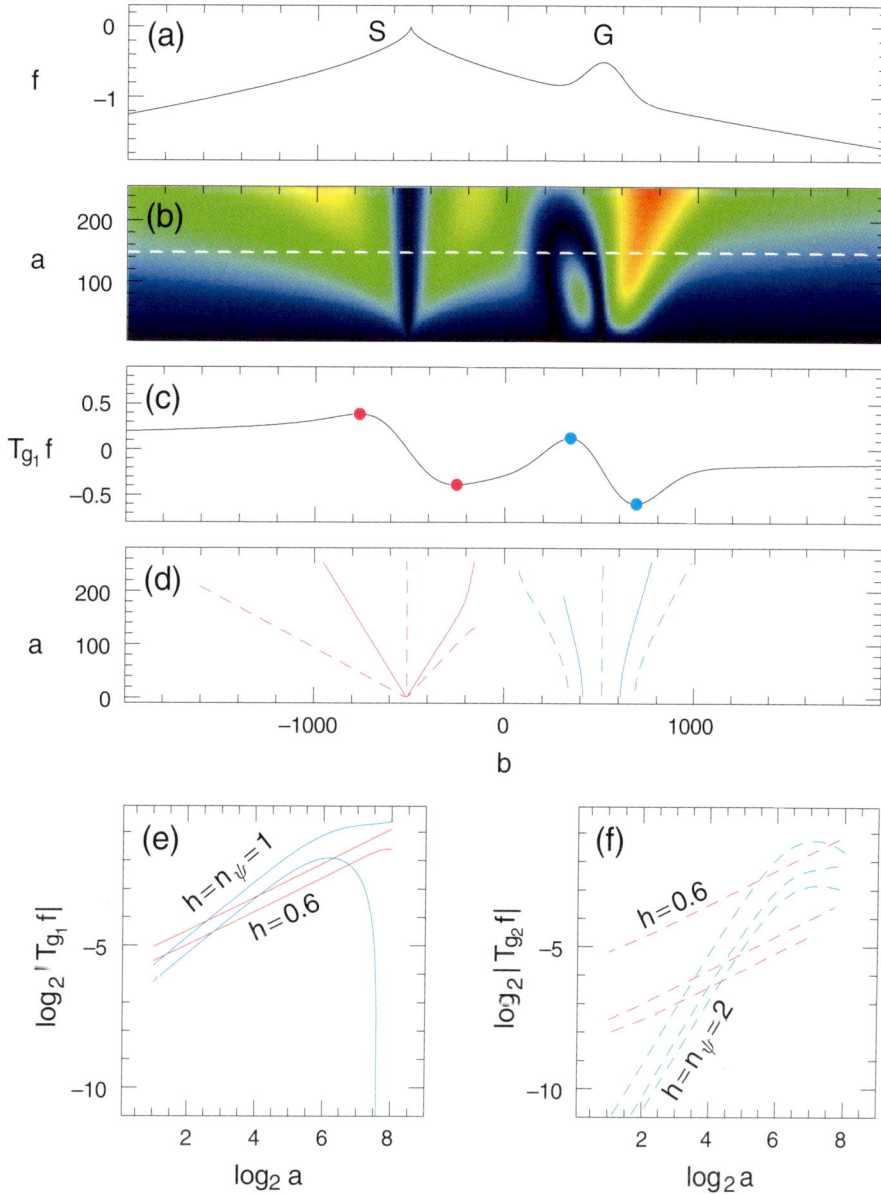

Fig. 2.1. How to estimate the Hölder exponent from the behavior of the WT modulus along the maxima lines. (**a**) Graph of the function $f(x) = -|\frac{x-x_0}{1024}|^{0.6} + \frac{1}{2}\exp(-\frac{1}{2}(\frac{x-x_1}{100})^2)$, which displays a singularity S of Hölder exponent $h = 0.6$ located at $x_0 = -512$ and a smooth localized behavior G at $x_1 = 512$. (**b**) The WT of $f(x)$ computed using the first-order derivative of the Gaussian function $g^{(1)}$ (2.20) as coded according to the natural order of the light spectrum from black ($|T_\psi| = 0$) to red ($\max_{b,a}|T_\psi|$). (**c**) Horizontal section of $T_\psi[f](b,a)$ at the scale $a = a_0$; the symbols (•) represent the modulus maxima. (**d**) Maxima lines of the WT in the (b,a) half-plane as computed when using the first-order (——) or the second-order (– – –) derivative of the Gaussian function. (**e**) Local measurement of the scaling exponent along the maxima lines when using $g^{(1)}$ ($n_\psi = 1$): $h(x_0) = 0.6$, $h(x_1) = n_\psi = 1$; the slope of the curve $\log_2|T_\psi f|$ vs. $\log_2 a$ provides an estimate of the scaling exponent. (**f**) Same computation as in (**e**) but when using a second-order analyzing wavelet $g^{(2)}$ ($n_\psi = 2$): $h(x_0) = 0.6$, $h(x_1) = n_\psi = 2$. The results in (**e**) and in (**f**) are therefore in good agreement with the theoretical predictions given by (2.14) and (2.15). Let us remark that the number of maxima lines that point to a given singularity is $n_\psi + 1$

2.2.2 Scanning Singularities
with the Wavelet Transform Modulus Maxima

As introduced in Sect. 2.1, the strength of a singularity of a function is quantified by the *Hölder exponent* (2.6). This definition of the singularity strength naturally leads to a generalization of the so-called $f(\alpha)$ singularity spectrum (2.8) introduced for fractal measures in [2.45, 2.46, 2.51]. As originally defined by Parisi and Frisch [2.33], we will denote by $D(h)$ the Hausdorff dimension of the set where the Hölder exponent is equal to h [2.31, 2.32, 2.36]:

$$D(h) = \dim_{\mathrm{H}}\{x \ , \ h(x) = h\} \ , \qquad (2.13)$$

where h can take, a priori, positive as well as negative real values (e.g. the Dirac distribution $\delta(x)$ corresponds to the Hölder exponent $h(0) = -1$).

The main interest in using the WT for analyzing the regularity of a function lies in its ability to be blind to polynomial behavior by an appropriate choice of the analyzing wavelet ψ. Indeed, let us assume that according to (2.6), f has, at the point x_0, a local scaling (Hölder) exponent $h(x_0)$; then, assuming that the singularity is not oscillating [2.40, 2.92, 2.93], one can easily prove that the local behavior of f is mirrored by the WT which locally behaves like [2.37, 2.39–2.42, 2.82, 2.83, 2.94]:

$$T_\psi[f](x_0, a) \sim a^{h(x_0)} \ , a \to 0^+ \ , \qquad (2.14)$$

provided $n_\psi > h(x_0)$, where n_ψ is the number of vanishing moments of ψ (2.12). Therefore one can extract the exponent $h(x_0)$ as the slope of a log-log plot of the WT amplitude vs. the scale a. On the contrary, if one chooses $n_\psi < h(x_0)$, the WT still behaves as a power-law but with a scaling exponent which is n_ψ :

$$T_\psi[f](x_0, a) \sim a^{n_\psi} \ , a \to 0^+ \ . \qquad (2.15)$$

Thus, around a given point x_0, the faster the WT decreases when the scale goes to zero, the more regular f is around that point. In particular, if $f \in C^\infty$ at x_0 ($h(x_0) = +\infty$), then the WT scaling exponent is given by n_ψ, i.e. a value which is dependent on the shape of the analyzing wavelet. According to this observation, one can hope to detect the points where f is smooth by just checking the scaling behavior of the WT when increasing the order n_ψ of the analyzing wavelet [2.32, 2.35, 2.36, 2.42]. Let us also remark that if $h(x_0)$ is negative, then the WT no longer decreases but instead increases when the scale a goes to zero; this remark has been of fundamental use in [2.95] for detecting vorticity filaments in turbulent pressure signals.

A very important point (at least for practical purposes) raised by Mallat and Hwang [2.41] is that the local scaling exponent $h(x_0)$ can be equally estimated by looking at the value of the WT modulus along a maxima line converging towards the point x_0. Indeed one can prove that (2.14) and (2.15) still hold when following a maxima line from large down to small scales [2.41, 2.42]. This is illustrated in Fig. 2.1 for the particular case of an isolated singularity 'interacting' with a localized smooth structure. The situation is somewhat more intricate when investigating fractal functions. Indeed the characteristic feature of these singular functions is the existence of a hierarchical distribution of singularities [2.31, 2.32, 2.35, 2.36, 2.42, 2.82, 2.83]. Locally, the Hölder exponent $h(x_0)$ is governed by the singularities which accumulate at x_0. This results in unavoidable oscillations around the expected power-law behavior of the WT amplitude [2.31, 2.32, 2.36, 2.42, 2.96]. Therefore the exact determination of h from log-log plots on a finite range of scales is somewhat uncertain [2.97–2.99]. Of course, there have been many attempts to circumvent these difficulties [2.31, 2.98]. Nevertheless there exist fundamental limitations (which are not intrinsic to the WT technique) to the estimate of Hölder exponents of fractal functions. Consequently, the determination of statistical quantities like the $D(h)$ singularity spectrum requires a method which is more feasible and more appropriate than a systematic investigation of the local scaling behavior of the WT. This is the purpose of the recently developed wavelet-based multifractal formalism [2.32, 2.35, 2.36, 2.42] that we will explicitly use in Sect. 2.3 to analyze the statistical scaling properties of DNA walks.

2.2.3 A Wavelet-Based Multifractal Formalism: The Wavelet Transform Modulus Maxima Method

A natural way of performing a multifractal analysis of fractal functions consists in generalizing the 'classical' multifractal formalism [2.43–2.51] using wavelets instead of boxes. By taking advantage of the freedom in the choice of the 'generalized oscillating boxes' that are the wavelets, one can hope to get rid of possible smooth behavior that could mask singularities or perturb the estimation of their strength h. But the major difficulty with respect to box-counting techniques [2.49, 2.61, 2.62, 2.64, 2.65] for singular measures, consists in defining a covering of the support of the singular part of the function with our set of wavelets of different sizes. A simple method would rely on the definition of the following *partition function* in terms of WT coefficients [2.82, 2.83]:

$$Z(q,a) = \int \left| T_\psi[f](x,a) \right|^q dx \,, \tag{2.16}$$

where $q \in \mathbb{R}$. This method based on a continuous covering of the real line would be, however, unstable for negative q values since nothing prevents the WT coefficients from vanishing at some point (x_0, a) of the space-scale half-plane.

The wavelet transform modulus maxima (WTMM) [2.32, 2.35, 2.36, 2.42] method implies that one changes the continuous sum over space in (2.16) into a discrete sum over the local maxima of $|T_\psi[f](x,a)|$:

$$Z(q, a) = \sum_{l \in \mathcal{L}(a)} \left(\sup_{\substack{(x,a') \in l \\ a' \le a}} |T_\psi[f](x, a')| \right)^q . \qquad (2.17)$$

As emphasized in [2.32, 2.35, 2.36, 2.42], the branching structure of the WT skeleton in the (x, a) half-plane enlightens the hierarchical organization of the singularities (see Fig. 2.3e and 2.3f). Thus the WT skeleton indicates how to position, at the considered scale a, the oscillating boxes in order to obtain a partition of the singularities of f. The sup in (2.17) can be regarded as a way to define a scale adaptive 'Hausdorff-like' partition. Now from the deep analogy that links the multifractal formalism to thermodynamics [2.36, 2.47, 2.50], one can define the exponent $\tau(q)$ from the power-law behavior of the partition function:

$$Z(q, a) \sim a^{\tau(q)} , \quad a \to 0^+ , \qquad (2.18)$$

where q and $\tau(q)$ again play respectively the role of the inverse temperature and the free energy. The main result of this wavelet-based multifractal formalism is that in place of the energy and the entropy (i.e. the variables conjugated to q and τ), one has h, the Hölder exponent, and $D(h)$, the singularity spectrum. This means that the singularity spectrum of f can be determined from the Legendre transform of the partition function scaling exponent $\tau(q)$ [2.42, 2.100]:

$$D(h) = \min_q(qh - \tau(q)) . \qquad (2.19)$$

From the properties of the Legendre transform, it is easy to see that *homogeneous* fractal functions that involve singularities of unique Hölder exponent $h = \partial\tau/\partial q$, are characterized by a $\tau(q)$ spectrum which is a *linear* function of q. On the contrary, a *nonlinear* $\tau(q)$ curve is the signature of nonhomogeneous functions that exhibit *multifractal* properties, in the sense that the Hölder exponent $h(x)$ is a fluctuating quantity that depends upon the spatial position x.

Remark 1: The partition function exponents $\tau(q)$ are much more than simply some intermediate quantities of a rather easy experimental access. For some specific values of q, they have well known meaning [2.31, 2.32]. In full analogy with standard box-counting arguments, $-\tau(0)$ can be identified to the capacity of the set of singularities of f: $-\tau(0) = d_C(\{x, h(x) < +\infty\})$. Similarly, $\tau(1)$ is related to the capacity of the graph $\mathcal{G}$ of the considered function: $d_C(\mathcal{G}) = \max(1, 1-\tau(1))$. Finally $\tau(2)$ is related to the scaling exponent β of the spectral density $S(k) = |\hat{f}(k)|^2 \sim k^{-\beta}$ with $\beta = 2 + \tau(2)$.

Remark 2: Since the WTMM method is mainly devoted to practical applications to stochastic systems, let us point out that the theoretical treatment of *random multifractal functions* requires special attention. A priori, there is no reason that all the realizations of the same stochastic multifractal process correspond to a unique $D(h)$ curve. Each realization has its own unique distribution of singularities and one crucial issue is to relate these distributions to some averaged versions computed experimentally. As emphasized by Hentschel [2.101], one can take advantage of the analogy that links the multifractal description to statistical thermodynamics [2.36, 2.45–2.47, 2.50], by using methods created specifically to study disorder in spin-glass theory [2.102]. When carrying out replica averages of the random partition function associated with a stochastic function, one gets multifractal spectra $\tau(q, n)$ that generally depend on the number n of members in the replica average (let us note that $n = 0$ and $n = 1$ respectively correspond to commonly used *quenched* and *annealed* averaging [2.101]). Then, by Legendre transforming $\tau(q, n)$, some type of average $D(h)$ spectra are being found [2.101]. Some care is thus required when interpreting these average spectra in order to avoid some misunderstanding of the underlying physics. We refer the reader to [2.42, 2.100] for rigorous mathematical results concerning the application of the WTMM method to stochastic functions.

2.2.4 Defining our Battery of Analyzing Wavelets

There are almost as many analyzing wavelets as applications of the continuous WT. In our original works [2.32, 2.36, 2.77, 2.84–2.86], we have mainly used the class of analyzing wavelets defined by the successive derivatives of the Gaussian function:

$$g^{(N)}(x) = \frac{d^N}{dx^N} e^{-x^2/2} , \qquad (2.20)$$

for which $n_\psi = N$. Throughout this study, we will rather use the set of compactly supported analyzing wavelets $\psi^{(n)}_{(m)}$ plotted in Fig. 2.2 [2.95, 2.103, 2.104]. They are constructed from Dirac distributions $(\psi^{(n)}_{(0)})$ via successive convolutions with the box function χ. The index m_ψ, corresponding to the number of convolutions, characterizes the smoothness of the analyzing wavelet. n_ψ is by definition the number of vanishing moments of ψ.

Let us note that the functions $\psi^{(0)}_{(m)}$ are not analyzing wavelets since they are not of zero mean. However, when using $\psi^{(0)}_{(1)}$ (which is nothing else than the box function χ) in the (continuous) partition function defined in (2.16), one recovers some variant of classical box-counting techniques [2.49, 2.61, 2.62, 2.64, 2.65] commonly used for multifractal analysis of singular measures. Let us also remark that when using $\psi^{(1)}_{(0)}(x) = \delta(x-1) - \delta(x)$, the WT is nothing but the local increment of the considered function and the (continuous) partition function (2.16) then reduces to the so-called structure function of order q (2.9) [2.15, 2.33, 2.67] . We refer the reader to [2.68] where a comparative study

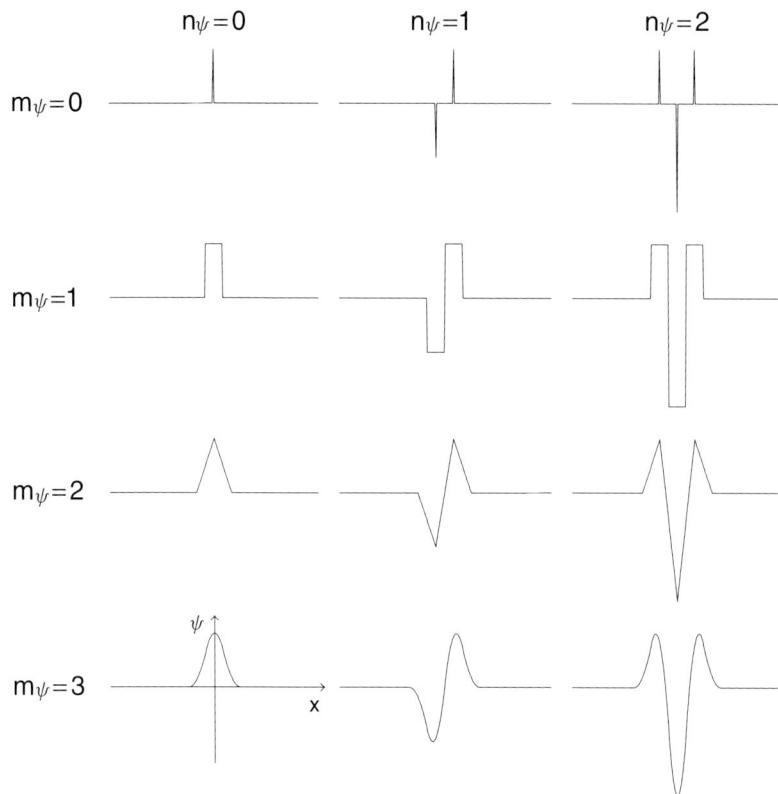

Fig. 2.2. Set of compactly supported analyzing wavelets $\psi_{(m)}^{(n)}$. n_ψ corresponds to the number of vanishing moments. The functions $\psi_{(m)}^{(n)}$ are smooth versions of $\psi_{(0)}^{(n)}$ obtained after m successive convolutions with the box function χ

of the structure function approach and the WTMM method is carried out. The main messages of this study are reported in the next section.

2.2.5 The Structure Function Approach vs. the WTMM Method

It is tempting to relate the exponents $\tau(q)$, defined from the WT partition functions (2.17), (2.18), to the scaling exponents ζ_p of the structure functions (2.9). A simple comparison of (2.10) and (2.19) gives immediately [2.31, 2.32]:

$$\tau(q) = \zeta_q - 1. \tag{2.21}$$

But this relationship does not hold for all values of q; this results from intrinsic limitations of the SF approach [2.68]. As pointed out just above [2.31,2.32,2.98], the local increment of a function can be seen as its WT computed at point x_0 and scale l, with the 'poor man's wavelet' $\Delta(x) = \psi_{(0)}^{(1)}(x) = \delta(x-1) - \delta(x)$. In this spirit, the SFs (2.9) are analogous to the naive partition functions defined in (2.16):

$$S_q(l) = \int \left| \delta f_l(x) \right|^q dx. \qquad (2.22)$$

However, there are two main differences between the WTMM partition functions (2.17) and the SFs (2.22), from which result the insufficiencies of the SF method [2.68].

(i) *The continuous integral used to define $S_q(l)$.* Indeed, there is no reason, a priori, that the increment PDF vanishes around $\delta f_l = 0$. Thus the SF $S_q(l)$ may diverge for $q \leq -1$. Consequently, the ζ_q spectrum estimate is unstable for $q < 0$ and from the Legendre transform properties, only the (increasing) part of the $D(h)$ singularity spectrum (e.g. the left part of the $D(h)$ curve in Fig. 2.4) corresponding to the strongest singularities is amenable to the SF approach. The first crucial advantage of the WTMM method is that the partition function (2.17) is computed using a discrete summation over the WT skeleton where the WT coefficients do not vanish (by definition of the WTMM); this keeps the calculation of $Z(q, a)$ stable for $q < 0$.

(ii) *The poorness of the analyzing wavelet $\Delta(x)$.* The second main disadvantage of the SF method is that the 'poor man's wavelet' $\Delta(x)$ [2.31, 2.32, 2.98] does not satisfy the criteria to be an efficient analyzing wavelet. First $\Delta(x)$ is only orthogonal to constants but not to higher order polynomials ($n_\psi = 1$). This precludes the detection of singularities with $h \geq 1$. Thus, as demonstrated in [2.68], if the maximum value of $D(h)$ is reached for some $h \geq 1$, the accessible range of singularities is even more truncated to values $h < h_{\mathrm{crit}} < 1$. Moreover, $\Delta(x)$ is a singular analyzing wavelet made of two Dirac distributions and therefore it cannot generally be integrated against tempered distributions. When using (2.22) to study some distribution which involves singularities with $h \leq 0$, one is generally faced with severe instabilities in the computation of the SFs. Therefore, the range of Hölder exponents accessible to the SF method is not only limited from above ($h < h_{\mathrm{crit}} \leq 1$), but also from below ($h > 0$). Because one has the freedom to select an analyzing wavelet ψ which is smooth and which has enough vanishing moments ($n_\psi > h_{\max}$), the WTMM method does not possess any of these drawbacks [2.68] . This is why the WTMM method is much more than a simple generalization of commonly used box-counting and structure function techniques [2.31, 2.32, 2.35, 2.36, 2.42, 2.77].

2.2.6 Discriminating Multifractal from Monofractal Synthetic Random 1D Signals

This section is devoted to test applications of the WTMM method to random functions generated either by *additive* models like fractional Brownian motions [2.21, 2.22] or by *multiplicative* models like random $\mathcal{W}$-cascades on wavelet dyadic trees [2.103–2.107]. For each model, we first wavelet transform 1 000 realizations of length $L = 65\,536$ with a first order ($n_\psi = 1$) analyzing wavelet. From the WT skeletons defined by the WTMM, we compute the mean partition function (2.17) from which we extract the annealed $\tau(q)$ (2.18) and, in turn, $D(h)$ (2.19) multifractal spectra. We systematically test the robustness of our estimates with respect to some change of the shape of the analyzing wavelet, in particular when increasing the number n_ψ of zero moments.

Fractional Brownian Signals: Since its introduction by Mandelbrot and van Ness [2.21], the fractional Brownian motion (fBm) has become a very popular model in signal and image processing [2.1–2.22, 2.27]. In 1D, fBm has proved useful for modeling various physical phenomena with long-range dependence, e.g. '$1/f$' noises. The fBm exhibits a power spectral density $S(k) \sim 1/k^\beta$, where the spectral exponent $\beta = 2H + 1$ is related to the Hurst exponent H. fBm has been extensively used as test stochastic signals for Hurst exponent measurements. The performances of classical methods (e.g. height-height correlation function, variance and power spectral methods, first return and multi-return probability distributions, maximum likelihood techniques) [2.29, 2.30, 2.108–2.115] have been recently competed by wavelet-based techniques [2.116–2.126]. A fBm $B_H(x)$ indexed by $H \in\,]0, 1[$ is a Gaussian process of mean value 0 and whose correlation function is given by

$$\left\langle B_H(x)B_H(y)\right\rangle = \frac{\sigma^2}{2}\left(\left|x\right|^{2H} + \left|y\right|^{2H} - \left|x - y\right|^{2H}\right),\qquad(2.23)$$

where $\langle \ldots \rangle$ represents the mean value. The variance of such process is

$$\mathrm{var}\left(B_H(x)\right) = \sigma^2 |x|^{2H}\,.\qquad(2.24)$$

The classical Brownian motion corresponds to $H = 1/2$ and to a variance $\mathrm{var}(B_{1/2}(x)) = \sigma^2 |x|$. On can easily show that the increments of a fBm, i.e $\delta B_{H,l} = B_H(x + l) - B_H(x)$ ($l \in \mathbb{R}^{+*}$ fixed), are stationary. Indeed, the correlation function depends only on $x - y$ and l:

$$\left\langle \delta B_{H,l}(x)\delta B_{H,l}(y)\right\rangle = \frac{\sigma^2}{2}\left(\left|x - y + l\right|^{2H} + \left|x - y - l\right|^{2H} - 2\left|x - y\right|^{2H}\right).\qquad(2.25)$$

For $H = 1/2$, we recover the fact that the increments of the classical Brownian motion are independent. For any other value of H, the increments are either positively correlated ($H > 1/2$: persistent random walk) or anti-correlated ($H < 1/2$: antipersistent random walk). Moreover, from (2.23) one gets:

$$B_H(x + \lambda y) - B_H(x) \simeq \lambda^H\left(B_H(x + y) - B_H(x)\right),\qquad(2.26)$$

where $\simeq$ stands for the equality in law. This means that fBms are self-affine processes (2.1) and that the Hurst exponent is H. The higher H, the more regular the motion. But since (2.26) holds for any x and y, this means that almost all realizations of the fBm are continuous, everywhere nondifferentiable with a unique Hölder exponent $h(x) = H$, $\forall x$ [2.1, 2.22, 2.32, 2.127]. Thus the fBms are homogeneous fractals characterized by a singularity spectrum which reduces to a single point:

$$D(h) = 1 \qquad \text{if } h = H , \qquad (2.27)$$
$$= -\infty \quad \text{if } h \neq H .$$

By Legendre transforming $D(h)$, one gets the following expression for the partition function exponents:

$$\tau(q) = qH - 1 . \qquad (2.28)$$

$\tau(q)$ is a linear function of q with a slope given by the index H of the fBm. Let us point out that $\tau(2) \neq 0$ (i.e. $H \neq 1/2$) indicates the presence of long-range correlations.

In Fig. 2.3 and 2.4, we report the results of a statistical analysis of fBms using the WTMM method [2.32, 2.35, 2.36, 2.68]. We mainly concentrate on $B_{1/3}$ since it has a $k^{-5/3}$ power spectrum similar to the spectrum of the multifractal stochastic signal we will study in the next section. Actually, our goal is to demonstrate that, where the power spectrum analysis fails, the WTMM method succeeds in discriminating unambiguously between these two fractal signals. The numerical signals were generated by filtering uniformly generated pseudo-random noise in Fourier space in order to have the required $k^{-5/3}$ spectral density. A $B_{1/3}$ fractional Brownian trail is shown in Fig. 2.3a. Figure 2.3c illustrates the WT coded, independently at each scale a, using 256 colors. The analyzing wavelet is $\psi^{(1)}_{(3)}$ ($n_\psi = 1$). Figure 2.4a displays some plots of $\log_2 Z(q, a)$ vs. $\log_2(a)$ for different values of q, where the partition function $Z(q, a)$ has been computed on the WTMM skeleton shown in Fig. 2.3e, according to the definition in (2.17). Using a linear regression fit, we then obtain the slopes $\tau(q)$ of these graphs. As shown in Fig. 2.4c, when plotted vs. q, the data for the exponents $\tau(q)$ consistently fall on a straight line that is remarkably fitted by the theoretical prediction $\tau(q) = q/3 - 1$ (2.28). As expected theoretically, we find from the numerical application of the WTMM method that the fBm $B_{1/3}$ is a nowhere differentiable homogeneous fractal signal with a unique Hölder exponent $h = H = 1/3$ as given by the slope of the linear $\tau(q)$ spectrum (the hallmark of homogeneous fractal scaling). Similar good estimates are obtained when using analyzing wavelets of different orders, and this whatever the value of the index H of the fBm.

Random $\mathcal{W}$-Cascades: Multiplicative cascade models have enjoyed increasing interest in recent years as the paradigm of multifractal objects [2.1, 2.45, 2.48, 2.53, 2.54, 2.101]. The notion of cascade actually refers to a self-similar process whose properties are defined multiplicatively from coarse to

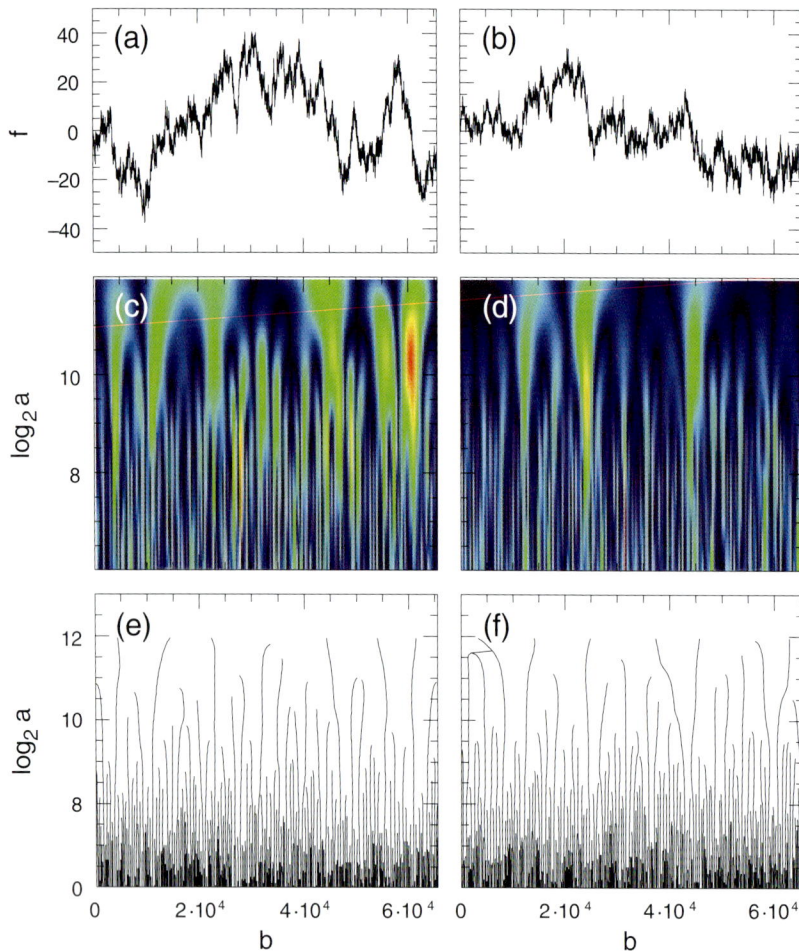

Fig. 2.3. WT of monofractal and multifractal stochastic signals. *Fractional Brownian motion:* (**a**) a realization of $B_{1/3}$ ($L = 65\,536$); (**c**) WT of $B_{1/3}$ as coded, independently at each scale a, using 256 colors from black ($|T_\psi| = 0$) to red ($\max_b |T_\psi|$); (**e**) WT skeleton defined by the set of all the maxima lines. *Log-normal random $\mathcal{W}$-cascades:* (**b**) a realization of the log-normal $\mathcal{W}$-cascade model ($L = 65\,536$) with the following parameter values $m = -0.355 \ln 2$ and $\sigma^2 = 0.02 \ln 2$ (see [2.104–2.106]); (**d**) WT of the realization in (**b**) represented with the same color coding as in (**c**); (**f**) WT skeleton. The analyzing wavelet is $\psi^{(1)}_{(3)}$ (see Fig. 2.2)

fine scales. In that respect, it occupies a central place in the statistical theory of turbulence [2.15, 2.54, 2.67]. Since Richardson's scenario [2.128], the turbulent cascade picture has often been invoked to account for the intermittency phenomenon observed in fully developed turbulent flows [2.15, 2.53, 2.54, 2.67]: energy is transferred from large eddies down to small scales (where it is dissipated) through a cascade process in which the transfer rate at a given scale is not spatially homogeneous, as supposed in the theory developed by Kolmogorov in 1941 [2.129], but undergoes local intermittent fluctuations [2.15, 2.53, 2.54, 2.67]. Over the past forty years, refined models including the log-normal model of Kolmogorov [2.130] and Obukhov [2.131], multi-

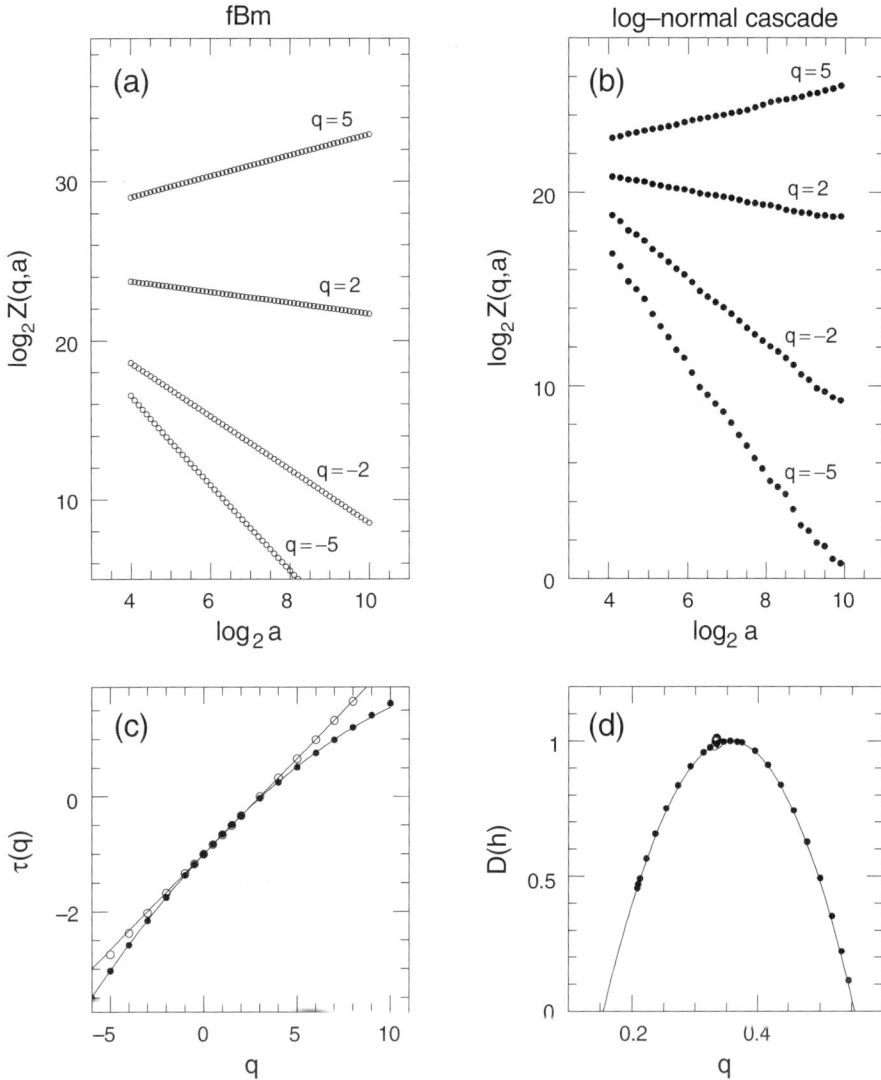

Fig. 2.4. Determination of the $\tau(q)$ and $D(h)$ multifractal spectra of fBm $B_{1/3}$ (circles) and log-normal random $\mathcal{W}$-cascades (dots) using the WTMM method. (**a**) $\log_2 Z(q,a)$ vs. $\log_2 a$: $B_{1/3}$. (**b**) $\log_2 Z(q,a)$ vs. $\log_2 a$: log-normal $\mathcal{W}$-cascades with the same parameters as in Fig. 2.3b. (**c**) $\tau(q)$ vs. q; the solid lines correspond respectively to the theoretical spectra (2.28) and (2.29). (**d**) $D(h)$ vs. h; the solid lines correspond respectively to the theoretical predictions (2.27) and (2.30). The analyzing wavelet is $\psi_{(3)}^{(1)}$. The reported results correspond to annealed averaging over 1 000 realizations of $L = 65\,536$

plicative hierarchical cascade models like the random β-model [2.43], the α-model [2.132], the p-model [2.133] (for a review, see [2.54]), the log-stable models [2.134–2.136] and more recently the log-infinitely divisible cascade models [2.137–2.141], have grown in the literature as reasonable models to mimic the energy cascading process in turbulent flows. On a very general ground, a self-similar cascade is defined by the way the scales are refined and by the statistics of the multiplicative factors at each step of the process [2.53, 2.54, 2.101, 2.136]. One can thus distinguish discrete cascades that involve discrete scale ratios leading to log-periodic corrections to scaling (*discrete scale invariance* [2.32, 2.85, 2.86, 2.96]), from continuous cascades without preferable scale factors (*continuous scale invariance*). More fundamentally, there are two main classes of self-similar cascade processes: *deterministic cascades* that generally correspond to solvable models and *random cascades* that are likely to provide more realistic models but for which some theoretical care is required as far as their multifractal limit and some basic multifractal properties (including multifractal phase transitions) are concerned [2.101]. As a notable member of the later class, the independent random cascades introduced by Mandelbrot (commonly called $\mathcal{M}$-*cascades* [2.53, 2.142, 2.143]) as a general model of random curdling in fully developed turbulence, have a special status since they are the main cascade model for which deep mathematical results have been obtained [2.59, 2.60]. Moreover, as pointed out by Schertzer and Lovejoy [2.144], this 'first generation' of cascade models is *static* in the sense that it accounts for the multiplicative hierarchical structure of the data in the spatial domain only [2.136]. With the specific goal to model temporal evolution, these authors have proposed a 'second generation' of space-time cascade models that take into account both the scaling anisotropy between space and time, and the breaking of the mirror symmetry along the temporal axis, i.e. causality [2.136, 2.144, 2.145].

Paradoxically, if there is a plethora of mono- and multifractal cascade models in the literature that generate deterministic as well as random singular measures in the small-scale limit, there are still only a handful of distinct algorithms for synthesizing 'rough' functions of a single variable with multifractal statistics. Beyond the problem of the multifractal description of singular functions that has been solved with the WTMM method [2.32, 2.35, 2.36, 2.42], there is thus the practical issue of defining in any concrete way how to build a multifractal function. Schertzer and Lovejoy [2.134] suggested a simple power-law filtering (fractional integration) of singular cascade measures. So, this model combines a multiplicative procedure with an additive one reminiscent of some algorithms to generate fBm [2.1, 2.27, 2.30]. In [2.34, 2.146], the midpoint displacement technique for building fBm was generalized to generate deterministic or random multi-affine functions. The same goal was achieved in [2.32, 2.42] by combining fractional or ordinary integration with signed measures obtained by recursive cascade like procedures. Several other attempts to simulate 'synthetic turbulence' that shares the intermittency properties of turbulent velocity data have partially succeeded [2.147–2.150]. More recently, the concept of self-similar cascades leading to multifractal measures has been generalized to the construction of scale-invariant signals using orthonormal wavelet basis [2.103–2.107]. Instead of re-distributing the measure over subintervals with multiplicative weights, one allocates the wavelet coefficients in a multiplicative way on the dyadic grid. This method has been implemented to generate multifractal functions from a given deterministic or probabilistic multiplicative process (see Sect. 2.6.2). From a mathematical point of view, the convergence of these $\mathcal{W}$-*cascades* and the regularity properties of the so-obtained deterministic or stochastic functions have been discussed in [2.106].

As inspired from the modeling of fully developed turbulent signals by log-infinitely divisible multiplicative processes [2.137–2.141], we will mainly concentrate here on the log-normal $\mathcal{W}$-cascades in order to calibrate the WTMM method. If m and σ^2 are respectively the mean and the variance of $\ln W$ (where W is a multiplicative random variable with log-normal probability distribution), then, as shown in [2.104–2.106], a straightforward computation leads to the following $\tau(q)$ spectrum:

$$\tau(q) = -\log_2 \langle W^q \rangle - 1 \ , \ \forall \ q \ \in \mathbb{R} \qquad (2.29)$$
$$= -\frac{\sigma^2}{2\ln 2} q^2 - \frac{m}{\ln 2} q - 1 \ ,$$

where $\langle \ldots \rangle$ means ensemble average. The corresponding $D(h)$ singularity spectrum is obtained by Legendre transforming $\tau(q)$ (2.29):

$$D(h) = -\frac{(h + m/\ln 2)^2}{2\sigma^2/\ln 2} + 1 \ . \qquad (2.30)$$

According to the convergence criteria established in [2.106], m and σ^2 have to satisfy the conditions:

$$m < 0 \ \text{ and } \ \frac{|m|}{\sigma} > \sqrt{2\ln 2} \ . \qquad (2.31)$$

Moreover, by solving $D(h) = 0$, one gets the following bounds for the support of the $D(h)$ singularity spectrum:

$$h_{\min} = -\frac{m}{\ln 2} - \frac{\sqrt{2}\sigma}{\sqrt{\ln 2}} \ , \ h_{\max} = -\frac{m}{\ln 2} + \frac{\sqrt{2}\sigma}{\sqrt{\ln 2}} \ . \qquad (2.32)$$

In Fig. 2.3b is illustrated a realization of a log-normal $\mathcal{W}$-cascade for the parameter values $m = -0.355\ln 2$ and $\sigma^2 = 0.02\ln 2$. The corresponding WT and WT skeleton as computed with $\psi_{(3)}^{(1)}$ are shown in Fig. 2.3d and 2.3f respectively. The results of the application of the WTMM method are reported in Fig. 2.4. As shown in Fig. 2.4b, when plotted vs. the scale parameter a in a logarithmic representation, the annealed average of the partition functions $Z(q, a)$ displays a well defined scaling behavior over a range of scales of about five octaves. Let us point out that scaling of quite good quality is found for a rather wide range of q values : $-5 \leq q \leq 10$. When processing to a linear regression fit of the data over the first four octaves, one gets the $\tau(q)$ spectrum shown in Fig. 2.4c. This spectrum is clearly a nonlinear function of q, the hallmark of multifractal scaling. Moreover, the numerical data are in remarkable agreement with the theoretical quadratic prediction (2.29). Similar quantitative agreement is observed on the $D(h)$ singularity spectrum in Fig. 2.4d which displays a single humped parabola shape that characterizes intermittent fluctuations corresponding to Hölder exponent values ranging from $h_{\min} = 0.155$ to $h_{\max} = 0.555$ (2.32). Unfortunately, to capture the strongest and the weakest singularities, one needs to compute the $\tau(q)$ spectrum for very large values of $|q|$. This requires the processing of many more realizations of the considered log-normal random $\mathcal{W}$-cascade.

The test applications reported in this section demonstrate the ability of the WTMM method to resolve multifractal scaling of 1D signals, a hopeless task for classical power spectrum analysis. They were used on purpose to calibrate and to test the reliability of our methodology and of the corresponding numerical tools with respect to finite-size effects and statistical convergence.

2.3 Wavelet Based Fractal Analysis of DNA Sequences

Applications of the WTMM method (for other examples see Chap. 7) to 1D signals have already provided insight into a wide variety of outstanding problems [2.77] such as fully developed turbulence [2.32, 2.35, 2.36, 2.95, 2.103–2.105, 2.151], finance [2.152, 2.153], medical time series analysis [2.154], and fractal growth phenomena [2.155–2.158]. In this section, we report recent developments in the context of statistical analysis of the complexity of DNA sequences [2.159–2.163].

The possible relevance of scale invariance and fractal concepts to the structural complexity of genomic sequences is the subject of considerable increasing interest [2.25, 2.26, 2.164]. During the past few years, there has been intense discussion about the existence, the nature, and the origin of *long-range correlations* in DNA sequences. Different techniques including mutual information functions [2.165, 2.166], autocorrelation functions [2.167–2.169], power spectra [2.24, 2.164, 2.166, 2.170–2.172], the *DNA walk* representation [2.24, 2.25, 2.173–2.178], and Zipf analysis [2.179–2.182] were used for statistical analysis of DNA sequences. But despite the effort spent, there is still some continuing debate on rather struggling questions. In that respect, it is of fundamental importance to corroborate the fact that the reported long-range correlations are not just an artefact of the compositional heterogeneity of the genome organization [2.167–2.169, 2.172, 2.173, 2.176–2.178]. Furthermore, it is still an open question whether the long-range correlation properties are different for *protein-coding* (exonic) and *noncoding* (intronic, intergenetic) sequences [2.24, 2.25, 2.164–2.171, 2.174, 2.175, 2.177, 2.179–2.183]. One of the main obstacles to long-range correlation analysis is the *mosaic structure* of DNA sequences which are well known to be formed of 'patches' ('strand bias') of different underlying compositions [2.24, 2.25, 2.173]. These patches appear as trends in the DNA walk landscapes and are likely to introduce some breaking of scale invariance [2.24, 2.25, 2.172, 2.173, 2.176, 2.177]. Most of the techniques used so far for characterizing the presence of long-range correlations are not well adapted to study patchy sequences. In a preliminary work [2.159, 2.160], we have emphasized the WT as a very powerful technique for fractal analysis of DNA sequences. By considering analyzing wavelets that make the WT microscope blind to low-frequency trends, one can reveal and quantify the scaling properties of DNA walks. Here we report on recent results obtained by applying the WTMM method to various genomic sequences selected in the human genome as well as in other genomes of the three kingdoms [2.159–2.163].

2.3.1 How to Make the WT Microscope Blind to Compositional Patchiness ?

A DNA sequence is a four-letter (A, C, G, T) text where A, C, G, and T are the bases adenine, cytosine, guanine, and thymine respectively. A popular method to graphically portray the genetic information stored in DNA sequences consists in mapping them in some variants of a n-dimensional random walk [2.184–2.186]. In this work, we will follow the strategy originally proposed by Peng et al. [2.24]. The so-called 'DNA walk' analysis requires first to convert the DNA text into a binary sequence. This can be done, for example, on the basis of purine $(Pu = A, G)$ vs. pyrimidine $(Py = C, T)$ distinction, by defining an incremental variable that associates to position i the value $\chi(i) = 1$ or -1, depending on whether the ith nucleotide of the sequence is Pu or Py. Each DNA sequence is thus represented as a string of purines and pyrimidines. Then the graph of the DNA walk is defined by the cumulative variables:

$$f(n) = \sum_{i=1}^{n} \chi(i) .$$

(2.33)

As an illustration, we show in Fig. 2.5a the graph of the DNA walk for the sequence of the bacteriophage λ. The patchiness of this coding sequence is patent; one clearly recognizes four regions of different strand bias, while the fluctuations about these main trends are hardly perceptible to the eye at the scale of the entire sequence. Actually, if one proceeds to some enlargements of this picture, one notices that the fluctuations involved are quite complex and possess some scale-invariant properties. The mosaic character of DNA consisting of patches of different composition (purine-rich regions, as compared to the average concentration over the entire strand, alternate with pyrimidine-rich regions corresponding to different trends in the DNA landscape shown in Fig. 2.5a) is generic of coding sequences. Nonconding sequences also display some patchiness which seems generally less obvious to distinguish from the bare fluctuations as already noticed in previous works [2.24, 2.25, 2.112, 2.173, 2.187].

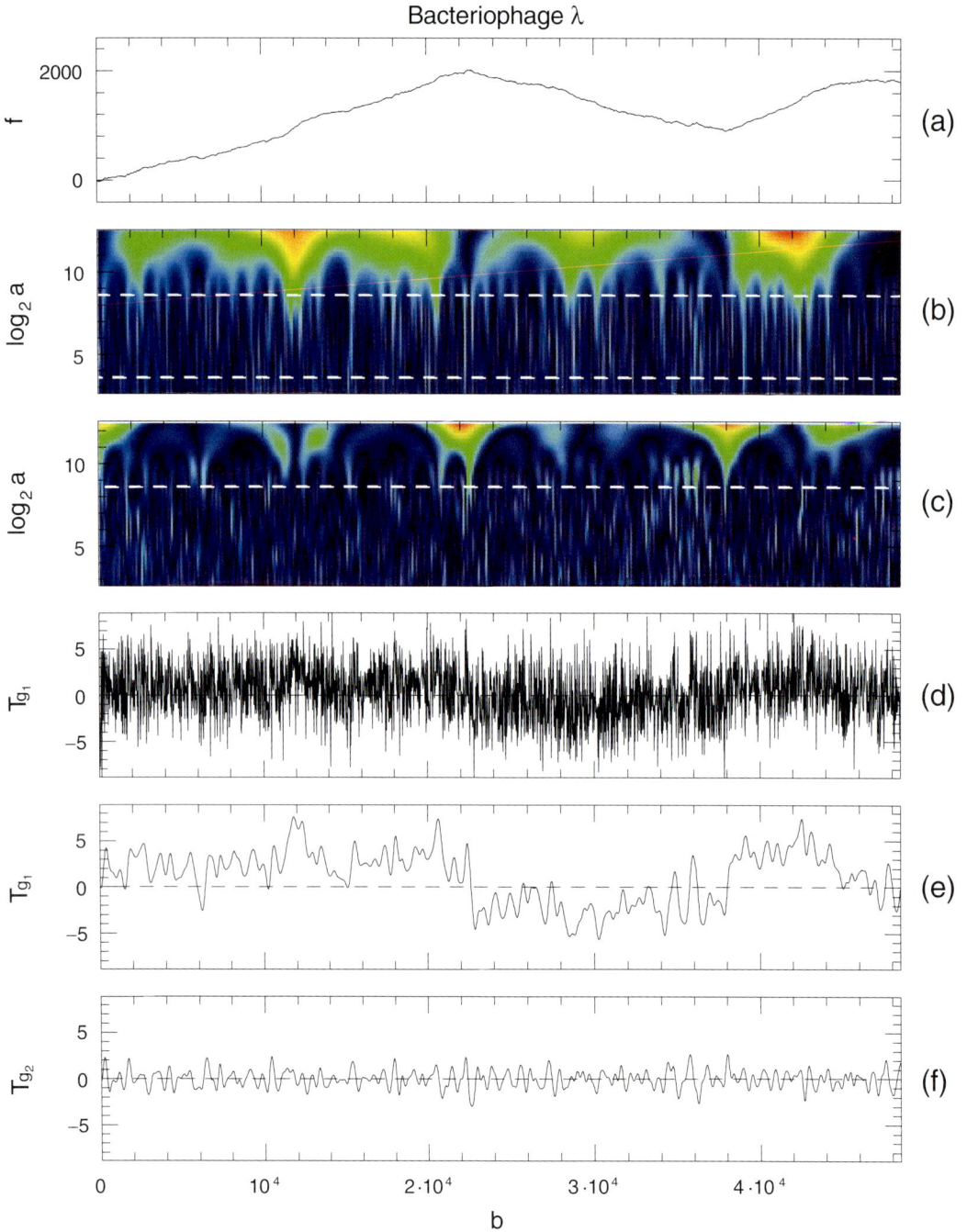

Fig. 2.5. WT analysis of the bacteriophage λ genome ($L = 48\,502$). (**a**) DNA walk displacement $f(x)$ based on purine-pyrimidine distinction, vs. nucleotide distance x. (**b**) WT of $f(x)$ computed with $g^{(1)}$ (2.20); $T_{g^{(1)}}(b,a)$ is coded, independently at each scale a, using 256 colors from black ($\min_b |T_{g^{(1)}}|$) to red ($\max_b |T_{g^{(1)}}|$). (**c**) Same analysis as in (**b**) but with the second-order analyzing wavelet $g^{(2)}$. (**d**) $T_{g^{(1)}}(b, a = a_1)$ vs. b for $a_1 = 12$ (nucleotides). (**e**) $T_{g^{(1)}}(b, a = a_2)$ vs. b for $a_2 = 384$ (nucleotides). (**f**) Same analysis as in (**e**) but with $g^{(2)}$

This mosaic structure is recovered whatever the coding rule one uses. We refer the reader to [2.160] for the use of alternative base pair codings. In the present study, we will also consider a representation inspired by the binary coding proposed by Voss [2.170, 2.171] and which consists in decomposing the nucleotide sequence into four sequences corresponding to A, C, T, or G, coding with $\chi(i) = 1$ at the nucleotide position and $\chi(i) = -1/3$ at other positions. The DNA walk obtained with the A mononucleotide coding for the yeast chromosome I is shown in Fig. 2.7a for illustration [2.162].

Even though the patchy structure of DNA sequences probably contains biological information of great importance, it is rather cumbersome as far as long-range correlation investigation is concerned. A lot of effort has been spent in order to master the presence of trends in DNA walks (and of course of trend changing). Several phenomenological methods have been proposed mainly by the Boston group. The 'min-max' method in the pioneering work of Peng et al. [2.24] has the major drawback that it requires the investigator to remove by hand the trends after identifying the local minima and maxima of the DNA landscape. The 'detrended fluctuation analysis' [2.112, 2.187] recently used by this group looks much more reliable since it does not require, a priori, any decision of the investigator. This method consists in partitioning the entire landscape into boxes of length l and in computing the 'detrended walk' as the difference between the original walk and the local trend (for some applications in the field of human physiology see Chaps. 7, p. 227, and 8, p. 264; for the analysis of climate data cf. Chap. 5, p. 178). But as emphasized by Voss [2.171], some care is required when attempting to remove bias.

The wavelet analysis of the bacteriophage λ sequence is shown in Fig. 2.5. Figure 2.5b shows the WT space-scale representation of this DNA signal when using the first derivative of the Gaussian function (2.20). When using a coding similar to the one prompted in previous studies, the WT is organized in a tree like structure from large to small scales, that looks qualitatively similar to the fractal branching observed in the WT representations of fBms (Fig. 2.3c) and log-normal random signals (Fig. 2.3d). In Fig. 2.5d and 2.5e, two horizontal cuts of $T_{g^{(1)}}(b, a)$ are shown at two different scales $a = a_1 = 2^2$ and $a_2 = 2^7$ that are represented by the dashed lines in Fig. 2.5b. When taking into account the characteristic size of the analyzing wavelet at the scale $a = 1$ corresponding to three nucleotides, these two scales correspond to looking at the fluctuations of the DNA walk over a characteristic length of the order of 12 and 384 nucleotides respectively. When focusing the WT microscope at the small scale $a = a_1$ in Fig. 2.5d, since $g^{(1)}$ is orthogonal to constants ($n_\psi = 1$), one filters out the low frequency component and reveals the local (high frequency) fluctuations of $f(x)$, i.e. the local fluctuations in purine and pyrimidine compositions over small size (~ 12 nucleotides) domains. When increasing the WT magnification in Fig. 2.5e, one realizes that these fluctuations actually occur around three successive linear trends; $g^{(1)}$ not being blind to linear behavior, the WT coefficients fluctuate about nonzero constant behavior that correspond to the slopes of those linear trends. Although this phenomenon is more pronounced when progressively increasing the scale parameter a towards a value that corresponds to the characteristic size ($\sim 15\,000$ nucleotides) of these strand biases, it is indeed present at all scales. In Fig. 2.5f, at the same coarse scale $a = a_2$ as in Fig. 2.5e, the fluctuations of the WT coefficients are shown as computed with the order-2 analyzing wavelet $g^{(2)}$ ($n_\psi = 2$). The WT microscope now being also orthogonal to linear behavior, the WT coefficients fluctuate about zero and one does not see the influence of the strand bias anymore, and this at all scales (Fig. 2.5c). Furthermore, by considering successively $g^{(3)}$, $g^{(4)}$, ..., one can hope not only to restore the stationarity of the increments of the DNA signal but also to eliminate more complicated nonlinear trends, that could be confused with the presence of long-range correlations in DNA sequences [2.159, 2.160].

2.3.2 Application of the WTMM Method to Human DNA Sequences

Demonstration of the Monofractality of DNA Walks: As a first application of the WTMM method, let us focus on the statistical analysis of coding and noncoding sequences in the human genome [2.159–2.161, 2.163]. The results reported in Fig. 2.6 correspond to the study of 2 184 introns of length $L \geq 800$ and of 226 exons of length $L \geq 600$ selected in the EMBL data bank [2.163]. The length criteria used to select these sequences result from some compromise between the control of finite-size effects in the WTMM scaling analysis (those sequences are rather short sequences: $\langle L \rangle_{\text{introns}} \simeq 800$, $\langle L \rangle_{\text{exons}} \simeq 150$) and the achievement of statistical convergence (2 184 introns and 226 exons are rich enough statistical samples). Figure 2.6a displays plots of the partition functions $Z(q, a)$, computed from the WT skeletons using $g^{(2)}$ as analyzing wavelet, vs. the scale parameter a in a logarithmic representation. These results correspond to quenched averaging over the corresponding 2 184 intron walks (red curves) and 226 exon walks (blue curves) generated using the G mononucleotide coding. For a rather wide range of q values: $-2 \leq q \leq 4$, scaling actually operates over a sufficiently large range of scales for the estimate of the $\tau(q)$ exponents to be meaningful. The $\tau(q)$ spectra obtained from linear regression fits of the intron and exon data are shown in Fig. 2.6b. For both these noncoding and coding sequences, the data points fall remarkably on a straight line which indicates that the considered intron walks and exon walks are likely to display monofractal scaling. But the slope of the straight line obtained for introns $H = \partial\tau(q)/\partial q = 0.60 \pm 0.02$ is definitely larger than the slope of the straight line derived for the exons $H = 0.53 \pm 0.02$. This is a clear indication that intron walks display long-range correlations ($H > 1/2$), while exon walks look much more like uncorrelated random walks ($H \simeq 1/2$). At first sight these results are in good agreement with the conclusions of pre-

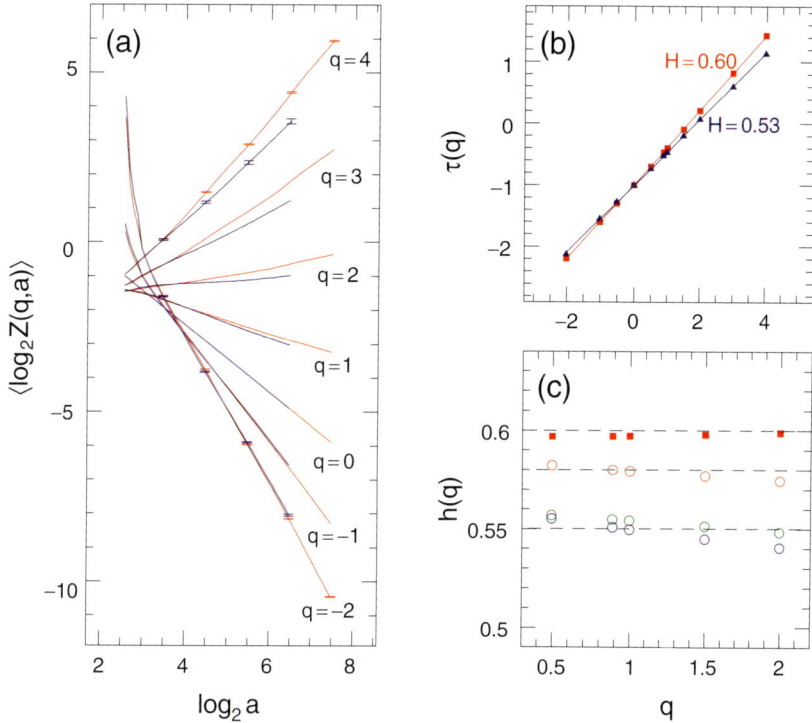

Fig. 2.6. Comparative WTMM analysis of the DNA walks of 2 184 introns ($L \geq 800$) and 226 exons ($L \geq 600$) in the human genome. The analyzing wavelet is $g^{(2)}$. The DNA walks are generated using the G mononucleotide coding for introns (red) and exons (blue). The reported results correspond to quenched averaging over our statistical samples of introns and exons respectively. (**a**) $\langle \log_2 Z(q,a) \rangle$ vs. $\log_2 a$ for various values of q. (**b**) $\tau(q)$ vs. q; the solid lines correspond to the theoretical spectrum $\tau(q) = qH - 1$ (2.28) for fBms with $H = 0.60 \pm 0.02$ (introns) and 0.53 ± 0.02 (exons). (**c**) $h(q) = (\tau(q) + 1)/q$ vs. q for the coding subsequences relative to position 1 (blue circles), 2 (green circles), and 3 (red circles) of the bases within the codons; the data for the introns (red squares) are shown for comparison; the horizontal broken-lines correspond to the following respective values of the Hurst exponent $H = 0.55$, 0.58, and 0.60

vious studies concerning the existence of long-range correlations in noncoding DNA sequences only [2.24, 2.25, 2.165, 2.183]. Similar observations are reported in [2.159, 2.163], when investigating the largest individual introns and exons found in the EMBL data bank.

One of the most striking results of our WTMM analysis in Fig. 2.6, is the fact that the $\tau(q)$ spectra extracted for both sets of introns and exons, are well fitted by (2.28), i.e. the analytical spectrum for fBms. Let us note that this remarkable finding is robust with respect to change in the considered mononucleotide coding used to generate the random walks.

About the Gaussian Character of the Fluctuations in DNA Walk Landscapes: Within the perspective of confirming the monofractality of DNA walks, we have studied the probability density function (PDF) of wavelet co-efficient values $\rho_a(T_{g^{(2)}}(\cdot, a))$, as computed at a fixed scale a in the fractal scaling range. According to the monofractal scaling properties, one expects

4

these PDFs to satisfy the self-similarity relationship:

$$a^H \rho_a(a^H T) = \rho(T) \, , \qquad (2.34)$$

where $\rho(T)$ is a 'universal' PDF (actually the PDF obtained at scale $a = 1$) that does not depend on the scale parameter a. In [2.159, 2.163], we have shown that when plotting $a^H \rho_a$ vs. $a^H T$, all the ρ_a curves corresponding to different scales actually collapse on a unique curve when using the exponent $H = 0.6$ for the set of human introns and $H = 0.53$ for the set of human exons. Moreover, independently of the coding or noncoding nature of the sequences, the so-obtained universal PDFs cannot be distinguished from a Gaussian distribution. We will see in Fig. 2.9c and 2.9d, that similar Gaussian WT coefficient statistics are observed, on a comparable range of (small) scales in between 10 and 200 nucleotides, for both the yeast and *Escherichia coli* genomes. Thus, as explored through the optics of the WT microscope, the basic fluctuations (about the low frequency trends due to compositional patchiness) in DNA walks are likely to have monofractal Gaussian statistics. The presence of long-range correlations in the human introns is in fact contained in the scale dependence of the variance of this distribution, $\sigma^2(a) \sim a^{2H}$, with $H = 0.60 \pm 0.02$ like for persistent random walk fBms, as compared to the uncorrelated random walk value $H = 0.53 \pm 0.02 \simeq 1/2$ obtained for the coding sequences.

Uncovering Long-Range Correlations in Coding DNA Sequences: Because of the 'period three' codon structure of coding DNA, it is natural to investigate separately the three subsequences relative to the position (1, 2, or 3) of the bases within their codons [2.161, 2.188]. We have built up these subsequences from our set of 226 human exons and we have repeated the WTMM analysis. The data obtained for the corresponding $\tau(q)$ spectra again fall on straight lines which corroborates monofractal scaling properties for the three subsequences. As shown in Fig. 2.6c for the subsequences relative to positions 1 and 2, one gets the same slope $h(q) = \partial\tau(q)/\partial q = 0.55 \pm 0.02$, which is indistinguishable from the value obtained for the overall coding sequences. Surprisingly, the data for the subsequence relative to position 3, exhibit a slope which is clearly larger ($H = 0.58 \pm 0.02$), i.e. a value which is very close to the exponent $H = 0.60 \pm 0.02$ estimated for the set of introns. This observation suggests that this third coding subsequence is likely to display the same degree of long-range correlations as noncoding sequences [2.161, 2.163]. We refer the reader to [2.161, 2.163] for the experimental demonstration that these long-range correlations are actually GC content dependent. Several mechanisms involved in the plasticity of genomes can be proposed to account for the observed long-range correlations. Among these mechanisms, one can exclude insertion-deletion events of DNA fragments of widely variable sizes which are very rare in exonic regions due to the strong constraints imposed by their coding properties. Selection pressure arguments can be further invoked to ex-

DNA A-Walk

DNA Bendability Walk

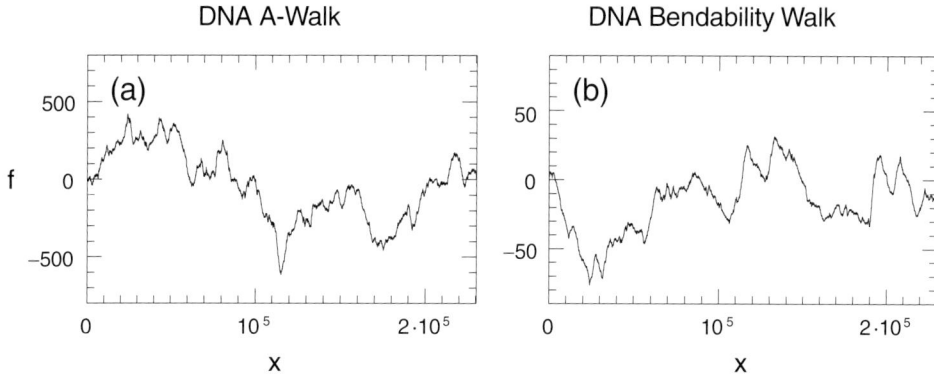

Fig. 2.7. DNA walk landscapes of the yeast chromosome I ($L = 230\,209$). (**a**) 'A' mononucleotide coding. (**b**) Trinucleotide (PNuc) coding proposed in [2.189] to account for the fluctuations of the local DNA curvature

plain the difference between the correlations observed for the third bases of the codons as compared to those for bases 1 and 2. Indeed it is well known that most of the degeneracy of the genetic code is contained in codon position three.

2.3.3 Towards a Structural Interpretation of the Long-Range Correlations in DNA Sequences

Recent success in DNA sequencing provides a wide range of investigations for statistical analysis of genomic sequences [2.190–2.196]. The availability of fully sequenced genomes offers the possibility to study scale invariance properties of DNA sequences over a wide range of scales extending from tens to thousands of nucleotides. The first completely sequenced eucaryotic genome *Saccharomyces cerevisiae* [2.194] provides an opportunity to perform a comparative wavelet analysis of the scaling properties displayed by each chromosome [2.162, 2.163] (Fig. 2.7a). When looking at the scale dependence of the root-mean square $\sigma(a)$ of the wavelet coefficient PDF computed with $g^{(2)}$ (2.20), over the DNA walks corresponding to 'A' in each of the 16 yeast chromosomes, one sees, in Fig. 2.8a, that they all present superimposable behavior. We notably observe the same characteristic scale that separates two different scaling regimes. Let us note that common behavior to the 16 yeast chromosomes has already been pointed out in [2.197, 2.198]. At small scales, $20 \lesssim a \lesssim 200$, weak power-law correlations (PLC) are observed as characterized by $H = 0.59 \pm 0.02$, a mean value which is significantly larger than 1/2. At large scales, $200 \lesssim a \lesssim 5\,000$, strong PLC with $H = 0.82 \pm 0.01$ become dominant with a cutoff around $10\,000$ bp (a number which is by no means accurate) where uncorrelated behavior is observed. (In this section the scale parameter is expressed in nucleotide units.) The existence of these two scaling regimes is confirmed in Fig. 2.9a, 2.9c and 2.9e [2.163], where the WT PDFs computed at different scales (Fig. 2.9a) are shown to collapse on a single curve, as predicted by the self-similarity relationship (2.34),

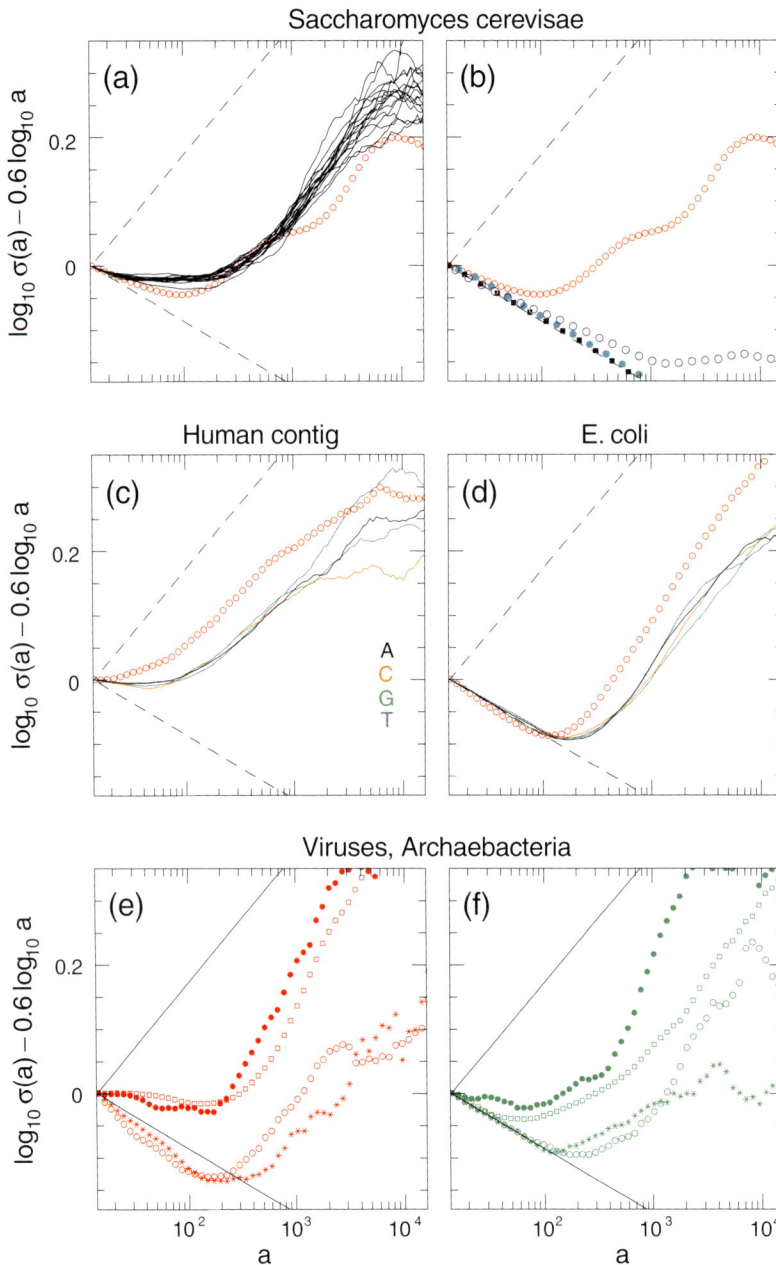

Fig. 2.8. Global estimate of the r.m.s. of WT coefficients of DNA walk landscapes as computed with $g^{(2)}$ (2.20). $\log_{10}\sigma(a) - 0.6\log_{10}a$ vs. $\log_{10}a$; the dashed lines corresponding to uncorrelated ($H = 1/2$) and strongly correlated ($H = 0.80$) regimes are drawn to guide the eye. *Saccharomyces cerevisiae*: (**a**) A DNA walks of the 16 *Saccharomyces cerevisiae* chromosomes (solid lines) and of the corresponding bending profiles when averaged over the 16 chromosomes (red circles). (**b**) True bending profile (red circles) and a synthetic random bending profile (blue circles). The results are averaged over the 16 yeast chromosomes. Similar computations after randomly shuffling the original DNA sequences for both the A DNA walks (black squares) and the bending profiles (green dots). *Human contig*: (**c**) A (black curve), C (orange curve), G (green curve), and T (blue curve) DNA walks and the corresponding bending profiles (red circles). *Escherichia coli*: (**d**) Same representation as in (**c**). Viruses, Archaebacteria: *Archaeglobus fulgidus* (squares), *Epstein Barr virus* (dots), *Melanoplus sanguinipes entomapoxvirus* (circles), and *T4 bacteriophage* (stars): (**e**) 'G' DNA walks; (**f**) bending profiles

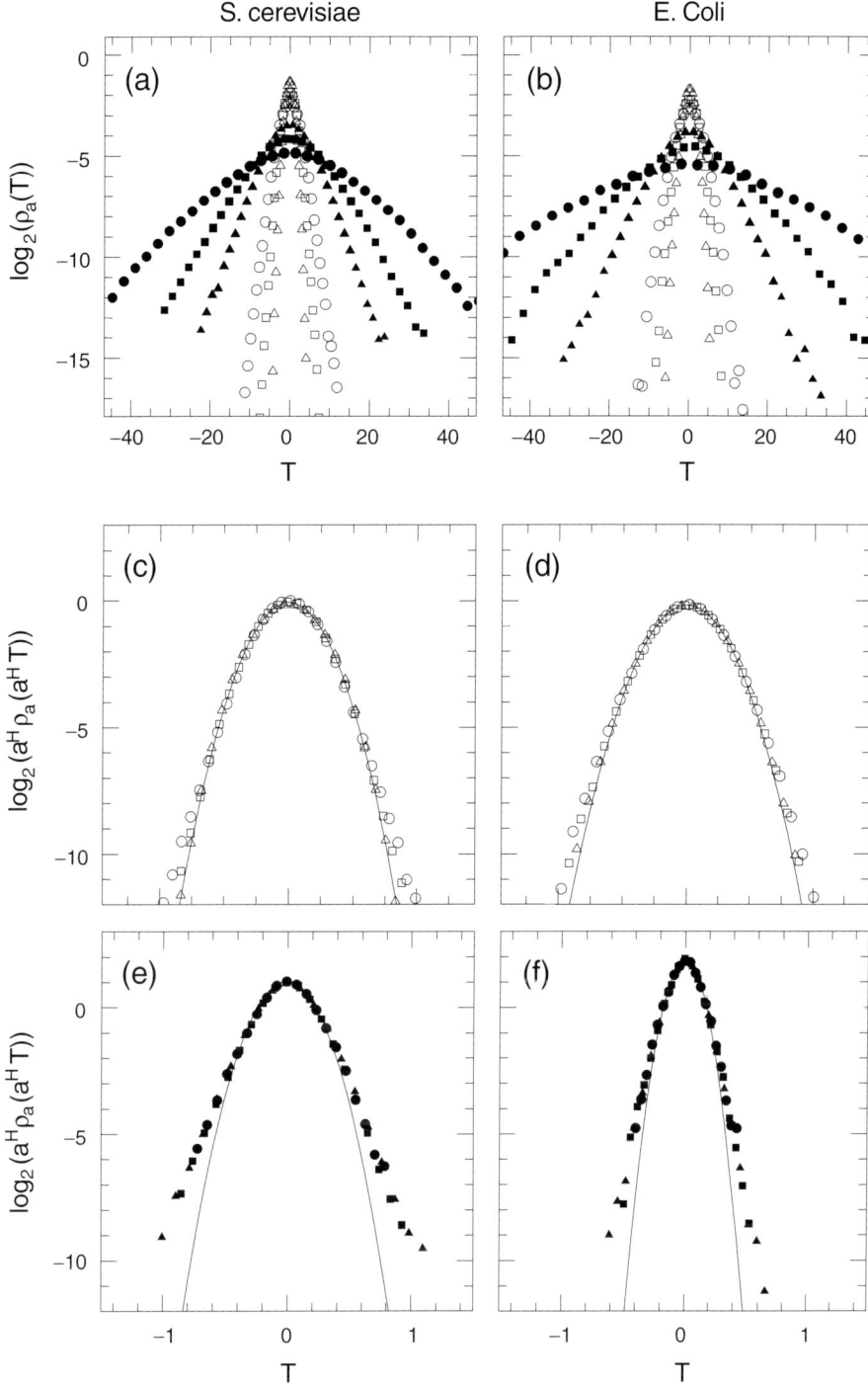

Fig. 2.9. Probability distribution functions of wavelet coefficient values of 'A' DNA walks. The analyzing wavelet is the Mexican hat $g^{(2)}$ (2.20). *Saccharomyces cerevisiae*: (**a**) $\log_2(\rho_a)$ vs. $T_{g^{(2)}}$ for the set of scales $a = 12$ ($\triangle$), 24 ($\square$), 48 ($\circ$), 192 ($\blacktriangle$), 384 ($\blacksquare$), and 768 ($\bullet$); (**c**) small-scale regime: $\log_2(a^H \rho_a(a^H T_{g^{(2)}}))$ vs. $T_{g^{(2)}}$ with $H = 0.59$; (**e**) large-scale regime: $\log_2(a^H \rho_a(a^H T_{g^{(2)}}))$ vs. $T_{g^{(2)}}$ with $H = 0.75$. *Escherichia coli*: (**b**) $\log_2(\rho_a)$ vs. $T_{g^{(2)}}$ for the set of scales $a = 24$ ($\triangle$), 48 ($\square$), 96 ($\circ$), 384 ($\blacktriangle$), 768 ($\blacksquare$), and 1536 ($\bullet$); (**d**) small-scale regime: $\log_2(a^H \rho_a(a^H T_{g^{(2)}}))$ vs. $T_{g^{(2)}}$ with $H = 0.50$; (**f**) large-scale regime: $\log_2(a^H \rho_a(a^H T_{g^{(2)}}))$ vs. $T_{g^{(2)}}$ with $H = 0.80$

56   Alain Arneodo et al.

provided one uses the scaling exponent value $H = 0.59$ in the scale range $10 \lesssim a \lesssim 100$ (Fig. 2.9c) and $H = 0.75$ in the scale range $200 \lesssim a \lesssim 1\,000$ (Fig. 2.9e). In the small scale regime, the PDFs are very well approximated by Gaussian distributions (Fig. 2.9c) which recalls the results of the WTMM analysis of noncoding eucaryotic sequences in Sect. 2.3.2. In the large scale regime, the PDFs have stretched exponential-like tails (Fig. 2.9e). In both regimes, the fact that (2.34) is verified, corroborates the monofractal nature of the roughness fluctuations of the yeast DNA walks [2.163]. We have also examined other eucaryotic contigs from different organisms (human, rodent, avian, plant, and insect) and we have observed the same characteristic features as those obtained with *Saccharomyces cerevisiae*. In Fig. 2.8c, a similar characteristic scale $a^* \simeq 100$ bp is clearly seen on a human contig. Moreover, the crossover from a $H = 0.62 \pm 0.01$ to a $H = 0.75 \pm 0.02$ PLC regime is remarkably robust since the data for the four A, C, G, and T DNA walks fall almost on the same curve in the range $20 \lesssim a \lesssim 2\,000$ [2.162,2.163]. The striking overall similarity of the results obtained with these different eucaryotic genomes prompted us to also examine the scale invariance properties of bacterial genomes. In Figs. 2.8d, 2.9b, 2.9d and 2.9f, are reported the results obtained for *Escherichia coli* [2.199] which are typical of what we have observed with 15 other bacterial genomes (data not shown) [2.162,2.163]. Again, there exists a well defined characteristic scale $a^* \simeq 100$–200 bp that delimits the transition to very strong PLC with $H = 0.85 \pm 0.02$ at large scales. In order to examine if these properties actually extend homogeneously over the whole genomes, $\sigma(a)$ was calculated over a window of width $l = 2\,000$, sliding along the DNA walk profiles. The results reported in Fig. 2.0 clearly reveal the existence of a characteristic scale $a^* \simeq 100$–200 bp which seems to be robust all along the corresponding DNA molecules and this for all investigated genomes [2.162,2.163]. There exists, however, an important difference between eucaryotic and bacterial genomes: no PLC are observed in the small-scale regime where uncorrelated $H = 1/2$ Brownian motion-like behavior is observed (Fig. 2.9d). At this point, it may seem that PLC are inherent to mostly noncoding genomes, but that is not the case. As shown in Fig. 2.8e for *Archaeglobus fulgidus*, the wavelet investigation of five archaean genomes (which are mostly coding) also reveals the presence of small-scale PLC as observed in eucaryotic genomes, although somewhat less pronounced [2.162,2.163]. But the striking feature of these data is that the strong large-scale PLC are present in all bacterial, archaeabacterial, and eucaryotic genomes [2.170,2.171].

What mechanism or phenomenon might explain the small-scale PLC in eucaryotic genomes? Their total absence in bacterial genomes raises the possibility that they could be related to certain nucleotide arrangements in the 150 bp long DNA regions which are wrapped around histone proteins to form the eucaryotic nucleosomes [2.200–2.202]. Indeed, eubacterial genomic DNA is associated with histone-like proteins (e.g. Hu), but no nucleosome-type structure has been detected in these organisms [2.203]. Following this hypothesis, we have extended the application of the WT microscope to the investigation of the multiscale properties of DNA bending signals that are likely to reflect the hierarchical organization of nucleoprotein complexes. To construct a bending profile that accounts for the fluctuations of the local DNA curvature, we use the trinucleotide (PNUC) model proposed in [2.189]. The results of the WTMM analysis of the bending profiles of the yeast chromosomes (Fig. 2.7b) are reported in Fig. 2.8a and 2.8b [2.162, 2.163]. This analysis reveals striking similarities with the curves resulting from the DNA walk analysis, in both the small-scale and the large-scale regimes. To ensure that these observations are not simply due to a 're-coding' of the DNA sequences, but due to the proper values of roll angles used to determine the bending profile, we have randomly changed the table that maps trinucleotides to roll-angle values. The new table is obtained using a Gaussian distribution of same mean, variance, and symmetries as the original table [2.189]. As shown in Fig. 2.8b, this results in the vanishing of these PLC (now $H = 0.51 \pm 0.01$ at small scales), establishing that these PLC do indeed reflect persistent scale-invariant structural properties. We have also examined a number of eucaryotic and eubacterial genomes with similar conclusions [2.162, 2.163]. As shown in Fig. 2.8c and 2.8d for a human contig and *Escherichia coli*, the structural information that is contained in the scaling properties of bending profiles can be directly extracted from the WTMM reading of the DNA texts [2.162, 2.163].

Following the nucleosomal interpretation of PLC in the small-scale regime, the observation of such correlations in archaean genomes in Fig. 2.8e and 2.8f [2.162, 2.163] is consistent with the presence in archaebacteria of structures similar to the eucaryotic nucleosomes [2.204–2.208]. This analysis has also been extended to viral genomes. Small-scale PLC are clearly detected in organelle genomes and in most viral double-stranded DNA genomes (*Herpes-, Adeno-, Papova-, Parvo-,* and *Hepadna viruses*) as shown for *Epstein Barr virus* in Fig. 2.8e and 2.8f. This further supports the hypothesis of nucleosome-based PLC since nucleosomes have been observed on several classes of double-stranded DNA viruses [2.209, 2.210]. The Poxviridae, which are the only animal DNA viruses replicating in the cytoplasm of their host cells, code for a bacterial-type of histone-like protein [2.211], and no PLC are found in this scale range as shown in Fig. 2.8e and 2.8f for *Melanoplus sanguinipes* virus. This observation is consistent with our hypothesis and suggests that the genomic DNA of these viruses is submitted to a packaging process different from other animal viruses. Finally, bacteriophage genomes do not present any small-scale PLC (Fig. 2.8e and 2.8f for *T4 bacteriophage* and data not shown) as already observed for their eubacterial hosts. Other classes of virus genomes like the single and double-strand RNA viruses (to the exception of the retroviruses) are very unlikely to be associated to nucleosomes. In all cases except retroviruses, we observe a total absence of small-scale PLC. In the case of retroviruses, it is known that the integrated viral DNA is associated to nucleosomes in the cell nucleus [2.212]; we clearly confirm the presence of small-scale PLC ($H \simeq 0.57 \pm 0.02$). These wavelet based fractal analyses of viral and cellular genomes of all three kingdoms sustain without failure the fact that small-scale PLC provide a reliable diagnostic of the existence of eucaryotic nucleosomes.

Several studies have established the presence in genomic sequences of repetitive DNA motifs related to bending properties [2.213, 2.214]. It is noteworthy that a 10.2 base periodic modulation of correlation functions (not to be misunderstood with strict periodicity) is observed in eucaryotic genomes, where it has been interpreted in relation to nucleosomal structures. However, there is a fundamental difference between this nucleosome diagnostic based on periodicity (i.e. invariance with respect to discrete translations) and our analysis based on scale invariance properties (i.e. invariance with respect to continuous dilations) which show up as a power-law behavior of the envelope of the correlation function. The main conclusion of our WT analysis is that the mechanisms underlying the nucleosomal structure of eucaryotic genomes are likely to be multiscale phenomena that involve the whole set of scales in the 1–200 bp range. In other words, the bending sites are not positioned in a regular sequential manner, as suggested by the periodic-pattern picture, but are fractally distributed. A number of experimental observations help us interpret our results. Nucleosomes can be regarded as mobile (nucleosome sliding [2.215]) and with inherently statistical positioning [2.216]. They also can be viewed as dynamical structures transiently exposing stretches of their DNA allowing access to regulatory proteins [2.217]. In that context, the understanding of small-scale PLC should provide further insight into the nucleosome structure and dynamics.

What mechanisms might explain the strongly correlated patterns observed on the DNA walks as well as on the corresponding bending profiles in all genomes, in the 200–5 000 bp range? As already suggested in [2.216], the signals involved in nucleosome binding and positioning may act collectively over large distances to the packing of nucleosomal arrays into high-order chromatin structures [2.200–2.202, 2.218]. Since DNA bending sites are key elements for nucleosomal structures, the detailed investigation of large-scale PLC observed in eucaryotic bending profiles should shed light on the compaction mechanisms at work in the hierarchical formation and dynamics of chromatin. An important clue provided by our studies is that similar long distance correlations in bending profiles are also observed in eubacteria and archaebacteria. It is then tempting to conjecture that the compaction of the eubacterial nucleoid in rosette-shaped chromosomal structures [2.219] is submitted to similar multiscale structural constraints. Indeed, it is well established that architectural proteins interacting with DNA bending sites participate to the stabilization of large nucleoprotein arrays, as for example, the eucaryotic high mobility group (HMG) and the eubacterial Hu histone-like proteins [2.220]. Moreover, the individual domains constituting the bacterial nucleoid are variable in size and position [2.221]. Actually eucaryotic, eubacterial, and archaebacterial chromosomes are submitted to dynamical constraints that might result in scale invariant properties of the DNA text. For instance, large-scale PLC may result from DNA structural features related to the modulation of transcription, replication, and recombination events [2.222–2.224]. A deep understanding of the large-scale PLC observed in DNA bending profiles as well as in DNA walks for all three kingdoms and their interpretation in terms of structural and dynamical properties remain challenging questions requiring further investigations.

2.4 The 2D Wavelet Transform Modulus Maxima Method for the Multifractal Analysis of Rough Surfaces

Ever since the explosive propagation of fractal ideas [2.1] throughout the scientific community in the late 1970s and early 1980s, there have been numerous applications to surface science [2.5–2.8, 2.10–2.13, 2.16–2.18]. Both real space imaging techniques (including all kinds of microscopy and optical imaging techniques) and diffraction techniques have been extensively used to study rough surfaces [2.18]. The characterization of surface roughness is an important problem from a fundamental point of view as well as for the wealth of potential applications in applied sciences. Indeed, a wide variety of natural and technological processes lead to the formation of complex interfaces [2.1–2.20]. Assigning a fractal dimension to those irregular surfaces has now become routine in various fields including topography, defect and fracture studies, growth phenomena, erosion and corrosion processes, catalysis, and many other areas in physics, chemistry, biology, geology, meteorology, and material sciences [2.1–2.20]. For rough surfaces which are well described by self-affine fractals with anisotropic scale invariance [2.1, 2.5, 2.10, 2.22, 2.22, 2.27, 2.28] various methods (e.g. divider, box, triangle, slit-island, power spectral, variogram and distribution methods) of computing D_F were shown to be very sensitive to limited resolution as well as finite-size effects [2.28, 2.29, 2.108, 2.109, 2.112, 2.225]. An alternative strategy consists in computing the so-called roughness exponent H [2.1, 2.5, 2.10] that describes the scaling of the width (or thickness) of the rough interface with respect to measurement scale. As for 1D signals (see Sects. 2.1 and 2.2.6), different methods [2.29, 2.30, 2.108, 2.109, 2.112, 2.114, 2.115] are available to estimate this exponent which is supposed to be related to the fractal dimension $D_F = d - H$ of self-affine surfaces embedded in a d-dimensional space. Again a number of artefacts may pollute the estimate of the roughness exponent. Since sensitivity and accuracy are method dependent, it is usually recommended to simultaneously use different tools in order to appreciate in a quantitative way, the level of confidence in the measured exponent.

The purpose of this section is to generalize the WTMM method described in Sect. 2.2 [2.32, 2.35, 2.36, 2.42] from 1D to 2D, with the specific goal to achieve multifractal analysis of rough surfaces with fractal dimension D_F anywhere between two and three. In recent years, increasing interest has been paid to the application of the WT to image processing [2.65, 2.77, 2.80, 2.86, 2.226–2.228]. In this context, Mallat and collaborators [2.40, 2.41] have extended the WTMM representation in 2D in a manner inspired from Canny's multiscale edge detectors commonly used in computer vision [2.229]. Our strategy will thus consist in using this representation to define a (3D) WT skeleton from which one can compute partition functions and ultimately extract multifractal spectra. A detailed description of this approach can be found in [2.230–2.234].

2.4.1 2D Continuous Wavelet Transform for Multiscale Edge Detection

The edges of the different structures that appear in an image are often the most important features for pattern recognition. Hence, in computer vision [2.235, 2.236], a large class of edge detectors look for points where the gradient of the image intensity has a modulus which is locally maximum in its direction. As originally noticed by Mallat and collaborators [2.40, 2.41], with an appropriate choice of the analyzing wavelet, one can recast the Canny's multiscale edge detector [2.229] in terms of a 2D WT. The general idea is to start smoothing the discrete image data by convolving it with a filter and then to compute the gradient of the smoothed signal.

Let us consider two wavelets that are, respectively, the partial derivatives with respect to x and y of a 2D smoothing function $\phi(x, y)$:

$$\psi_1(x, y) = \frac{\partial \phi(x, y)}{\partial x} \quad \text{and} \quad \psi_2(x, y) = \frac{\partial \phi(x, y)}{\partial y} \ . \tag{2.35}$$

We will assume that ϕ is a well localized (around $x = y = 0$) isotropic function that depends on $|\mathbf{x}|$ only. In this study, we will mainly use the Gaussian function:

$$\phi(x, y) = \mathrm{e}^{-(x^2+y^2)/2} = \mathrm{e}^{-|\mathbf{x}|^2/2} \ , \tag{2.36}$$

as well as the isotropic Mexican hat:

$$\phi(\mathbf{x}) = (2 - \mathbf{x}^2)\mathrm{e}^{-|\mathbf{x}|^2/2} \ . \tag{2.37}$$

The corresponding analyzing wavelets ψ_1 and ψ_2 are illustrated in Fig. 2.10. They have one and three vanishing moments when using respectively the Gaussian function (2.36) and the Mexican hat (2.37) as smoothing function.

For any function $f(x, y) \in L^2(\mathbb{R})$, the WT with respect to ψ_1 and ψ_2 has two components and therefore can be expressed in a vectorial form:

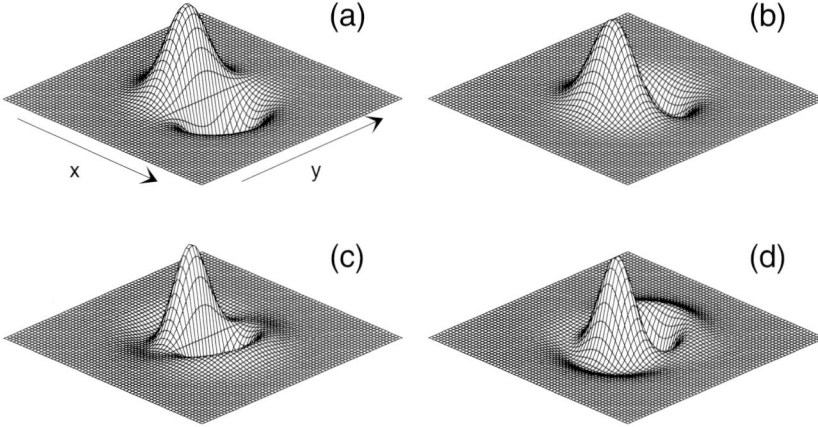

Fig. 2.10. The analyzing wavelets ψ_1 and ψ_2 defined in (2.35). First-order analyzing wavelets obtained from a Gaussian smoothing function ϕ (2.36): (**a**) ψ_1; (**b**) ψ_2. Third-order analyzing wavelets obtained from the isotropic Mexican hat smoothing function ϕ (2.37): (**c**) ψ_1; (**d**) ψ_2

$$\mathbf{T}_{\boldsymbol{\psi}}[f](\mathbf{b}, a) = \begin{pmatrix} T_{\psi_1}[f] = a^{-2} \int d^2\mathbf{x}\, \psi_1\big(a^{-1}(\mathbf{x} - \mathbf{b})\big) f(\mathbf{x}) \\ T_{\psi_2}[f] = a^{-2} \int d^2\mathbf{x}\, \psi_2\big(a^{-1}(\mathbf{x} - \mathbf{b})\big) f(\mathbf{x}) \end{pmatrix} . \tag{2.38}$$

Then, after a straightforward integration by parts, one gets:

$$\begin{aligned} \mathbf{T}_{\boldsymbol{\psi}}[f](\mathbf{b}, a) &= a^{-2} \boldsymbol{\nabla} \left\{ \int d^2\mathbf{x}\, \phi\big(a^{-1}(\mathbf{x} - \mathbf{b})\big) f(\mathbf{x}) \right\} \\ &= \boldsymbol{\nabla} \big\{ T_\phi[f](\mathbf{b}, a) \big\} \\ &= \boldsymbol{\nabla} \big\{ \phi_{\mathbf{b},a} * f \big\} . \end{aligned} \tag{2.39}$$

If $\phi(\mathbf{x})$ is simply a smoothing filter like the Gaussian function (2.36), then (2.39) amounts to define the 2D wavelet transform as the gradient vector of $f(\mathbf{x})$ smoothed by dilated versions $\phi(a^{-1}\mathbf{x})$ of this filter. If $\phi(\mathbf{x})$ has some vanishing moments, then $T_\phi[f](\mathbf{b}, a)$ in (2.39) is nothing but the continuous 2D WT of $f(\mathbf{x})$ as defined by Murenzi [2.237, 2.238], provided $\phi(\mathbf{x})$ is an isotropic analyzing wavelet.

As far as notations are concerned, we will mainly use the representation involving the modulus and the argument of the WT:

$$\mathbf{T}_{\boldsymbol{\psi}}[f](\mathbf{b}, a) = \big(\mathcal{M}_{\boldsymbol{\psi}}[f](\mathbf{b}, a), \mathcal{A}_{\boldsymbol{\psi}}[f](\mathbf{b}, a) \big) , \tag{2.40}$$

with

$$\mathcal{M}_{\boldsymbol{\psi}}[f](\mathbf{b}, a) = \left[\big(T_{\psi_1}[f](\mathbf{b}, a) \big)^2 + \big(T_{\psi_2}[f](\mathbf{b}, a) \big)^2 \right]^{1/2} , \tag{2.41}$$

and

$$\mathcal{A}_{\boldsymbol{\psi}}[f](\mathbf{b}, a) = \arg\big(T_{\psi_1}[f](\mathbf{b}, a) + \mathrm{i} T_{\psi_2}[f](\mathbf{b}, a) \big) . \tag{2.42}$$

2.4.2 Characterizing the Local Regularity Properties
of Rough Surfaces with the Wavelet Transform Modulus Maxima

In the present study, we will use the term *rough surface* for an irregular surface on which there are no overhanging regions. This means that the surface can be correctly described by a function $z = f(\mathbf{x})$, which specifies the height of the surface at the point $\mathbf{x} = (x, y)$. Of particular interest are the rough surfaces corresponding to self-affine fractals in $\mathbb{R}^3$ [2.1, 2.5, 2.10, 2.22, 2.27, 2.28]. Moreover, we will assume that the considered functions possess only cusp-like singularities [2.232]. Under these assumptions, the situation is nevertheless trickier than in 1D [2.230–2.233]. Indeed, one has to distinguish two main cases depending on whether scale invariance is under isotropic or anisotropic dilations [2.1, 2.29, 2.136, 2.144, 2.145, 2.239].

Isotropic Dilations: Local scale invariance under isotropic dilations means that locally, around the point $\mathbf{x}_0$, the function f behaves as:

$$f(\mathbf{x}_0 + \lambda \mathbf{u}) - f(\mathbf{x}_0) \simeq \lambda^{h(\mathbf{x}_0)} \Big(f(\mathbf{x}_0 + \mathbf{u}) - f(\mathbf{x}_0) \Big) , \qquad (2.43)$$

where $\lambda > 0$ and $\mathbf{u}$ is a unit vector. If the scaling exponent $h(\mathbf{x}_0)$ does not depend upon the direction of $\mathbf{u}$, then f displays isotropic local scale invariance around $\mathbf{x}_0$ and the corresponding singularity is of Hölder exponent $h(\mathbf{x}_0)$. If, on the contrary, the scaling exponent depends upon the direction of $\mathbf{u}$, then the Hölder exponent is the minimum value of h over all the possible orientations of $\mathbf{u}$. Thus f displays anisotropic scale-invariance around $\mathbf{x}_0$ with one, several or a continuum of privileged directions along which the variation of f defines the Hölder exponent of the singularity located at $\mathbf{x}_0$.

Anisotropic Dilations: Local scale invariance under anisotropic dilations means that locally around the point $\mathbf{x}_0$, the function f behaves as [2.134, 2.136, 2.144, 2.145, 2.239]:

$$f\Big(\mathbf{x}_0 + \Lambda_\alpha(\lambda) r_\theta \mathbf{u} \Big) - f(\mathbf{x}_0) \simeq \lambda^{h(\mathbf{x}_0)} \Big(f(\mathbf{x}_0 + \mathbf{u}) - f(\mathbf{x}_0) \Big) , \qquad (2.44)$$

where $\lambda > 0$ and $\mathbf{u}$ is a unit vector. r_θ is a rotation matrix and $\Lambda_\alpha(\lambda)$ is a positive diagonal 2×2 matrix that accounts for anisotropic self-affine scale transformation in the θ-rotated referential with origin $\mathbf{x}_0$:

$$\Lambda_\alpha(\lambda) = \begin{pmatrix} \lambda & 0 \\ 0 & \lambda^\alpha \end{pmatrix} . \qquad (2.45)$$

The function f thus displays anisotropic scale invariance around $\mathbf{x}_0$ and the Hölder exponent is given by the behavior of f in the direction θ ($\alpha < 1$) or $\theta + \pi/2$ ($\alpha > 1$).

Very much like the WT analysis of cusp singularities in 1D (Sect. 2.2.2), in order to recover the Hölder exponent $h(\mathbf{x}_0)$ of a function f from $\mathbb{R}^2$ to $\mathbb{R}$, one needs to study the behavior of the WT modulus inside a cone $|\mathbf{x} - \mathbf{x}_0| < Ca$ in the space-scale half-space [2.230, 2.231, 2.239]. As originally proposed by Mallat and collaborators [2.40, 2.41], a very efficient way to perform point-wise regularity analysis is to use the wavelet transform modulus maxima (WTMM). In the spirit of Canny edge detection [2.229], at a given scale a, the WTMM are defined as the points $\mathbf{b}$ where the WT modulus $\mathcal{M}_\psi[f](\mathbf{b}, a)$ (2.41) is locally maximum along the gradient direction given by the argument $\mathcal{A}_\psi[f](\mathbf{b}, a)$ (2.42). These modulus maxima are inflection points of $f * \phi_a(\mathbf{x})$. As illustrated in the example just below, these WTMM lie on connected chains hereafter called *maxima chains* [2.230–2.233]. In theory, one only needs to record the position of the local maxima of $\mathcal{M}_\psi$ along the maxima chains together with the value of $\mathcal{M}_\psi[f]$ and $\mathcal{A}_\psi[f]$ at the corresponding locations. At each scale a, our wavelet analysis thus reduces to store those WTMM maxima (WTMMM) only. They indicate locally the direction where the signal has the sharpest variation. This orientation component is the main difference between 1D and 2D WT analysis. These WTMMM are disposed along connected curves across scales called *maxima lines* [2.231, 2.232]. We will define the WT skeleton as the set of maxima lines that converge to the (x, y)-plane in the limit $a \to 0^+$. This WT skeleton is likely to contain all the information concerning the local Hölder regularity properties of the function f under consideration.

Let us first illustrate the above definitions on the function f shown in Fig. 2.11a:

$$f(\mathbf{x}) = A \mathrm{e}^{-(\mathbf{x}-\mathbf{x}_1)^2/2\sigma^2} + B \left| \mathbf{x} - \mathbf{x}_0 \right|^{0.3}. \tag{2.46}$$

This function is C^∞ everywhere except at $\mathbf{x} = \mathbf{x}_0$ where f is isotropically singular with a Hölder exponent $h(\mathbf{x}_0) = 0.3$. Its 2D WT (2.38) with a first-order analyzing wavelet (the smoothing function $\phi(\mathbf{x})$ is the isotropic Gaussian function) is shown in Fig. 2.11b for a given scale $a = 2^3 \sigma_W$, where $\sigma_W = 13$ is the width (in pixel units) of the analyzing wavelet at the smallest scale where it is still well enough resolved. Actually, Fig. 2.11b illustrates the behavior of $\mathcal{M}_\psi[f]$. From a visual inspection of this figure, on can convince oneself that the modulus is radially symmetric around $\mathbf{x}_0$ where is located the singularity S. This is confirmed by the behavior of $\mathcal{A}_\psi[f]$ which rotates uniformly from 0 to 2π around $\mathbf{x}_0$. The WTMM as well as the WTMMM are shown in Fig. 2.11c for a discrete set of scales which allows us to reconstruct a three-dimensional representation in the space-scale half-space. At small scale, there exist mainly two maxima chains. One is a closed curve around $\mathbf{x}_0$ where is located the singularity S. The other one is an open curve which partially surrounds the localized smooth structure G. On each of these maxima chains, one finds only one WTMMM ($\bullet$) whose corresponding argument is such that the gradient vector points to S and G respectively. As far as the singularity S is concerned,

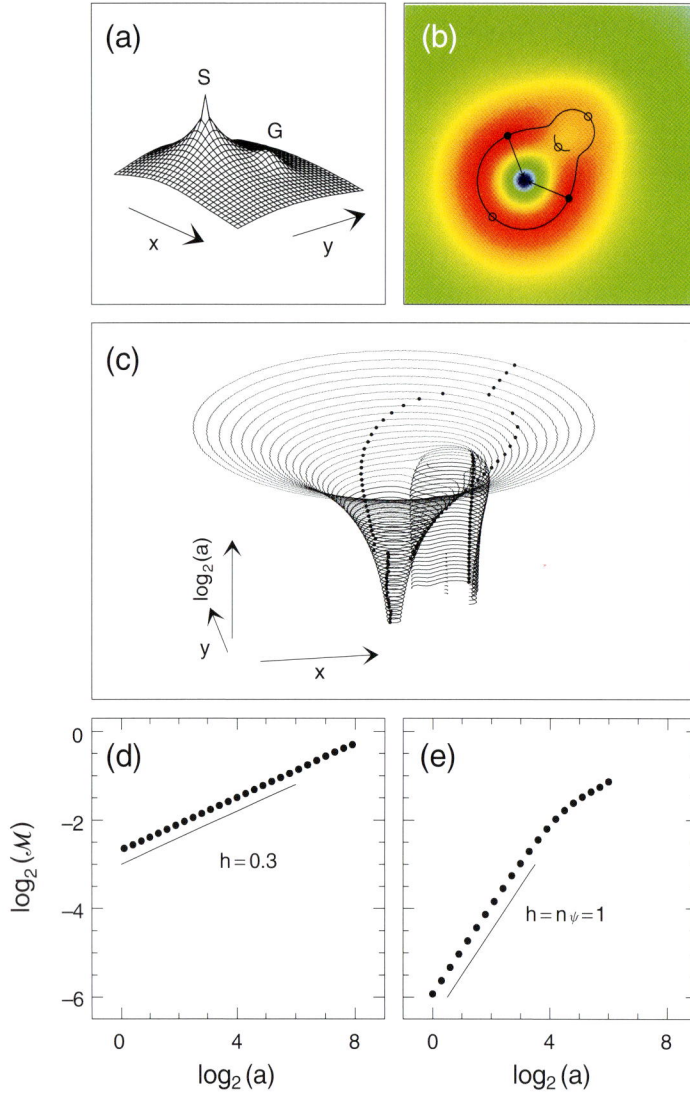

Fig. 2.11. Estimating the Hölder exponent from the behavior of the WT modulus along the maxima lines. (**a**) Graph of the function $f(\mathbf{x})$ defined in (2.46): the isotropic singularity S is located at $\mathbf{x}_0 = (-256, -256)$; the Gaussian localized structure G of width $\sigma = 128$ is located at $\mathbf{x}_1 = (256, 256)$; the parameters are $A = 1$ and $B = -1$. (**b**) $\mathcal{M}_\psi[f]$ coded from black ($\mathcal{M}_\psi = 0$) to red (max $\mathcal{M}_\psi$) with 256 colors as computed at scale $a = 2^3 \sigma_W$ where $\sigma_W = 13$ (pixels) is the characteristic size of the first-order analyzing wavelet ψ (see Fig. 2.10) at the smallest resolved scale; the solid lines correspond to the maxima chains defined by the WTMM; the local maxima (minima) along these chains are indicated by (●) (○) from which originates an arrow whose length is proportional to $\mathcal{M}_\psi[f]$ and its direction (with respect to the x-axis) is given by the WTMM argument $\mathcal{A}_\psi[f]$. (**c**) Three-dimensional representation of the topological evolution of the WTMM chains of f in the space-scale half-hyperplane. The WTMMM (●) are disposed on connected curves called maxima lines. These maxima lines are obtained by linking each WTMMM computed at a given scale to the nearest WTMMM computed at the scale just above. There exist two maxima lines, $\mathcal{L}_{\mathbf{x}_0}$ and $\mathcal{L}_{\mathbf{x}_1}$, pointing respectively to the singularity S and to the smooth localized structure G in the limit $a \mapsto 0^+$. (**d**) Evolution of $\mathcal{M}_\psi[f]$ along $\mathcal{L}_{\mathbf{x}_0}$. (**e**) Evolution of $\mathcal{M}_\psi[f]$ along $\mathcal{L}_{\mathbf{x}_1}$

this means that the direction of largest variation of f around S is given by $\theta_{\mathbf{x}_0} = \mathcal{A}_\psi[f] + \pi$, where $\mathcal{A}_\psi[f]$ is the argument of the corresponding WTMMM. When increasing the scale parameter, the maxima chains evolve; in particular the closed maxima chain around S swells (its characteristic size behaves like a) until it connects with the maxima chain associated with G to form a single closed curve surrounding both S and G. The topological evolution of the maxima chains in the space-scale half-space in Fig. 2.11c enlightens the existence of two maxima lines obtained by linking the WTMMM step by step (i.e. as continuously as possible) from small to large scales. One of these maxima lines points to the singularity S in the limit $a \to 0^+$. (To understand the bubble structure of $\mathcal{L}_{\mathbf{x}_0}(a)$ in Fig. 2.11c, we refer the reader to [2.232, 2.234] where a detailed description of the algorithm to construct the whole 3D skeleton in general situations is provided.) As shown in Fig. 2.11d, along this maxima line $(\mathcal{L}_{\mathbf{x}_0}(a))$, the WT modulus behaves as [2.40, 2.41]

$$\mathcal{M}_\psi[f]\Big(\mathcal{L}_{\mathbf{x}_0}(a)\Big) \sim a^{h(\mathbf{x}_0)} \;,\; a \to 0^+ \;, \tag{2.47}$$

where $h(\mathbf{x}_0) = 0.3 < n_\psi$ is the Hölder exponent of S. Moreover, along this maxima line, the WT argument evolves towards the value [2.232]:

$$\mathcal{A}_\psi[f]\Big(\mathcal{L}_{\mathbf{x}_0}(a)\Big) = \pi + \theta_{\mathbf{x}_0} \;, \tag{2.48}$$

in the limit $a \to 0^+$, where $\theta_{\mathbf{x}_0}$ is nothing but the direction of the largest variation of f around $\mathbf{x}_0$, i.e. the direction to follow from $\mathbf{x}_0$ to cross the maxima line at a given (small) scale. From the maxima line $\mathcal{L}_{\mathbf{x}_0}(a)$, one thus gets the required amplitude and directional information to characterize the local Hölder regularity of f at $\mathbf{x}_0$. Note that along the other maxima line $\mathcal{L}_{\mathbf{x}_1}(a)$ which points to $\mathbf{x}_1$ where is located the smooth localized structure G, the WT modulus behaves as (Fig. 2.11e):

$$\mathcal{M}_\psi[f]\Big(\mathcal{L}_{\mathbf{x}_1}(a)\Big) \sim a^{n_\psi} \;,\; a \to 0^+ \;, \tag{2.49}$$

where $n_\psi = 1$ is the order of the analyzing wavelet. Equations (2.47) and (2.49) can thus be seen as the analogs of (2.14) and (2.15) in 1D. We refer the reader to [2.232] where a similar WTMMM treatment is also shown to apply to anisotropic as well as self-affine singularities.

2.4.3 The 2D Wavelet Transform Modulus Maxima Method

Methodology: Our strategy will consist in mapping the methodology developed in Sect. 2.2.3 for multifractal analysis of irregular 1D landscapes, to the statistical characterization of roughness fluctuations of 2D surfaces [2.230–2.234]. The 2D WTMM method relies upon the space-scale partitioning given by the WT skeleton. As discussed in Sect. 2.4.2, this skeleton (see

Fig. 2.12e) is defined by the set of maxima lines which point to the singularities of the considered function and therefore is likely to contain all the information concerning the fluctuations of point-wise Hölder regularity. Let us define $\mathcal{L}(a)$ as the set of all maxima lines that exist at the scale a and which contain maxima at any scale $a' \leq a$. As discussed in Sect. 2.4.2, the important feature is that each time the analyzed image has a Hölder exponent $h(\mathbf{x}_0) < n_\psi$, there is at least one maxima line pointing towards $\mathbf{x}_0$ along which (2.47) is expected to hold. In the case of fractal functions, we thus expect that the number of maxima lines will diverge in the limit $a \to 0^+$, as the signature of the hierarchical organization of the singularities. The WTMM method consists in defining the following partition function directly from the WTMMM that belong to the WT skeleton:

$$Z(q,a) = \sum_{\mathcal{L} \in \mathcal{L}(a)} \left(\mathcal{M}_\psi[f](\mathbf{x}, a) \right)^q , \qquad (2.50)$$

where $q \in \mathbb{R}$. As in 1D (Sect. 2.2.3), from the scaling behavior of this partition function:

$$Z(q,a) \sim a^{\tau(q)} , \quad a \to 0^+ , \qquad (2.51)$$

one can extract the $D(h)$ singularity spectrum (as defined as the Hausdorff dimension of the set of points such that the Hölder exponent of f is h) from a simple Legendre transform:

$$D(h) = \min_q \left(qh - \tau(q) \right) . \qquad (2.52)$$

Probability Density Functions: From the definition of the partition function in (2.50), one can transform the discrete sum over the WTMMM into a continuous integral over $\mathcal{M}_\psi[f]$:

$$Z(q,a)/Z(0,a) = \langle \mathcal{M}^q \rangle(a) = \int d\mathcal{M} \, \mathcal{M}^q P_a(\mathcal{M}) . \qquad (2.53)$$

The multifractal description thus consists in characterizing how the moments of the PDF $P_a(\mathcal{M})$ of $\mathcal{M}$ behave as a function of the scale parameter a. The power-law exponents $\tau(q)$ in (2.51) therefore quantify the evolution of the shape of the $\mathcal{M}$ PDF across scales. At this point, let us remark that one of the main advantages of using the WT skeleton is the fact that, by definition, $\mathcal{M}$ is different from zero and consequently that $P_a(\mathcal{M})$ generally decreases exponentially fast to zero at zero. This observation is at the heart of the WTMM method since, for this reason, one cannot only compute the $\tau(q)$ spectrum for $q > 0$ but also for $q < 0$ [2.230–2.233]. From the Legendre transform of $\tau(q)$ (2.52), one is thus able to compute the whole $D(h)$ singularity spectrum, i.e. its increasing left part ($q > 0$) as well as its decreasing right part ($q < 0$).

But, although we have decided to mainly use isotropic analyzing wavelets, we have seen in Sect. 2.4.2 that from the analysis of the WT skeleton, one is able to also extract directional information via the computation of $\mathcal{A}_\psi[f](\mathbf{x}, a)$. It is thus very instructive to extend our statistical analysis to the investigation of the joint PDF $P_a(\mathcal{M}, \mathcal{A})$ [2.231, 2.232]. Two main situations have to be distinguished:

(i) $\mathcal{M}$ and $\mathcal{A}$ are independent. This means that, whatever the scale a, the joint PDF factorizes:

$$P_a(\mathcal{M}, \mathcal{A}) = P_a(\mathcal{M})P_a(\mathcal{A}) . \tag{2.54}$$

In other words, the Hölder exponent h is statistically independent of the direction $\theta = \mathcal{A} + \pi$ to which it is associated. This implies that the $D(h)$ singularity spectrum is decoupled from the angular information contained in $P_a(\mathcal{A})$. If this angle PDF is flat, this means that the rough surface under study displays isotropic scale invariance properties. If, on the contrary, this PDF is a nonuniform distribution on $[0, 2\pi]$, this suggests that some anisotropy is present in the analyzed image. The possible existence of privileged directions can then be revealed by investigating the correlations between the values of $\mathcal{A}$ for different maxima lines. Furthermore, $P_a(\mathcal{A})$ may evolve when varying the scale parameter a. The way its shape changes indicates whether (and how) anisotropy is enhanced (or weakened) when going from large scales to small scales. Even though we are mainly interested in the scaling properties in the limit $a \to 0^+$, the evolution of the shape of $P_a(\mathcal{A})$ across scales is likely to enlighten possible deep structural changes.

(ii) $\mathcal{M}$ and $\mathcal{A}$ are dependent. If (2.54) definitely does not apply, this means that the rough surface under consideration is likely to display anisotropic scale invariance properties. By conditioning the statistical analysis of $\mathcal{M}$ to a given value of $\mathcal{A}$, one can then investigate the scaling properties of the conditioned partition function:

$$Z_\mathcal{A}(q, a) = Z_\mathcal{A}(0, a) \int d\mathcal{M} \; \mathcal{M}^q P_a(\mathcal{M}|\mathcal{A}) \\ \sim a^{\tau_\mathcal{A}(q)} . \tag{2.55}$$

Then by Legendre transforming $\tau_\mathcal{A}(q)$, one gets the singularity spectrum $D_\mathcal{A}(h)$ conditioned to the value of the angle $\mathcal{A}$ $(= \theta - \pi)$. The investigation of the $\mathcal{A}$-dependence of the singularity spectrum $D_\mathcal{A}(h)$ can be rich of information concerning anisotropic multifractal scaling properties.

Remark 3: There have been previous attempts in the literature to carry out anisotropic multifractal analysis. In the context of geophysical (fracture and faulting) data analysis, Ouillon et al. [2.240–2.242] have used an optimized anisotropic wavelet coefficient method to detect and characterize the different levels of mineral organization via the changes of statistical scale invariance.

From a mathematical point of view, Ben Slimane [2.239] has recently proposed a way to generalize the multifractal formalism to anisotropic self-similar functions. His strategy consists in modifying the definition of the 2D WT so that anisotropic zooming is operationally integrated in the optics of this mathematical microscope.

2.4.4 Application of the 2D WTMM Method to Synthetic Isotropic Monofractal and Multifractal Rough Surfaces

Fractional Brownian Surfaces: The generalization of Brownian motion to more than one dimension was first considered by Lévy [2.243]. The generalization of fBm follows along similar lines. A 2D fBm $B_H(\mathbf{x})$ indexed by $H \in \,]0,1[$, is a process with stationary zero-mean Gaussian increments and whose correlation function is given by [2.1, 2.22, 2.243]:

$$\langle B_H(\mathbf{x})B_H(\mathbf{y})\rangle = \frac{\sigma^2}{2}\Big(|\mathbf{x}|^{2H} + |\mathbf{y}|^{2H} - |\mathbf{x}-\mathbf{y}|^{2H}\Big) , \qquad (2.56)$$

where $\langle \ldots \rangle$ represents the ensemble mean value. The variance of such a process is

$$\mathrm{var}\Big(B_H(\mathbf{x})\Big) = \sigma^2|\mathbf{x}|^{2H} , \qquad (2.57)$$

from which one recovers the classical behavior $\mathrm{var}(B_{1/2}(\mathbf{x})) = \sigma^2|\mathbf{x}|$ for Brownian motion with $H = 1/2$. 2D fBms are self-affine processes that are statistically invariant under isotropic dilations:

$$B_H(\mathbf{x}_0 + \lambda\mathbf{u}) - B_H(\mathbf{x}_0) \simeq \lambda^H\Big[B_H(\mathbf{x}_0 + \mathbf{u}) - B_H(\mathbf{x}_0)\Big] , \qquad (2.58)$$

where $\mathbf{u}$ is a unitary vector and $\simeq$ stands for the equality in law. The index H corresponds to the Hurst exponent; the higher the exponent H, the more regular the fBm surface. But since (2.58) holds for any $\mathbf{x}_0$ and any direction $\mathbf{u}$, this means that almost all realizations of the fBm process are continuous, everywhere nondifferentiable, isotropically scale-invariant as characterized by a unique Hölder exponent $h(\mathbf{x}) = H$ [2.1, 2.22, 2.126], $\forall \mathbf{x}$. Thus fBm surfaces are the representation of homogeneous stochastic fractal functions characterized by a singularity spectrum which reduces to a single point

$$\begin{aligned} D(h) &= 2 \quad \text{if } h = H , \\ &= -\infty \quad \text{if } h \neq H . \end{aligned} \qquad (2.59)$$

By Legendre transforming $D(h)$ according to (2.52), one gets the following expression for the partition function exponent (2.51):

$$\tau(q) = qH - 2 . \qquad (2.60)$$

$\tau(q)$ is a linear function of q, the signature of monofractal scaling, with a slope given by the index H of the fBm.

We have tested the 2D WTMM method described in Sect. 2.4.3 [2.232] on fBm surfaces generated by the so-called fast Fourier transform filtering method [2.22, 2.27]. We have used this particular synthesis method because of its implementation simplicity. Indeed it amounts to a fractional integration of a 2D 'white noise' and therefore it is expected to reproduce quite faithfully the expected isotropic scaling invariance properties (2.58). We have investigated fBms for various values of the index H. Here we report, for illustration, the results obtained on 32 realizations of a 2D fBm process with $H = 1/3$. Along the lines of the numerical implementation procedure described in Sect. 2.4.2 [2.232], we have wavelet transformed 32 ($1\,024 \times 1\,024$) images of $B_{H=1/3}$ with an isotropic first-order analyzing wavelet. To master edge effects, we then restrain our analysis to the 512×512 central part of the WT of each image. In Fig. 2.12c is illustrated the computation of the maxima chains and the WTMMM for an individual image at a given scale. In this figure is also shown the convolution of the original image (Fig. 2.12a) with the isotropic Gaussian smoothing filter ϕ (2.39). According to the definition of the WTMM, the maxima chains correspond to well defined edge curves of the smoothed image. The local maxima of $\mathcal{M}_\psi$ along these curves are located at the points where the sharpest intensity variation is observed. The corresponding arrows clearly indicate that locally, the gradient vector points to the direction (as given by $\mathcal{A}_\psi$) of maximum change of the intensity surface. When going from large to small scale, the average distance between two nearest neighbor WTMMM decreases like a. This means that the number of WTMMM, and in turn, the number of maxima lines, proliferates across scales like a^{-2}. The corresponding WT skeleton is shown in Fig. 2.12e. As confirmed just below, when extrapolating the arborescent structure of this skeleton to the limit $a \rightarrow 0^+$, one recovers the theoretical result that the support of the singularities of a 2D fBm has a dimension $D_F = 2$, i.e. $B_{H=1/3}(\mathbf{x})$ is nowhere differentiable [2.1, 2.22, 2.243].

In Fig. 2.13 are reported the results of the computation of the $\tau(q)$ and $D(h)$ spectra using the 2D WTMM method described in Sect. 2.4.3. As shown in Fig. 2.13a, for $q \in [-4, 6]$, the annealed average partition function $Z(q, a)$ (over 32 images of $B_{1/3}(\mathbf{x})$) displays a well defined scaling behavior over more than three octaves. When proceeding to a linear regression fit of the data over the first two octaves, one gets the $\tau(q)$ spectrum (2.51) shown in Fig. 2.13c. The data systematically fall on a straight line which is in remarkable agreement with the theoretical $\tau(q)$ spectrum (2.60) with $H = 1/3$. As shown in Fig. 2.13d, it is therefore not surprising that the $D(h)$ spectrum obtained by Legendre transforming the $\tau(q)$ data reduce to a single point $D(h = H = 1/3) = 2.00 \pm 0.02$ (2.59).

In Fig. 2.14 are shown the PDFs $P_a(\mathcal{M}) = \int d\mathcal{A} P_a(\mathcal{M}, \mathcal{A})$ and $P_a(\mathcal{A}) = \int d\mathcal{M} P_a(\mathcal{M}, \mathcal{A})$, for four different values of the scale parameter. As seen in

fBm rough surface   log–normal cascade

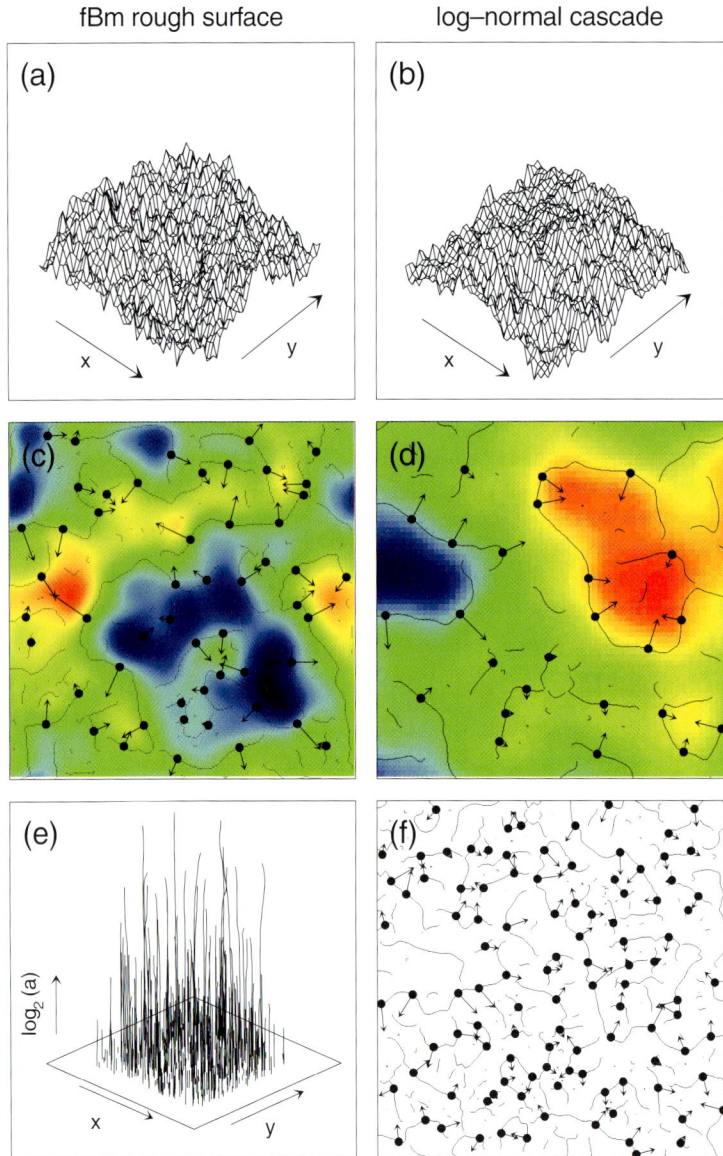

Fig. 2.12. Continuous WT of monofractal and multifractal synthetic rough surfaces. *Fractional Brownian rough surfaces:* (**a**) A $(1\,024 \times 1\,024)$ realization of a $H = 1/3$ fBm rough surface; (**c**) maxima chains defined by the WTMM as computed at the scale $a = 2^2 \sigma_W$; the WTMMM (•) are defined by the local maxima of $\mathcal{M}_\psi$ along the chains; (**e**) the WT skeleton defined by the maxima lines obtained after linking the WTMMM detected at different scales. *Log-normal random 2D $\mathcal{W}$-cascades:* (**b**) A $(1\,024 \times 1\,024)$ rough surface generated using the log-normal $\mathcal{W}$-cascade model with the parameter values $m = -0.35 \ln 2$ and $\sigma^2 = 0.03 \ln 2$; (**d**) maxima chains and WTMMM as computed at the scale $a = 2^3 \sigma_W$; (**f**) same as in (**d**) but at the scale $a = 2\sigma_W$. In (**c**) and (**d**), the smoothed image $\phi_{\mathbf{b},a} \star f$ is shown as a color coded background from black (min) to red (max). ψ is the first-order radially symmetric wavelet shown in Fig. 2.10

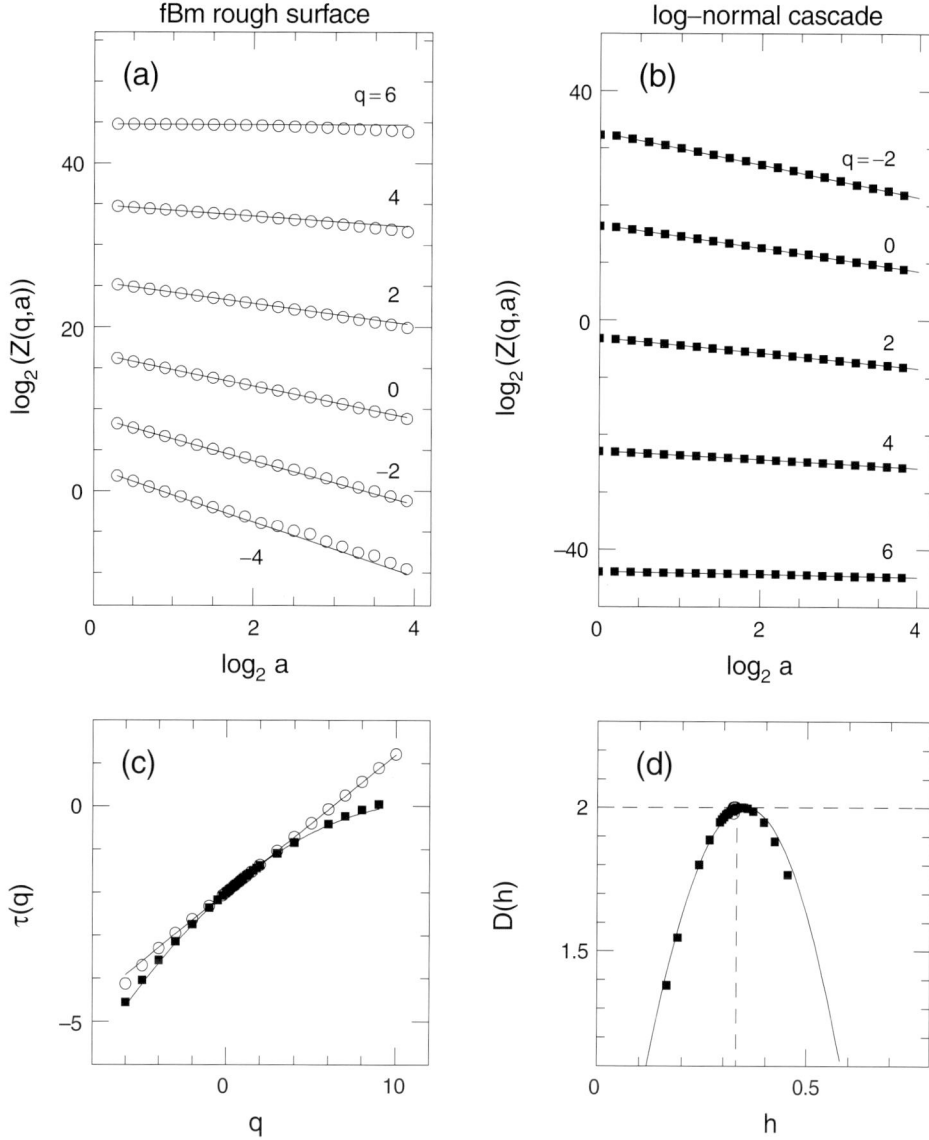

Fig. 2.13. Determination of the $\tau(q)$ and $D(h)$ multifractal spectra of fractional Brownian rough surfaces with $H = 1/3$ (circles) and log-normal random 2D $\mathcal{W}$-cascades (squares) using the 2D WTMM method. (**a**) $\log_2 Z(q,a)$ vs. $\log_2 a$: $B_{1/3}(\mathbf{x})$. (**b**) $\log_2 Z(q,a)$ vs. $\log_2 a$: log-normal $\mathcal{W}$-cascade with the same parameters as in Fig. 2.12b. (**c**) $\tau(q)$ vs. q; the solid lines correspond respectively to the theoretical spectra given by (2.60) and (2.61). (**d**) $D(h)$ vs. h; the solid lines correspond respectively to the theoretical predictions given by (2.59) and (2.62). The analyzing wavelet is the same as in Fig. 2.12. The reported results correspond to annealed averaging over 32 ($1\,024 \times 1\,024$) realizations

Fig. 2.14a, $P_a(\mathcal{M})$ is not a Gaussian (in contrast to the PDF of the continuous 2D wavelet coefficients), but decreases very fast to zero at zero as expected when using the WTMM method. The corresponding PDFs $P_a(\mathcal{A})$ are represented in Fig. 2.14c. $P_a(\mathcal{A})$ clearly does not evolve across scales. Moreover, except for some small amplitude fluctuations observed at the largest scale, $P_a(\mathcal{A}) = 1/2\pi$ is a flat distribution as expected for statistically isotropic scale-

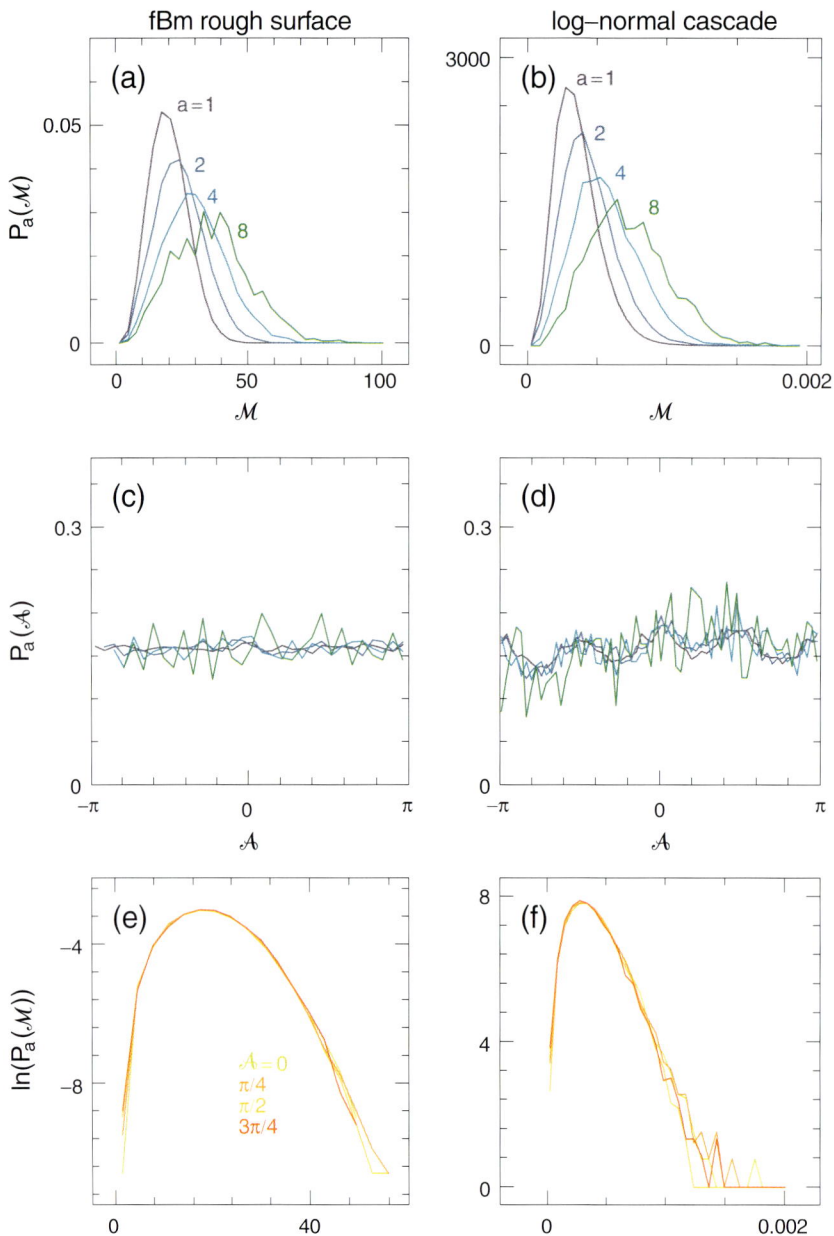

Fig. 2.14. PDFs of the WTMMM coefficients as computed at the scales $a = 1, 2, 4,$ and 8 (in σ_W units). *Fractional Brownian rough surfaces:* (**a**) $P_a(\mathcal{M})$ vs. $\mathcal{M}$; (**c**) $P_a(\mathcal{A})$ vs. $\mathcal{A}$; (**e**) PDFs of $\mathcal{M}$ when conditioned by $\mathcal{A}$, at the scale $a = \sigma_W$. *Log-normal random 2D $\mathcal{W}$-cascades:* (**b**) $P_a(\mathcal{M})$ vs. $\mathcal{M}$; (**d**) $P_a(\mathcal{A})$ vs. $\mathcal{A}$; (**f**) PDFs of $\mathcal{M}$ when conditioned by $\mathcal{A}$, at the scale $a = 2^{0.1}\sigma_W$. In (**e**) and (**f**) the different curves correspond to fixing $\mathcal{A}$ (mod π) to $0 \pm \pi/8$, $\pi/4 \pm \pi/8$, $\pi/2 \pm \pi/8$, and $3\pi/4 \pm \pi/8$. Same 2D WTMM computations as in Fig. 2.13

invariant rough surfaces. The results reported in Fig. 2.14e not only corroborate statistical isotropy but they bring unambiguous evidence for the independence of $\mathcal{M}$ and $\mathcal{A}$ (2.54). For two different scales, the PDF of $\mathcal{M}$, when conditioned by the argument $\mathcal{A}$, is shown to be shape invariant.

2D Random $\mathcal{W}$-Cascades: As originally introduced in [2.233], the concept of $\mathcal{W}$-cascade [2.103–2.107] (Sect. 2.2.6) can be generalized in 2D in order to generate synthetic multifractal functions of two variables. A 2D random $\mathcal{W}$-cascade is built recursively on the two-dimensional square grid of separable wavelet orthogonal basis, involving only scales that range between a given large scale L and the scale 0 (excluded). Thus the corresponding fractal function $f(\mathbf{x})$ will not involve scales greater than L. For that purpose, we will use compactly supported wavelets defined by Daubechies [2.73]. In Fig. 2.12b is shown a rough surface generated with the log-normal $\mathcal{W}$-cascade model proposed in [2.233]. The multifractal spectra have the following analytical expression:

$$\tau(q) = -\log_2\langle W^q\rangle - 2 \;,\; \forall q \in \mathbb{R}$$
$$= -\frac{\sigma^2}{2\ln 2}q^2 - \frac{m}{\ln 2}q - 2 \;, \tag{2.61}$$

and

$$D(h) = -\frac{(h + m/\ln 2)^2}{2\sigma^2/\ln 2} + 2 \;, \tag{2.62}$$

which are reminiscent of (2.29) and (2.30) derived for 1D log-normal $\mathcal{W}$-cascades. Let us mention that (2.31) and (2.32) are slightly modified in 2D (indeed one has simply to replace σ^2 by $2\sigma^2$ in these formulas).

In Figs. 2.12, 2.13, and 2.14, we mainly report results obtained with the first-order radially symmetric analyzing wavelets shown in Fig. 2.10. Let us point out that quite robust results are obtained with the third-order analyzing wavelet. In Fig. 2.12d and 2.12f is illustrated the computation of the maxima chains and the WTMMM for an individual image (Fig. 2.12b) of a multifractal rough surface generated with the log-normal $\mathcal{W}$-cascade model described in [2.233]. The model parameters are $m = -0.35\ln 2$ and $\sigma^2 = 0.03\ln 2$. From the WTMMM defined on these maxima chains, one constructs the WT skeleton according to the procedure described in Sects. 2.4.2 and 2.4.3. From the WT skeletons of 32 ($1\,024 \times 1\,024$) images like the one in Fig. 2.12b, one computes the annealed average of the partition functions $Z(q,a)$. As shown in Fig. 2.13b, when plotted vs. the scale parameter a in a logarithmic representation, these annealed average partition functions display a well defined linear behavior over a range of scales of about four octaves (i.e. $\sigma_W \lesssim a \lesssim 16\sigma_W$, where $\sigma_W = 13$ pixels). Let us point out that scaling of quite good quality is found for a rather wide range of values of q: $-6 \lesssim q \lesssim 8$. When processing to a linear regression fit of the data over the first four octaves, one gets the $\tau(q)$ spectrum (■) shown in Fig. 2.13c. For the range of investigated q values, the numerical data are in remarkable agreement with the theoretical nonlinear $\tau(q)$ spectrum given by (2.61). Similar quantitative agreement is observed on the $D(h)$ singularity spectrum in Fig. 2.13d. This is the quantitative demonstration that the 2D WTMM method is able to resolve multifractality. From the theoretical spectra, the multifractal rough surfaces under study display intermittent fluctuations corresponding to Hölder exponent values ranging from $h_{\min} = 0.005$ to $h_{\max} = 0.700$. Unfortunately, to capture the strongest and weakest singularities, one needs to compute the $\tau(q)$ spectrum for very large values of $|q|$. This requires the processing of many more images of much larger size, which is out of our current computer capabilities. Note that with the statistical sample studied here, one has $D(h(q = 0) = 0.35) = 2.00 \pm 0.02$, which allows us to conclude that the rough surfaces under consideration are singular everywhere.

Remark 4: Note that we have chosen on purpose the parameters m and σ^2 so that the spectral exponent $\beta = 4 - \tau(2) = 2.66$ is the same for the $B_{1/3}$ monofractal rough surfaces and the log-normal $\mathcal{W}$ rough surfaces. The results reported in this section show that the 2D WTMM method succeeds in distinguishing monofractal and multifractal surfaces that exhibit exactly the same density spectrum power law decay.

From the construction rule of these synthetic log-normal rough surfaces [2.233, 2.234], the multifractal nature of these random functions is expected to be contained in the way the shape of the WT modulus PDF $P_a(\mathcal{M})$ evolves when varying the scale parameter a, as shown in Fig. 2.14b. Indeed the joint PDF $P_a(\mathcal{M}, \mathcal{A})$ is expected to factorize, as the signature of the implicit decoupling of $\mathcal{M}$ and $\mathcal{A}$ in the construction process. This decoupling is numerically retrieved in Fig. 2.14f, where, for two different scales, the PDF of $\mathcal{M}$, when conditioned by the argument $\mathcal{A}$, is shown to be shape invariant. Moreover, as seen in Fig. 2.14d, $P_a(\mathcal{A})$ does not exhibit any significant change when increasing a, except some loss in statistical convergence at large scales due to the rarefaction of the maxima lines. Let us point out that, even though $P_a(\mathcal{A})$ looks globally rather flat, one can notice some small amplitude almost periodic oscillations at the smallest scales which reflects the existence of privileged directions in the wavelet cascading process. These oscillations are maximum for $\mathcal{A} = 0$, $\pi/2$, π, and $3\pi/2$, as the witness to the square lattice anisotropy underlying the 2D wavelet tree decomposition.

2.5 Application of the 2D WTMM Method
to High-Resolution Satellite Images of Cloud Structure

The problematic of nonlinear variability over a wide range of scales has been considered for a long time with respect to the highly intermittent nature of turbulent flows in fluid dynamics [2.15, 2.67]. Special attention has been paid to their asymptotic and possibly universal behavior when the dissipation length goes to zero, i.e. when the Reynolds number goes to infinity. Besides wind-tunnel and laboratory (grid, jet, ...) experiments, the atmosphere is a huge natural laboratory where high Reynolds number (fully developed) turbulent dynamics can be studied. Clouds, which are at the source of the hydrological cycle, are the most obvious manifestation of the Earth's turbulent atmospheric dynamics [2.16, 2.244–2.246]. By modulating the input of solar radiation, they play a critical role in the maintenance of the Earth's climate [2.247]. They are also one of the main sources of uncertainty in current climate modeling [2.248], where clouds are assumed to be homogeneous media lying parallel to the Earth's surface; at best, a linear combination of cloudy and clear portions according to cloud fraction is used to account for horizontal inhomogeneity when predicting radiative properties. During many years, the lack of data hindered our understanding of cloud microphysics and cloud-radiation interactions. Nowadays, it is well-recognized that clouds are variable in all directions and that fractal [2.244, 2.245, 2.249–2.254] and multifractal [2.16, 2.246, 2.255–2.257] concepts are likely to be relevant to the description of their complex 3D geometry. Until quite recently, the internal structure of clouds was probed by balloons or aircrafts that penetrated the cloud layer, revealing an extreme variability of 1D cuts of some cloud fields [2.257–2.265]. In particular, in situ measurements of cloud liquid water content (LWC) were performed during many intensive field programs (FIRE [2.266], ASTEX [2.267], SOCEX [2.268], ...). Indeed, during the past fifteen years, vast amounts of data on the distribution of atmospheric liquid water from a variety of sources were collected and analyzed in many different ways. All these data contain information on spatial and/or temporal correlations in cloudiness, enabling the investigation of scale invariance over a range from a few centimeters to hundreds of kilometers. An attractive alternative to in situ probing is to use *high-resolution satellite imagery* that now provides direct information about the fluctuations in liquid water concentration in the depth of clouds [2.245, 2.250, 2.252, 2.254, 2.269–2.274]. These rather sophisticated remote sensing systems called 'millimeter radars' are actually sensitive not only to precipitating rain drops but also to suspended cloud droplets. Spectral analysis of the recorded 2D radiance field [2.269–2.274] confirms previous 1D findings that make it likely that cloud scenes display structures over a wide range of scales.

One has to give credit to Lovejoy and co-workers [2.132, 2.134, 2.136, 2.144, 2.145, 2.255, 2.256, 2.275, 2.276] for applying the multifractal description to atmospheric phenomena. Using the trace moment and double trace moment techniques [2.134, 2.136, 2.275, 2.276] they have brought experimental evidence for multiple scaling (or in other words, the existence of a continuum of scaling exponent values) in various geophysical fields. More recently, Davis and co-workers [2.246, 2.257, 2.265] have used the structure function method to study LWC data recorded during ASTEX and FIRE programs. Both these analyses lead to the conclusion that the internal marine stratocumulus (Sc) structure is multifractal over at least three decades in scales. Similar multifractal behavior has been reported by Wiscombe et al. [2.274] when analyzing liquid water path (LWP) data (i.e. column integrated LWC) from the Atmospheric Radiation Measurement (ARM) archive. Even though all these studies seem to agree, at least as far as their common diagnostic of multifractal scaling of the cloud structure is concerned, they all address 1D data. To our knowledge, the structure function method has been also applied to 1D cuts of high-resolution satellite images [2.270, 2.277], but we are not aware of any results coming out from a specific 2D analysis. Our goal here is to take advantage of the 2D WTMM method to carry out a multifractal analysis of high-resolution satellite images of cloudy scenes [2.230, 2.231, 2.234, 2.278]. Beyond the issue of improving statistical characterization of in situ and remotely sensed data, there is a most challenging aspect which consists in extracting structural information to constraint stochastic cloud models which in turn will be used for radiative transfer simulations [2.246, 2.253, 2.255, 2.270, 2.279–2.285]. Then by comparing the multifractal properties of the numerically generated artificial radiation fields with those of actual measurements, one can hope to achieve some degree of closure.

2.5.1 Landsat Data of Marine Stratocumulus Cloud Scenes

Over the past fifteen years, Landsat imagery has provided the remote sensing community at large with a very attractive and reliable tool for studying the Earth's environment [2.246, 2.250, 2.252–2.254, 2.269–2.273, 2.286, 2.287]. One of the main advantages of high-resolution satellite imagery is its rather low effective cost as compared to outfitting and flying research aircraft. Moreover, this instrument is well calibrated and it offers the possibility to reach unusual high spatial, spectral, and radiometric resolutions [2.270, 2.287]. Mainly two types of statistical analysis have been applied so far to Landsat imagery: spectral analysis of the 2D radiance field [2.269–2.273, 2.287] and joint area and perimeter distributions for ensembles of individual clouds [2.250, 2.252–2.254] defined by some threshold in radiance. One of the most remarkable properties of Landsat cloud scenes is their statistical 'scale invariance' over a rather large range of scales, which justifies why fractal and multifractal concepts have progressively gained more acceptance in the atmospheric scientist community [2.16, 2.246].

Off all cloud types, marine stratocumulus are without any doubt the ones which have attracted the most attention, mainly because of their first-order effect on the Earth's energy balance [2.16, 2.245, 2.246, 2.287, 2.288]. Being at once very persistent and horizontally extended, marine Sc layers carry considerable weight in the overall reflectance (albedo) of the planet and, from there, command a strong effect on its global climate [2.247]. Furthermore, with respect to climate modeling [2.248] and the major problem of cloud-radiation interaction [2.246, 2.255, 2.269, 2.270, 2.279–2.281, 2.283, 2.289], they are presumably at their simplest in marine Sc which are relatively thin ($\sim$ 300-500 m), with well-defined (quasi-planar) top and bottom, thus approximating the plane-parallel geometry where radiative transfer theory is well developed [2.245, 2.255, 2.270, 2.280, 2.281, 2.283]. However, because of its internal homogeneity assumption, plane-parallel theory shows systematic biases in large-scale average reflectance [2.281, 2.290] relevant to general circulation model (GCM) energetics and large random errors in small-scale values [2.283, 2.291] relevant to remote-sensing applications. Indeed, marine Sc have huge internal variability [2.257, 2.265], not necessarily apparent to the remote observer (see Fig. 2.15a).

In the next section, we challenge previous analysis [2.250, 2.252–2.254, 2.269–2.273, 2.286, 2.287] of Landsat imagery using the 2D WTMM methodology [2.230, 2.231, 2.278] described in Sect. 2.4. Our specific goal will be to improve statistical characterization of the highly intermittent radiance fluctuations of marine Sc, a prerequisite for developing better models of cloud structure. For that purpose, we analyze a ($\simeq 196 \times 168\,\mathrm{km}^2$) original cloudy Landsat 5 scene captured with the TM camera (1 pixel $-$ 30 m) in the 0.6–0.7 μm channel (i.e. reflected solar photons as opposed to their counterparts emitted in the thermal infrared) during the first ISCCP (International Satellite Cloud Climatology Project) Research Experiment (FIRE) field program, which took place over the Pacific Ocean off San Diego in summer 1987. For computational convenience, we actually select 32 overlapping $1\,024 \times 1\,024$ pixels^2 subscenes in this cloudy region. The overall extent of the explored area is about $7\,840\,\mathrm{km}^2$ [2.278]. Figure 2.15a shows a typical ($1\,024 \times 1\,024$) portion of the original image where the eight-bit gray scale coding of the quasi-nadir viewing radiance clearly reveals the presence of some anisotropic texture induced by convective structures which are generally aligned to the wind direction.

2.5.2 WTMM Multifractal Analysis of Landsat Images of Stratocumulus

We systematically follow the numerical implementation procedure previously used in Sect. 2.4.4 for synthetic rough surfaces. We first wavelet transform the

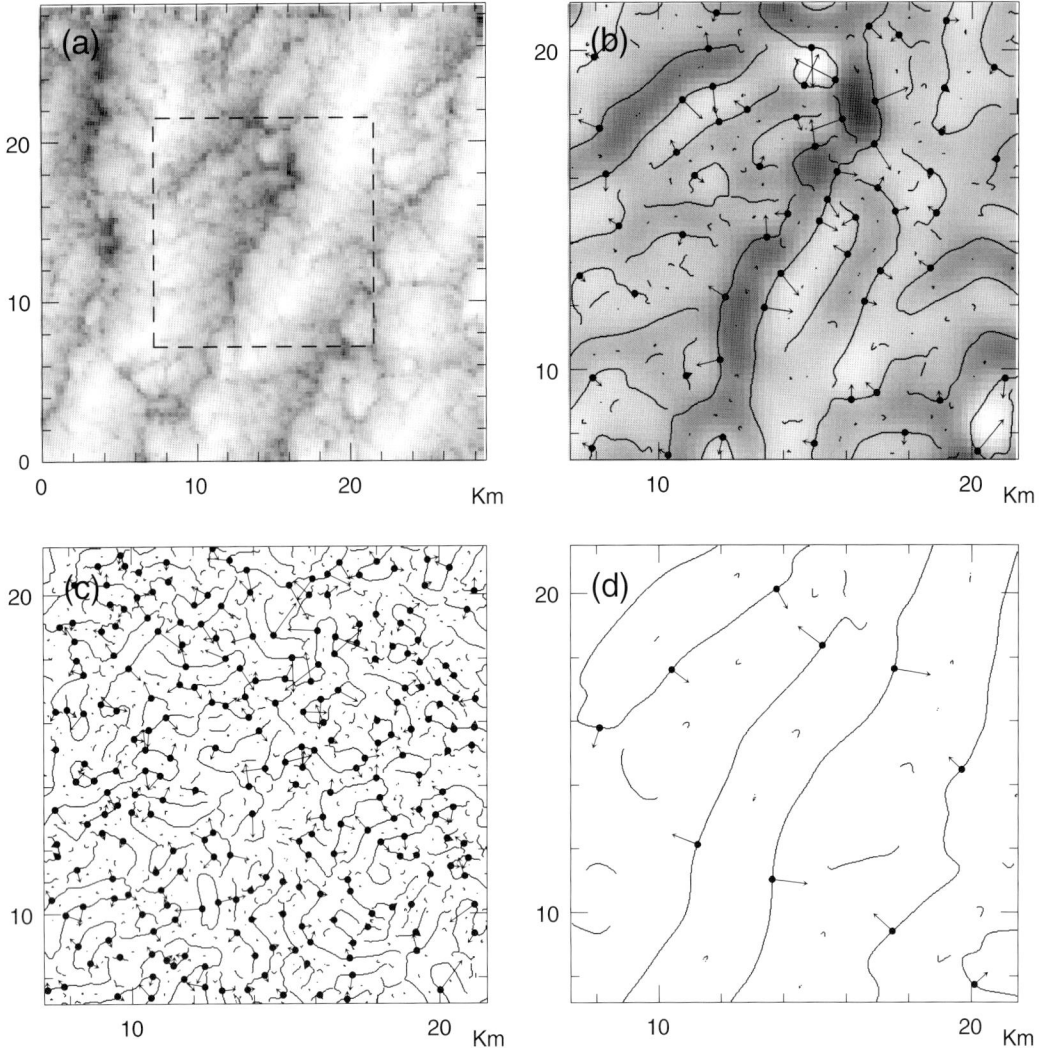

Fig. 2.15. 2D WT analysis of a Landsat image of marine Sc clouds captured at $l = 30$ m resolution on July 7, 1987, off the coast of San Diego (CA) [2.266]. (**a**) 256 gray-scale coding of a $(1\,024 \times 1\,024)$ portion of the original radiance image. In (**b**) $a = 2^{2.9}\sigma_W$, (**c**) $a = 2^{1.9}\sigma_W$ and (**d**) $a = 2^{3.9}\sigma_W$ where $(\sigma_W = 13 \text{ pixels} \simeq 390 \text{ m})$, are shown the maxima chains; the local maxima of $\mathcal{M}_\psi$ along these chains are indicated by ($\bullet$) from which originates an arrow whose length is proportional to $\mathcal{M}_\psi$ and its direction (with respect to the x-axis) is given by $\mathcal{A}_\psi$; only the central (512×512) part delimited by a dashed square in (**a**) is taken into account to define the WT skeleton. In (**b**), the smoothed image $\phi_{\mathbf{b},a} \star I$ is shown as a gray-scale coded background from white (min) to black (max). $\psi(\mathbf{x})$ is the first-order radially symmetric analyzing wavelet shown in Fig. 2.10

32 overlapping $(1\,024 \times 1\,024)$ images, cut out of the original image, with the first-order $(n_\psi = 1)$ radially symmetric analyzing wavelet defined in Fig. 2.10. From the WT skeleton defined by the WTMMM, we compute the partition functions from which we extract the $\tau(q)$ and $D(h)$ multifractal spectra as explained in Sect. 2.4.3. We systematically test the robustness of our estimates with respect to some change in the shape of the analyzing wavelet, in particular when increasing the number of zero moments.

Numerical Computation of the Multifractal $\tau(q)$ and $D(h)$ Spectra:
In Fig. 2.15 is illustrated the computation of the maxima chains and the WT-
MMM for the marine Sc subscene. After linking these WTMMM across scales,
one constructs the WT skeleton from which one computes the partition functions $Z(q, a)$ (2.50). As reported in Fig. 2.16a, the annealed average partition
functions ($\bullet$) display some well-defined scaling behavior over the first three
octaves, i.e. over the range of scales $390\,\mathrm{m} \lesssim a \lesssim 3\,120\,\mathrm{m}$, when plotted vs.
a. Indeed, the scaling deteriorates progressively from the large scale side when
one goes to large values of $|q| \gtrsim 3$. As discussed in [2.278], besides the fact that
we are suffering from insufficient sampling, the presence of localized Dirac like
structures is likely to explain the fact that the observed crossover to a steeper
power-law decay occurs at a smaller and a smaller scale when one increases
$q > 0$. Actually for $q \gtrsim 3$, the crossover scale $a^* \lesssim 1\,200\,\mathrm{m}$ becomes significantly smaller than the so-called integral scale which is approximately given
by the characteristic width $\lambda \simeq 5$–$6\,\mathrm{km}$ of the convective rolls (Fig. 2.15a).
When processing to a linear regression fit of the data in Fig. 2.16a over the
first octave and a half (in order to avoid any bias induced by the presence of
the observed crossover at large scales), one gets the $\tau(q)$ spectrum ($\bullet$) shown
in Fig. 2.16b. In contrast to the fBm rough surfaces studied in Sect. 2.4.4, this
$\tau(q)$ spectrum unambiguously deviates from a straight line. When Legendre
transforming this nonlinear $\tau(q)$ curve, one gets the $D(h)$ singularity spectrum
reported in Fig. 2.16c. Its characteristic single humped shape over a finite range
of Hölder exponents is a clear signature of the multifractal nature of the marine
Sc radiance fluctuations.

In Fig. 2.16 are also shown for comparison the results ($\circ$) obtained when
applying the 2D WTMM method with a third-order ($n_\psi = 3$) radially symmetric analyzing wavelet (the smoothing function ϕ being the isotropic 2D
Mexican hat). As seen in Fig. 2.16a, the use of a wavelet which has more zero
moments seems to somehow improve scaling. For the range of q-values investigated, the crossover scale turns out to be rejected at a larger scale, enlarging
by some amount the range of scales over which scaling properties can be mea-
sured, especially for the largest values of $|q|$. The fact that one improves scaling
when increasing the order of the analyzing wavelet suggests that perhaps some
smooth behavior unfortunately deteriorates our statistical estimate of the multifractal spectra of the original Landsat radiance image. Let us recall that, as
explained in Sect. 2.4.2, smooth C^∞ behavior may give rise to maxima lines
along which $\mathcal{M}_\psi \sim a^{n_\psi}$ (see Fig. 2.11); hence the larger n_ψ, the smaller is
the overall contribution of those 'spurious' maxima lines in the partition func-
tion summation over the WT skeleton. As seen in Fig. 2.15a, the anisotropic
texture induced by the convective streets or rolls might well be at the origin of the relative lack of well defined scale invariance. When looking at the
corresponding $\tau(q)$ spectrum ($\circ$) extracted from the data in Fig. 2.16b, one
gets quantitatively the same estimates for $q \gtrsim -1$. For more negative values

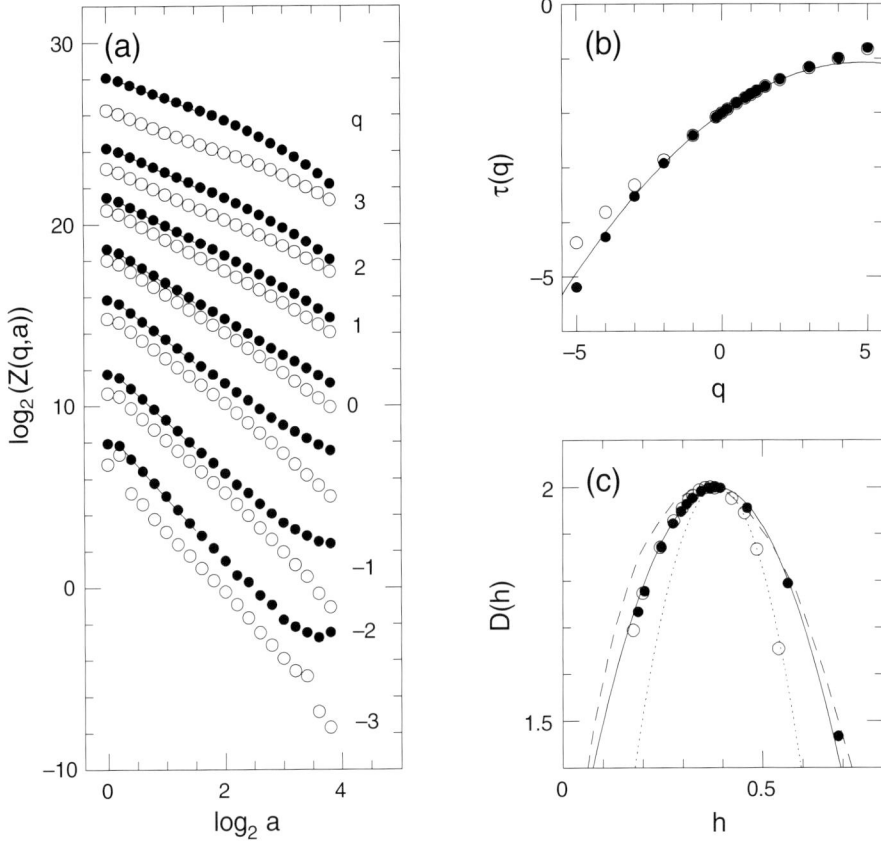

Fig. 2.16. Determination of the $\tau(q)$ and $D(h)$ spectra of radiance Landsat images of marine Sc. The 2D WTMM method is used with either a first-order ($\bullet$) or a third-order ($\circ$) radially symmetric analyzing wavelet (see Fig. 2.10). (**a**) $\log_2 Z(q,a)$ vs. $\log_2 a$; the solid lines correspond to linear regression fits of the data over the first and a half octave. (**b**) $\tau(q)$ vs. q obtained from a linear regression fit of the data in (**a**). (**c**) $D(h)$ vs. h, after Legendre transforming the $\tau(q)$ curve in (**b**). In (**b**) and (**c**), the solid lines correspond to the theoretical multifractal spectra for the log-normal $\mathcal{W}$-cascades with parameter values $m = -0.38 \ln 2$ and $\sigma^2 = 0.07 \ln 2$ ((2.61) and (2.62)). The $D(h)$ singularity spectrum of velocity (dotted line) and temperature (dashed line) fluctuations in fully developed turbulence are shown for comparison in (**c**)

of q, the data obtained with the third-order analyzing wavelet clearly depart from the previous estimates with the first-order wavelet. The slope of the new $\tau(q)$ spectrum is somehow weakened which implies, from the Legendre transform properties, that the corresponding values of $h(q) = \partial\tau/\partial q$ are reduced. The computation of the $D(h)$ singularity spectrum ($\circ$) in Fig. 2.16c enlightens this phenomenon: while the increasing left-hand branch (which corresponds to the strongest singularities) of the $D(h)$ curve appears to be quite robust with respect to the choice of $\boldsymbol{\psi}$, the decreasing right-hand branch (associated to the weakest singularities) is modified when increasing the number of zero moments of $\boldsymbol{\psi}$. As shown in Fig. 2.16b and 2.16c, the $\tau(q)$ spectrum as well as the $D(h)$ spectrum data, are very well fitted by the theoretical quadratic spectra of log-normal random $\mathcal{W}$-cascades (2.61) and (2.62). However, with the first-order analyzing wavelet, the best fit is obtained with the parameter values $m = -0.38\ln 2 = -0.263$ and $\sigma^2 = 0.07\ln 2 = 0.049$, while for the third-order wavelet these parameters take slightly different values, namely $m = -0.366\ln 2 = -0.254$ and $\sigma^2 = 0.06\ln 2 = 0.042$. The variance parameter σ^2 which characterizes the intermittent nature of marine Sc radiance fluctuations is therefore somehow reduced when going from $n_\psi = 1$ to $n_\psi = 3$. Actually, it is the lack of statistical convergence because of insufficient sampling which is the main reason for this uncertainty in the estimate of σ^2 [2.278]. As previously experienced in [2.233] for synthetic multifractal rough surfaces, an accurate estimate of the exponents $\tau(q)$ for $q \lesssim -3$ requires more than 32 $(1\,024 \times 1\,024)$ images. With the statistical sample of Landsat images we have at our disposal, one gets $D(h(q = 0) = 0.37 \pm 0.02) = 2.00 \pm 0.01$, which is a strong indication that the radiance field is singular everywhere. From the estimate of $\tau(q = 2) = -1.38 \pm 0.02$, one gets the following estimate of the spectral exponent: $\beta = \tau(2) + 4 = 2.62 \pm 0.02$, i.e. a value which is in good agreement with previous estimates [2.258–2.262, 2.264, 2.269–2.273, 2.287].

WTMMM Probability Density Functions: This subsection is mainly devoted to the analysis of the joint probability distribution function $P_a(\mathcal{M}, \mathcal{A})$ (see Sect. 2.4.3) as computed from the WT skeletons of the 32 $(1\,024 \times 1\,024)$ radiance images with the first-order radially symmetric analyzing wavelet $(n_\psi = 1)$. In Fig. 2.17a and 2.17b are respectively shown the PDFs $P_a(\mathcal{M}) = \int d\mathcal{A}\, P_a(\mathcal{M}, \mathcal{A})$ and $P_a(\mathcal{A}) = \int d\mathcal{M}\, P_a(\mathcal{M}, \mathcal{A})$, for three different values of the scale parameter $a = 2^{0.3}\sigma_W$ (480 m), $2^{1.3}\sigma_W$ (960 m), and $2^{2.3}\sigma_W$ (1 920 m). First let us focus on the results shown in Fig. 2.17b for $P_a(\mathcal{A})$. This distribution is clearly scale dependent with some evidence of anisotropy enhancement when going from small to large scales, in particular when one reaches scales which become comparable to the characteristic width of the convective structures (i.e. a few kilometers wide). Two peaks around the values $\mathcal{A} \simeq -\pi/6$ and $5\pi/6$ become more and more pronounced as the signature of a privileged direction in the analyzed images. As one can check from a visual inspection of Fig. 2.15a, this direction is nothing but the perpendicular to the mean direction of the convective rolls that are generally aligned to the wind direction. This is another clear indication that, at large scales, the WT microscope is sensitive to the convective roll texture, a rather regular modulation superimposed to the background radiance fluctuations [2.231, 2.278]. Another important message which comes out from our analysis is illustrated in Fig. 2.17c and 2.17d. When conditioning the PDF of $\mathcal{M}$ by the argument $\mathcal{A}$, the shape of this PDF is shown to be independent of the considered value of $\mathcal{A}$, as long as the value of the scale parameter a remains small as compared to the characteristic width of the convective structures. The observation that the joint probability distribution actually factorizes, i.e. satisfies (2.54), is the signature that $\mathcal{M}$ and $\mathcal{A}$ are likely to be independent [2.231, 2.278]. This implies that all the multifrac-

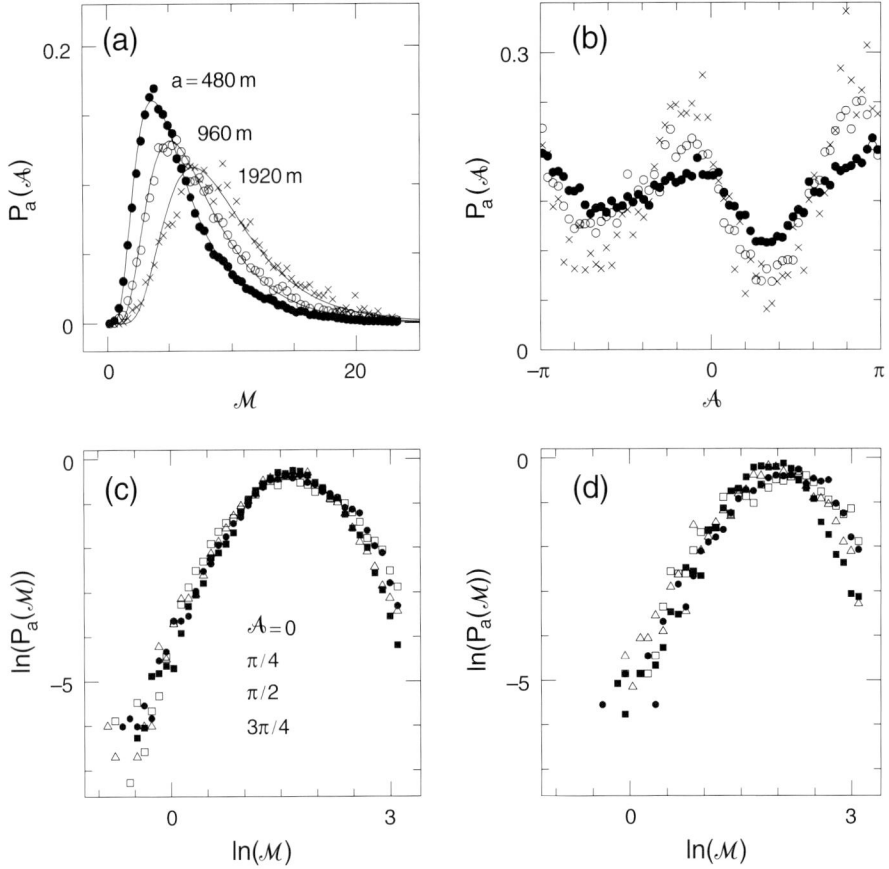

Fig. 2.17. PDFs of the WTMMM coefficients of 32 (1 024 × 1 024) radiance Landsat images as computed with the first-order radially symmetric analyzing wavelet. (**a**) $P_a(\mathcal{M})$ vs. $\mathcal{M}$; (**b**) $P_a(\mathcal{A})$ vs. $\mathcal{A}$; the symbols correspond to the following scales $a = 2^{0.3}\sigma_W = 480\,\mathrm{m}$ (•), $2^{1.3}\sigma_W = 960\,\mathrm{m}$ (○), and $2^{2.3}\sigma_W = 1\,920\,\mathrm{m}$ (×). PDFs of $\mathcal{M}$ when conditioned by $\mathcal{A}$ at the scales (**c**) $a = 2^{0.3}\sigma_W = 480\,\mathrm{m}$ and (**d**) $a = 2^{1.3}\sigma_W = 960\,\mathrm{m}$. The different symbols in (**c**) and (**d**) correspond to fixing $\mathcal{A}$ (mod π) to $0 \pm \pi/8$ (•), $\pi/4 \pm \pi/8$ (□), $\pi/2 \pm \pi/8$ (△), and $3\pi/4 \pm \pi/8$ (■)

tal properties of the marine Sc radiance fluctuations are contained in the way the shape of the PDF of $\mathcal{M}$ evolves when one decreases the scale parameter a. As shown in Fig. 2.17a, for any scale significantly smaller than the integral scale ($\sim$ 5–6 km, as given by the characteristic width of the convective structures) all the data points fall, within a good approximation, on a log-normal curve [2.231, 2.278].

2.5.3 Comparative WTMM Multifractal Analysis of Landsat Radiance Field and Velocity and Temperature Fields in Fully Developed Turbulence

Let us point out that a similar 1D WTMM analysis of the velocity fluctuations in high Reynolds number turbulence has come to conclusions very close to those of the present study [2.88, 2.103–2.105, 2.292]. Besides the presence of rare localized Dirac like structures that witness to the probing of vorticity filaments [2.77, 2.95, 2.97, 2.103], the multifractal nature of turbulent velocity is likely to be understood in terms of a log-normal cascading process which is expected to be scale-invariant in the limit of very high Reynolds numbers [2.88, 2.292]. In Fig. 2.16c are shown for comparison the results obtained for the $D(h)$ singularity spectrum of the radiance Landsat images together with the $D(h)$ data extracted from the 1D analysis of a turbulent velocity signal recorded at the Modane wind tunnel ($R_\lambda \simeq 2\,000$) [2.88, 2.105] (indeed $D(h) + 1$ is represented for the latter in order to compare 1D to 2D data). The turbulent velocity $D(h)$ spectrum significantly differs from the results obtained for the marine Sc cloud. They have a common feature, i.e. the Hölder exponent the most frequently encountered in the radiance field $h = m/\ln 2 = h(q = 0) = \partial\tau/\partial q|_{q=0} = 0.38 \pm 0.01$ is indistinguishable from the corresponding exponent $h = h(q = 0) = 0.39 \pm 0.01$ found for the turbulent velocity field. Note that these values are significantly larger than the theoretical value $h = 1/3$ predicted by Kolmogorov in 1941 [2.129] to account for the observed $k^{-5/3}$ power spectrum behavior. The main difference comes from the intermittency parameter which is much stronger for the cloud, $\sigma^2/\ln 2 = 0.07 \pm 0.01$ ($n_\psi = 1$) or $\sigma^2/\ln 2 = 0.06 \pm 0.01$ ($n_\psi = 3$) than for the turbulent velocity, $\sigma^2/\ln 2 = 0.036 \pm 0.004$. This is the signature that the radiance field is much more intermittent than the velocity field: the $D(h)$ singularity spectrum for the former is unambiguously wider than the corresponding spectrum for the latter. For the sake of comparison, we have also reported in Fig. 2.16c the multifractal $D(h)$ spectrum of the temperature fluctuations recorded in a $R_\lambda = 400$ turbulent flow [2.293]. The corresponding single humped curve is definitely much wider than the velocity $D(h)$ spectrum and it is rather close to the data corresponding to the marine Sc radiance field. It is well recognized, however, that liquid water is not really passive and that its identification with a passive component in atmospheric dynamics offers limited insight into cloud structure since, by definition, near-saturation conditions prevail and latent heat production affects buoyancy [2.294]. So cloud microphysical processes are expected to interact with the circulation at some, if not all, scales [2.295]. Nevertheless, our results in Fig. 2.16c tell us that from a multifractal point of view, the intermittency captured by the Landsat satellite looks statistically equivalent to the intermittency of a passive scalar in fully developed 3D turbulence. The fact that the internal structure of Sc cloud somehow reflects some statistical properties of atmospheric turbulence is not such a surprise in this highly turbulent environment. The investigation of different sets of Landsat data is urgently required in order to test the degree of generality of the results reported in this first WTMM analysis of high-resolution satellite images. In particular, one may wonder up to which extent the marine Sc Landsat data collected off the coast of San Diego on July 7, 1987 under specific observation conditions, actually reflect the specific internal structure of Sc clouds. Work in this direction is currently in progress.

Finally, with respect to the issue of cloud modeling, it comes out quite naturally from the WTMM analysis of marine Sc Landsat data, that the 2D random $\mathcal{W}$-cascade models introduced in [2.233, 2.234], are much more realistic hierarchical models than commonly used multifractal models like the fractionally integrated singular cascade [2.134, 2.136, 2.275] or the bounded cascade [2.288, 2.296] models. We are quite optimistic in view of using the log-normal $\mathcal{W}$-cascade models with realistic parameter values for radiation transfer simulations. In our opinion, random $\mathcal{W}$-cascade models are a real breakthrough, not only for the general purpose of image synthesis, but more specifically for cloud modeling. It is likely that better cloud modeling will make further progress in our understanding of cloud-radiation interaction possible.

2.6 Beyond Multifractal Analysis with Wavelet-Based Space-Scale Correlation Functions: Revealing a Causal Information Cascade in Stock Market Data

2.6.1 Space-Scale Correlation Functions from Wavelet Analysis

Correlations in multifractals have already been experienced in the literature [2.297–2.299]. However, all these studies rely upon the computation of the scaling behavior of some partition functions involving different points; they thus mainly concentrate on spatial correlations of the local singularity exponents. The approach developed in [2.87] is different since it does not focus on (nor suppose) any scaling property but rather consists in studying the correlations of the *logarithms* of the amplitude of a space-scale decomposition of the signal. For that purpose, the WT is a natural tool to perform space-scale analysis. More specifically, if $\chi(x)$ is a bump function such that $||\chi||_1 = 1$, then by taking

$$\varepsilon(x,a) = a^{-2} \int \chi\big((x-y)/a\big)\big|T_\psi[f](y,a)\big|^2 dy, \qquad (2.63)$$

one has:

$$||f||_2^2 = \int\int \varepsilon(x,a)dx\,da\,, \qquad (2.64)$$

and thus $\varepsilon(x,a)$ can be interpreted as the *local space-scale energy density* of the considered signal f [2.300]. Since $\varepsilon(x,a)$ is a positive quantity, we can define the *magnitude* of the field f at point x and scale a as:

$$\omega(x,a) = \frac{1}{2}\ln\varepsilon(x,a)\,. \qquad (2.65)$$

Our aim in this section is to show that a cascade process can be studied through the correlations of its space-scale magnitudes [2.87]:

$$C(x_1, x_2, a_1, a_2) = \overline{\tilde{\omega}(x_1, a_1)\tilde{\omega}(x_2, a_2)} \,, \tag{2.66}$$

where the overline stands for ensemble average and $\tilde{\omega}$ for the centered process $\omega - \overline{\omega}$.

2.6.2 Analysis of Random $\mathcal{W}$-Cascades Using Space-Scale Correlation Functions

As discussed in Sect. 2.2.6, cascade processes can be defined in various ways. Periodic wavelet orthogonal bases [2.70, 2.73] provide a general framework in which $\mathcal{W}$-cascades can be constructed easily [2.103–2.107]. Let us consider the following wavelet series:

$$f(x) = \sum_{j=0}^{+\infty} \sum_{k=0}^{2^j - 1} c_{j,k} \psi_{j,k}(x) \,, \tag{2.67}$$

where the set $\{\psi_{j,k}(x) = 2^{j/2}\psi(2^j x - k)\}$ is an orthonormal basis of $L^2([0, L])$ and the coefficients $c_{j,k}$ correspond to the WT of f at scale $a = L2^{-j}$ (L is the 'integral' scale that corresponds to the size of the support of $\psi(x)$) and position $x = ka$. The above sampling of the space-scale half-plane defines a dyadic tree [2.70, 2.73]. If one indexes by a dyadic sequence $\{\epsilon_1, ..., \epsilon_j\}$ ($\epsilon_k = 0$ or 1) each of the 2^j nodes at depth j of this tree, the cascade is defined by the multiplicative rule: $c_{j,k} = c_{\epsilon_1...\epsilon_j} = c_0 \prod_{i=1}^{j} W_{\epsilon_i}$. The law chosen for the weights W determines the nature of the cascade and the multifractal (regularity) properties of f [2.106, 2.107]. From the above multiplicative structure (Fig. 2.18), if one assumes that there is no correlation between the weights at a given cascade step, then it is easy to show that for $a_p = L2^{-j_p}$ and $x_p = k_p a_p$ ($p = 1$ or 2), the correlation function (2.66) is nothing but the variance $V(j)$ of $\ln c_{j,k} = \sum \ln W_{\epsilon_i}$, where (j, k) is the deepest common ancestor to the nodes (j_1, k_1) and (j_2, k_2) on the dyadic tree [2.87, 2.106]. This *ultrametric* structure of the correlation function shows that such a process is not stationary. However, we will generally consider uncorrelated consecutive realizations of length L of the same cascade process, so that, in good approximation, C depends only on the space lag $\Delta x = x_2 - x_1$ and one can replace ensemble average by space average. In that case, $C(\Delta k, j_1, j_2) = \langle C(k_1, k_1 + \Delta k, j_1, j_2)\rangle$ can be expressed as [2.87, 2.106]:

$$C(\Delta k, j_1, j_2) = 2^{-(j-n)} \sum_{p=1}^{j-n} 2^{j-n-p} V(j - n - p) \,, \tag{2.68}$$

where $j = \sup(j_1, j_2)$ and $n = \log_2 \Delta k$.

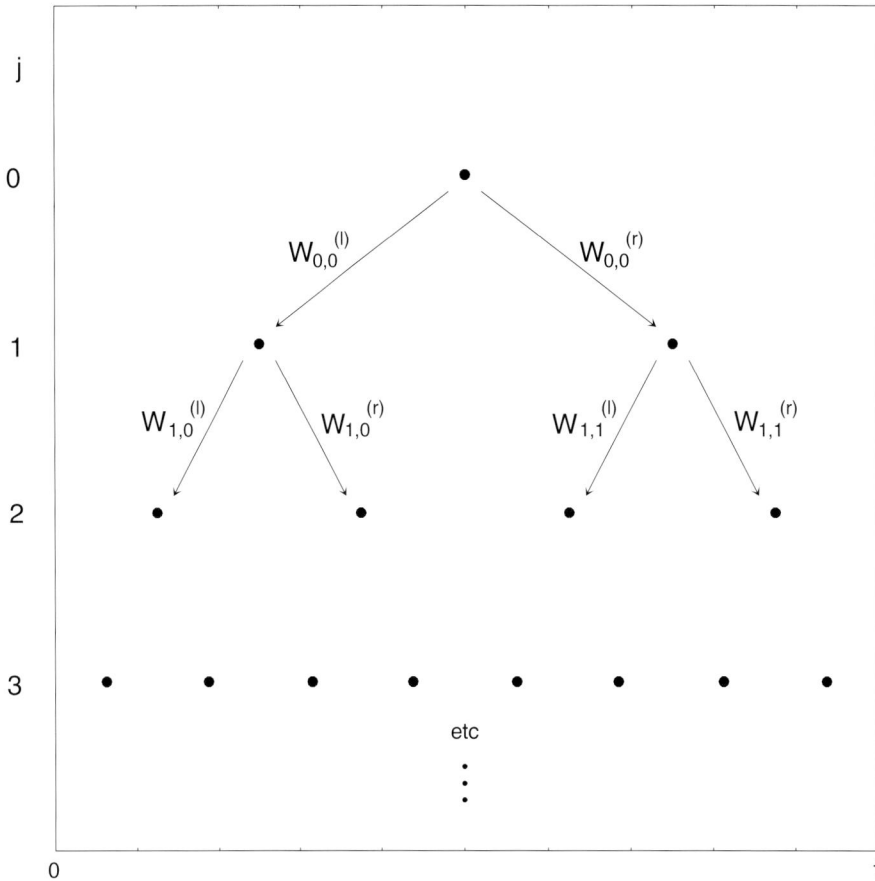

Fig. 2.18. Sketch of the construction rule of a $\mathcal{W}$-cascade. The wavelet coefficients $\{c_{j,k}\}_{j,k}$ lie on a dyadic grid. At each scale $a_j = 2^{-j}$, the grid displays 2^j coefficients with abscissa $x_{j,k} = 2^{-j}k$. The value of the wavelet coefficient $c_{j,2k}$ $(c_{j,2k+1})$ is obtained from the value of the wavelet coefficient $c_{j-1,k}$ by multiplying it by $W_{j-1,k}^{(l)}$ $(W_{j-1,k}^{(r)})$ where $W_{j-1,k}^{(\epsilon)}$ are i.i.d. real valued random variables

Let us illustrate these features on some simple case of random $\mathcal{W}$-cascades (e.g. log-normal cascades) [2.103–2.107]. As in classical cascades, at each step of the cascade, one chooses i.i.d. random variables $\ln W_{\epsilon_i}$ of variance σ^2. Then $V(j) = \sigma^2 j$ and it can be established [2.87, 2.106] that, for $\sup(a_1, a_2) \leq \Delta x < L$,

$$C(\Delta x, a_1, a_2) = \sigma^2 \left(\log_2 \left(\frac{L}{\Delta x} \right) - 2 + 2\frac{\Delta x}{L} \right) . \qquad (2.69)$$

Thus, the correlation function decreases very slowly, independently of a_1 and a_2, as a *logarithm function* of Δx. This behavior is illustrated in Fig. 2.20a and 2.20b where a log-normal cascade (Fig. 2.19a) has been constructed using Daubechies' compactly supported wavelet basis (D-5) [2.73]. The correlation functions of the magnitudes of $f(x)$ have been computed as described above (2.66) using a simple box function for $\chi(x)$. Let us note that all the results

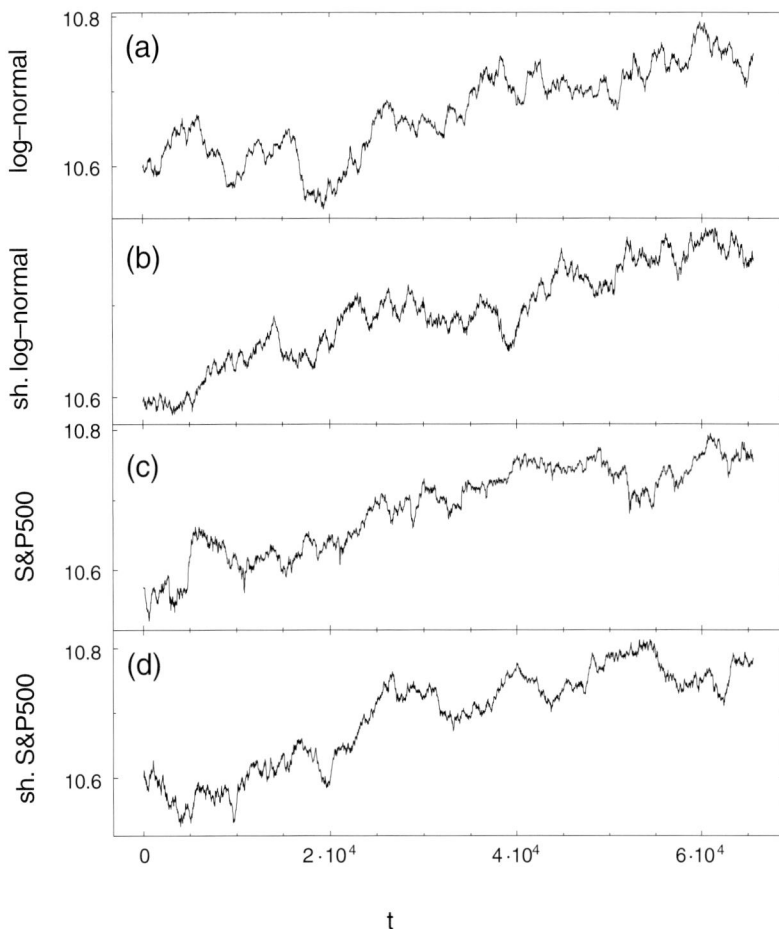

Fig. 2.19. Financial time series as compared to synthetic multifractal signals: (**a**) Realization of a random function generated by a log-normal $\mathcal{W}$-cascade using the 'Daubechies 5' compactly supported orthogonal wavelet basis. The law of $\ln|W|$ is Gaussian with mean $m = -H \ln 2 = -0.6 \ln 2$ and variance $\sigma^2 = 0.02 \ln 2 = 0.0077$. (**b**) 'White noise' version of the log-normal $\mathcal{W}$-cascade realization shown in (**a**) after randomly shuffling the wavelet coefficients at each scale j. (**c**) Time evolution of $\ln P(t)$, where $P(t)$ is the S&P 500 index, sampled with a time resolution $\delta t = 5$ min in the period October 1991-February 1995. The data have been preprocessed in order to remove 'parasitic' daily oscillatory effects. (**d**) Same as in (**c**) but after having randomly shuffled the increments of the signal in (**c**)

reported in this section concern the increments (we use $\psi_{(0)}^{(1)}$) of the considered signal and that we have checked that they are actually independent of the specific choice of the analyzing wavelet $\psi_{(m)}^{(n)}$ (Fig. 2.2). In Fig. 2.20a are plotted the 'one-scale' ($a_1 = a_2 = a$) correlation functions for three different scales $a = 4$, 8, and 32. One can see that, for $\Delta x > a$, all the curves collapse to a single one, which is in perfect agreement with the expression (2.69): in semi-log-coordinates, the correlation functions decrease almost linearly (with slope σ^2) up to the integral scale L that is of order 2^{16} points. In Fig. 2.20b are displayed these correlation functions when the two scales a_1 and a_2 are

88   Alain Arneodo et al.

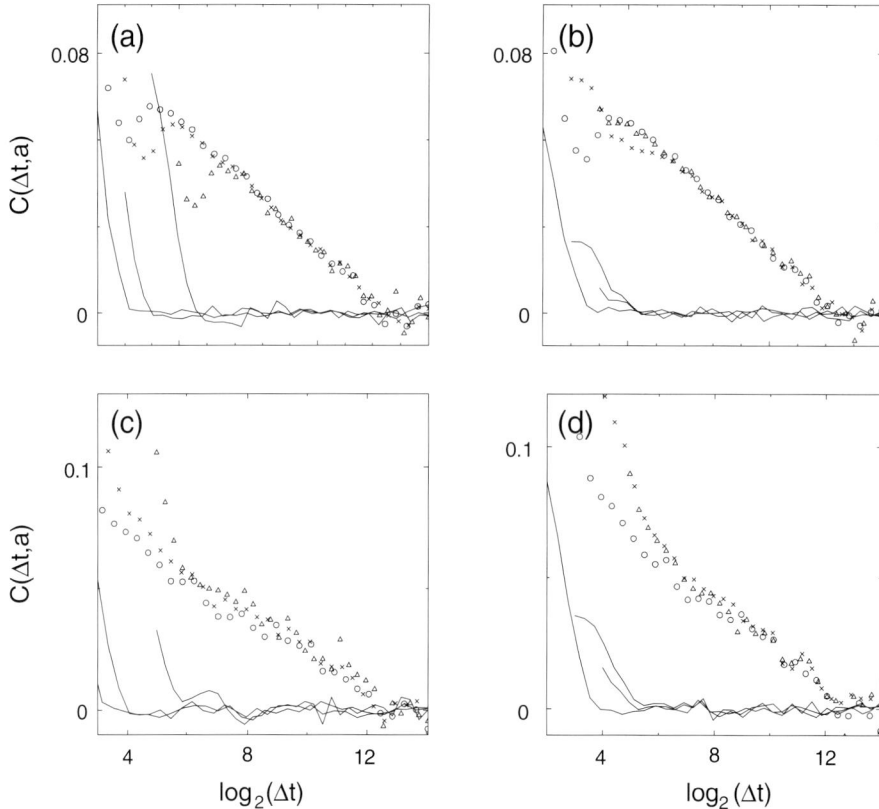

Fig. 2.20. Space-scale correlation functions $C(t, t + \Delta t, a, a')$. *Log-normal random $\mathcal{W}$-cascade:* (**a**) 'one-scale' correlation functions ($a = a'$) for $a = 4$ ($\circ$), 8 ($\times$), and 32 ($\triangle$); (**b**) 'two-scale' correlation functions for $a = 4$, $a' = 8$ ($\circ$), $a = 8$, $a' = 32$ ($\times$), and $a = 4$, $a' = 32$ ($\triangle$); the solid curves correspond to similar computations performed on the 'white noise' log-normal $\mathcal{W}$-cascade realizations. S&P 500 index: (**c**) 'one-scale' correlation functions of the log-volatility for various scales a corresponding to $a = 30$ ($\circ$), 120 ($\times$), and 480 ($\triangle$) min; (**d**) 'two-scale' correlation functions for various pairs of scales corresponding to $a = 30$, $a' = 120$ ($\circ$), $a = 120$, $a' = 480$ ($\times$), and $a = 30$, $a' = 480$ ($\triangle$) in min units; the solid curves correspond to similar computations performed on the randomly shuffled increment version of the S&P 500 index

different. One can check that, as expected, they still do not depend on the scales provided $\Delta x \geq \sup(a_1, a_2)$; moreover, they are again very well fitted by the above theoretical curve (except at very large Δx where finite size effects show up). The linear behavior of $C(\Delta x, a_1, a_2)$ vs. $\ln(\Delta x)$ is characteristic for 'classical' scale-invariant cascades for which the random weights are uncorrelated [2.87, 2.88, 2.106].

2.6.3 Distinguishing 'Multiplicative' from 'Additive' Processes

The two previous examples illustrate the fact that magnitudes in random cascades are correlated over very long distances. Moreover, the slow decay of the correlation functions is independent of scales for large enough space lags

($\Delta x > a$). This is reminiscent of the multiplicative structure along a space-scale tree. These features are not observed in 'additive' models like fBms (Sect. 2.2.6) whose long-range correlations originate from the sign of their variations rather than from the amplitudes. In Fig. 2.20a and 2.20b are plotted the correlation functions of an 'uncorrelated' log-normal model constructed using the same parameters as in the first example but without any multiplicative structure (the coefficients $c_{j,k}$ have, at each scale j, the same log-normal law as before but are independent). Let us note that from the point of view of both the multifractal formalism and the increment PDF scale properties, the 'uncorrelated' (Fig. 2.19b) and 'multiplicative' (Fig. 2.19a) log-normal models are indistinguishable since their one-point statistics at a given scale are identical. As far as the magnitude space-scale correlations are concerned, the difference between the cascade and the other models is striking: for $\Delta x > a$, the magnitudes of the 'white noise' log-normal model are found to be uncorrelated. Let us emphasize that similar uncorrelated behavior is observed for fBms and Lévy processes [2.87, 2.106]. For their ability to reveal the existence of a multiplicative hierarchical structure underlying fractal landscapes or turbulent signals, the wavelet-based space-scale correlation functions are a definite step beyond statistical multifractal analysis [2.87, 2.106].

2.6.4 Analysis of Stock Market Data
Using Space-Scale Correlations

Modeling accurately financial price variations is an essential step underlying portfolio allocation optimization, derivative pricing and hedging, and fund management and trading [2.19, 2.20, 2.301–2.303]. The observed complex price fluctuations guide and constrain our theoretical understanding of agent interactions and of the organization of the market. The Gaussian paradigm of independent normally distributed price increments [2.304, 2.305] has long been known to be incorrect with many attempts to improve it. Mandelbrot first proposed to use Lévy distributions [2.23, 2.306], which are characterized by a fat tail decaying as a power-law with index between 0 and 2. His suggestion arrived at an epoch when Markovitz famous mean-variance portfolio and Black and Scholes option pricing theories were being developed and widely applied. For main stream economists, the econometric nonlinear autoregressive models with conditional heteroscedasticity (ARCH) [2.307] and their generalizations [2.308] are more natural because they keep the volatility (standard deviation of price variations) as the main descriptor. Recall that heteroscedasticity refers to the fact that the variance (or volatility) is itself a stochastic variable (the exact definition is the following: when the errors for different dates (or points) have different variances but are unrelated with each other, then the errors are said to exhibit heteroscedasticity. If the variances are related, the heteroscedasticity is said to be correlated). These models address volatility clustering and partly

the observed 'fat tails' of distributions. The problem, however, is that these
GARCH models capture only imperfectly the volatility correlations and the
fat tails of the probability density function (PDF) of price variations. Moreover, as far as changes in time scales are concerned, the so-called 'aggregation'
properties of these models are not easy to control.

Recently, physicists have characterized more precisely the distribution of
market price variations [2.19, 2.20] and found that a power-law truncated by
an exponential provides a reasonable fit at short time scales (less than one day),
while at larger time scales the distributions may cross over progressively to the
Gaussian distribution which becomes approximately correct for monthly and
larger scale price variations [2.19, 2.20, 2.153, 2.309]. Alternatively, Ghashghaie
et al. [2.310] proposed a 'multiplicative' cascade model based on an analogy between price dynamics and hydrodynamic turbulence. According to this model,
the return r at a given time scale $a < T$, is given by:

$$r_a(t) \equiv \ln P(t+a) - \ln P(t) = \varepsilon(t,a)^{1/2} u(t) , \qquad (2.70)$$

where $u(t)$ is some scale independent random variable, T is some coarse 'integral' time scale and $\varepsilon(t,a)^{1/2}$ is the local r.m.s. of the return that can be
multiplicatively decomposed, for any decreasing sequence of scales $\{a_i\}_{i=0,..,n}$
with $a_0 = T$ and $a_n = a$, as [2.88, 2.103–2.106, 2.140, 2.311–2.313]

$$\varepsilon(t,a)^{1/2} = \prod_{i=0}^{n-1} W_{a_{i+1},a_i} \varepsilon(t,T)^{1/2} . \qquad (2.71)$$

Equation (2.70) together with (2.71) show that the logarithm of the price is
a multiplicative process. But, this is different from the classical multiplicative
processes studied in finance literature, due to the tree-like structure of the correlations that are added by the hierarchical construction of the multiplicands
as discussed in Sect. 2.6.2.

In turbulence the field ε is related to the energy while in finance ε is called
the *volatility*. Recall that the volatility has fundamental importance in finance
since it provides a measure of the amplitude of price fluctuations, hence of the
market risk. Using $\omega(t,a) = \frac{1}{2} \ln \varepsilon(t,a)$ as a natural variable, if one supposes
that W_{a_{i+1},a_i} depends only on the scale ratio a_{i+1}/a_i, one can easily show, by
choosing the a_i as a geometric series Ts^i ($s < 1$), that (2.71) implies that the
PDF of ω at scale $a = Ts^n$ can be written as:

$$P_a(\omega) = (G_s^{\otimes n} \otimes P_T)(\omega) , \qquad (2.72)$$

where $\otimes$ means the convolution product, G_s is the PDF of $\ln W_{sa,a}$ and P_T is the
PDF of $\omega(t,T)$. The symbol $G_s^{\otimes n} \otimes P_T$ means that G_s has been convoluted with
itself n times before being convoluted with P_T. Equation (2.72) is the exact
reformulation (in log variables) of the paradigm that Ghashghaie et al. [2.310]

used to fit foreign exchange (FX) rate data at different scales. Recall that it simply means that the distributions of the logarithm of the *absolute value* of the price variations can be represented by a superposition of elementary laws G_s. In this formalism, G can be proven to be the PDF of an infinitely divisible random variable [2.103–2.106, 2.140, 2.311] (hence ε is called 'log-infinitely divisible'). In [2.310], G is assumed to be normal (the cascade is called 'log-normal') of variance $-\sigma^2 \ln s$.

Note that a cascade model does not necessarily imply the existence of correlations between returns. As pointed out in Remark 1, if the exponent $\tau(2) = 0$, then the power spectral density behaves as $1/k^2$, i.e. as the power spectral density of Brownian motions. This is why the $1/k^2$ shape of the power spectrum of financial time series cannot be invoked as an argument against a cascade model [2.153, 2.314]. Moreover, as far as scaling properties of price fluctuations are concerned, it is easy to deduce from (2.72) that, if $H \ln s$ is the mean of G_s and $-\sigma^2 \ln s$ its variance, then the maximum of the PDF of $\varepsilon(t, a)^{1/2}$ varies as $a^{H-\sigma^2/2}$ (H plays the same role as the Lévy index in truncated Lévy flight models [2.19, 2.20, 2.309] with $H = 1/\mu$), while its standard deviation behaves as $a^{(H-\sigma^2)/2}$. These features are observed in both turbulence ($H \simeq 0.38$ and $\sigma^2 \simeq 0.03$) [2.88, 2.103–2.106, 2.311–2.313] and finance ($H \simeq 0.6$ and $\sigma^2 \simeq 0.02$) [2.310]. Therefore, as advocated in [2.310], (2.72) accounts reasonably well for one-point statistical properties of financial times series. However, because of the relatively small statistics available in finance, it is very difficult to demonstrate that (2.72) is more pertinent to fit the data than a 'truncated Lévy' distribution [2.19, 2.20, 2.153, 2.309, 2.314].

At this point, let us emphasize that (2.71) imposes much more constraints on the statistics than (2.72) that only refers to one point statistics. The main difference between the *multiplicative* cascade model and the truncated Lévy *additive* model is that the former predicts strong correlations in the volatility while the latter assumes no correlation. It is then tempting to compute the correlations of the log-volatility $\omega(t, a)$ at different time scales a [2.315]. In Fig. 2.19 are shown time series for which we study increment time correlations (i.e. the WT coefficient correlation as computed with the analyzing wavelet $\psi_{(0)}^{(1)}$ shown in Fig. 2.2). Figure 2.19c represents the logarithm of the S&P 500 index. Figure 2.19d is the same as Fig. 2.19c but after having randomly shuffled the increments $\ln P(i+1) - \ln P(i)$. In Fig. 2.20c and 2.20d are reported the one-scale and two-scale correlation functions $C^\omega(\Delta t, a_1, a_2) = \overline{\tilde{\omega}(t, a_1)\tilde{\omega}(t + \Delta t, a_2)}$ of the log-volatility as functions of $\ln \Delta t$. As compared to the absence of correlation found for the randomly shuffled S&P 500 signal, the S&P 500 index log-volatility clearly displays correlations that are well fitted by an equation analogous to (2.69):

$$C^\omega(\Delta t, a_1, a_2) = \sigma^2 \left(\log_2 \left(\frac{T}{\Delta t} \right) - 2 + 2\frac{\Delta t}{T} \right) + \sigma_T^2 , \qquad (2.73)$$

provided $\sup(a_1, a_2) \leq \Delta t < T$, where σ_T^2 is the variance of $\omega(t, T)$. One actually observes in Fig. 2.20c a slow linear decay of the one-scale correlation coefficient with a slope $\sigma^2 \simeq 0.0077$ (a value which can be obtained independently from the fit of the PDFs), with only one adjustable parameter $T \simeq 3$ months. Similar results are obtained on the two-scale correlation functions in Fig. 2.20d. As expected from (2.73), all the data collapse on a single curve which is nearly linear up to some integral time T of order 3 months, provided $\Delta t > \sup(a_1, a_2)$.

Let us point out that volatility at large time intervals that cascades to smaller scales cannot do so instantaneously. From causality properties of financial signals, the 'infrared' towards 'ultraviolet' cascade must manifest itself in a time asymmetry of the cross-correlation coefficients $\rho_a^\omega(\Delta t, \Delta a) = C^\omega(\Delta t, a, a + \Delta a)/\sqrt{C^\omega(0, a)C^\omega(0, a + \Delta a)}$. In particular, one expects that $\rho_a^\omega(\Delta t, \Delta a) \leq \rho_a^\omega(-\Delta t, \Delta a)$ if $\Delta t > 0$ and $\Delta a > 0$. From the observed nearly Gaussian properties of $\omega(t, a)$ [2.315], one can derive the following expression for the mean mutual information of the variables $\omega(t, a)$ and $\omega(t + \Delta t, a + \Delta a)$:

$$I_a(\Delta t, \Delta a) = -0.5 \log_2\left(1 - (\rho_a^\omega(\Delta t, \Delta a))^2\right) . \tag{2.74}$$

Since the process is causal, this quantity can be interpreted as the information contained in $\omega(t, a)$ that 'propagates' to $\omega(t + \Delta t, a + \Delta a)$. In Fig. 2.21, we have computed $I_a(\Delta t, \Delta a)$ for the S&P 500 index (Fig. 2.21c) and its randomly shuffled version (Fig. 2.21d) [2.315]. One can see on the bottom right picture that there is no well defined structure that emerges from the noisy background. Except in a small domain at small scales around $\Delta t = 0$, the mutual information is in the noise level as expected for uncorrelated variables. In contrast, two features are clearly visible on the bottom left representation. First, the mutual information at different scales is mostly important for equal times. This is not so surprising since there are strong localized structures in the signal that are 'coherent' over a wide range of scales.

The extraordinary new fact is the appearance of a nonsymmetric propagation cone of information showing that the volatility at large scales influences causally (in the future) the volatility at shorter scales. Although one can also detect some information that propagates from past fine to future coarse scales, it is clear that this phenomenon is weaker than past coarse/future fine flux. As compared to the symmetric cone structure observed in Fig. 2.21a for log-normal random $\mathcal{W}$-cascades, the dissymmetry observed in Fig. 2.21c in the mutual information of the S&P 500 index is thus a clear demonstration of the pertinence of the notion of causal cascade in market dynamics. Let us point out that similar features are found on FX rates and other stock market data.

There are several mechanisms that can be invoked to rationalize our observations, such as the heterogeneity of traders and their different time horizon [2.316] leading to an 'information' cascade from large time scales to short time scales, the lag between stock market fluctuations and long-run movements

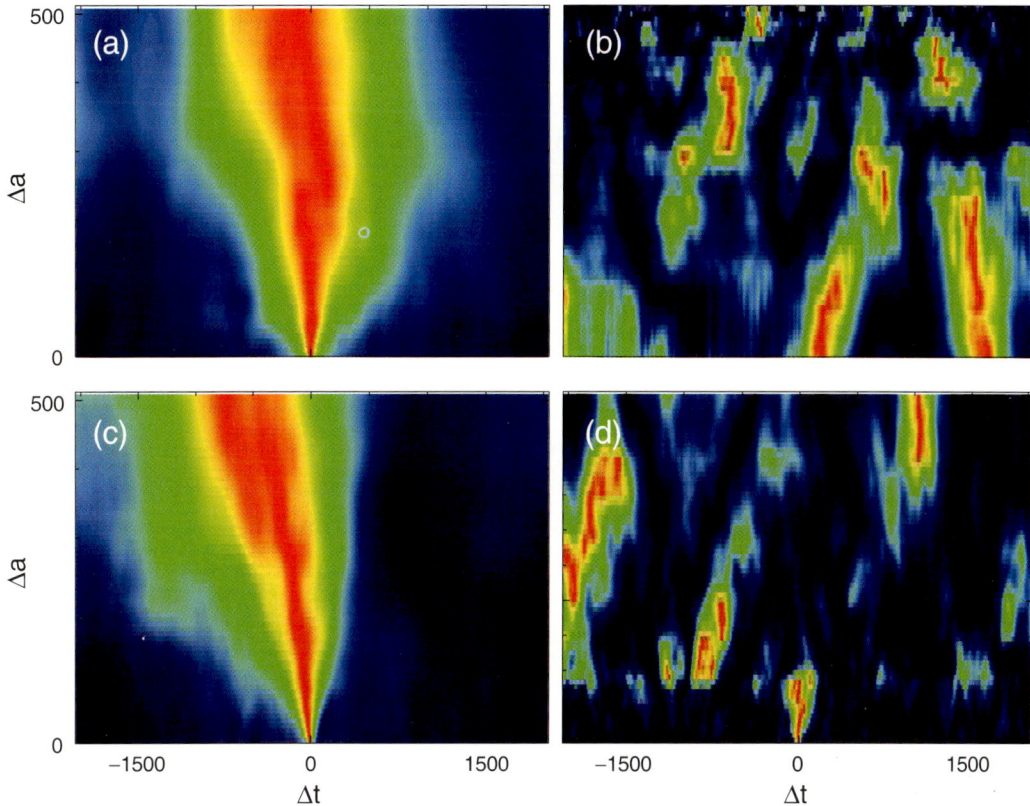

Fig. 2.21. The mutual information $I_a(\Delta t, \Delta a)$ of the variables $\omega(t, a)$ and $\omega(t + \Delta t, a + \Delta a)$ is represented in the $(\Delta t, \Delta a)$ half-plane. The small scale $a = 4$ is fixed. The amplitude of $I_a(\Delta t, \Delta a)$ is coded from black for zero values to red for maximum positive values ('heat code'), independently at each scale lag Δa. (**a**) Log-normal random $\mathcal{W}$-cascade; same parameters values as in Fig. 2.19a; (**b**) 'white noise' version of the log-normal $\mathcal{W}$-cascade; (**c**) S&P 500 index; (**d**) randomly shuffled increment version of the S&P 500. Note that, for middle scale lag values, the maxima (red spots) of the mutual information in (**a**) and (**c**) are two orders of magnitude larger than the corresponding maxima in (**b**) and (**d**) respectively. For the S&P 500, $\Delta a = 1$ corresponds to 5 min

in dividends [2.317], the effect of the regular release (monthly, quarterly) of major economic indicators which cascades to fine time scale. Correlations of the volatility have been known for a while and have been partially modeled by mixtures of distributions [2.318], ARCH/GARCH models [2.307], and their extensions [2.308]. However, as previously pointed out, because they are constructed to fit the fluctuations at a given time interval, these models are not adapted to account for the above described multiscale properties of financial time series. We have performed the same correlation analysis for simulated GARCH(1,1) processes and obtained structureless pictures similar to the one corresponding to the shuffled S&P 500 in Fig. 2.21d. More recently, Muller et al. [2.316] have proposed the HARCH model in which the variance at time t is a function of the realized variances at different scales. By construction, this model captures the lagged correlation of the volatility from the large to the small time scales. However, it does not contain the notion of cascade and involves only a few time scales. Moreover, it suffers from the same deficiencies as ARCH-type models concerning the difficulties to control and interpret parameters at different scales. Let us also mention some recent work by Mandelbrot et al. [2.319] that introduces and tests a multifractal model of asset prices. Their key idea is to construct a subordinated Brownian motion using a multifractal trading time. From simple arguments, one can show that there are strong similarities with the cascade model approach advocated in this section.

Putting together the evidence provided by the logarithmic decay of the log-volatility correlations and the volatility cascade from the infrared to the ultraviolet, we have revisited the analogy with turbulence, albeit on the *volatility* and not on the price variations. Our main finding is the exhibition of this information cascading process: the fact that variations of prices over a few month scale influence in the future the daily price variations is extraordinarily rich of consequences. This is not so only for the fundamental understanding of the nature of financial markets but also, and maybe more important, for practical applications. Indeed, the nature of the correlations that are implied by this cascade across scales, has profound implications on the market risk, a problem of utmost concern for all financial institutions as well as individuals. In particular, these correlations are likely to have strong consequences on derivative pricing and hedging.

2.7 Conclusion

To summarize, we have presented a first step towards a statistical theory of multifractal 1D and 2D signals based on wavelet analysis. Indeed we believe that the WTMM method (and its generalization to 2D), for determining the $D(h)$ singularity spectrum of a fractal landscape, a turbulent signal, or the image of a fractal object, is likely to become as useful as the well-known phase portrait reconstruction, Poincaré section, and first return map techniques for the analysis of chaotic time series [2.320, 2.321]. The reported results of previous analysis of DNA walks (Sect. 2.3) and satellite images of fractal clouds (Sect. 2.5), together with former wavelet-based statistical studies of fully developed turbulent velocity signals [2.32, 2.35, 2.36, 2.95, 2.103–2.105, 2.151, 2.292] as well as stock market data [2.152, 2.153], show that this method is readily applicable to experimental situations. We have also shown that one can further use the WT to go beyond this thermodynamic description of fractal objects and eventually to reveal from the branching structure of the WT skeleton, the existence of an underlying multiplicative hierarchical structure. As explained in [2.32, 2.36, 2.157, 2.322], in some situations, one can hope to solve the 'inverse fractal problem' by extracting from the data some dynamical system which accounts for its internal self-similar structure. The application of a wavelet-based tree matching algorithm to characterize the fractal properties of DLA azimuthal Cantor sets in [2.155–2.158] has revealed the existence of a predominant Fibonacci multiplicative process in the apparently disordered arborescent morphology of diffusion-limited aggregates. This discovery is a spectacular manifestation of the statistical relevance of the golden mean arithmetic to Laplacian growth phenomena. As shown in Sect. 2.6, in some situations, the use of the space-scale correlation functions computed from the WT representation can provide deep insight on the underlying random cascade structure [2.87]. Recent applications of this methodology in the context of fully developed turbulence have revealed the existence of a (nonscale-invariant) log-normal cascading process underlying the turbulent velocity fluctuations [2.87, 2.88, 2.103–2.105, 2.151, 2.292]. More surprising are the results reported in Sect. 2.6, of a similar investigation of financial time series [2.315]. Underlying the fluctuations of the volatility of the price variations, there exists a causal information cascade from large to small time scales that can be visualized with the WT representation. We are convinced that further applications of this wavelet based methodology (WTMM method, wavelet based tree matching algorithm for solving the inverse fractal problem, space-scale correlation functions) will lead to similar major breakthroughs in various fields where multiscale phenomena are ubiquitous.

Acknowledgements. The material reported in this review paper relies on collaborative works with Y. d'Aubenton-Carafa, E. Bacry, S. Manneville, S.G. Roux, D. Sornette, and C. Thermes. These works were supported by the GIP GREG (project 'Motif dans les Séquences'), by the Ministère de l'Education Nationale, de l'Enseignement Supérieur, de la Recherche et de l'Insertion Professionnelle ACC-SV (project 'Génétique et Environnement'), and by NATO (Grant no. CRG 960176). We are indebted to R.F. Cahalan, A. Davis, Y. Gagne, Y. Malecot, and S. Ciliberto for the permission to use their experimental data.

References

2.1 B.B. Mandelbrot, *The Fractal Geometry of Nature* (Freeman, San Francisco, 1982).

2.2 H.E. Stanley and N. Ostrowski (eds.), *On Growth and Form: Fractal and Non-Fractal Patterns in Physics* (Martinus Nijhof, Dordrecht, 1986).

2.3 L. Pietronero and E. Tosatti (eds.), *Fractals in Physics* (North-Holland, Amsterdam, 1986).

2.4 H.E. Stanley and N. Ostrowski (eds.), *Random Fluctuations and Pattern Growth* (Kluwer, Dordrecht, 1988).

2.5 J. Feder, *Fractals* (Pergamon, New York, 1988).

2.6 T. Vicsek, *Fractal Growth Phenomena* (World Scientific, Singapore, 1989).

2.7 D. Avnir (ed.), *The Fractal Approach to Heterogeneous Chemistry: Surfaces, Colloids, Polymers* (Wiley, New York, 1989).

2.8 A. Aharony and J. Feder (eds.), *Fractals in Physics, Essays in Honour of B.B. Mandelbrot*, Physica D, **38** (North-Holland, Amsterdam, 1989).

2.9 B.J. West, *Fractal Physiology and Chaos in Medicine* (World Scientific, Singapore, 1990).

2.10 F. Family and T. Vicsek, *Dynamics of Fractal Surfaces* (World Scientific, Singapore, 1991).

2.11 A. Bunde and S. Havlin (eds.), *Fractals and Disordered Systems* (Springer, Berlin, 1991).

2.12 T. Vicsek, M. Schlesinger, and M. Matsushita (eds.), *Fractal in Natural Science* (World Scientific, Singapore, 1994).

2.13 A. Bunde and S. Havlin (eds.), *Fractals in Science* (Springer, Berlin, 1994).

2.14 B.J. West and W. Deering, Phys. Rep. **254**, 1 (1994).

2.15 U. Frisch, *Turbulence* (Cambridge University Press, Cambridge, 1995).

2.16 G.G. Wilkinson, J. Kanellopoulos, and J. Megier (eds.), *Fractals in Geoscience and Remote Sensing, Image Understanding Research Series, Vol.1, ECSC-EC-EAEC* (Brussels, Luxemburg, 1995).

2.17 A.L. Barabási and H.E. Stanley, *Fractal Concepts in Surface Growth* (Cambridge University Press, Cambridge, 1995).

2.18 F. Family, P. Meakin, B. Sapoval, and R. Wool (eds.), *Fractal Aspects of Materials, Material Research Society Symposium Proceeding, Vol. 367* (MRS, Pittsburg, 1995).

2.19 J.-P. Bouchaud and M. Potters, *Théorie des Risques Financiers* (Eyrolles, Aléa-Saclay, 1997).

2.20 R.N. Mantegna and H.E. Stanley, *An Introduction to Econophysics* (Cambridge University Press, Cambridge, 2000).

2.21 B.B. Mandelbrot and J.W. Van Ness, S.I.A.M. Rev. **10**, 422 (1968).

2.22 H.O. Peitgen and D. Saupe (eds.), *The Science of Fractal Images* (Springer, New York, 1987).

2.23 B.B. Mandelbrot, J. Bus. Univ. Chicago **40**, 393 (1967).

2.24 C.-K. Peng, S.V. Buldyrev, A.L. Goldberger, S. Havlin, F. Sciortino, M. Simons, and H.E. Stanley, Nature **356**, 168 (1992).

2.25 H.E. Stanley, S.V. Buldyrev, A.L. Goldberger, S. Havlin, S.M. Ossadnik, C.-K. Peng, and M. Simons, Fractals **1**, 283 (1993).

2.26 W. Li, T.G. Marr, and K. Kaneko, Physica D **75**, 392 (1994).

2.27 R.F. Voss, Physica D **38**, 362 (1989).

2.28 B. Dubuc, J.F. Quiniou, C. Roques-Carmes, C. Tricot, and S.W. Zucker, Phys. Rev. A **39**, 1500 (1989).

2.29 J. Schmittbuhl, J.P. Violette, and S. Roux, Phys. Rev. E **51**, 131 (1995).

2.30 M.S. Taqqu, V. Teverovsky, and W. Willinger, Fractals **3**, 785 (1995).

2.31 J.-F. Muzy, Ph.D. Thesis, University of Nice, 1993.

2.32 J.-F. Muzy, E. Bacry, and A. Arneodo, Int. J. Bifurcat. Chaos **4**, 245 (1994).

2.33 G. Parisi and U. Frisch, in *Turbulence and Predictability in Geophysical Fluid Dynamics and Climate Dynamics, Proc. Int. School*, edited by M. Ghil, R. Benzi, and G. Parisi (North-Holland, Amsterdam, 1985), p. 84.

2.34 A.L. Barabási and T. Vicsek, Phys. Rev. A **44**, 2730 (1991).

2.35 J.-F. Muzy, E. Bacry, and A. Arneodo, Phys. Rev. Lett. **67**, 3515 (1991).

2.36 A. Arneodo, E. Bacry, and J.-F. Muzy, Physica A **213**, 232 (1995).

2.37 S. Jaffard, C. R. Acad. Sci. Paris Sér. I **308**, 79 (1989).

2.38 S. Jaffard, Publ. Mat. **35**, 155 (1991).

2.39 M. Holschneider and P. Tchamitchian, in [2.71], p. 102.

2.40 S. Mallat and S. Zhong, IEEE Trans. Pattern Anal. Mach. Intell. **14**, 710 (1992).

2.41 S. Mallat and W.L. Hwang, IEEE Trans. Inf. Theory **38**, 617 (1992).

2.42 E. Bacry, J.-F. Muzy, and A. Arneodo, J. Statist. Phys. **70**, 635 (1993).

2.43 R. Benzi, G. Paladin, G. Parisi, and A. Vulpiani, J. Phys. A **17**, 3521 (1984).

2.44 E.B. Vul, Y.G. Sinai, and K.M. Khanin, J. Russ. Math. Surv. **39**, 1 (1984).

2.45 T.C. Halsey, M.H. Jensen, L.P. Kadanoff, I. Procaccia, and B.I. Shraiman, Phys. Rev. A **33**, 1141 (1986).

2.46 P. Collet, J. Lebowitz, and A. Porzio, J. Statist. Phys. **47**, 609 (1987).

2.47 R. Badii, Ph.D. Thesis, University of Zurich, 1987.

2.48 G. Paladin and A. Vulpiani, Phys. Rep. **156**, 148 (1987).

2.49 P. Grassberger, R. Badii, and A. Politi, J. Statist. Phys. **51**, 135 (1988).

2.50 T. Bohr and T. Tèl, in *Direction in Chaos*, Vol. 2, edited by B.L. Hao (World Scientific, Singapore, 1988), p. 194.

2.51 D. Rand, Ergod. Theory. Dyn. Syst. **9**, 527 (1989).

2.52 P. Meakin, in *Phase Transition and Critical Phenomena*, Vol. 12, edited by C. Domb and J.L. Lebowitz (Academic, Orlando, 1988), p. 355.

2.53 B.B. Mandelbrot, J. Fluid Mech. **62**, 331 (1974).

2.54 C. Meneveau and K.R. Sreenivasan, J. Fluid Mech. **224**, 429 (1991).

2.55 Y.G. Sinai, J. Russ. Math Surv. **27**, 21 (1972).

2.56 R. Bowen, in *Lect. Notes Maths.* **470**, 1 (Springer, New York, 1975).

2.57 D. Ruelle, *Thermodynamic Formalism* (Addison Wesley, Reading, 1978).

2.58 P. Grassberger and I. Procaccia, Physica D **13**, 34 (1984).

2.59 J.P. Kahane and J. Peyrière, Adv. Math. **22**, 131 (1976).

2.60 G.M. Molchan, Commun. Math. Phys. **179**, 681 (1996).

2.61 J.D. Farmer, E. Ott, and J.A. Yorke, Physica D **7**, 153 (1983).

2.62 P. Grassberger and I. Procaccia, Physica D **9**, 189 (1983).

2.63 R. Badii and A. Politi, J. Statist. Phys. **40**, 725 (1985).

2.64 G. Grasseau, Ph.D. Thesis, University of Bordeaux I (1989).

2.65 F. Argoul, A. Arneodo, J. Elezgaray, G. Grasseau, and R. Murenzi, Phys. Rev. A **41**, 5537 (1990).

2.66 L.V. Meisel, M. Jonhson, and P.J. Cote, Phys. Rev. A **45**, 6989 (1992).

2.67 A.S. Monin and A.M. Yaglom, *Statistical Fluid Mechanics: Mechanics of Turbulence*, Vol. 2 (MIT, Cambridge, 1975).

2.68 J.F. Muzy, E. Bacry, and A. Arneodo, Phys. Rev. E **47**, 875 (1993).

2.69 J.M. Combes, A. Grossmann, and P. Tchamitchian (eds.), *Wavelets* (Springer, Berlin, 1989).

2.70 Y. Meyer, *Ondelettes* (Herman, Paris, 1990).

2.71 P.G. Lemarié (ed.), *Les Ondelettes en 1989* (Springer, Berlin, 1990).

2.72 Y. Meyer (ed.), *Wavelets and Applications* (Springer, Berlin, 1992).

2.73 I. Daubechies, *Ten Lectures on Wavelets* (S.I.A.M., Philadelphia, 1992).

2.74 M.B. Ruskai, G. Beylkin, R. Coifman, I. Daubechies, S. Mallat, Y. Meyer, and L. Raphael (eds.), *Wavelets and their Applications* (Jones and Barlett, Boston, 1992).

2.75 C.K. Chui, *An Introduction to Wavelets* (Academic, Boston, 1992).

2.76 Y. Meyer and S. Roques (eds.), *Progress in Wavelets Analysis and Applications* (Editions Frontières, Gif-sur-Yvette, 1993).

2.77 A. Arneodo, F. Argoul, E. Bacry, J. Elezgaray, and J.-F. Muzy, *Ondelettes, Multifractales et Turbulences: de l'ADN aux Croissances Cristallines* (Diderot Editeur, Art et Sciences, Paris, 1995).

2.78 G. Erlebacher, M.Y. Hussaini, and L.M. Jameson (eds.), *Wavelets: Theory and Applications* (Oxford University Press, Oxford, 1996).

2.79 M. Holschneider, *Wavelets: An Analysis Tool* (Oxford University Press, Oxford, 1996).

2.80 S. Mallat, *A Wavelet Tour in Signal Processing* (Academic, New York, 1998).

2.81 B. Torresani, *Analyse Continue par Ondelettes* (Editions de Physique, Les Ulis, 1998).

2.82 M. Holschneider, J. Statist. Phys. **50**, 963 (1988).

2.83 M. Holschneider, Ph.D. Thesis, University of Aix-Marseilles II (1988).

2.84 A. Arneodo, G. Grasseau, and M. Holschneider, Phys. Rev. Lett. **61**, 2281 (1988), in [2.69], p. 182.

2.85 A. Arneodo, F. Argoul, J. Elezgaray, and G. Grasseau, in *Nonlinear Dynamics*, edited by G. Turchetti (World Scientific, Singapore, 1989), p. 130.

2.86 A. Arneodo, F. Argoul, E. Bacry, J. Elezgaray, E. Freysz, G. Grasseau, J.-F. Muzy, and B. Pouligny, in [2.72], p. 286.

2.87 A. Arneodo, E. Bacry, S. Manneville, and J.-F. Muzy, Phys. Rev. Lett. **80**, 708 (1998).

2.88 A. Arneodo, S. Manneville, J.-F. Muzy, and S.G. Roux, Philos. Trans. R. Soc. Lond. A **357**, 2415 (1999).

2.89 P. Goupillaud, A. Grossmann, and J. Morlet, Geoexploration **23**, 85 (1984).

2.90 A. Grossmann and J. Morlet, S.I.A.M. J. Math. Anal. **15**, 723 (1984).

2.91 A. Grossmann and J. Morlet, in *Mathematics and Physics, Lectures on Recent Results*, edited by L. Streit (World Scientific, Singapore, 1985), p. 135.

2.92 A. Arneodo, E. Bacry, and J.-F. Muzy, Phys. Rev. Lett. **74**, 4823 (1995).

2.93 A. Arneodo, E. Bacry, S. Jaffard, and J.-F. Muzy, J. Statist. Phys. **87**, 179 (1997).

2.94 S. Jaffard and Y. Meyer, J. Math. Pures Appl. **68**, 95 (1989).

2.95 S.G. Roux, J.-F. Muzy, and A. Arneodo, Eur. Phys. J. B **8**, 301 (1999).

2.96 D. Sornette, in *Scale Invariance and Beyond*, edited by B. Dubrulle, F. Graner, and D. Sornette (EDP Sciences and Springer, Les Ulis and Berlin, 1997), p. 235.

2.97 E. Bacry, A. Arneodo, U. Frisch, Y. Gagne, and E. Hopfinger, in *Turbulence and Coherent Structures*, edited by M. Lesieur and O. Metais (Kluwer, Dordrecht, 1991), p. 203.

2.98 M. Vergassola and U. Frisch, Physica D **54**, 58 (1991).

2.99 M. Vergassola, R. Benzi, L. Biferale, and D. Pisarenko, J. Phys. A **26**, 6093 (1993).

2.100 S. Jaffard, SIAM J. Math. Anal. **28**, 944 (1997).

2.101 H.G.E. Hentschel, Phys. Rev. E **50**, 243 (1994).

2.102 S.F. Edwards and P.W. Anderson, J. Phys. F **5**, 965 (1975).

2.103 S.G. Roux, Ph.D. Thesis, University of Aix-Marseille II, 1996.

2.104 A. Arneodo, J.-F. Muzy, and S.G. Roux, J. Phys. II France **7**, 363 (1997).

2.105 A. Arneodo, S. Manneville, J.-F. Muzy, and S.G. Roux, Eur. Phys. J. B **1**, 129 (1998).

2.106 A. Arneodo, E. Bacry, and J.-F. Muzy, J. Math. Phys. **39**, 4142 (1998).

2.107 R. Benzi, L. Biferale, A. Crisanti, G. Paladin, M. Vergassola, and A. Vulpiani, Physica D **65**, 352 (1993).

2.108 T. Higuchi, Physica D **46**, 254 (1990).

2.109 N.P. Greis and H.P. Greenside, Phys. Rev. A **44**, 2324 (1991).

2.110 G. Wornell and A.V. Oppenheim, IEEE Trans. Signal Proc. **40**, 611 (1992).

2.111 J. Beran, *Statistics for Long-Memory Process* (Chapman and Hall, New York, 1994).

2.112 C.-K. Peng, S.V. Buldyrev, S. Havlin, M. Simons, H.E. Stanley, and A.L. Goldberger, Phys. Rev. E **49**, 1685 (1994).

2.113 A. Scotti, C. Meneveau, and S.G. Saddoughi, Phys. Rev. E **51**, 5594 (1995).

2.114 A.R. Mehrabi, H. Rassamdana, and M. Sahimi, Phys. Rev. E **56**, 712 (1997).

2.115 B. Pilgram and D.T. Kaplan, Physica D **114**, 108 (1998).

2.116 P. Flandrin, IEEE Trans. Inf. Theory **35**, 197 (1989).

2.117 P. Flandrin, IEEE Trans. Inf. Theory **38**, 910 (1992).

2.118 E. Masry, IEEE Trans. Inf. Theory **39**, 260 (1993).

2.119 P. Abry, P. Goncalvès, and P. Flandrin, Lect. Notes Statist. **105**, 15 (1995).

2.120 L. Delbeke and P. Abry, submitted to: Applied and Computational Harmonic Analysis (1998).

2.121 P. Abry, P. Flandrin, M.S. Taqqu, and D. Veitch, in *Self-Similarity in Network Traffic*, edited by Parks and W. Willinger (Wiley, New York, 1998).

2.122 A.H. Tewfik and M. Kim, IEEE Trans. Inf. Theory **38**, 904 (1992).

2.123 J. Pando and L.Z. Fang, Phys. Rev. E **57**, 3593 (1998).

2.124 I. Simonsen, A. Hansen, and O.M. Nes, Phys. Rev. E **58**, 2779 (1998).

2.125 C.L. Jones, G.T. Lonergan, and D.E. Mainwaring, J. Phys. A: Math. Gen. **29**, 2509 (1996).

2.126 B. Audit, E. Bacry, J.-F. Muzy, and A. Arneodo, submitted to: IEEE Trans. Info. Theory (2000).

2.127 G. Samorodnisky and M.S. Taqqu, *Stable Non-Gaussian Random Processes* (Chapman and Hall, New York, 1994).

2.128 L. Richardson, Proc. R. Soc. Lond., Ser. A **110**, 709 (1926).

2.129 A.N. Kolmogorov, C. R. Acad. Sci. USSR **30**, 301 (1941).

2.130 A.N. Kolmogorov, J. Fluid Mech. **13**, 82 (1962).

2.131 A.M. Obukhov, J. Fluid Mech. **13**, 77 (1962).

2.132 D. Schertzer and S. Lovejoy, in *Turbulence and Chaotic Phenomena in Fluids*, edited by T. Tatsumi (North-Holland, Amsterdam, 1984), p. 505.

2.133 C. Meneveau and K.R. Sreenivasan, Phys. Rev. Lett. **59**, 1424 (1987).

2.134 D. Schertzer and S. Lovejoy, J. Geophys. Res. **92**, 9693 (1987).

2.135 S. Kida, J. Phys. Soc. Jpn. **60**, 5 (1990).

2.136 D. Schertzer, S. Lovejoy, F. Schmitt, Y. Ghigirinskaya, and D. Marsan, Fractals **5**, 427 (1997).

2.137 E.A. Novikov, Phys. Fluids A **2**, 814 (1990).

2.138 B. Dubrulle, Phys. Rev. Lett. **73**, 959 (1994).

2.139 Z.-S. She and E.C. Waymire, Phys. Rev. Lett. **74**, 262 (1995).

2.140 B. Castaing and B. Dubrulle, J. Phys. II France **5**, 895 (1995).

2.141 B. Dubrulle, J. Phys. II France **6**, 1825 (1996).

2.142 B.B. Mandelbrot, C. R. Acad. Sci. Paris Ser. A **278**, 289 (1974).

2.143 B.B. Mandelbrot, C. R. Acad. Sci. Paris Ser. A **278**, 355 (1974).

2.144 D. Schertzer and S. Lovejoy, Phys. Chem. Hydrol. J. **6**, 623 (1985).

2.145 S. Lovejoy and D. Schertzer, in [2.16], p. 102.

2.146 T. Vicsek and A.L. Barabási, J. Phys. A: Math. Gen. **24**, L845 (1991).

2.147 J. Eggers and S. Grossmann, Phys. Rev. A **45**, 2360 (1992).

2.148 J.A.C. Humphrey, C.A. Schuler, and B. Rubinski, Fluid Dyn. Res. **9**, 81 (1992).

2.149 A. Juneja, D.P. Lathrop, K.R. Sreenivasan, and G. Stolovitzky, Phys. Rev. E **49**, 5179 (1994).

2.150 L. Biferale, G. Boffetta, A. Celani, A. Crisanti, and A. Vulpiani, Phys. Rev. E **57**, R6261 (1998).

2.151 A. Arneodo, B. Audit, E. Bacry, S. Manneville, J.-F. Muzy, and S.G. Roux, Physica A **254**, 24 (1998).

2.152 A. Arneodo and J.-F. Muzy, Tech. Rept., Science and Finance (unpublished).

2.153 A. Arneodo, J.P. Bouchaud, R. Cont, J.-F. Muzy, M. Potters, and D. Sornette, Comment on Turbulent cascades in foreign exchange markets (1996); [available from `xxx.lanl.gov/abs/cond-mat/9607120`].

2.154 P.C. Ivanov, L.A. Nines Amaral, A.L. Goldberger, S. Havlin, M.G. Rosenblum, Z.K. Struzik, and H.E. Stanley, Nature **399**, 461 (1999).

2.155 A. Arneodo, F. Argoul, E. Bacry, J.-F. Muzy, and M. Tabard, Phys. Rev. Lett. **68**, 3456 (1992).

2.156 A. Arneodo, F. Argoul, J.-F. Muzy, and M. Tabard, Physica A **188**, 217 (1992).

2.157 A. Arneodo, F. Argoul, E. Bacry, J.-F. Muzy, and M. Tabard, Fractals **1**, 629 (1993).

2.158 A. Khun, F. Argoul, J.-F. Muzy, and A. Arneodo, Phys. Rev. Lett. **73**, 2998 (1994).

2.159 A. Arneodo, E. Bacry, P.V. Graves, and J.-F. Muzy, Phys. Rev. Lett. **74**, 3293 (1995).

2.160 A. Arneodo, Y. d'Aubenton-Carafa, E. Bacry, P.V. Graves, J.-F. Muzy, and C. Thermes, Physica D **96**, 291 (1996).

2.161 A. Arneodo, Y. d'Aubenton-Carafa, B. Audit, E. Bacry, J.-F. Muzy, and C. Thermes, Eur. Phys. J. B **1**, 259 (1998).

2.162 B. Audit, C. Thermes, C. Vaillant, Y. d'Aubenton-Carafa, J.-F. Muzy, and A. Arneodo, Phys. Rev. Lett. **86**, 2471 (2001).

2.163 B. Audit, Ph.D. Thesis, University of Paris VI, 1999.

2.164 S.V. Buldyrev, A.L. Goldberger, S. Havlin, C.-K. Peng, and H.E. Stanley, in [2.13], p. 49.

2.165 W. Li and K. Kaneko, Europhys. Lett. **17**, 655 (1992).

2.166 W. Li, Int. J. Bifurcat. Chaos **2**, 137 (1992).

2.167 B. Borštnik, Int. J. Quantum Chem. **52**, 457 (1994).

2.168 M.Y. Azbel', Phys. Rev. Lett. **75**, 168 (1995).

2.169 H. Herzel and I. Große, Physica A **216**, 518 (1995).

2.170 R.F. Voss, Phys. Rev. Lett. **68**, 3805 (1992).

2.171 R.F. Voss, Fractals **2**, 1 (1994).

2.172 B. Borštnik, D. Pumpernik, and D. Lukman, Europhys. Lett. **23**, 389 (1993).

2.173 S. Nee, Nature **357**, 450 (1992).

2.174 V.V. Prabhu and J.-M. Claverie, Nature **357**, 782 (1992).

2.175 P.J. Munson, R.C. Taylor, and G.S. Michaels, Nature **360**, 635 (1992).

2.176 S. Karlin and V. Brendel, Science **259**, 677 (1993).

2.177 C.A. Chatzidimitriou-Dreismann and D. Larhammar, Nature **361**, 212 (1993).

2.178 D. Larhammar and C.A. Chatzidimitriou-Dreismann, Nucl. Acids Res. **21**, 5167 (1993).

2.179 R.N. Mantegna, S.V. Buldyrev, A.L. Goldberger, S. Havlin, C.-K. Peng, M. Simons, and H.E. Stanley, Phys. Rev. Lett. **73**, 3169 (1994).

2.180 A. Czirók, R.N. Mantegna, S. Havlin, and H.E. Stanley, Phys. Rev. E **52**, 446 (1995).

2.181 R.N. Mantegna, S.V. Buldyrev, A.L. Goldberger, S. Havlin, C.-K. Peng, M. Simons, and H.E. Stanley, Phys. Rev. E **52**, 2939 (1995).

2.182 A. Czirók, H.E. Stanley, and T. Vicsek, Phys. Rev. E **53**, 6371 (1996).

2.183 W. Li and K. Kaneko, Nature **360**, 635 (1992).

2.184 C.L. Berthelsen, J.A. Glazier, and M.H. Skolnick, Phys. Rev. A **45**, 8902 (1992).

2.185 G.A. Dietler and Y.C. Zhang, Fractals **2**, 473 (1994).

2.186 J.A. Glazier, S. Raghavachari, C.L. Berthelsen, and M.H. Skolnick, Phys. Rev. E **51**, 2665 (1995).

2.187 S.V. Buldyrev, A.L. Goldberger, S. Havlin, R.N. Mantegna, M.E. Matsa, C.-K. Peng, M. Simons, and H.E. Stanley, Phys. Rev. E **51**, 5084 (1995).

2.188 P. Allegrini, M. Barbi, P. Grigolini, and B.J. West, Phys. Rev. E **52**, 5281 (1995).

2.189 D.S. Goodsell and R.E. Dickerson, Nucl. Acids Res. **22**, 5497 (1994).

2.190 J.D. Watson, M. Gilman, J. Witkowski, and M. Zoller, *Recombinant DNA* (Freeman, New York, 1992).

2.191 A. Danchin, La Recherche **24**, 222 (1993).

2.192 A. Bernot, *L'Analyse des Génomes* (Nathan, Paris, 1996).

2.193 R.D. Fleischmanm et al., Science **269**, 496 (1995).

2.194 B. Dujon, Trends Genet. **12**, 263 (1996).

2.195 J. Carol et al., Science **273**, 1058 (1996).

2.196 T. Gaasterland, S. Andersson, and C. Sensen, Magpie genome sequencing project list; [available from `www-fp.mcs.anl.gov/~gaasterland/genomes.html`].

2.197 W. Li, G. Stolovitzky, P. Bernaola-Galván, and J.L. Oliver, Genome Res. **8**, 916 (1998).

2.198 G.M. Viswanathan, S.V. Buldyrev, S. Havlin, and H.E. Stanley, Physica A **249**, 581 (1998).

2.199 F.R. Blattner et al., Science **277**, 1453 (1997).

2.200 T.J. Richmond, J.T. Finch, B. Rushton, D. Rhodes, and A. Klug, Nature **311**, 532 (1984).

2.201 J. Widom and A. Klug, Cell **43**, 207 (1985).

2.202 Y. Saitoh and U.K. Laemmli, Cold Spring Harb. Symp. Quant. Biol. **58**, 755 (1993).

2.203 L.D. Murphy and S.B. Zimmerman, J. Struct. Biol. **119**, 336 (1997).

2.204 J.N. Reeve, K. Sandman, and C.J. Daniels, Cell **89**, 999 (1997).

2.205 M.R. Starich, K. Sandman, J.N. Reeve, and M.F. Summers, J. Mol. Biol. **255**, 187 (1996).

2.206 S.L. Pereira, R.A. Grayling, R. Lurz, and J.N. Reeve, Proc. Natl. Acad. Sci. USA **94**, 12 633 (1997).

2.207 K. Luger, A.W. Mader, R.K. Richmond, D.F. Sargent, and T.J. Richmond, Nature **389**, 251 (1997).

2.208 J.R. Brown and W.F. Doolittle, Microbiol. Mol. Biol. Rev. **61**, 456 (1997).

2.209 G.J. Olsen and C.R. Woese, Cell **89**, 991 (1997).

2.210 M. Coca-Prados, H.Y. Yu, and M.T. Hsu, J. Virol. **44**, 603 (1982).

2.211 M.V. Borca, P.M. Irusta, G.F. Kutish, C. Carillo, C.L. Afonso, A.T. Burrage, J.G. Neilan, and D.L. Rock, Arch. Virol. **141**, 301 (1996).

2.212 S.A. Stanfield-Oakley and J.D. Griffith, J. Mol. Biol. **256**, 503 (1996).

2.213 E.N. Trifonov, Physica A **249**, 511 (1998).

2.214 H. Herzel, O. Weiss, and E.N. Trifonov, Bioinformatics **15**, 187 (1999).

2.215 G. Meersseman, S. Pennings, and E.M. Bradbury, EMBO J. **11**, 2951 (1992).

2.216 P.T. Lowary and J. Widom, Proc. Natl. Acad. Sci. USA **94**, 1183 (1997).

2.217 K.J. Polach and J. Widom, J. Mol. Biol. **254**, 130 (1995).

2.218 A. Grosberg, Y. Rabin, S. Havlin, and A. Neer, Europhys. Lett. **23**, 373 (1993).

2.219 D.E. Pettijohn, in *Escherichia coli and Salmonella*, edited by F. Neidhardt et al. (AMS, Washington DC, 1996), p. 158.

2.220 A.M. Segall, S.D. Goodman, and H.A. Nash, EMBO J. **13**, 4536 (1994).

2.221 N.P. Higgins, X. Yang, Q. Fu, and J.R. Roth, J. Bacteriol. **178**, 2825 (1996).

2.222 G. Pruss and K. Drlica, Cell **56**, 521 (1989).

2.223 H.A. Nash, Trends Biochem. Sci. **15**, 222 (1990).

2.224 R. Kanaar, A. Klippel, E. Shekhtman, J.M. Dungan, R. Khamann, and N.R. Cozzarelli, Cell **27**, 353 (1990).

2.225 B. Lea-Cox and J.S.Y. Wang, Fractals **1**, 87 (1993).

2.226 E. Freysz, B. Pouligny, F. Argoul, and A. Arneodo, Phys. Rev. Lett. **64**, 745 (1990).

2.227 A. Arneodo, F. Argoul, J.-F. Muzy, B. Pouligny, and E. Freysz, in [2.74], p. 241.

2.228 J.P. Antoine, P. Carette, R. Murenzi, and B. Piette, Signal Proc. **31**, 241 (1993).

2.229 J. Canny, IEEE Trans. Pattern Anal. Mach. Intell. **8**, 679 (1986).

2.230 J. Arrault, A. Arneodo, A. Davis, and A. Marshak, Phys. Rev. Lett. **79**, 75 (1997).

2.231 A. Arneodo, N. Decoster, and S.G. Roux, Phys. Rev. Lett. **83**, 1255 (1999).

2.232 A. Arneodo, N. Decoster, and S.G. Roux, Eur. Phys. J. B **15**, 567 (2000).

2.233 N. Decoster, S.G. Roux, and A. Arneodo, Eur. Phys. J. B **15**, 739 (2000).

2.234 N. Decoster, Ph.D. Thesis, University of Bordeaux I, 1999.

2.235 D. Marr, *Vision* (Freemann, San-Francisco, 1982).

2.236 A. Rosenfeld and M. Thurston, IEEE Trans. Comput. C 29 (1971).

2.237 R. Murenzi, Ph.D. Thesis, University of Louvain la Neuve, 1990.

2.238 R. Murenzi, in [2.69], p. 239.

2.239 M. Ben Slimane, Ph.D. Thesis, E.N.P.C., France, 1996.

2.240 G. Ouillon, D. Sornette, and C. Castaing, Nonlin. Proc. Geophys. **2**, 158 (1995).

2.241 G. Ouillon, C. Castaing, and D. Sornette, J. Geophys Res. **101**, 5477 (1996).

2.242 P. Gaillot, J. Darrozes, M. de Saint Blanquat, and G. Ouillon, Geophys. Res. Lett. **24**, 1819 (1997).

2.243 P. Lévy, *Processus Stochastiques et Mouvement Brownien* (Gauthier-Villars, Paris, 1965).

2.244 S. Lovejoy, Science **216**, 185 (1982).

2.245 R.F. Cahalan, in *Advances in Remote Sensing and Retrieval Methods*, edited by A. Deepak, H. Fleming, and J. Theon (Deepak, Hampton, 1989), p. 371.

2.246 A. Davis, A. Marshak, R.F. Cahalan, and W.J. Wiscombe, (1997) unpublished.

2.247 V. Ramanatahn, R.D. Cess, E.F. Harrison, P. Minnis, B.R. Barkston, E. Ahmad, and D. Hartmann, Science **243**, 57 (1989).

2.248 R.D. Cess et al., Science **245**, 513 (1989).

2.249 F.S. Rys and A. Waldvogel, in [2.3], p. 461.

2.250 R.M. Welch and B.A. Wielicki, Clim. Appl. Meteorol. **25**, 261 (1986).

2.251 J.I. Yano and Y. Takeuchi, J. Meteorol. Soc. Jpn. **65**, 661 (1987).

2.252 R.M. Welch, K.S. Kuo, B.A. Wielicki, S.K. Sengupta, and L. Parker, J. Appl. Meteorol. **27**, 341 (1988).

2.253 R.F. Cahalan and J.H. Joseph, Mon. Weather Rev. **117**, 261 (1989).

2.254 G. Sèze and L. Smith, in *Proc. Seventh Conf. Atmospheric Radiation* (Am. Meteorol. Soc., San Francisco, 1990).

2.255 A. Davis, S. Lovejoy, and D. Schertzer, in *Scaling, Fractals and Nonlinear Variability in Geophysics*, edited by S. Lovejoy and D. Schertzer (Kluwer, Dordrecht, 1991), p. 303.

2.256 Y. Tessier, S. Lovejoy, and D. Schertzer, J. Appl. Meteorol. **32**, 223 (1993).

2.257 A. Davis, A. Marshak, W.J. Wiscombe, and R.F. Cahalan, J. Geophys. Res. **99**, 8055 (1994).

2.258 W.D. King, C.T. Maher, and G.A. Hepburn, J. Appl. Meteorol. **20**, 195 (1981).

2.259 C. Duroure and B. Guillemet, Atmos. Res. **25**, 331 (1990).

2.260 B. Baker, J. Atmos. Sci. **49**, 387 (1992).

2.261 S.P. Malinowski and I. Zawadski, J. Atmos. Sci. **50**, 5 (1993).

2.262 A.V. Korolev and I.P. Mazin, J. Appl. Meteorol. **32**, 760 (1993).

2.263 S.P. Malinowski, M.Y. Leclerc, and D.G. Baumgardner, J. Atmos. Sci. **51**, 397 (1994).

2.264 A. Davis, A. Marshak, W.J. Wiscombe, and R.F. Cahalan, J. Atmos. Sci. **53**, 1538 (1996).

2.265 A. Marshak, A. Davis, W.J. Wiscombe, and R.F. Cahalan, J. Atmos. Sci, **54**, 1423 (1997).

2.266 S. Cox, D. McDougal, D. Randall, and R. Schiffer, Bull. Am. Meteorol. Soc. **68**, 114 (1987).

2.267 B.A. Albrecht, C.S. Bretherton, D. Johnson, W.H. Schubert, and A.S. Frisch, Bull. Am. Meteorol. Soc. **76**, 889 (1995).

2.268 R. Boers, J.B. Jensen, P.B. Krummel, and H. Gerber, Q. J. R. Meteorol. Soc. **122**, 1307 (1996).

2.269 H.W. Baker and J.A. Davies, Remote Sens. Env. **42**, 51 (1992).

2.270 A. Davis et al., J. Atmos. Sci. **54**, 241 (1997).

2.271 R.F. Cahalan and J.B. Snider, Remote Sens. Env. **28**, 95 (1989).

2.272 S. Lovejoy, D. Schertzer, P. Silas, Y. Tessier, and D. Lavallée, Ann. Geophys. **11**, 119 (1993).

2.273 S.M. Gollmer, M. Harshvardan, and R.F. Cahalan, J. Atmos. Sci. **52**, 3013 (1995).

2.274 W.J. Wiscombe, A. Davis, A. Marshak, and R.F. Cahalan, Proc. Fourth Atmospheric Radiation Measurement (ARM) Science Team Meeting, Charleston, U.S. Dept. of Energy 11 (1995).

2.275 D. Schertzer and S. Lovejoy, in *Fractals: their Physical Origin and Properties*, edited by L. Pietronero (Plenum, New York, 1989), p. 49.

2.276 D. Schertzer and S. Lovejoy, in [2.10], p. 11.

2.277 A. Davis, A. Marshak, W.J. Wiscombe, and R.F. Cahalan, Proc. 2nd Workshop on Nonstationary Random Processes and their Applications (1995), preprint.

2.278 S.G. Roux, A. Arneodo, and N. Decoster, Eur. Phys. J. B **15**, 765 (2000).

2.279 L.M. Romanova, Izv. Acad. Sci. USSR Atmos. Oceanic Phys. **11**, 509 (1975).

2.280 A. Davis, Ph.D. Thesis, McGill University, Montreal, 1992.

2.281 R.F. Cahalan, W. Ridgway, W.J. Wiscombe, T.L. Bell, and J.B. Snider, J. Atmos. Sci. **51**, 2434 (1994).

2.282 K. Stamnes, S.-C. Tsay, W.J. Wiscombe, and K. Jayaweera, Appl. Opt. **27**, 2502 (1988).

2.283 R.F. Cahalan, W. Ridgway, W.J. Wiscombe, S. Gollmer, and M. Harshvardan, J. Atmos. Sci. **51**, 3776 (1994).

2.284 A. Marshak, A. Davis, W.J. Wiscombe, and R.F. Cahalan, J. Geophys. Res. **100**, 26 247 (1995).

2.285 M. Tiedke, Mon. Weather Res. **124**, 745 (1996).

2.286 M. Harshvardan, B.A. Wielicki, and K.M. Ginger, J.Clim. **7**, 1987 (1994).

2.287 A. Davis, A. Marshak, W.J. Wiscombe, and R.F. Cahalan, in *Current Topics in Nonstationary Analysis*, edited by G. Treviño et al. (World Scientific, Singapore, 1996), p. 97.

2.288 R.F. Cahalan, M. Nestler, W. Ridgway, W.J. Wiscombe, and T.L. Bell, in *Proc. 4th International Meeting on Statistical Climatology*, edited by J. Sansom (New Zealand Meteorological Service, Wellington, 1990), p. 28.

2.289 R.D. Cess, M.H. Zhang, Y. Zhou, X. Jing, and V. Dvortsov, J. Geophys. Res. **101**, 23299 (1996).

2.290 A. Davis, S. Lovejoy, and D. Schertzer, SPIE Proc. **1558**, 37 (1991).

2.291 A. Marshak, A. Davis, W.J. Wiscombe, and G. Titov, Remote Sens. Env. **52**, 72 (1995).

2.292 A. Arneodo, S. Manneville, J.-F. Muzy, and S.G. Roux, Appl. Comput. Harmonic Anal. **6**, 374 (1999).

2.293 G. Ruiz-Chavarria, C. Baudet, and S. Ciliberto, Physica D **99**, 369 (1996).

2.294 A. Davis, A. Marshak, H. Gerber, and W.J. Wiscombe, J. Geophys. Res. (1998), to appear.

2.295 C.H. Meong, W.R. Cotton, C. Bretherton, A. Chlond, M. Khairoutdinov, S. Krueger, W.S. Lewellen, M.K. McVean, J.R.M. Pasquier, H.A. Rand, A.P. Siebesma, B. Stevens, and R.I. Sykes, Bull. Am. Meteorol. Soc. **77**, 261 (1996).

2.296 A. Marshak, A. Davis, R.F. Cahalan, and W.J. Wiscombe, Phys. Rev. E **49**, 55 (1994).

2.297 M.E. Cates and J.M. Deutsch, Phys. Rev. A **35**, 4907 (1987).

2.298 A.P. Siebesma, in *Universality in Condensed Matter*, edited by R. Julien, L. Peliti, R. Rammal, and N. Boccara (Springer, Heidelberg, 1988), p. 188.

2.299 J. O'Neil and C. Meneveau, Phys. Fluids A **5**, 158 (1993).

2.300 F. Morel-Bailly, M. Chauve, J. Liandrat, and P. Tchamitchian, C.R. Acad. Sci. Paris Sér. II **313**, 591 (1991).

2.301 H. Markowitz, *Portfolio Selection: Efficient Diversification of Investment* (Wiley, New York, 1959).

2.302 R. Merton, *Continuous-Time Finance* (Blackwell, Cambridge, 1990).

2.303 J.C. Hull, *Futures, Options and other Derivatives* (Prentice-Hall, Upper Saddle River, 1997).

2.304 M.L. Bachelier, *Thérie de la Spéculation* (Gauthiers-Villars, Paris, 1900).

2.305 P.A. Samuelson, *Collected Scientific Papers* (MIT, Cambridge, 1972).

2.306 B.B. Mandelbrot, *Fractal and Scaling in Finance: Discontinuity, Concentration, Risk* (Springer, New York, 1997).

2.307 R.F. Engle, Econometrica **50**, 987 (1982).

2.308 T. Bollersev, R.Y. Chous, and K.F. Kroner, J. Economet. **52**, 5 (1992).

2.309 R. Mantegna and H.E. Stanley, Nature **376**, 46 (1995).

2.310 S. Ghashghaie, W. Breymann, J. Peinke, P. Talkner, and Y. Dodge, Nature **381**, 767 (1996).

2.311 B. Castaing, Y. Gagne, and E.J. Hopfinger, Physica **46 D**, 177 (1990).

2.312 B. Castaing, Y. Gagne, and M. Marchand, Physica D **68**, 387 (1993).

2.313 Y. Gagne, M. Marchand, and B. Castaing, J. Phys. II France **4**, 1 (1994).

2.314 R. Mantegna and H.E. Stanley, Nature **383**, 587 (1996).

2.315 A. Arneodo, J.-F. Muzy, and D. Sornette, Eur. Phys. J. B **2**, 277 (1998).

2.316 U. Muller, M. Dacorogna, R.D. Davé, R.B. Olsen, and O.V. Pictet, First Int, Conf. High Frequency Data in Finance, HFDF I, Zurich (1995).

2.317 R.B. Barsky and J.B. De Long, Q. J. Econ. **CVIII**, 291 (1993).

2.318 S.J. Kon, J. Finance **39**, 147 (1984).

2.319 B.B. Mandelbrot, A. Fisher, and L. Calvet, Tech. Rept. (unpublished), working paper.

2.320 P. Cvitanovic (ed.) *Universality in Chaos* (Helger, Bristol, 1984).

2.321 B. Hao (ed.), *Chaos* (World Scientific, Singapore, 1984).

2.322 A. Arneodo, E. Bacry, and J.-F. Muzy, Europhys. Lett. **25**, 479 (1994).

Part II

Climate Systems

Köppen Climate Classification

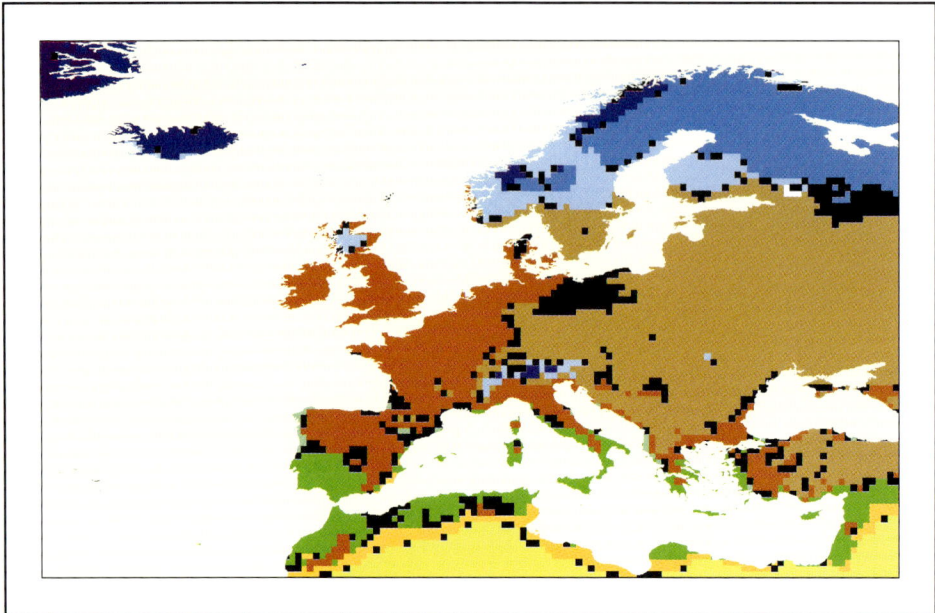

Climate types 1901-1995 · **Change in 1891-1995**

Ar Am Aw As BS BW Cr Cs Cw Do Dc Eo Ec FT Fl

3. Space-Time Variability of the European Climate

Klaus Fraedrich and Christian-D. Schönwiese

Observational datasets are analyzed in four steps to provide a comprehensive view of climate and climate variability in Europe. The *first* step towards an overall description of climate is the analysis of the spatial distributions of means and variances of basic climate elements (surface air temperature, sea level pressure, and precipitation). The *second* step identifies the climate generating dynamical processes; here, cyclone paths and storm tracks provide climatologies from quantitative Lagrangian and Eulerian analyses of weather data; Grosswetterlagen and climate zones, on the other hand, represent climatologies based on qualitative patterns (regimes) embedding the climate into its geographical and biospheric environment. A *third* step of analysis is necessary because these climatologies indicate recent climate changes at the end of the last century. This is documented in terms of the spatial distribution of time averaged climate elements and of various measures of their variability (see preceding steps). The *fourth* step evaluates the causes of the observed trends, analyzing the effects of external forcing (volcanism, solar, and anthropogenic) by empirical methods.

3.1 Introduction

Climate describes the relatively long-term behavior of the climate system whose components comprise the atmosphere, the hydrosphere (ocean and fresh water of land areas), the cryosphere (land and sea ice), the pedosphere, the lithosphere (both representing the land surface), and the biosphere (especially vegetation). This system is observed specifying the space-time variations of climate elements like temperature, air humidity, precipitation, air pressure,

◄ **Fig. 3.0.** Climate zones in Europe (1901–1995) and their shifts in the period 1981–1995 (in black): tropical (A), dry (B), subtropical (C), temperate (D), boreal (E), snow (F) with tundra and perpetual frost (T, I)

wind, cloudiness etc., most of them characterizing the subsystem atmosphere. Information from prehistoric time is provided by reconstructions (palaeoclimatology). Consequently, climate is described using proper statistics of all types of variability and for all relevant space-time scales. This variability is forced by internal interactions within the climate system and external influences (cf. also Chap. 6). Related process studies and modeling may lead to an understanding of the climate system behavior, which includes deterministic and statistical modeling and verification against observations.

Climate and climate change gain public interest for the following reasons: (i) human welfare can be linked to climate variability and, in the future, mankind may also benefit or suffer from natural or anthropogenically induced climate variations. (ii) In historical and, much intensified, in industrial time mankind is becoming a climate forcing factor, so that climate protection (UN Framework Convention on Climate Change) appears to be an urgent task (cf. the discussions in Chap. 4).

The scientific knowledge about climate and climate variability is still incomplete: the climate database, the understanding of the processes participating in climate change, and the role of the anthroposphere with its socio-economic response as a feedback process. Future scenarios and predictions of the climate system have an extremely important presupposition: description and understanding of past climate variability need to be both as exact and comprehensive as possible. This requires intensive cooperation of scientists.

As a contribution to the very basis of this task, that is the awareness of our present knowledge, some aspects of space-time statistics of climate variability are systematically addressed focusing our attention on Europe. Time and space scales are discussed (Sect. 3.2); we deal with the quantitative and qualitative measures of climate (Sect. 3.3, steps one and two of the climate analysis) in terms of means and variabilities, including storm tracks, cyclone paths, Grosswetterlagen, and climate zones. Finally, trends of these measures are presented and the effects of external forcing are discussed (Sect. 3.4, steps three and four of the analysis).

3.2 Time and Space Scales: Peaks, Gaps, and Scaling

In each climate compartment the life-span and intensity of the largest energy containing eddy characterizes the predictability and variability in terms of the decay period of a perturbation. Useful estimates of their memories and predictabilities can be obtained from the residence times of water as an important carrier of latent energy (Table 3.1). The observed large differences of the memories of the interacting systems challenge monitoring and modeling of climate variability.

Table 3.1. Time-scale estimates

Subsystem	Time-scale
Atmosphere	< 10 days (weather)
Ocean/land	1 month (upper layers) to 10^3 yr (deep ocean)
Cryosphere	< 10 yr (sea ice) to > 10^3 yr (ice shields)

The scale range of atmospheric phenomena, for example, spans from raindrops to planetary waves in space and covers events between very short turbulent outbursts and climate change. Power spectra of time series of climate observables, their extrema and power-law slopes (Fig. 3.1) provide further information on the climate dynamics and variability.

Spectral *maxima* (or peaks) represent dominating time scales. Such peaks occur in the following period-bands: 12 and 24 hr (diurnal and half-daily cycle), three days to one month (weather variability ranging from turbulent eddies to planetary waves), half-annual and annual cycle, the quasi-biennial oscillation (QBO) of the tropical stratosphere, the El Niño/Southern Oscillation (ENSO) (3–7 yr), solar cycles (10–20 yr), 100–1 500 yr (little ice age like events), 22, 41, and 100 thousand years (astronomical cycles) and, finally, geological scales related to orogenesis and continental drift. An idealized spectrum of a time series of a hypothetical climate element is presented in Fig. 3.1.

Spectral *minima* (or gaps) identify periods useful for time-averaging and sampling (see Fig. 3.1a, points 1–4): these averaging periods range from a 3 hr time scale, which characterizes synoptic sampling, via monthly periods describing the march of the seasons and intra-annual variability, to decadal periods, which correspond to climate fluctuations, while millions of years represent tectonic changes.

Spectral *scaling* characterizes the nonlinearity of atmospheric dynamics. A scale range, which exhibits scale invariance, relates fluctuations at smaller scales to those of larger ones by the same power-law scaling without a preferred mode of excitation. In this sense, a power-law or scaling behavior within a frequency band

$$S(\omega) \sim \omega^{-b} \qquad (3.1)$$

is another characteristic of the variability of the dynamical system (Fig. 3.1b). Analysis of rainfall at European stations reveals a regime of climate fluctuations ($b \sim 0.7$) for periods > 3 yr; this differs from the $b \sim 2$ red noise power-law spectrum connecting the level of small scale white noise forcing with that of the white large scale response. A spectral plateau $b \sim 0$ represents variability of the large scale circulation (3 yr to 1 month) which is linked through a

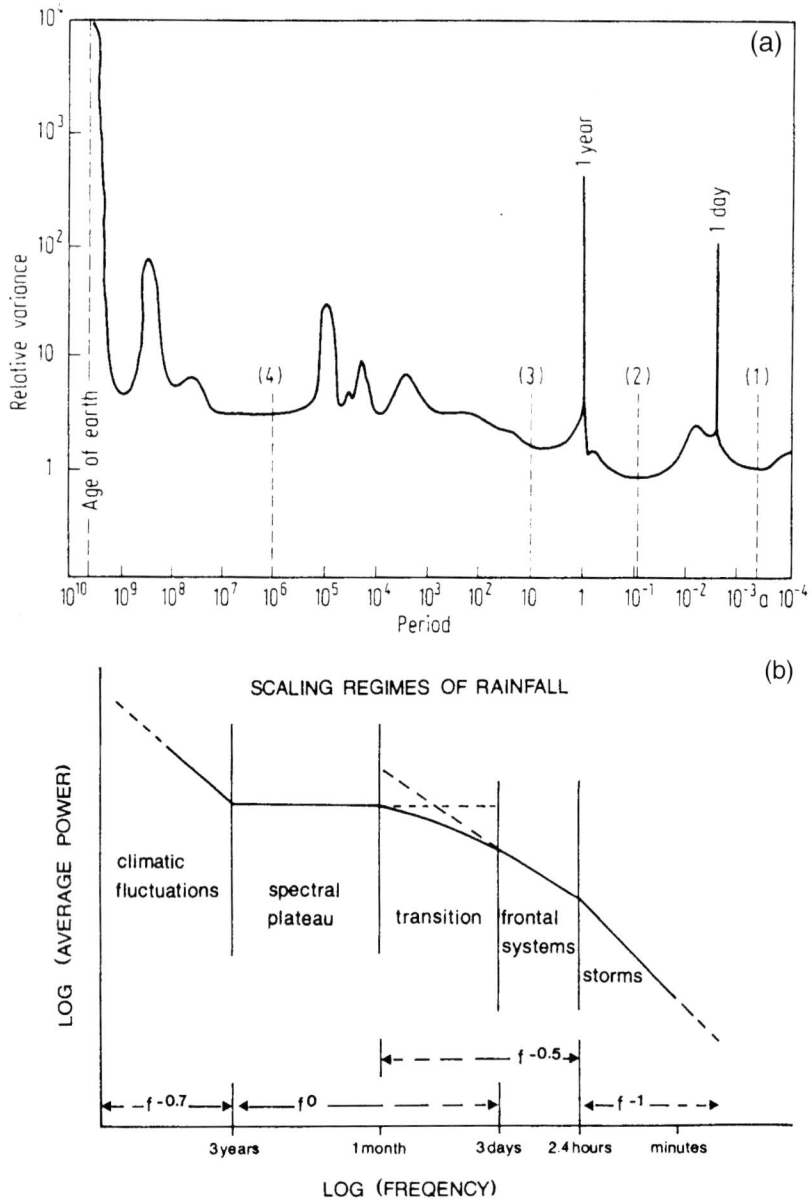

Fig. 3.1. Spectra of climate variability: (**a**) Schematic climate spectrum [3.43,3.50]; spectral maxima represent dominant atmospheric and climatic phenomena and their associated periods; spectral minima (points 1–4) identify periods of time-averaging. (**b**) Schematic scaling regimes of continental European rainfall [3.17]; spectral power-law scaling (scale invariance) characterizes the nonlinear dynamics of regimes of atmospheric variability

transition zone with a regime of frontal systems ($b \sim 0.5$, lasting less than 3 days). The break at 2.4 hr and the standard meteorological interpretation of the associated meso-scale regime remain an open question.

3.2.1 Time Scales: Power Spectrum Analysis

Power spectrum analysis transforms any time series, e.g. of climate elements like temperature or precipitation, into a spectrum where the contributions of variance are related to a particular sequence of frequency or period bands, respectively. Spectral peaks point to possible cyclical variance components whereas the background spectrum, especially its behavior in respect to low frequencies, reveals whether the process implies persistence (leading to so-called 'red noise') or not ('white noise').

There are a number of algorithms available providing spectral density estimates (see e.g. [3.42, 3.64]). The easiest way is to compute the Fourier transform of the autocorrelation function, called autocorrelation spectrum analysis (ASA). Corresponding confidence tests (χ^2-test) are available and easy to handle. An alternative method is based on the information entropy theory and called maximum entropy spectrum analysis (MESA). Compared to ASA it reveals a better resolution of the frequency bands (especially at low frequencies) but involves confidence test problems. The singular spectrum analysis (SSA) is again, like ASA, based on the computation of the autocovariance matrix but introduces an EOF (empirical orthogonal function) transformation by the computation of the eigenvalues of this matrix. In this case it needs to be decided for all EOFs whether the confidence intervals of the eigenvalues overlap. Then the related frequencies are isolated using ASA or MESA estimates. In addition, wavelet analysis (for a general introduction and some applications, see Chap. 2) is some type of a combination of spectral analysis and numerical filter techniques where these filter techniques themselves are able to suppress variations within the relatively high-frequency part of the spectrum (low-pass filter) or low-frequency part (high-pass filter) or both, so that a particular frequency band remains (band-pass filter). Usually, the result is shown in terms of the filtered time series.

As an example, the annual surface air temperature variations 1758–1998 at Frankfurt/Main (Germany, see Fig. 3.2a) are shown including the 10 yr low-pass filtered (smoothed) data suppressing variations of periods < 10 yr. Because of an observation gap and some uncertainties in earlier years we use only the period since 1857. ASA and MESA, see Fig. 3.2b, indicate some peaks, where in the case of ASA only two peaks exceed the 90% confidence level, 2.1 and 2.3 yr (quasi-biennial oscillation), whereas in the case of MESA the most prominent peak is 7.8 yr although a pronounced 2.3 yr peak is also indicated. It should be mentioned that an approximately 8 yr peak is also found in the time series data of the North Atlantic Oscillation (NAO). ASA shows a 'red noise' background spectrum with variance increasing towards low frequencies/long periods, which is reflected in the confidence level; MESA does not. Figure 3.2c

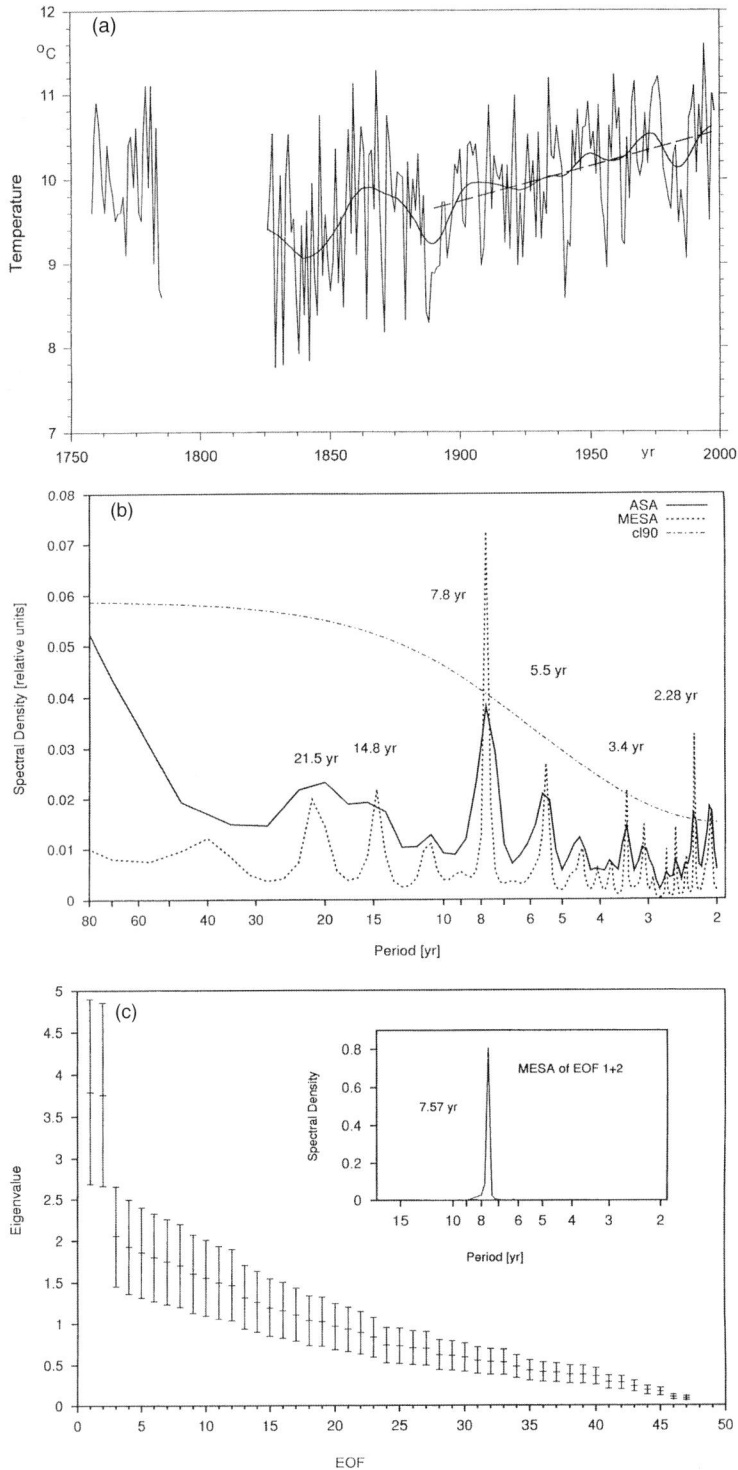

Fig. 3.2. (a) Observed annual surface air temperature variations 1758–1998 (observation gap 1786–1825) with 10 yr low-pass filtered data (smooth line) and +1.1 K linear trend (dashed line) at Frankfurt/Main, Germany (1850–1998). (b) Related power spectrum analysis 1857–1996 (ASA = autocorrelation spectrum analysis, MESA = maximum entropy spectrum analysis, cl90 = 90% confidence level). (c) Related singular spectrum analysis (SSA) including MESA of EOF1 + EOF2

shows the result employing SSA analysis. Note that the eigenvalue (95%) confidence intervals of the first two EOFs do not overlap with the other. A MESA of these first two EOFs indicates a 7.6 yr cycle, which is close to the dominant MESA peak of the original time series. The eigenvalues of the first two EOFs are almost identical indicating a harmonic oscillation.

3.2.2 Space Scales: Correlations and Representativeness

Climate time series statistics varying in space are represented by their correlations from station to station or grid point to grid point. This leads to information about the number of stations or grid points necessary to derive proper space-related statistics. An example may illustrate this problem. Figure 3.3a (from [3.38]) shows a relationship where, related to the Nordic station Helsinki (1881–1980), the annual surface air temperature time series correlation coefficients (linear Pearson correlation) are plotted against the station distance (182 stations considered, 114 from Europe); confidence levels 90%–99.9% are also indicated. It appears that these correlations drop to values < 0.7 (corresponding to about 50% of the common variance) approximately at a distance of 1 000 km and that even at a 2 000 km distance correlations are still significant. At a distance of 6–8 000 km, correlations again turn positive (and significant); these teleconnections are related to the rainbearing frontal systems which, associated with the Rossby wave pattern of the atmospheric circulation, dominate Europe's climate. Mean sea level pressure shows a similar behavior [3.53]. For summer precipitation, however, correlations drop to values < 0.7 at distances near 50 km, Fig. 3.3b (for 250 stations surrounding Frankfurt/Main, 1891–1990; [3.53]). In winter the correlation distance changes to about 200 km. These distance numbers are a measure of representativeness: in the case of temperature and pressure (large numbers) we have a fair representativeness and do not need so much information in space; this, however, is not the case for precipitation (small numbers). Note that the typical representativeness distance defined by any reasonable correlation range does not only depend on climate elements and seasons but also on the data resolution in time and on the climate regime. Moreover, there are also long-term changes associated with Grosswetter or climate trends. The subsequent analysis of climate variablity is extended by deriving the degrees of freedom of Grosswetterlagen and climate zones.

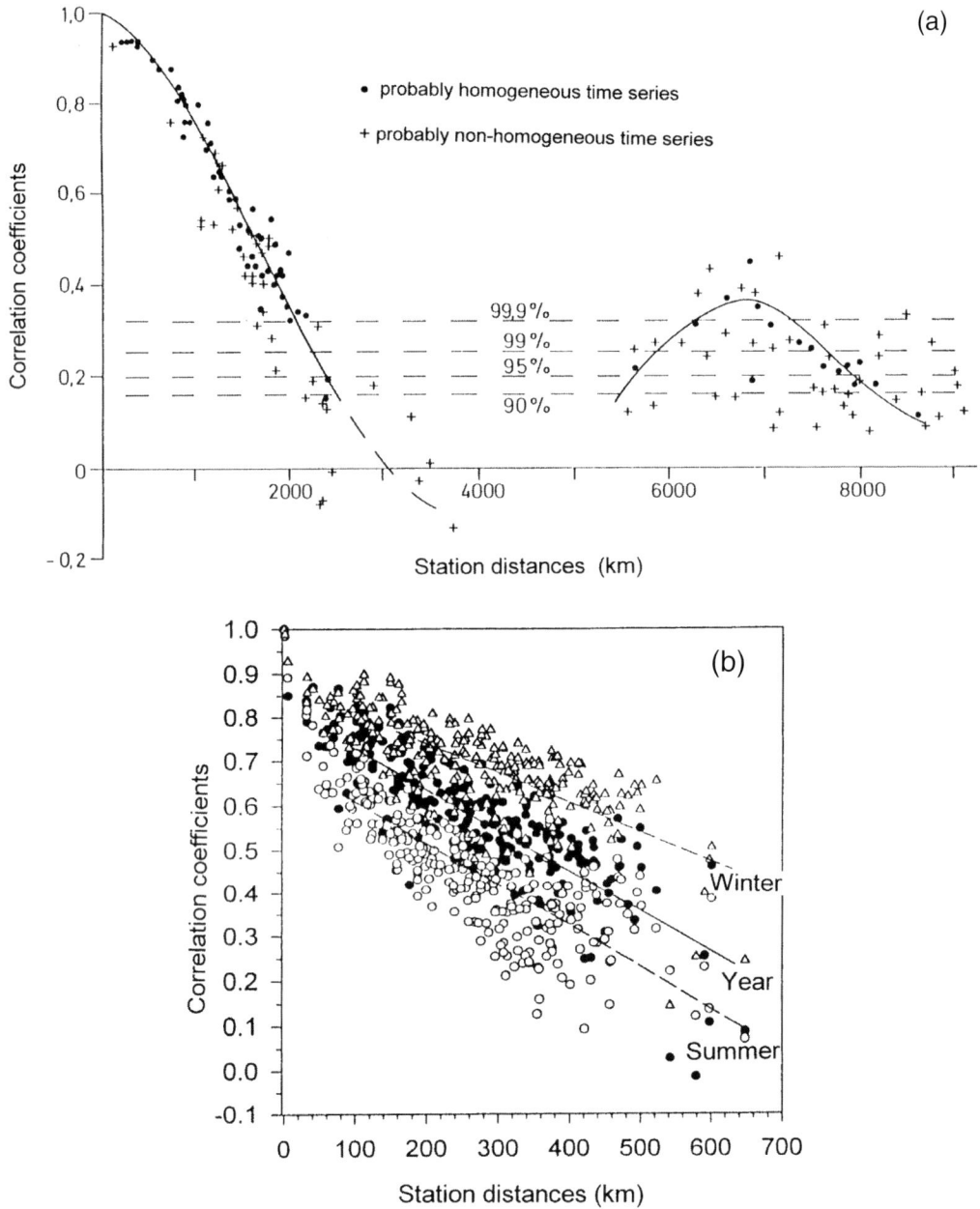

Fig. 3.3. (a) Surface air temperature correlations (1881–1980) changing with distances between reference station Helsinki (Finland) and other stations in Europe and North America; dashed lines specify confidence levels [3.38]. (b) As (a) but for precipitation correlations in Germany (reference station Frankfurt/Main, 1891–1990): annual (full circles), summer (Jun., Jul., Aug., open circles) and winter (Dec., Jan., Feb., triangles) [3.53]

3.3 Europe's Climate: Storm Tracks, Grosswetterlagen, and Climate Zones

Weather and climate observations are analyzed to obtain quantitative physical information in terms of energy and momentum budgets of the atmospheric dynamics, their space and time filtered properties in the space-time or wavenumber-frequency domain, and nonlinear interactions. Regional climates and the associated variability are described by spatial distributions of time means and anomalies. Another type of data analysis which is qualitative or phenomenological, identifies and describes structurally stable patterns. That is, large scale circulation systems which are related to real weather (highs and lows) processes or climate zones (associated with typical biota). Both types of analysis describe the same physical processes but contribute different aspects to it. For both approaches to represent climatologically meaningful space-time behavior, the underlying quantitative datasets have to be subjected to prior determination of the spatial degrees of freedom (or dimension) in order to substantiate the number of phenomenological patterns selected for the analysis: the linear approach is based on the superposition of independent modes and leads to estimates of the spatial degrees of freedom; the nonlinear method of scaling is more dynamically oriented. This section presents a description of European climate in terms of its space-time variability which is based on suitable physical and phenomenological datasets; climate change issues are addressed in the subsequent section.

3.3.1 Climate Means and Variability Patterns: Pressure, Temperature, and Precipitation

The local surface climate at a grid point or single station is conventionally described by three variables (climate elements), which imply a suitable time average (say, a winter month) $C(t) = (P, T, N)$: the monthly mean sea level pressure P, temperature T, and precipitation N. Variability is suitably described by anomalies $C(t)' = (C(t) - \langle C \rangle)/\sigma_i$, from the long-term ensemble mean of, say, thirty years $\langle C \rangle$, and which are normalized by the respective standard deviations, σ_i. In this sense the North Atlantic/European climate is represented by the fields of the ensemble means $\langle C \rangle$, while the space-time variability is comprised in the remaining anomalies $C'(t)$. The information in this anomaly dataset can be suitably reduced by EOF analysis which separates the space and the time variability. The eigenvectors of the covariance matrix of the anomalies describe spatial patterns which are commonly ranked by their contribution to the total variance (denoted by their eigenvalues). These North Atlantic/Europe eigenvectors (see [3.34] for a climate analysis of North America) represent contributions to regional climate anomalies containing information on the simultaneous distribution of the sea level pressure, temperature, and precipitation. They are used to reconstruct the original anomaly state vector by their respective principal components (PCs or amplitudes $E1(t), E2(t), ...$).

Space and Time Variability (North Atlantic Oscillation): Forty years of NCAR (National Center for Atmospheric Research) re-analyses (1958–1997, representing the synoptic scale with a 250 km resolution) reveal a dominating eigenvector, EOF1, which contributes almost a quarter (24.4%) to the total variance. Its spatial pattern provides a first indication that precipitating frontal systems determine the variability of the North Atlantic/European climate (Fig. 3.4). The following two describe 14.3% and 11.9% of the variance (not shown), so that a hypothetical degeneracy of EOF2 and 3, whose eigenvalues should differ by the estimated standard deviation of either, cannot be excluded (see [3.47]).

EOF1 reveals a marked north-south anomaly pressure difference between Iceland and the Azores/Iberian Peninsula, which is known as the North Atlantic Oscillation (NAO). It represents, for positive NAO index, an enhanced zonal flow across the North Atlantic, which is associated with a deep Iceland low and a strong Azores high generating higher (lower) temperatures and more (less) precipitation over the eastern (western) part of the North Atlantic and northern (southern) parts of the European continent. These situations influence the climate variability between Newfoundland and the Black Sea.

The other EOFs (not shown) modify the EOF1-variability; EOF2, for example, represents situations with enhanced anticyclonic 'Grosswetter' over the European continent, which are associated with a shift of the tail end of the cross Atlantic storm track. In low pressure situations, a negative pressure anomaly extends over the European continent. The related more southward position and zonally oriented cyclone track is associated with frontal systems which lead to enhanced precipitation and positive temperature anomalies in the central and southern parts of Europe.

As the European climate is dominated by the NAO, it is not surprising that suitable NAO circulation indices are subjected to climate change analysis. These are the PCs of the dominating first EOF of the surface pressure or combined fields projecting the vector time series $C_i(\mathrm{t})$ onto the EOF or eigenvectors, or simply the pressure difference between stations on the Azores and on Iceland. About a hundred years of observations show the following results: (i) the first combined EOF indicates a significant bimodality (see [3.16]). (ii) The observed NAO station index varies on interannual up to interdecadal time scales, indicating nonstationarity of the NAO (Fig. 3.5). A standard wavelet analysis exhibits active and passive NAO phases with low and high frequency variability, respectively. The hemispheric circulation during the *active* phase is dominated by the North Atlantic Oscillation, whose index fluctuates with

NCAR 1958-1997 DJF EOF 1   24.4%

Sea Level Pressure

2m-Temperature

Precipitation

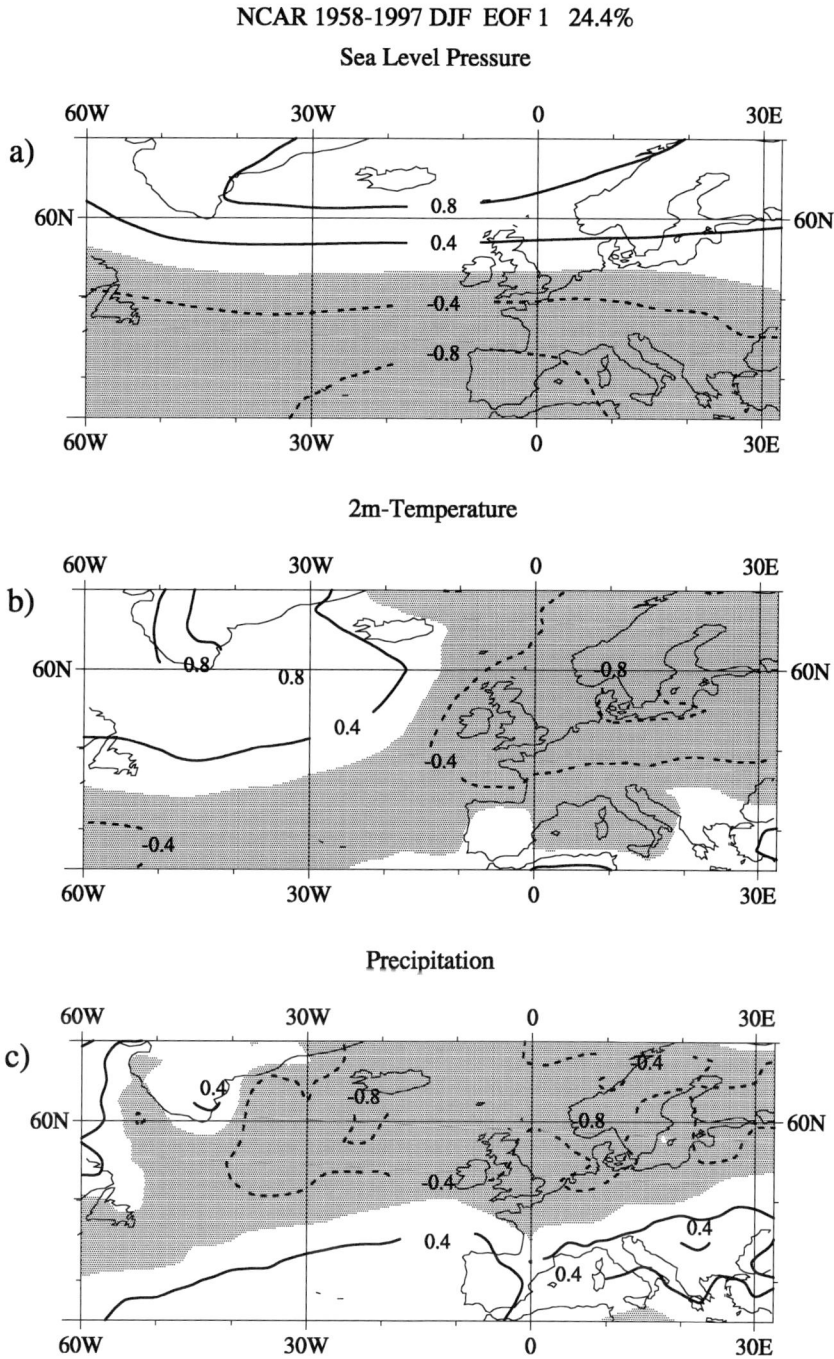

Fig. 3.4. Joint spatial variability of the winter climate in the North Atlantic/European sector (winter seasons 1958–1997): the dominating EOF1 describes the North Atlantic Oscillation (NAO) in terms of (**a**) monthly mean sea level pressure P, (**b**) temperature T and (**c**) precipitation N

periods between 15 and 7 yr, shows enhanced low-frequency variability and large extremes (see [3.1] and, for the 1962–1976 winter months, [3.66]). The *passive* phase reveals a minimum in low-frequency variability characterized by enhanced variance on shorter time scales (about 4 yr period). The hemispheric

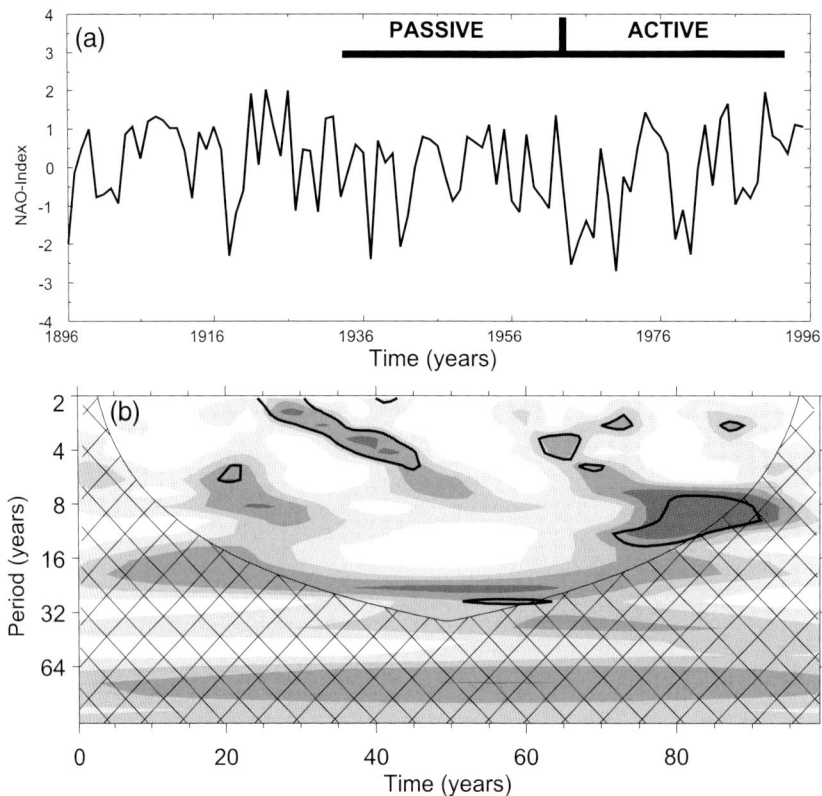

Fig. 3.5. North Atlantic Oscillation index time series: (**a**) winter mean surface pressure differences Azores minus Iceland (from [3.27]), and (**b**) wavelet power spectrum (see also [3.1])

circulation is dominated by the Pacific/North America teleconnection pattern (PNA); it evolves in the tropical/subtropical Pacific and has a strong influence on the southwestern part of the North Atlantic (exciting the Aleutian Low and the North American High).

Extrema: The analysis so far is confined to the first and second moments (means and variances) and needs to be extended to higher moments or, equivalently, to the complete probability distribution, in order to obtain additional information about its intensity fluctuations, in particular extrema or intermittency (and thus the higher moments). They are represented by the tail end of the cumulative distribution. For example, the 5-min rainfall totals, N, of a single station (Potsdam, Germany) reveal the following structure (Fig. 3.6, [3.17]). The linear drop of the tail end of the probability distribution is not approached smoothly from the curve of the 'bulk' of the distribution, but rather abruptly as a break from the rest of the distribution. Quantitatively, this tail, if approximated by a power-law of the type

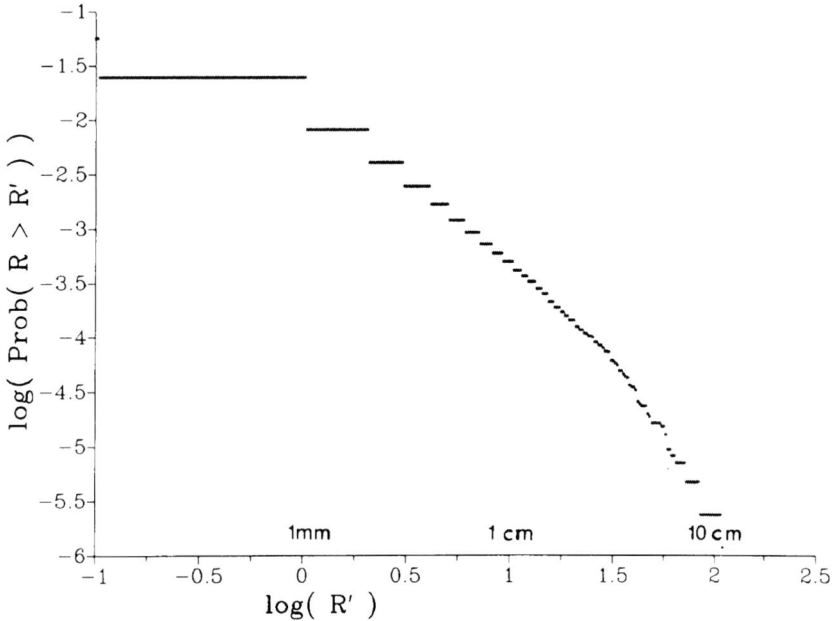

Fig. 3.6. Probability distribution of rainfall in a log-log diagram: single station (Potsdam, Germany) 5-min data, which are distribution averaged over eight summers (after [3.17])

$$F(N) = \text{Prob } \{N > n\} \sim n^{-a}, \tag{3.2}$$

shows a lesser reduction in probability for increasing intensity fluctuations in the 1 mm to 2 cm regime ($a \sim 1.7$), than for intensities > 2 cm ($a \sim 3.0$). That is, the recurrence time (or return period) of the same hypothetically large event is shorter for the smaller power-law slope (and vice versa). For example: an event > 10 cm, which occurs with probability 10^{-6} in the $a \sim 3.0$ regime, would, in the $a \sim 1.7$ regime, occur with probability 10^{-5} and thus return an order of magnitude earlier, a more pessimistic scenario.

A *biased coin flip* model is the basis of this analysis of the probability distribution [3.24]. Here, the continuous rainfall variable is subjected to dichotomy utilizing the occurrence of observations smaller than the fixed rainfall threshold $n : q = 1 - F(n) = \text{Prob } \{N < n\}$. A subsequent Bernoulli experiment determines the first exceedance of the threshold p at the jth trial; its outcome is the geometric distribution, $\omega(j) = Fq^{j-1}$, with mean $\langle j \rangle = \tau = 1/F$, and standard deviation $s = (\tau^2 - \tau)^{1/2}$. Multiplying by the sampling time, the mean, $\langle j \rangle = \tau$, defines the expected return period of (the occurrence of) an intermittent event (that is, threshold exceedance or $N > n$), for which to happen once, $\tau = 1/F$ trials are necessary in the average; if n is the median, then the event $N > n$ returns on the average every second trial, $\tau = 2$; if n is the upper quartile: $\tau = 4$, etc.

3.3.2 Storm Tracks and Cyclone Paths: Eulerian and Lagrangian View

As indicated by the eigenvector analysis, synoptic weather systems form an integral part of the space-time variability of European climate. They undergo a life cycle during which they follow more or less well defined paths known as cyclone or storm tracks; they are associated with rainbearing frontal weather systems affecting regional climates, in particular the water cycle and extreme weather events. The European weather and climate are affected by the variability of the cross Atlantic cyclone track, in particular by its sensitive tail, which is associated with the final stages of a cyclone's life cycle. About 70-80% of the winter precipitation in continental Europe originates from about 15 frontal cyclones based on single station (Berlin) composite analysis [3.19]. These storm tracks can also be influenced by distant atmospheric events, like the ENSO system in the tropical Pacific (see, for example, [3.16]). Furthermore, relatively fast fluctuations in the North Atlantic storm track interact with the oceanic conveyor belt through the freshwater flux (or rainfall) possibly leading to low frequency variability in the ocean and, likewise, in the atmosphere. Space-time variability of atmospheric dynamics is conveniently described by Eulerian statistics of model or observational datasets to comprise the physical information contained in the sequence of weather maps. The height (or geopotential) of a pressure surface is traditionally used to describe mid-latitude synoptic scale systems.

Storm Tracks: The storm track is a common measure for the mean intensity of mid-latitude disturbances. It is defined as a region with enhanced standard deviation of the variability of the band-pass filtered (2.5–6 days) 500 hPa geopotential height to identify regions of strongest baroclinic activity [3.36]. The axis of the storm track is located along the jetstream as determined in the 500 hPa geopotential height. There are two distinct regions recognized over the oceanic basins of the Northern Hemisphere with a magnitude of 60 m (Fig. 3.7a, deduced from the European Center for Medium Range Weather Forecasting (ECMWF) re-analyses; [3.56]): the North Atlantic and the North Pacific. The North Atlantic storm track is considerably stronger and extends from the center of the North American continent to Northern Europe, whose climate is considerably affected by the eddy activity of these synoptic scale systems; the Pacific storm track is restricted to the ocean.

Cyclone Paths (Scaling and Climatology): The cyclone track is the region of an enhanced density of synoptic cyclones. The comparison between storm track and cyclone track reveals some discrepancies: anticyclones are included in the statistics defining a storm track, and its variability is not confined to geopotential height minima [3.65]. For example, the occurrence of minima in

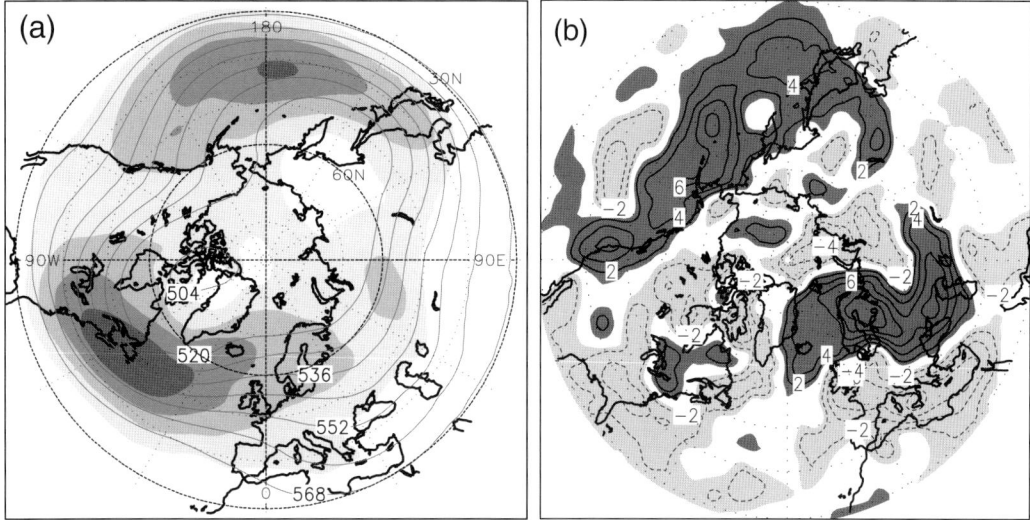

Fig. 3.7. Storm tracks (1979–1997): (**a**) the r.m.s. of the 500 hPa band pass filtered geopotential height anomaly; the climatological distribution of the winter 500 hPa height averages (in gpdm) is also included. (**b**) The decadal trend of the storm track. The contour intervals of the stippled area are (**a**) 10 m, (**b**) 2 m, dark shade denotes positive, light shade negative anomalies (after [3.56])

the superposition of a wave with a zonal flow depends on the flow intensity, whereas the variability depends on the wave only. Cyclone paths are sensitive to the particular tracking method; they are identified subjectively or by a search algorithm (as, for example, developed by Blender et al. [3.7]), the simplest of which consists of two steps: (i) a cyclone is detected as surface pressure low and therefore as a local minimum from the 1 000 hPa reference field in an area covering 3 × 3 grid points. (ii) A sufficient intensity of the low requires a positive mean gradient of the 1 000 hPa height over an area characterized by the Rossby deformation radius (1 000 km). The trajectory $X_j(t)$ of the individual cyclone j is a sequence of cyclone positions $(x_j(t), y_j(t))$ for 6 hr time steps $t = 0, \ldots, T = 3$ days; the high temporal resolution is required to guarantee the cyclone traces of the sequentially identified lows. Cyclone locations are further analyzed relative to their initial position, $dX_j(t) = X_j(t) - X_j(t = 0)$,

$$X_j(t = T) = \Big[x_j(t = 0), y_j(t = 0); \ldots; x_j(T), y_j(T) \Big], \text{ and} \qquad (3.3)$$

$$dX_j(t = T) = \Big[dx_j(t = 0), dy_j(t = 0); \ldots; dx_j(T), dy_j(T) \Big], \qquad (3.4)$$

where (3.3) and (3.4) indicate the cyclone and its relative track. Formally, the trajectory or the relative displacement of an individual cyclone is characterized by a single point in a time-delay coordinate phase space spanned by the consecutive (relative) cyclone positions. Applying cluster analysis to this phase space leads to three dominating centroids of mean relative displacement, which represent Lagrangian-type regimes of stationary, north-eastward and zonally

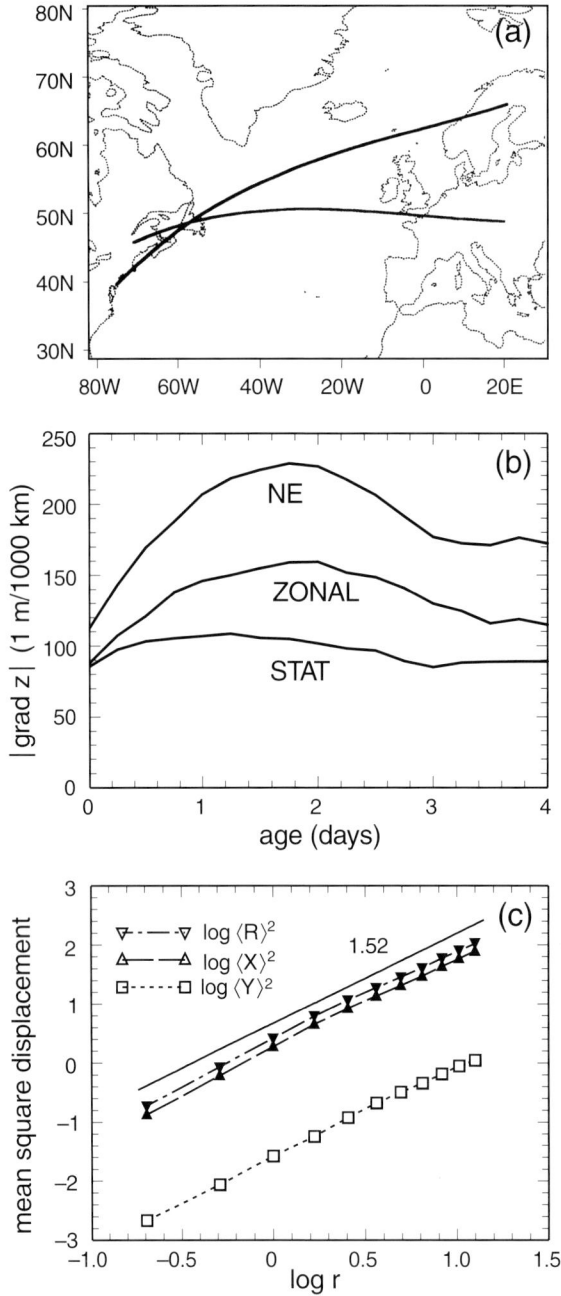

Fig. 3.8. Paths of North Atlantic cyclones: (**a**) mean trace of cyclone positions in the north-eastward and zonal clusters; (**b**) the average life cycle of the 1 000 hPa height gradient, and (**c**) their scaling (after [3.7])

traveling cyclones (see Fig. 3.8). The latter two centroids correspond to the extremes of the NAO which characterizes the Iceland low and Azores high seesaw and is associated with low frequency (decadal) variability (see Sect. 3.3.1). An ensemble averaged ($\langle \ldots \rangle$) position of the propagating cyclone clusters and the life-cycle of the horizontal pressure gradient are shown in Fig. 3.8a, b. *Cyclone track scaling* or, more generally, scaling in ocean and atmosphere dynamics, originates from diffusion processes. In this sense Fraedrich and Leslie [3.20] adopt Richardson's experiment [3.49] and analyze tropical and mid-latitude cyclones (for a review see [3.61]). Treating traveling cyclones as large scale diffusive elements, they obey a fractional power-law scaling of the time behavior of the mean square displacements, $dX^2(t) = dx^2(t) + dy^2(t)$ (normalized by standard deviations). This leads to the structure function which, for stationarity, corresponds to a power-law spectrum:

$$\langle dx^2(t) + dy^2(t) \rangle \sim t^q. \tag{3.5}$$

For geophysical flows $1 < q < 2$ is observed. The mean square displacement is calculated for the whole set of cyclones and for the different clusters. Figure 3.8c shows a power-law scaling of the total, zonal, and meridional displacements of the cyclones in the North Atlantic/European sector in log-log format: $\langle dx^2 + dy^2 \rangle$, $\langle x^2 \rangle$, $\langle y^2 \rangle$ for the upper, middle, and lower curves. The average of all cyclones yields $q \sim 1.52$, (the north-eastward $q \sim 1.67$, zonal $q \sim 1.59$). The stationary cyclones follow purely diffusive behavior with $q = 1$. The results for the zonal and north-eastward propagating cyclones can be compared with $q \sim 1.67$ of a simple point vortex model for the mid-latitude circulation [3.63]; ocean drifter trajectories follow $q \sim 1.5$ [3.51]. This behavior, observed in many geophysical flows, characterizes a trajectory of fractal or self-similar dimension $d = 2/q$; that is, cyclone motion is neither purely linear ($q = 2$) nor diffusive ($q = 1$).

A *cyclone climatology* is provided by cyclone densities following Köppen's [3.33] first analysis of North Atlantic/European cyclone paths. The 1979–1997 cyclone density (Fig. 3.9a, see also [3.56]) shows the temporal occupation of cyclones per $1\,000\,\mathrm{km}^2$. The cyclone trajectories are mainly located in the storm track area; they originate over the western parts of the oceans and travel eastward bending northward until the end of their life cycles. The cyclone density reaches 0.1 in the storm track areas. Similar to the storm track, the cyclone density over the North Atlantic extends westwards onto the American continent. Distinct density maxima within the storm tracks are found southeast of Greenland in the Denmark Strait, in the Barents Sea, the Canadian Sea, and the Gulf of Alaska. Further maxima are detected in the Mediterranean Sea. Cyclogenesis is located in the baroclinic zones at the western or upstream parts of the storm tracks. While the North Pacific cyclones arise over sea, North Atlantic cyclones originate partly over the North American continent; the Mediterranean and the Caspian Sea are other regions of cyclogenesis. There are areas of secondary cyclogenesis east of Greenland and over northern Europe. Cyclolysis occurs in the north-eastern end of the storm tracks with pronounced maxima in the Gulf of Alaska and the Denmark Strait. Secondary cyclones over northern Europe end in the Arctic sea while the Mediterranean cyclones end in south-eastern Europe.

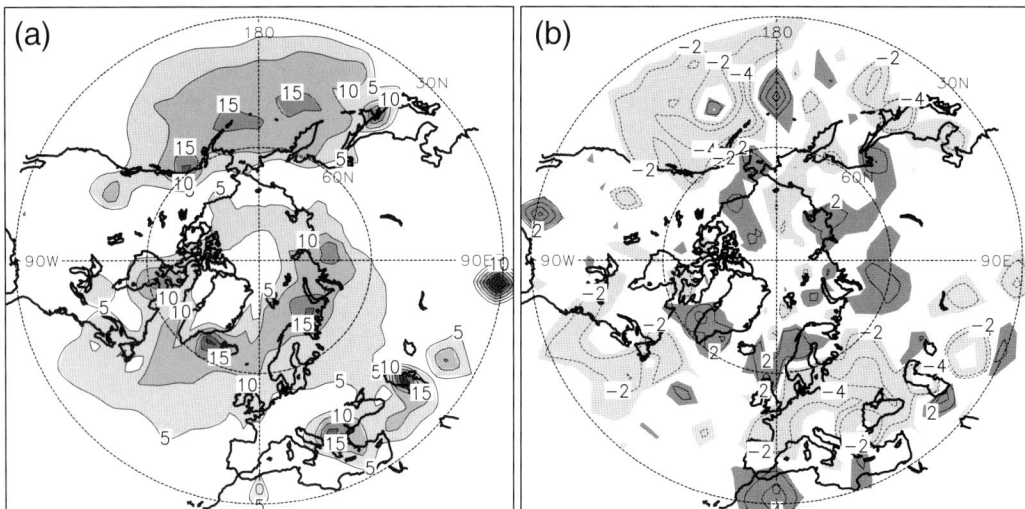

Fig. 3.9. Cyclone density (1979–1997): (**a**) occurrence and (**b**) decadal trend. The density gives the ratio (times 100) of an occupation of $1\,000\,\mathrm{km}^2$ by cyclones at an observation time. Contour intervals are 5% in (**a**), 2% in (**b**). The trend in (**b**) is positive for dark shade, negative for light shade (after [3.56])

3.3.3 Grosswetterlagen and Climate Zones

Early analyses of weather and climate were based on phenomenological classifications of the underlying synoptic scale processes and of the climatic zones realizing that large scale patterns of recurrent weather episodes or of vegetation-climate coherences are fundamental properties of the atmosphere and the climate system. Before utilizing these phenomenological datasets (types of Grosswetterlagen or climate zones) for further statistical analysis estimates of the appropriate number of spatial degrees of freedom are required. This is necessary, because the number of independent and distinct patterns (types, modes, or regimes) often obtained from practical experience may not be supported by the underlying datasets representing the spatial fields (daily surface pressure maps for Grosswetterlagen, monthly climate mean temperatures and precipitation for climate classifications; see also Sect. 3.2 on representativeness).

Degrees of Freedom (Spatial Embedding): Degrees of freedom (DOF) of a system are estimated from a random sample of state vectors of an M-dimensional space. That is, T observations at M grid points are stored in the $M \times T$ data matrix, $X_i(t_j), i = 1, \ldots, M; j = 1, \ldots, T$. Now, DOF-estimates are obtained by fitting the computed distribution of the standardized χ^2-state variable, $\chi^2_M = \sum_{i=1}^{M}(X_i)^2/M$, to the theoretical χ^2-distribution $((\chi_{\mathrm{DOF}})^2 = \chi^2/\mathrm{DOF}$, see [3.60]). Using the computed variance $\mathrm{var}\{\sum_{i=1}^{N} X_i^2\}$ as a sample estimate, the equality $\mathrm{var}(\chi^2_M) = \mathrm{var}(\chi^2/\mathrm{DOF})$, and chi-squared distributed variables with mean M and variance $2M$, the degrees of freedom are

$$\mathrm{DOF} = 2M^2/\mathrm{var}\sum_{i=1}^{N} X_i^2 \,. \tag{3.6}$$

The variance and, therefore, the DOF-estimate can be improved by transformation of the original state variables, $\mathbf{X}(t) = S\Lambda^{1/2}\xi(t)$, to M new independent variables $\xi(t) = \Lambda^{1/2}S^{-1}\mathbf{X}(t)$. They are the projections of the original data sample onto the orthogonal eigenvectors of the symmetric $M \times M$ correlation matrix $C = S\Lambda S^{-1} = \langle \mathbf{X}^* \mathbf{X}\rangle$, determined by singular value decomposition of the original $M \times T$ dataset $\mathbf{X}(t)$. The matrix Λ contains M eigenvalues λ_i along the main diagonal; the orthogonal matrix S is composed by the eigenvectors of C with $S^{-1} = S^*$ (superscript * denotes the transpose operator). Some algebra finally leads to $\mathrm{DOF} = M^2/\sum_{i=1}^{M}\lambda_i^2$. The relation $M = \mathrm{trace}\,C = \sum_{i=1}^{M}\lambda_i$ shows that the degrees of freedom and the equivalent number of statistically independent observations introduced by Megreditchian [3.39] are identical. In this sense the following DOF-relations can be deduced for the correlation (or covariance) matrices, C (or K), and their respective eigenvalues (λ_c or λ_k); the total variance is $\langle X^2\rangle = \sum_{k=1}^{M}\lambda_k$. With $(\mathrm{trace}\,C)^2 = (\sum_{c=1}^{M}\lambda_c)^2$ and $\mathrm{trace}\,(C^2) = \sum_{c=1}^{M}\lambda_c^2$ one obtains the degrees of freedom and the confidence limits associated with the correlation (or covariance) matrix C (or K):

$$\mathrm{DOF}(C) = (\mathrm{trace}\,C)^2/\mathrm{trace}\,(C^2)\,, \tag{3.7}$$

$$\pm\Delta\mathrm{DOF} = \pm\frac{\partial\mathrm{DOF}}{\partial\lambda_m}\Delta\lambda_m = \pm4\mathrm{DOF}\left(\frac{2}{T'}\right)^{1/2}. \tag{3.8}$$

Estimates of DOF-confidence limits can be related to those of the eigenvalues λ_m. The rule of thumb derived by [3.47], $\Delta\lambda_m = \pm2\lambda_m(2/T')^{1/2}$, depends on the number T' of independent realizations. Error propagation leads to the DOF confidence limits. That is, $T' = 1\,000$ independent realizations, corresponding to 10 seasons of daily data with an integral time scale of $\tau = 3$-4 days, lead to confidence limits $\mathrm{DOF} \pm \Delta\mathrm{DOF} = \mathrm{DOF}\{1 \pm 4(2/T')^{1/2}\}$ of about 18%. Two applications are presented following Fraedrich et al. [3.21].

Climate Zones: Climate classifications follow by two basic principles: *genetic* classifications consider the influence of the general circulation of the atmosphere (see, for example, [3.11, 3.29]), the surface energy fluxes, and the tropospheric air masses on climate; further differentiation is possible when utilizing more information: frequency and tracks of cyclones and anticyclones, the intensity and location of quasi-stationary upper troughs, frontal passages, position up- and downstream of mountainous terrain, of coasts, the soil properties which regulate evaporation, albedo etc. *Effective* classifications describe regional climate states by observables and their link with flora, fauna, soil, agricultural use etc. The primary information (or climate elements) on which the effective schemes are based are long-term monthly means of temperature and precipitation; threshold values, connected to the vegetation growth, for example, are also included. Such classifications appear to be most suitable for practical applications. A well-known effective scheme has been introduced by Köppen [3.32] which comprises a spatial analysis of the global climate of the continents. Utilizing monthly ensemble means of temperature and precipitation, the Köppen climate regimes are identified by linking maps of vegetation to the climate state variables which leads to world wide main climate zones. With modifications they characterize the tropical (A), dry (B), subtropical (C), temperate (D), boreal (E), snow (F) climates which are, in parts, subdivided further into wet, summer- and winter-dry, or monsoon-type, oceanic or continental (r, s, w, m, o, c), or tundra and perpetual frost (T, I). The success of this classification lies in its simplicity and its relation to the biosphere; it may be considered as the first step towards a biome oriented analysis of the Earth's climate.

The *degrees of freedom* are estimated prior to an application of a classification scheme using monthly means (NCAR, 1958–1997) of the near surface temperature and precipitation which, for the European continent, gives DOF = 3. This corresponds with the observed number of climate types employing the modified Köppen scheme: subtropical (C), temperate (D), and boreal (E); the snow area (F) covers only a small portion of the continent while the dry climate (B) occurs at the south-eastern boundaries in and near Asia. Thus the classification scheme appears to be a useful analysis tool of regional climates and their change analysing the period 1901–1995 (Climate Research Unit, University of East Anglia, Norwich [3.46], $0.5° \times 0.5°$ grid resolution): (i) The area cover of the dominating climate zones (Fig. 3.0, in million km^2) is for the subtropical zone (C = 1.5), temperate (D = 8.4), boreal (E = 2.6); snow (F = 0.3) and dry zone (B ~ 1.4). (ii) A significant correlation between NAO fluctuations and the European climate zones can only be determined for the area covered by boreal Eo climate (Fig. 3.10), applying a 15 yr averaging window to both the annual NAO index and the surface climate variables entering the classification scheme.

Grosswetterlagen: Grosswetterlagen classifications are introduced to describe the large scale circulation from hemispheric down to regional scales utilizing the evolution of surface pressure fields as the primary information source: (a) the whole Northern Hemisphere extratropics (polewards of $30°$ N [3.9]) is the largest area classified by circulation types (46 including an indeterminable one), which are effectively reduced to 36 individual in summer or winter seasons. These types characterize the hemispheric movements of cyclones and

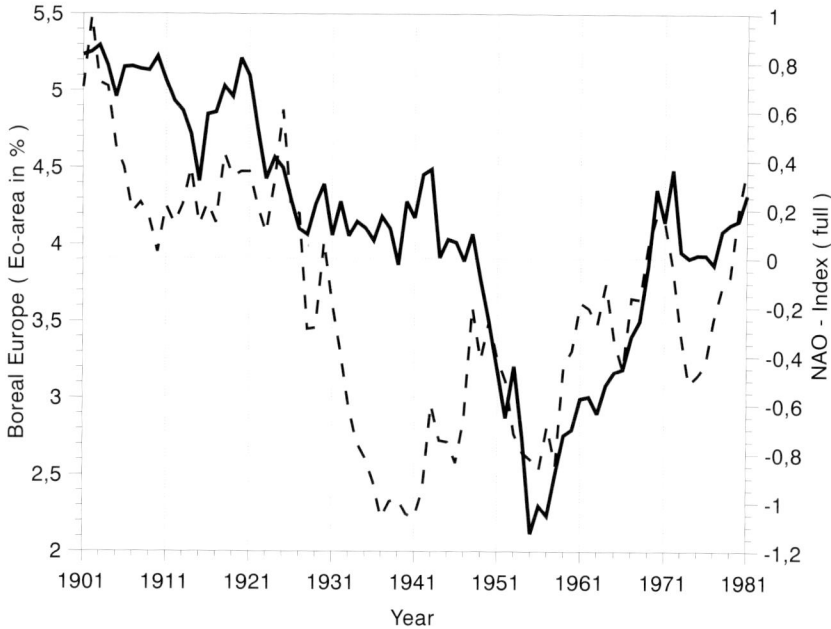

Fig. 3.10. The 1901–1995 time series of the North Atlantic Oscillation index (full line) and the relative area occupied by the boreal European Eo climate (dashed)

anticyclones which are combined to form 13 elementary circulation patterns which, in turn can be categorized into four principal groups (zonal mean, zonally asymmetric, meridional, and mixed). (b) The Grosswetterlagen catalogue characterizes the circulation of the eastern North Atlantic/European sector ([3.22, 3.26]; Fig. 3.11a) by centers of action leading to 29 (plus one indeterminable) weather types; they are arranged to 10 major classes which can be combined to three basic large scale circulation regimes (zonal, meridional, and mixed). (c) On a smaller scale (from 50° to 60° N and from 10° W to 2° E) the British Isles weather is based on seven circulation types ([3.35], see also [3.31]). Other regional classification schemes are also known [3.3]. All catalogues provide long time series of large scale circulation systems identified by often subjective 'pattern recognition' applied to the daily surface pressure fields. They have been used, for example, to analyze regime transition probabilities [3.59], to identify far distant teleconnections like the possible link between the ENSO in the tropical Pacific and Europe [3.13, 3.15, 3.70], and explain calendar singularities [3.5] and climate trends [3.69].

A *degree of freedom* analysis is performed prior to a statistical evaluation of North Atlantic/European Grosswetter, utilizing the daily surface pressure field. The data are the geopotential heights of the 1 000 hPa pressure level (ECMWF 1980–1989 with 500 km resolution) which characterize the large scale circulation. The following results are noted (see also [3.21]): (i) The *Northern Hemisphere* (from 30° to 75° latitude) represented by $M \sim 510$ grid points, yields $DOF \sim 30$ for unfiltered, $DOF \sim 40$ for band-pass, and $DOF \sim 15$–20 for low-

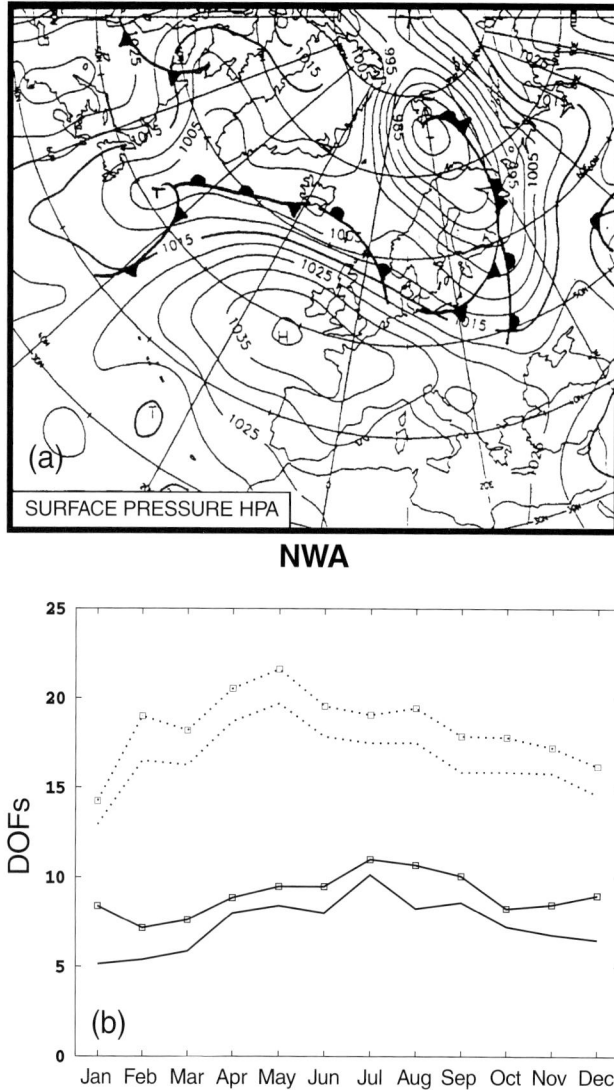

Fig. 3.11. Circulation regimes in the North Atlantic European sector: (**a**) an example from the Hess-Brezowsky Grosswetterlagen catalogue [3.26]. (**b**) Estimates of degrees of freedom based on observed anomalies from monthly (squares) and climate means (no squares) using low-pass (full line) and band-pass (dotted line) filtered data (from [3.21])

pass filtered data. This corresponds well with the Dzerdzeevskii catalogue identifying 36 different large scale circulation patterns, which are suitably combined to 13 weather regimes. General circulation model simulations with T-21 resolution reveal systematically larger values (plus 5). It appears that the model (although of limited resolution) is not able to generate the sufficiently small number of dominating and active modes which is necessary to adequately simulate the atmosphere's large scale dynamics. (ii) The *North Atlantic/European Sector* with $M \sim 160$ grid points gives, for the unfiltered data, DOF ~ 10–12; DOF ~ 7–10 are obtained for low-pass (10–90 days) and DOF ~ 15–20 for band-pass (2.5–6 days) filtered data (Fig. 3.11b). These DOF-estimates can be associated with the number of distinct large scale circulation patterns. Hess and Brezowsky define 36 Grosswetterlagen for the Atlantic/European sector [3.26]. This number reduces to 10 Grosswetter-types which corresponds with the low-pass filtered DOF-estimates. The large number of 36 Grosswetterlagen, however, appears to be too large and does not even compare with the band-pass filtered DOFs. In summarizing, the spatial DOFs indicate that the dataset analyzed does not allow a resolution of as many details as the large number of phenomenological Grosswetterlagen indicates. Thus, the number of (independent) Grosswetterlagen being substantially larger than the spatial degrees of freedom of the underlying dataset requires a considerable reduction of that number to a few regimes, if further statistical analysis is to be employed. Therefore, it is not surprising that analyses of singularities, climate, and climate change (as shown in the following) are based on a few regimes only, which comprise dynamically similar Grosswetter types.

The basic *time statistics* of the Grosswetter types describe the North Atlantic/European climate in terms of its synoptic genesis: (i) three meteorologically distinct large scale circulation regimes are selected. They show that the zonal (or westerly, 31%) and meridional (35%) flow patterns determine Europe's winter climate, while a set of mixed circulation patterns (34%) can be interpreted as a transition regime with a smaller mean residence time of five days compared to the six and seven day values attained by the zonal and meridional regimes. This interpretation is further supported by the mean first passage time describing regime alternation: 11 or 17 days from meridional to zonal or vice versa, while the transitory mixed state is occupied for about three or five days during the regime passage (see also [3.59]). (ii) NAO fluctuations and the North Atlantic/European Grosswetter (in particular the zonal regime) are closely related. Discarding the common Eulerian frame (of station data), Glowienka introduced a dynamically more relevant Lagrangian NAO index attributed to the changing positions of the mean Icelandic Low and Azores High [3.23, 3.48]. In this sense, the flow's seasonal mean occupation times in the zonal (or westerly) Grosswetter state appear to be a more suitable measure of the fluctuating North Atlantic Oscillation than the conventional Eulerian index measure (see Sect. 3.4.1 and Fig. 3.14).

3.3.4 Singularities

The annual cycle of, for example, single station surface air temperature at a time resolution of one or a few days (see Fig. 3.12) is not as smooth as may be expected for astronomical reasons (annual cycle of insolation). Particular deviations towards warmer or cooler conditions appear, even after ensemble averaging for many years. These phenomena of more or less regular temperature deviations from the smooth mean annual cycle are called 'singularities'

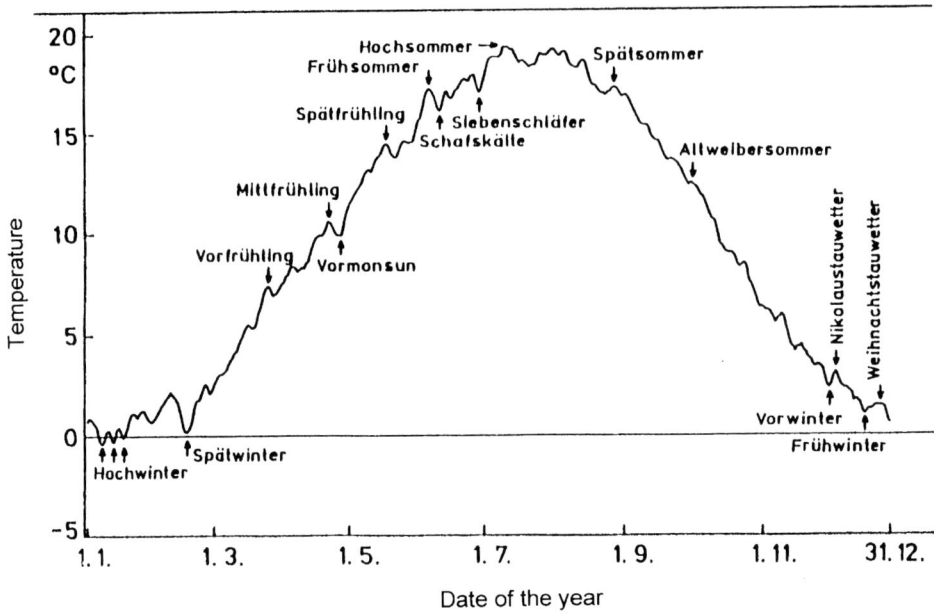

Fig. 3.12. Ensemble mean annual cycle of surface air temperature (1949–1985, three-day averages) at Frankfurt/Main; arrows indicate 'singularities' with their German names (from [3.5])

(or calendar singularities). Although related to a corresponding behavior of the 'Grosswetterlagen' (regimes), there is no steady relation to precipitation. Four types of singularities can be found associated with the following regime change: warm-moist (WM), warm-dry (WD), cold-moist (CM), and cold-dry (CD).

In continental Europe, some of these singularities have been known by experience for a long time, especially in the context of agriculture ('Bauernregeln'); some of them originated in the Roman Empire about two thousand years ago. Traditional German names of singularities are very popular, for instance 'Eisheilige' for a CD singularity in May, 'Schafskälte' for a CM singularity in June or 'Altweibersommer' for a WD singularity near the end of September (corresponding to the 'Indian summer' in the USA). Table 3.2 lists some prominent singularities, and their mean calender dates of occurrence and their frequencies. As every climatic phenomenon, also singularities change in time. For example, Table 3.2 does not indicate the CD singularity 'Eisheilige', because a statistical analysis shows that, nowadays, this singularity very rarely occurs and is replaced by a WD singularity around 7–18 May. In contrast to that, based on the 1891–1925 observation period, the CD singularity 'Eisheilige' appeared around 6–20 May with a frequency of 97%. Such a change is connected with a regime change of the observed frequency of 'Grosswetterlagen'.

Table 3.2. Type (WM = warm/moist, WD = warm/dry, CM = cold/moist, SD = cold/dry), mean calender data of occurrence and frequency (temperature-related, if not applicable precipitation-related in parentheses) of some Central European 'singularities' (simplified from [3.5], observation period 1946–1986, *not significant)

Type	Date	Freq. [%]	Type	Date	Freq. [%]	Type	Date	Freq. [%]
CD	4.–9.1.	83	CN	24.4.–2.5.	84	WD	25.–26.9.	(76)
CD	13.–14.1.	*	WD	7.–18.5.	95	WD	3.–10.10.	(73)
CD	17.–20.1.	58	CM	20.–31.5.	70	CD	13.–16.10.	87
WM	22.–30.1.	49	WD	2.–8.6.	84	WM	23.–25.10.	*
WM	3.–12.2.	60	CM	10.–12.6.	81	CD	28.10.–1.11.	(76)
CD	14.–20.2.	60	CM	6.–1.7.	70	WM	9.–11.11.	49
CD	27.2.	49	WD	3.–14.7.	89	CD	16.–18.11.	62
CD	7.3.	(76)	CM	16.–24.7.	60	CD	25.–26.11.	87
CD	11.–14.3.	*	WD	28.7.–7.8.	84	WM	27.–28.11.	92
CD	18.–20.3.	(95)	WD	13.–16.8.	*	CD	30.11.–2.12.	(76)
WD	23.–27.3.	62	CM	17.–24.8.	(60)	WM	4.–5.12.	56
WD	3.–4.4.	97	WD	29.8.–5.9.	62	CD	17.–21.12.	53
CD	7.–12.4.	60	WD	11.–12.9.	(81)	WM	24.–29.12.	53
WD	17.–22.4.	73	CM	19.9.	95	CD	4.–9.1.	83

3.4 Climate Trends: Europe at the End of the Twentieth Century

The increase of the global mean temperature observed in the last 100 years raises two questions [3.30]: can signals of climate change be detected in the North Atlantic/European sector and does this change exceed that of the natural variability (cf. Chap. 5)? A systematic treatment of climate change requires identification of climate variables or elements which characterize climate states and circulation patterns changing its long-term variability. Dynamical processes which are possibly associated with climate change, and indicators, which point to abrupt changes of the atmospheric circulation, need to be found, analyzed, and tested. For example, the ice cores in Central Greenland reveal temperature changes of about seven degrees within a few decades [3.10]. Although such abrupt temperature changes have not been observed since the Pleistocene-Holocene transition (after the Younger Dryas), appropriate climatological parameters incorporating the atmosphere's complexity may show such a behavior. For example, the North Atlantic Oscillation (see Fig. 3.5, [3.27]) reveals decadal fluctuations affecting weather and climate in Europe [3.68]. The following sections present analyses of climate trends in Europe at the end of the twentieth century utilizing a set of climate elements and indicators which characterize the atmospheric circulation dynamics in a statistically comprehensive manner: trends of the averages (first moments, Sect. 3.4.1) and of the variability (higher moments, Sect. 3.4.2) are followed by an introduction to the empirical analysis of the external forcing mechanisms.

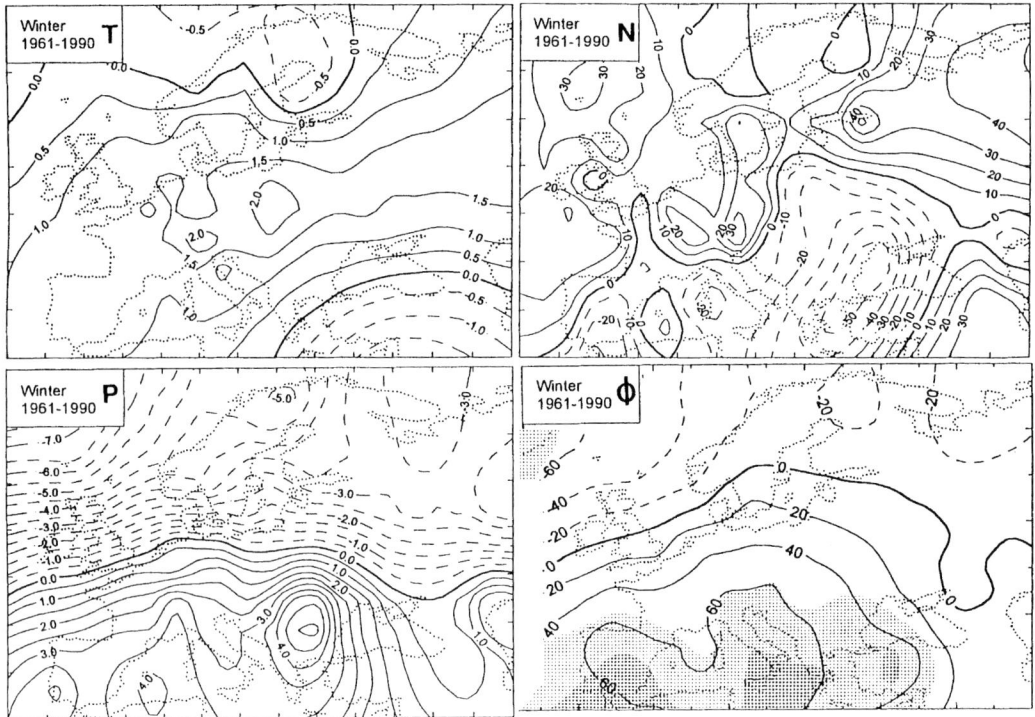

Fig. 3.13. Linear trend patterns (1961–1990) in Europe, T = surface air temperature in K, N = precipitation in percent, P = mean sea level pressure in hPa, and Φ = 500 hPa geopotential height in gpm (modified from [3.53])

3.4.1 Trend Patterns of Climate

The climate trend patterns 1961–1990 depending on both season and observation period are indicated concerning changes of the mean fields of the climate elements, surface air temperature, precipitation, mean sea level (MSL) pressure, and 500 hPa geopotential height in Europe [3.53]. Figure 3.13 shows that, except for northern Scandinavia and the SE Mediterranean area, all regions have experienced warming with maximum values up to 2 K in Central Europe. The precipitation trend pattern is more complicated; the precipitation increase in most areas of west, central and eastern Europe, which is contrasted by a decrease in the Mediterranean area. Note that the most pronounced precipitation decrease (up to 50%) coincides with the maximum of MSL pressure increase. The pressure fields both in MSL and 500 hPa show zonally oriented change patterns with a decrease in the North and an increase in the South (in coincidence with NAO behavior). On a secular time scale the trend patterns for temperature remain similar but not for precipitation and pressure (for details, including confidence tests, see [3.53]).

3.4.2 Storm Tracks and Cyclone Paths;
Grosswetterlagen and Climate Zones

Trends of means (that is, monthly, seasonal, or annual averages) need to be supplemented by trends of the intrinsic climate variability (second or higher moments) to characterize changes in storm tracks, cyclone densities, residence times of circulation regimes etc., which will be discussed in the following. Note that the effect of the recent changes of the North Atlantic Oscillation (Figs. 3.5, 3.10) is also reflected in the trends of storm tracks or cyclone paths, the changing Grosswetterlagen statistics, and the shifts of climate zones (see Sect. 3.3.3).

Trends of Storm Tracks and Cyclone Paths: The storm tracks trends are determined by the slope of a linear regression normalized per decade during 1979–1997. The North Atlantic storm track shifts to the Northeast, and, over Europe, the variability is displaced eastwards (Fig. 3.7b). The trend of the cyclone density (Fig. 3.9b) shows a north-eastward shift with a decrease in the western Atlantic and an increase over Scandinavia, which is very similar to the storm track trend (Fig. 3.7b). The reduction of the storm track over central Europe can be found in the trend of the cyclone density. In eastern Europe, however, the enhancement of the storm track is not due to cyclone density, which shows a distinct decrease in this area (see change of climate zones Fig. 3.9b). Note that, simultaneously with the north-eastward shift of the cyclone density, the numbers of short-lived cyclones (without minimal lifetime, hence including lows which exist only one time step) and the cyclones with at least 3 days lifetime decrease both with a rate of about two per decade [3 56].

Changing Grosswetterlagen Statistics (Frequency and Residence Time): The variability associated with the climate of the North Atlantic/European sector is suitably described in terms of large scale circulation regimes. Characterizing the underlying dynamics requires appropriate measures of the variability which are useful climate signals to identify trends and abrupt changes. Here, the statistics of the occupation time (frequency) or residence time (duration) will be applied, first to two univariate regimes (the cold- and the warm-type regimes) and then to a binary process (the zonal flow regime and its complement comprising all remaining other Grosswetter types). A trend analysis of the *frequency or occupation time* of warm- and cold-type Grosswetterlagen is summarized in Table 3.3. It reveals that in summer and winter a frequency increase of the warm regime is observed, corresponding to a decrease of the cold regime. This change is very pronounced in summer (47% increase of warm-type Grosswetterlagen).

Residence times characterize the persistence of circulation regimes of, for example, the zonal Grosswetter, which is closely associated with the cross At-

Table 3.3. Frequency of relatively warm- and cold-type European Grosswetterlagen (W or C) in summer and winter (from [3.28])

Period	Summer, W [%]	Summer, C [%]	Winter, W [%]	Winter, C [%]
1901–1930	31.6	51.0	39.3	27.1
1931–1960	40.4	49.3	35.8	33.3
1961–1990	46.7	42.3	43.1	31.2

lantic storm track (see Figs. 3.4, 3.7, 3.9) and, due the existance of the Gulf stream, determines the temperate climate of the European continent (see Fig. 3.0). Confining the dynamics to a dichotomous (or binary) process of two circulation states, the zonal regime $Z(t)$ and its complement (no Z), decadal mean residence times of the state Z can be estimated; it shows an abrupt increase near the beginning of the 1970s (Fig. 3.14a, b). An outlier test (Thompson rule, see [3.45]) of this climate signal (decadal winter mean duration of the zonal regimes) identifies the 1981–1990 decade as the onset of climate change in the North Atlantic/European sector; the trend commences in the beginning of the 1970s. Beyond that decade the observed peaks do not belong to the previous centennial sample (1881–1980) on the 99% significance level (denoted by triangles). The observed climate change in the North Atlantic/European sector can be attributed to the surface temperature anomalies of the Earth's continents and to the increasing intensity of the North Atlantic Oscillation (see Fig. 3.5). Their decadal sequences from 1901 to 1997 show a simultaneous rise from the 1971–1980 decade onwards, which is another indication of a global climate mode. It describes the 'warm continent-cold ocean' state (and vice versa) dominating the end of this century. Other parameters (ENSO variability, North Atlantic sea surface temperatures, Northern Hemisphere temperatures) have been excluded to be significantly related to the observed climate change in Europe (see [3.69] for more details). Applying the same analysis to the Lamb circulation patterns, that is, the type 16 (west) plus 26 (cyclonal west), leads to the same results (not shown). Another support comes from an analysis of Central European Grosswetter, precipitation, and temperature [3.2]. Furthermore, the mean occupation times of the circulation states reveal an analogous time evolution but there are no statistically safe outliers on the required level of significance.

Shift of Climate Zones: The analysis of the climate classification scheme (from 1901 to 1995, see Sect. 3.3.3) is extended to the changes occurring near the end of this century. Shifts observed during the last 15 yr are presented in Fig. 3.0 (covering a 30 year span by the sliding 15 yr window [3.41]). It appears that Europe realises the largest change of climate zones (relative to its area) compared with all other continents. Here it is the maritime temperate climate

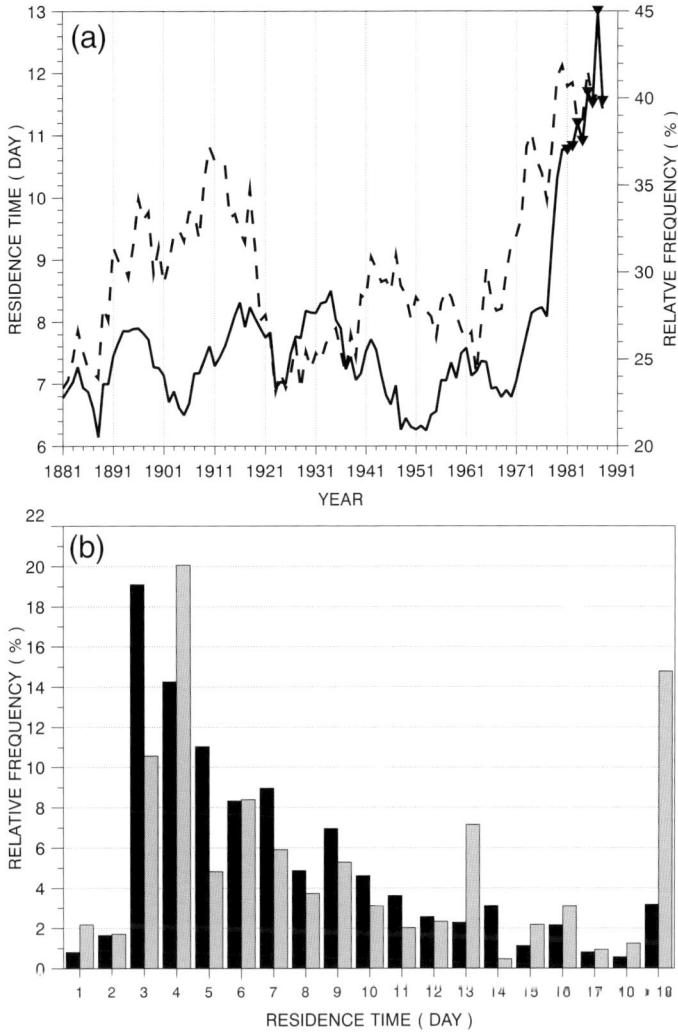

Fig. 3.14. (a) Time series of the decadal means of the residence times (full line, left axis) and occupation time (dashed, right axis) of the zonal (or westerly) Grosswetter state in winter (1881/82–1997/98); triangles define statistically safe outliers ($\beta = 0.05$); (b) residence time distributions (relative frequencies) for periods 1881 to 1970 (black columns) and 1971 to 1997 (gray; from [3.69])

(Do) which increases its area by $289 \times 10^3 \, \mathrm{km}^2$ replacing mainly the continental climate (Dc), which loses $222 \times 10^3 \, \mathrm{km}^2$.

3.4.3 External Forcing: Natural and Anthropogenic

External forcing mechanisms of the climate system are defined to operate without direct interactions. They can be identified in terms of radiation anomalies (radiative forcing) within the atmosphere surface subsystem leading to the corresponding temperature effects as they are simulated, for example, by energy balance models. However, they are also involved with indirect effects, often due to modifications caused by a change of the atmospheric composition (cloud effects) and circulation. This indirect forcing, which involves the response of all climate elements, is very uncertain and a matter of continuing research.

Causes: For example, explosive *volcanism* effects climate anomalies of a few years but climate change does not trigger volcanic eruptions (as far as we know). The climatic effect is that the additional volcanogenic aerosol layer which has a typical residence time of 1–3 yr after eruption, absorbs and scatters some part of the solar irradiation, leading to a warming of the stratosphere, whereas the insolation transmission into the lower part of the atmosphere (troposphere) is decreased leading to cooling effects near the Earth's surface. This radiation forcing which is dominated by sulfate particles can be quantified. Following [3.40] the tropospheric radiation forcing was $2.4\,\mathrm{W\,m^{-2}}$ in the year of the Pinatubo eruption (1991), $3.2\,\mathrm{W\,m^{-2}}$ one year later (maximum effect) and $0.9\,\mathrm{W\,m^{-2}}$ two years later. For earlier decades without direct observation of the stratosphere such assessments are not possible. However, some authors have tried to evaluate annual explosive volcanism parameter time series where this parameter is based on historical volcano chronologies (such as by [3.57]. One of the most recent assessments used in the following is from [3.25]) where both sulfate particle building via gas-to-particle conversion and sedimentation are taken into account.

Solar forcing of climate is a matter of controversial discussion for a long time. On a decadal to secular time scale the question has to be answered whether solar activity by solar flares etc. and indicated by sunspots has a significant influence on climate or not. Based on a number of recent publications [3.30] states a fluctuative forcing of $0.1\text{–}0.5\,\mathrm{W\,m^{-2}}$ where the related time series mostly are closely related to sunspot relative numbers (SRN, see [3.12]). So SRN time series may be used as a solar forcing proxy. Alternatively, solar forcing may have also contributed to a secular trend [3.37], however, at a similar magnitude as quantified above. We use this assumption, too. All other solar hypotheses, e.g. concerning an influence of the length of the solar cycle or solar diameter variations seem to be speculative [3.55].

Anthropogenic sources provide the most important secular forcing of global relevance. It is due to the emission of greenhouse gases (GHG: CO_2, CH_4, CFCs, N_2O, tropospheric O_3 etc.) in the context of human activities like energy use, traffic, deforestation, agriculture, and some other. This problem was described and discussed at length by [3.30] (see also Chap. 4 in this book). The most important point is that the corresponding atmospheric concentration increase since approximately 1850 (not only due to CO_2 but so-called CO_2 equivalents) is quantified as a $2.1\text{–}2.8\,\mathrm{W\,m^{-2}}$ radiative forcing and that in the case of a CO_2 doubling ($4.4\,\mathrm{W\,m^{-2}}$ forcing) a global mean surface air temperature increase of $2.1\text{–}4.6\,\mathrm{K}$ (equilibrium response) is simulated by coupled atmosphere ocean general circulation models (AOGCM, [3.30]). The same models attribute to this forcing so far roughly a $1\,\mathrm{K}$ temperature increase (transient

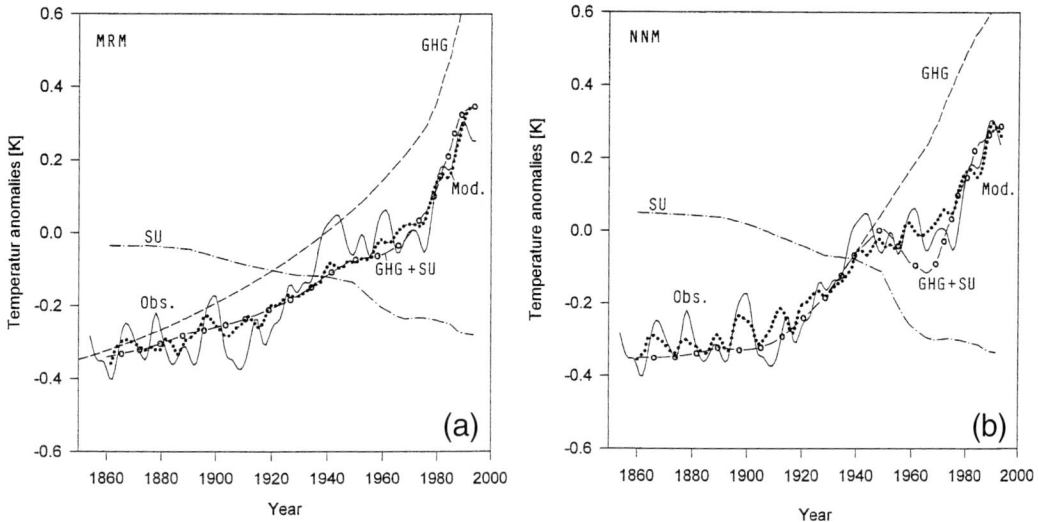

Fig. 3.15. (**a**) Observed global mean surface air temperature anomalies 1854–1994, 10 yr low-pass filtered (data from [3.30]), solid line, reproduction by means of a multiple regression model (MRM) using greenhouse gas (GHG), sulfate aerosol (SU, both anthropogenic) such as volcanic, solar, and ENSO forcing, dotted line, in addition related GHG (dashed line) and SU (dashed-dotted line) signal time series (from [3.52]). (**b**) As in (**a**) but neural network model (NNM, backpropagation) simulation

response, i.e. time lags implied). A second anthropogenic but negative forcing is due to tropospheric sulfate aerosol particles (SU) which may have led to a cooling of approximately 0.4–0.6 K during the same industrial time and as simulated by AOGCM as well [3.30]. Figure 3.15 shows the time series of all natural and anthropogenic forcing mechanisms mentioned above and related to in the following (including ENSO using the Southern Oscillation Index SOI).

Effects: Statistical techniques are used to check how much of the observed surface air temperature variance can be reproduced by a combination of GHG, SU, volcanic, solar, and ENSO forcing. It arises that a simple multiple regression model (MRM) (logarithmic GHG-temperature relationship, all other relationships are linear; the GHG-temperature time lag of 20 yr has been assessed by cross-correlation analysis; the volcanic delay is 1 yr and all other forcings are without delay) is able to explain 73% of this variance on a global mean annual basis. A neural network model (NNM, backpropagation architecture; [3.58]) which is trained to fit nonlinear relationships (and allows forcing factor interactions) comes out with 83%. The next step is to try to attribute particular isolated forcing factors to observed variation components (the sum of these components or climate 'signals' is the total variance explained). The results of this signal analysis are very similar in case of MRM and NNM and proportional to the radiative forcing postulated by [3.30] (see also Table 3.4).

Table 3.4. Global mean tropospheric radiative forcing of the specified influence factors since approximately 1850 (from [3.30]) and related surface air temperature signals as assessed by neural network and multiple regression models (NNM and MRM [3.52, 3.54, 3.67]; *only after major eruptions like Pinatubo: 1991 → 2.4, 1992 → 3.2, 1993 → 0.9 W m^{-2} [3.40]; **GCM [3.30]: equilibrium 2.1–4.6 K; transient 1.3–3.8 K)

Influence	Forcing [Wm^2]	NNM [K]	MRM [K]	Time structure
Greenhouse gases (GHG)	+2.1–2.8	0.9–1.3	0.8–1.2	Nonlinear trend
Sulfate aerosols (SU)	−0.4–1.5	0.2–0.4	0.1–0.4	Nonlinear var. trend
Combined (GHG + SU)	+(0.6–2.4)	0.5–0.7	0.6–0.8	Nonlinear var. trend
Volcanism (VIG)	−1.0–3.0*	0.1–0.2	0.1–0.4	Episodic (1–3 yr)
Solar activity (SRN)	+0.1–0.5	0.1–0.2	0.1–0.2	Fluctuative (+ trend?)
El Niño (ENSO)	+ (internal)	0.2–0.3	0.2–0.3	Episodic (0.5 yr)
CO$_2$ doubling, equilibrium	+4.4**	2.1	2.6–3.9	Nonlinear trend
CO$_2$ doubling, transient	+4.4**	1.7	1.8–2.6	Nonlinear trend
Explained variance		83%	73%	Annual data

On a regional-seasonal scale the statistical signal assessments suffer from a very low amount of explained variance. Nevertheless, some preliminary results for Europe, from a related MRM analysis, are listed in Table 3.5. Note that in this case the maximum GHG signals are found in summer. However, in line with the small amount of explained variance, the observed trends and the simulated signal sums differ considerably. In case of related NNM signals some strange results arise which seem to contradict the physical background. It appears that the neural network technique fails to reproduce reasonable signal results if the total amount of explained variance is relatively small.

Table 3.5. Statistical assessments of anthropogenic European surface air temperature signals 1892–1991 and related predictions using MRM (multiple regressions) or AOGCM (atmosphere ocean general circulation models) from [3.54]; *GHG trend scenario IS92A, 'business-as-usual' [3.30]; AOGCM from [3.8])

Influence	Spring [K]	Summer [K]	Autumn [K]	Winter [K]	Year [K]
Greenhouse gases (GHG)	1.2	1.4	1.2	0.8	1.2
Sulfate aerosols (SU)	−0.6	−0.9	−0.7	−0.3	−0.6
Combined (GHG + SU)	0.7	0.5	0.5	0.5	0.6
Sum of all signals simulated	0.7	0.5	0.5	0.5	0.6
Observed trends	0.4	0.2	0.3	0.4	0.4
Explained variance (MRM)	20%	23%	18%	11%	28%
MRM prediction 1991–2040*	1.7	2.0	1.7	1.1	1.7
AOGCM prediction 1935–2084*	1.9	2.2	1.9	3.6	2.3

3.5 Conclusion

The space-time variability of the European climate is analyzed which, due to its position at the tail end of the North Atlantic storm track, depends strongly on the oceanic circulation. This privileged situation allows Europe to support a population of 300 million people polewards from the $40°$ N latitude circle which, compared to the 30 million people living in North America or Eastern Asia north of the same latitude, is unique for the Earth. Europe's climate analysis described in this review follows four steps employing both physical observations and phenomenological datasets:

The *first* step characterizes the climate by the spatial distributions of long-term means and variances of the fundamental physical climate elements (surface air temperature, sea level pressure, and precipitation).

The *second* step requires the analysis of the underlying processes generating the climate. In the European sector these are the rain bearing frontal systems which, in terms of cyclone paths and storm tracks, create dynamically meaningful quantitative climatologies employing Lagrangian and Eulerian analyses of weather maps. The analysis of the climate generating processes is supplemented by the statistics of qualitative phenomenological patterns (or regimes) like Grosswetterlagen and climate zones which embed the regional climate into its geographical and biospheric environment.

The climate state associated with its space-time variability may undergo long-term changes associated with the global change of natural or anthropogenic origin; this constitutes the *third* step of climate analysis. The climatologies described above document a recent climate change at the end of this century after employing trend analyses of the spatial distribution of time averaged climate elements and of various measures of their variability (storm tracks or second moments and cyclone densities, occupation and residence times of circulation regimes).

The *fourth* and final step consists of an analysis of the climate trend generating processes. Dynamical models or empirical analyses (the latter approach is adopted here) are used to determine the effects of external forcing of the climate due to volcanism, solar variability, and anthropogenic activities.

Acknowledgements. The authors gratefully acknowledge F.-W. Gerstengarbe, P.C. Werner, and P. Hupfer for the continuing discussions, cooperation, and support on the climate and climate change issues raised in this paper.

References

3.1 C. Appenzeller, T.F. Stocker, and M. Anklin, Science **282**, 446 (1998).

3.2 A. Bardossy and H.J. Caspary, Theor. Appl. Climatol. **42**, 155 (1990).

3.3 R.G. Barry and A.H. Perry (eds.), *Synoptic Climatology, Methods and Applications* (Methuen, London, 1973), p. 555.

3.4 P. Bissolli and C.-D. Schönwiese, Meteorol. Rdsch. **40**, 147 (1987).

3.5 P. Bissolli and C.-D. Schönwiese, Naturwiss. Rdsch. **44**, 169 (1991).

3.6 M.L. Blackmon, J. Atmos. Sci. **33**, 1607 (1976).

3.7 R. Blender, K. Fraedrich, and F. Lunkeit, Q. J. R. Meteorol. Soc. **123**, 727 (1997).

3.8 U. Cubasch, G.C. Hegerl, A. Hellbach, H. Hoeck, U. Mikolajewicz, B.D. Santer, and E. Voss, Clim. Dyn. **11**, 71 (1995).

3.9 B.L. Dzerdzeevskii (ed.), *The Observed Circulation of the Atmosphere and Climate* (Nauka, Moscow, 1975), p. 285.

3.10 W. Dansgaard, J.W.C. White, and S.J. Johnsen, Nature **339**, 532 (1989).

3.11 H. Flohn, Erdkunde **4**, 141 (1950).

3.12 P.V. Foukal and J. Lean, Science **247**, 556 (1990).

3.13 K. Fraedrich, Int. J. Climatol. **10**, 21 (1990).

3.14 K. Fraedrich and K. Müller, Int. J. Climatol. **12**, 25 (1992).

3.15 K. Fraedrich, E.R. Kuglin, and K. Mueller, Tellus **44A**, 33 (1992).

3.16 K. Fraedrich, C. Bantzer, and U. Burkhardt, Clim. Dyn. **8**, 161 (1993).

3.17 K. Fraedrich and C. Larnder, Tellus **45A**, 289 (1993).

3.18 K. Fraedrich, Tellus **46A**, 541 (1994).

3.19 K. Fraedrich, R. Bach, and G. Naujokat, Contrib. Atmos. Phys. **59**, 54 (1986).

3.20 K. Fraedrich and L.M. Leslie, Q. J. R. Meteorol. Soc. **115**, 79 (1989).

3.21 K. Fraedrich, C. Ziehmann, and F. Sielmann, J. Clim. **8**, 361 (1995).

3.22 F.W. Gerstengarbe and P.C. Werner, Ber. Dtsch. Wetterdienstes **113**, 249 (1993).

3.23 R. Glowienka, Contrib. Atmos. Phys. **58**, 160 (1985).

3.24 E. J. Gumbel, *Statistics of Extremes* (Columbia University Press, New York, 1958).

3.25 J. Grieser and C.-D. Schönwiese, Atmosfera **12**, 111 (1999).

3.26 P. Hess and H. Brezowsky, Ber. Dtsch. Wetterdienstes **15**, 54 (1977).

3.27 J.W. Hurrel, Science **269**, 676 (1995).

3.28 P. Hupfer and C.-D. Schönwiese, Wiss. Auswertungen and GEO, Hamburg, 99 (1998).

3.29 P. Hupfer (ed.), *Das Klimasystem der Erde* (Akademie, Berlin, 1991) p. 464.

3.30 IPCC, *Climate Change* (Cambridge University Press, Cambridge, 1996), p. 572.

3.31 P.D. Jones, M. Hulme, and K.R. Briffa, Int. J. Climatol. **13**, 655 (1993).

3.32 W. Köppen, *Das geographische System der Klimate* (Bornträger, Berlin, 1936).

3.33 W. Köppen, Mitteil. Geogr. Gesell. Hamburg **1**, 76 (1881).

3.34 J.E. Kutzbach, J. Appl. Meteorol. **6**, 791 (1967).

3.35 H.H. Lamb, Geophys. Mem. 116, HMSO, London, 85 (1972).

3.36 N.C. Lau, J. Atmos. Sci. **45**, 2718 (1988).

3.37 J. Lean, J. Beer, and R. Bradley, Geophys. Res. Lett. **22**, 3195 (1995).

3.38 J. Malcher and C.-D. Schönwiese, Theor. Appl. Climatol. **38**, 157 (1987).

3.39 G. Megreditchian, Comp. Statist. Data Anal. **9**, 57 (1990).

3.40 M.P. McCormick, L.W. Thomason, and C.R. Trepte, Nature **373**, 399 (1995).

3.41 K. Fraedrich, F.W. Gerstengarbe and P.C. Werner, Clim. Change **45**, 405 (2001).

3.42 J.M. Mitchell, B. Dzerdzeevskii, H. Flohn, W.L. Hofmeyr, H.H. Lamb, K.N. Rao, and C.C. Wallen, WMO Tech. Note **79**, Geneva, 79 (1966).

3.43 J.M. Mitchell, *Das Klima. Analysen und Modelle, Geschichte und Zukunft* (Springer, Berlin, 1980), p. 296.

3.44 F. Molteni and S. Tibaldi, Q. J. R. Meteorol. Soc. **116**, 1263 (1990).

3.45 P.H. Mueller, P. Neumann, and R. Storm, *Tafel der mathematischen Statistik* (VEB, Leipzig, 1973).

3.46 M. New and M. Hulm, Physics of Climate Conference, R. Meteorol. Soc., Lond. (1997).

3.47 G.R. North, T.L. Bell, R.F. Cahalan, and F.J. Moeng, Mon. Weather Rev. **110**, 699 (1982).

3.48 H. Paeth, A. Hense, R. Glowienka-Hense, R. Voss, and U. Cubasch, Clim. Dyn. **15**, 953 (1999).

3.49 L.F. Richardson, Proc. Roy. Soc., Lond. **A110**, 709 (1926).

3.50 B. Saltzman, Adv. Geophys. **25**, 173 (1983).

3.51 B.G. Sanderson and D.A. Booth, Tellus **43A**, 334 (1991).

3.52 C.-D. Schönwiese, M. Denhard, J. Grieser, and A. Walter, Theor. Appl. Climatol. **57**, 119 (1997).

3.53 C.-D. Schönwiese and J. Rapp, *Climate Trend Atlas of Europe, Based on Observations* (Kluwer, Dordrecht, 1997), p. 228.

3.54 C.-D. Schönwiese, A. Walter, J. Rapp, S. Meyhöfer and M. Denhard, Ber. 102, Inst. Meteorol. Geophys. Univ. Frankfurt/Main, 156 (1998).

3.55 C.-D. Schönwiese, R. Ullrich, and F. Beck, Clim. Change **27**, 259 (1994).

3.56 M. Sickmöller, R. Blender, and K. Fraedrich, Q. J. R. Meteorol. Soc. **126**, 291 (2000).

3.57 T. Simkin, L. Siebert, L. McClelland, D. Bridge, C.G. Newhall, and J.H. Latter (eds.), *Volcanoes of the World* (Hutchinson, Stroudsbourg (and unpublished updates), 1981), p. 232.

3.58 M. Smith, *Neural Networks for Statistical Modelling* (Van Nostand Reinhold, New York, 1993), p. 235.

3.59 A. Spekat, B. Heller-Schulze, and M. Lutz, Meteorol. Rdsch. **36**, 243 (1983).

3.60 Z. Toth, Tellus **47A**, 457 (1995).

3.61 A. Tsinober, Nonlinear Proc. Geophys. **1**, 80 (1994).

3.62 R. Vautard, Mon. Weather Rev. **118**, 2056 (1990).

3.63 J.A. Viecelli, J. Atmos. Sci. **51**, 337 (1994).

3.64 H. von Storch and F.W. Zwiers, *Statistical Analysis in Climate Research* (Cambridge University Press, Cambridge, 1999), p. 484.

3.65 J.M. Wallace, X. Cheng, and D. Sun, Tellus **43A**, 16 (1991).

3.66 J.M. Wallace and D.S. Gutzler, Mon. Weather Rev. **109**, 782 (1981).

3.67 A. Walter, M. Denhard, and C.-D. Schönwiese, Meteorol. Z. **NF7**, 171 (1998).

3.68 H. Wanner, R. Rickli, E. Salvisberg, and C. Schmutz, *Klimawandel im Schweizer Alpenraum* (Hochschulverlag AG der ETH Zürich, Zürich, 2000), p. 285.

3.69 P.C. Werner, F.-W. Gerstengarbe, K. Fraedrich, and H. Oesterle, Int. J. Climatol. **20**, 463 (2000).

3.70 R. Wilby, Weather **48**, 234 (1993).

Atmospheric Carbon dioxide concentrations
from the last ice age until today

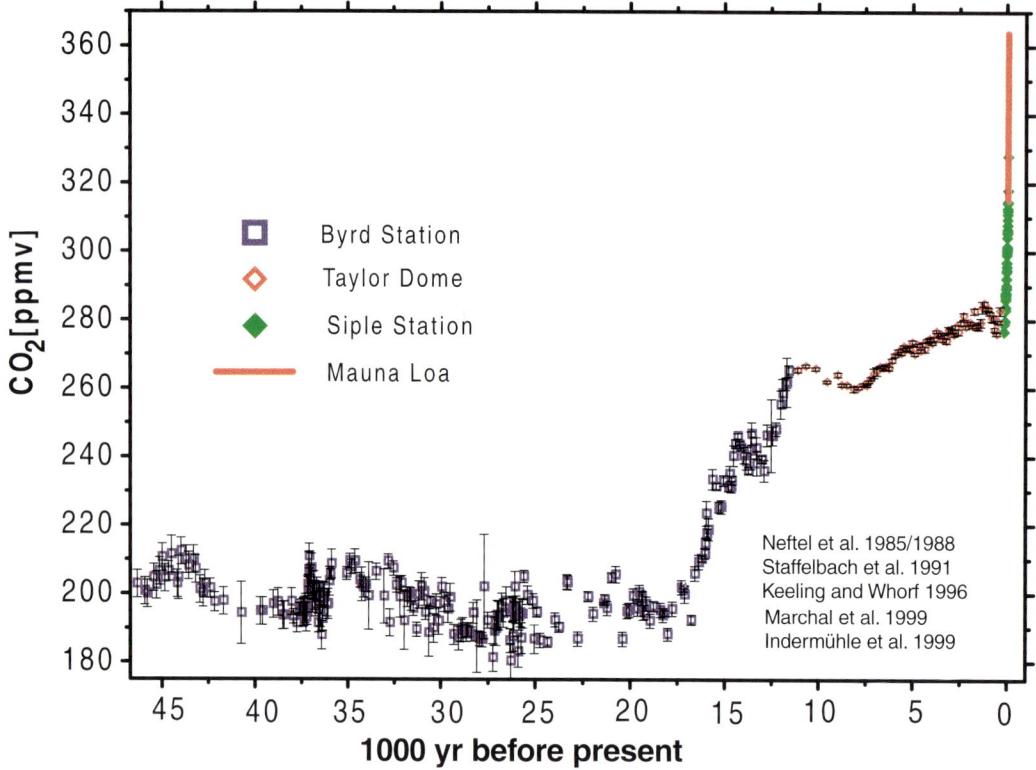

Legend:
- □ Byrd Station
- ◇ Taylor Dome
- ◆ Siple Station
- — Mauna Loa

Neftel et al. 1985/1988
Staffelbach et al. 1991
Keeling and Whorf 1996
Marchal et al. 1999
Indermühle et al. 1999

Y-axis: CO$_2$[ppmv]

X-axis: 1000 yr before present

4. Is Climate Predictable?

Klaus Hasselmann

Two forms of climate prediction are considered: the forecast of natural climate fluctuations such as El Niño, and the determination of the effect of human activities on the future climate. Although climate as a chaotic system is sometimes believed to be basically unpredictable, both forms of prediction are feasible, but with restrictions. These are determined by the degree of instability of the climate system with respect to the initial state, the stochastic forcing by short-term weather variability, the possible existence of bifurcation points, and, in the case of anthropogenic forcing, by the natural climate variability superimposed on the anthropogenic climate signal. The different forms of prediction and their present limitations are illustrated by a number of examples.

4.1 Introduction

It may appear surprising that climatologists apply models to compute the evolution of climate a hundred years or longer into the future, while meteorologists have difficulties, using rather similar models, in forecasting the development of weather beyond just a few days. The limitations of weather forecasting are often misinterpreted to imply that the climate system is basically unpredictable. However, there are basic differences between forecasting weather and predicting climate, both regarding the initial and boundary conditions and the relevant time scales and variables.

The meteorologist Edward Lorenz [4.36], one of the founders of chaos theory, distinguished between two kinds of prediction. Predictions of the first kind concern the time-dependent evolution of a system as a function of the initial conditions, with fixed boundary conditions. Predictions of the second kind

◄

concern the response of a system to changes in the boundary conditions, with fixed initial conditions. Weather forecasting is clearly a prediction problem of the first kind, while the prediction of the climate change due to human influences is normally regarded as a prediction problem of the second kind.

However, the computation of the climate change caused by the emission of greenhouse gases, for example, does not classify as a pure prediction problem of the second kind, since one is concerned with the change in climate as a function of time, and the dependence of the evolving climate state on the initial conditions becomes negligible only asymptotically, for large times. Thus, the computation of anthropogenic climate change represents a mixed prediction problem of the first and second kind. Moreover, even in the asymptotic limit, which is independent of the initial state, the climate state cannot be represented as a unique function of the boundary conditions, since climate exhibits natural, internally generated fluctuations superimposed on the externally generated mean climate response (cf. Chap. 5). The implications of this admixture of initial-value and boundary-value input together with superimposed natural climate variability has been the source of considerable debate on the predictability of anthropogenic climate change.

Although the prediction of anthropogenic climate change has become the centre of public concern, the prediction of natural climate fluctuations such as El Niño or the so called North Atlantic Oscillation (NAO) is also an important application of climate research (see Chap. 3). Both phenomena are examples of extensive climate anomaly patterns varying on annual to decadal time scales that have a major impact on weather regimes over large areas of the globe. As in the case of weather forecasting, the prediction of natural climate variations is a prediction of the first kind. The difference between weather and climate forecasting lies in this case not in the initial and boundary conditions, but in the time scales of the problem and the associated prediction variables.

4.2 Weather and Climate

By definition, weather and climate prediction are mutually exclusive. As a chaotic system, the atmosphere is unstable with respect to small perturbations of the initial conditions. This sets a natural limit to the predictability of detailed weather properties. Theoretically, the prediction limit is of the order of 20 days [4.35]. This period is sufficiently long that even very small-amplitude, small-scale perturbations in the initial state of the atmosphere have enough time to cascade through nonlinear interactions into larger scales and to amplify to a level at which reliable predictions are no longer feasible. In practice, limitations in the procurement of accurate initial data further reduce the predictability limit of weather to five to 10 days.

Climate, in contrast, is defined in terms of the statistics of weather. The data ensemble used to form the statistics is required, in the modern definition of climate (cf. GARP [4.10]), to cover a period at least as great as the theoretical limit of weather prediction[1]. Climate prediction is therefore concerned with slow changes in the statistical properties of weather on time scales of 20 days and longer. In other words, climate prediction begins where weather prediction ends.

The distinction between weather and climate prediction is analogous to the difference between predicting the molecule motions and the temperature of a gas. Apart from the practical difficulty of obtaining the initial data and performing the computation, the prediction of individual molecule motions is feasible (for given finite accuracy of the initial states of the molecules) only for microphysically short time spans of the order of a collision interval. However, the prediction of the local temperature of a gas, given by the average kinetic energy of a large local cluster of molecules, is feasible over much longer periods and is limited only by the macrophysical flow properties of the gas. Similarly, while weather prediction is limited by the dynamics of short-term weather fluctuations to maximally 20 days, climate predictability is governed by the long-term dynamics of the climate system and can extend (depending on the particular problem) from months to centuries, millenia, or even longer.

However, the analogy breaks down in one important aspect. In principle, the evolution of the temperature of a gas can be determined by computing the individual paths and kinetic energies of all gas molecules. Although the computed paths of individual molecules become meaningless after one or two collisions, the computation of the statistically averaged kinetic energy (temperature) of the gas would nevertheless remain valid. However, this brute force approach would be prohibitively expensive, and no physicist would seriously consider applying it, as there exists a much more efficient rigorous alternative: the kinetic theory of gases, which yields closed evolution equations for the macrophysical gas properties (temperature, density, and flow velocity) one is interested in. Unfortunately, however, this approach is not available for the climatologist. The strong nonlinearity of the climate system prohibits the application of the classical weak-interaction methods of kinetic gas theory, which require a clear separation of temporal and spatial interaction scales between microphysical and macrophysical processes. Thus, there exists no generally accepted closure theory of climate: it has not been possible to derive from first principles a set of closed equations for the evolution of the climate state which depends only on the climate variables one is interested in, independent of the rapidly varying weather variables.

---

[1] Prior to the modern view of climate as a dynamic rather than a static system, climate was defined in terms of 30 yr averages. This had the conceptual disadvantage of introducing a spectral gap between weather and climate prediction.

Climate modelers are accordingly faced with two alternatives: they can either compute the evolution of climate by computing the complete weather trajectories, in analogy with the brute force approach to computing the temperature of a gas; or they can invoke some ad hoc, necessarily approximate closure hypothesis. In the first approach, one runs a standard atmospheric general circulation model (AGCM) of the type used for weather forecasting, which resolves the detailed weather dynamics. However, the AGCM is run at reduced resolution to achieve the longer integration periods required for climate simulations. In addition, the AGCM must be coupled to an ocean general circulation model (OGCM) and further models of the biosphere, the cryosphere (snow and ice fields) and other slow components of the climate system, that can be treated as constant boundary conditions for weather forecasting, but whose dynamics become important on climatic time scales. The complexity and computing demands of these explicitly resolving climate models explain the insatiable desire of climatologists for ever larger supercomputers.

The second, closure approach is less demanding in computer resources, but is inherently speculative and open to criticism. Most climate modelers therefore pursue a combined strategy in which a smaller number of simulations with expensive high resolution models are used as reference standards, against which lower-resolution closure models can be calibrated. These can then applied for more extensive simulation series.

In practice, however, the hierarchy of models developed by climate modelers, although strongly bimodal, is better represented as a continuum of models of all resolutions, all of which invoke closure hypotheses at some level. For even the highest resolution models are unable to resolve processes of scale smaller than the finite model grid dictated by the available computer power, and these processes must then also be parameterized by appropriate statistical closure hypotheses.

4.3 Climate Prediction of the First Kind: ENSO

High-resolution climate models have been successfully applied for climate predictions of the the first kind. To extend the prediction limits of standard AGCMs used for weather prediction, the AGCMs are coupled to OGCMs. Although atmosphere ocean general circulation models (AOGCMs) are still unable to overcome the inherent stability limitation of an AGCM in the prediction of individual, rapidly varying weather systems, a AOGCM should in principle be able to predict the evolution of the state of the ocean over longer periods, since this is governed by the slow dynamics of the ocean circulation. The surface temperature of the ocean, in turn, acts back on the slowly varying statistical properties of the atmosphere that define the climatic state. The

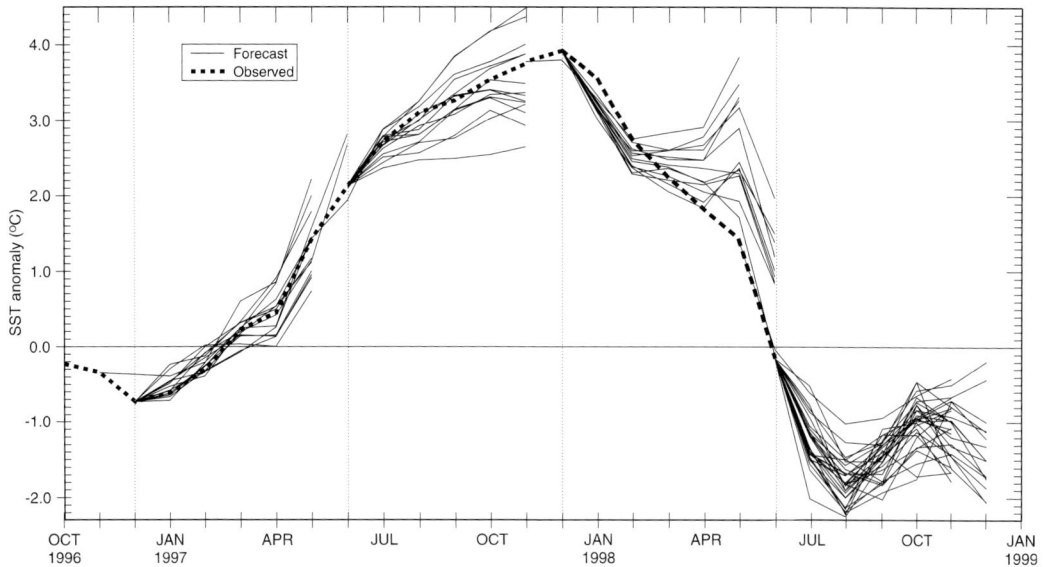

Fig. 4.1. Ensemble predictions of the 1997/1998 El Niño with subsequent La Nina by the European Centre for Medium Range Weather Forecasting

success of climate predictions using AOGCMs therefore depends ultimately on the ability to determine the initial state of the ocean and construct ocean circulation models that can reliably predict the future evolution of the ocean given its initial state.

So far, significant advances in climate prediction of the first kind have been achieved only in the forecasting of El Niño, the most pronounced and extensive climate anomaly in the range of short time scales. El Niño occurs irregularly every three to six years and normally lasts about 18 months. It begins as a warming of the eastern equatorial Pacific, accompanied by a drop in the East-West pressure gradient in the South Pacific. The change in the pressure gradient is referred to as the Southern Oscillation, and the coupled El Niño/Southern Oscillation phenomenon as ENSO. At its height, ENSO affects the entire tropical belt and many adjacent regions, including the United States and Australia [4.30].

Useful ENSO forecasts can be achieved on time scales of months to maximally about a year. Figure 4.1 shows as example forecasts of the 1997/1998 record El Niño produced by the European Centre for Medium Range Weather Forecasting (ECMWF). For clarity, only every sixth forecast of the monthly forecasts distributed by ECMWF on the Internet is shown. The predicted variable is the mean sea surface temperature averaged over the Eastern equatorial Pacific, which is highly correlated with anomalies in precipitation, temperature, and other climate variables in the regions affected by ENSO. For each forecast period an ensemble of forecasts was performed by varying the initial conditions in accordance with the variations in daily weather. The divergence of the plume of forecasts indicates the instability of the prediction with respect

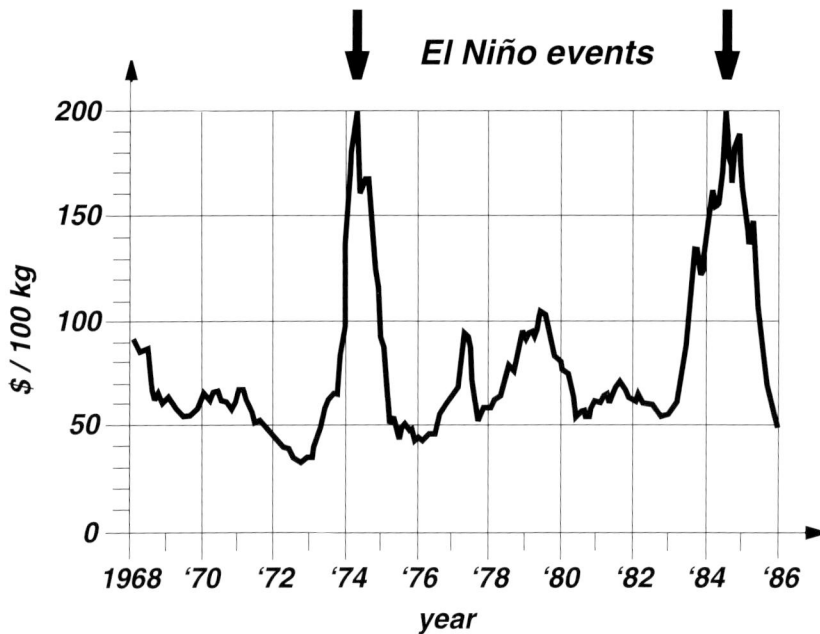

Fig. 4.2. Correlation of El Niño with the price of cocoa-nut oil

to the initial conditions and provides a useful measure of the reliability of the prediction. Although the spread of the plumes grows with time, meaningful predictions are clearly possible over periods up to about six months.

ENSO forecasts have a major impact on the economies and welfare of many countries worldwide. The successful prediction of the 1997/1998 El Niño yielded estimated savings of several hundred million US dollars in the USA alone. Examples of the impact of El Niño on the economy and health are illustrated in Figs. 4.2 and 4.3 showing, respectively, the correlation between El Niño and the price of cocoa-nut oil and the incidence of malaria in Columbia. Although public attention has focused on climate predictions of the second kind in relation to global warming, climate prediction of the first kind is therefore also an important application of climate research promising large benefits for society. As discussed below, the two kinds of prediction are in fact strongly interrelated, as many of the prediction problems posed by anthropogenic climate change are directly coupled to questions on the nature and predictability of natural climate variability.

Attempts to apply AOGCMs to predict natural climate variations also on decadal and longer time scales have so far proved less successful. In principle, the memory of the ocean is sufficiently long that the AOGCM modeling approach should be feasible also on these time scales, and the benefits of such predictions would be at least comparable to those of the prediction of El Niño. The North Atlantic Oscillation, for example, which exhibits variability on time scales of several years to one or two decades, has a major impact on the climate

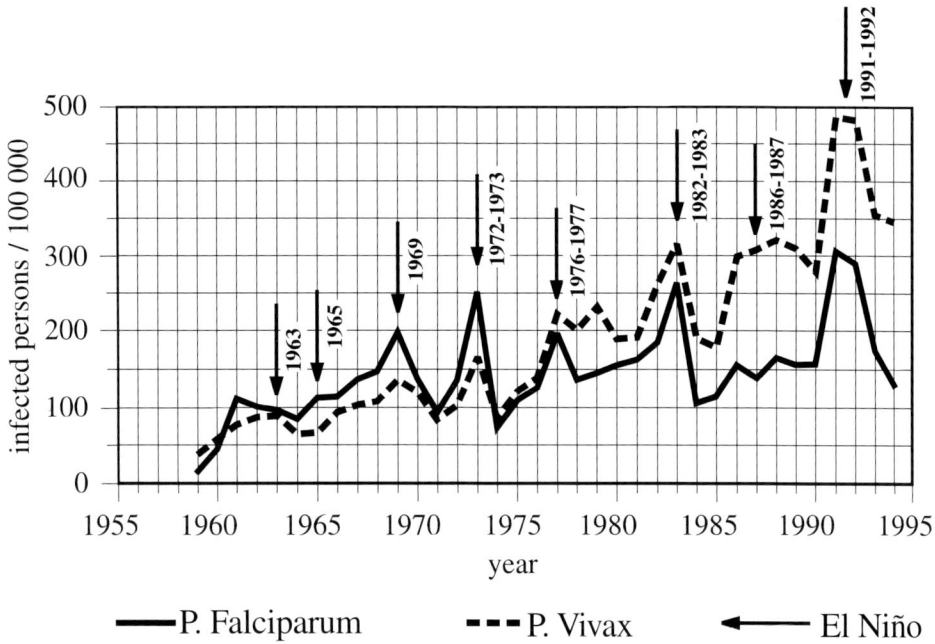

Fig. 4.3. Correlation of El Niño events with the incidence of malaria in Columbia

of Europe (cf. Chap. 3), and the value of El Niño forecasts would be greatly enhanced if the forecast period could be extended to include not only the event itself but also its irregular 3-to-6-year recurrence period.

The main difficulties in extending the present limit of short-range climate forecasts lie firstly in the determination of the initial anomalous state of the ocean, and secondly in constructing reliable ocean models that capture not only the shorter-period equatorial ocean dynamics but also the slower higher-latitude processes governing the global thermohaline circulation (THC) (see also Chap. 6). However, both problems are not insurmountable, and it is to be expected that the enhanced research efforts currently expended on the extension of the useful range of climate predictions of the first kind will ultimately bear fruit.

4.4 Stochastic Climate Models

Since the predictability of the ENSO phenomenon resides in the dynamics of the ocean rather than the atmosphere, one could consider constructing a simpler ENSO prediction model in which only the ocean is represented by a full OGCM, while the AGCM component of the AOGCM is replaced by a computationally less demanding diagnostic atmospheric-response model that determines the quasi-equilibrium mean anomaly state of the atmosphere for a given sea-surface-temperature anomaly field. The mean response can be determined empirically by comparison with observed data, or, more conveniently, by calibration against a simulation with a full AOGCM. The mean atmospheric response then acts back as a forcing function on the ocean in a closed loop system. Such hybrid models have indeed been constructed and yield ENSO forecasts of comparable skill to that of a AOGCM [4.30].

If the simulation ensemble of Fig. 4.1 is repeated using a hybrid model instead of a AOGCM, one finds that while the means of the prediction plumes are not modified significantly, the spread of the plumes is strongly reduced. This yields an important insight into the origin of the instabilities of ENSO predictions using AOGCMs: they cannot result primarily from interactions between the ocean and the slowly responding mean state of the atmosphere, as they would then also be seen in the predictions with hybrid models. They must be generated instead by the internal variability of the atmosphere, which is suppressed in a hybrid model, but retained in the AGCM-component of the AOGCM, in which the weather processes are explicitly simulated.

Thus, the application of a full AOGCM (at significant computational overhead) brings no significant advantage for the ENSO forecast itself, but provides useful information only on the reliability of the forecast. The role of the AGCM is to provide a continual noise input that degrades the reliability of the forecast with time. One may ask then whether the simulation of the noise generator cannot be achieved more efficiently than by running an expensive high-resolution AGCM. This is indeed the case. The occurrence of basically unpredictable weather-generated variability superimposed on the predictable ENSO signal is a particular manifestation of the general concept of stochastic models, introduced originally by Mitchell [4.42], developed independently by Hasselmann [4.11] and applied subsequently in a series of papers, see, e.g. [4.8, 4.9, 4.19, 4.31, 4.32, 4.41, 4.45] and others (see also [4.24]). Stochastic climate models represent special forms of closure models. They are based on a separation of the climate system into a fast subsystem y, the atmosphere, and a slow subsystem x, consisting of the oceans, cryosphere, biosphere, biogeochemical cycles, etc. Separated into these climate state components, the prognostic equations

$$\frac{dz}{dt} = P_z(z) \tag{4.1}$$

for the full climate state vector $z = (x, y)$ can be decomposed into the two coupled equations, with associated prognostic functions (propagators) $P_z = (P_x, P_y)$,

$$\frac{dx}{dt} = P_x(x, y) \,, \tag{4.2}$$

$$\frac{dy}{dt} = P_y(x, y) \,. \tag{4.3}$$

To avoid integrating the full system (4.1) or (4.2), (4.3) at the spatial and temporal resolution dictated by the dynamics of the fast system y, closure hypotheses are normally introduced. These postulate a set of closed prognostic equations for the climate state variables $x, \hat{y}$ that depends only on the climate state variables $x, \hat{y}$, where $\hat{y}$ denotes, as described above, the statistical properties of y (means, higher moments, probability distributions).

The standard closure approach involves two steps. First, the prognostic function $P_x(x, y)$ (4.2) is replaced by a function $\hat{P}_x(x, \hat{y})$ in which the dependence on the full atmospheric state y is reduced to a dependence on the statistical atmospheric state $\hat{y}$ (in most cases, this is reduced further to a dependence only on the mean atmospheric state $\bar{y}$). Second, the statistical atmospheric state $\hat{y}$ is assumed to adjust quasi-instantaneously (on the relevant climatic time scales) to the slow-system state: $\hat{y} = \hat{y}(x)$. In the hybrid ENSO model discussed above, for example, a linear response relation for the mean atmospheric state was assumed: $\bar{y} = Rx$, with an empirical response matrix R. The net effect of the two assumptions is to reduce the pair of coupled equations (4.2), (4.3) to a single closed prognostic equation

$$\frac{dx}{dt} = \hat{P}_x(x, \hat{y}(x)) \tag{4.4}$$

for the slow system x.

Although closure models of the form (4.4) may be able to simulate the mean evolution of the slow system, they are generally less successful, as pointed out for the case of hybrid ENSO models, in reproducing the climatic fluctuations about the mean. The origin of the deficiency is the elimination of all explicit weather variability in the prognostic equation (4.4). Although one is concerned in climate prediction only with slow changes in the variable x and the associated weather statistics $\hat{y}$, it is not correct to assume that short-time-scale weather variability has no impact on the low-frequency evolution of these variables. For although the fast system exhibits variability mainly on fast time scales, its variance spectrum - and, more importantly, the variance spectrum of the driving function P_x - normally extends down to very low frequencies. It is these very low frequencies that generate statistical fluctuations of the slow system on climatic time scales. It was shown in the papers cited above that much of the observed climate variability over a broad range of time scales can be explained as the response of the slow components of the climate system to this short-time scale forcing by atmospheric weather variability.

To allow for these forcing terms, the deterministic prognostic equation (4.4) should be replaced by the stochastic equation

$$\frac{dx}{dt} = \hat{P}_x(x, \hat{y}(x)) + \tilde{P}(x), \tag{4.5}$$

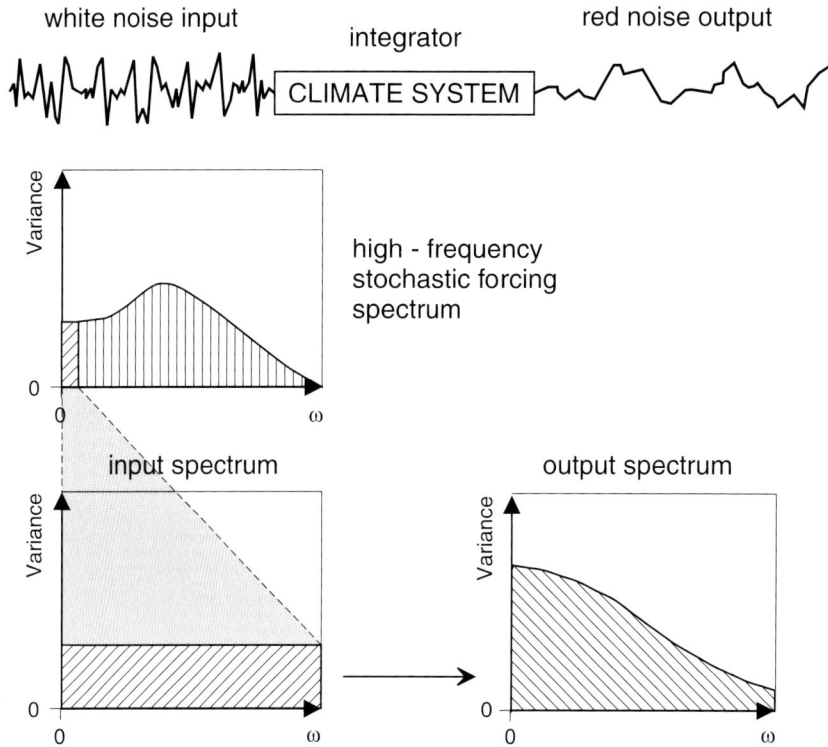

Fig. 4.4. Relation between stochastic forcing by atmospheric weather variability and climate response. At the low frequencies relevant for climate variability, the forcing spectrum is essentially white, while the integrating climate response spectrum is red

where $\hat{P}_x(x,\hat{y}(x))$ is the mean force arising from the coupling between the statistical atmospheric state $\hat{y}(x)$ and the slow system x and $\tilde{P}(x)$ is a stochastic forcing function generated by the weather variability.

The fact that only the very low frequencies of the stochastic forcing function $\tilde{P}_x$ generate natural climate variability greatly simplifies the representation of the forcing $\tilde{P}(x)$. In the low-frequency limit $\omega \to 0$, the variance spectrum $F_p(\omega)$ of the stochastic forcing $\tilde{P}$ can be represented as white noise, $F_p(\omega) = F_p(0) = \text{const}$ (Fig. 4.4), in analogy with the standard model of Brownian motion [4.50]. The details of the climate response to the white noise forcing depend on the dynamics of the slow system. Figure 4.5 shows three examples, corresponding to a straightforward integrator without feedback, linear feedback damping, and a linear damped oscillator. The integrator produces a nonintegrable, nonstationary stochastic process yielding an inverse-square variance spectrum $F_x \sim \omega^{-2}$ for x. The model represents a useful approximation for an intermediate range of frequencies between the low-frequency limit in which the the internal feedback dynamics of the slow system become important and the high-frequency limit above which the forcing spectrum can no

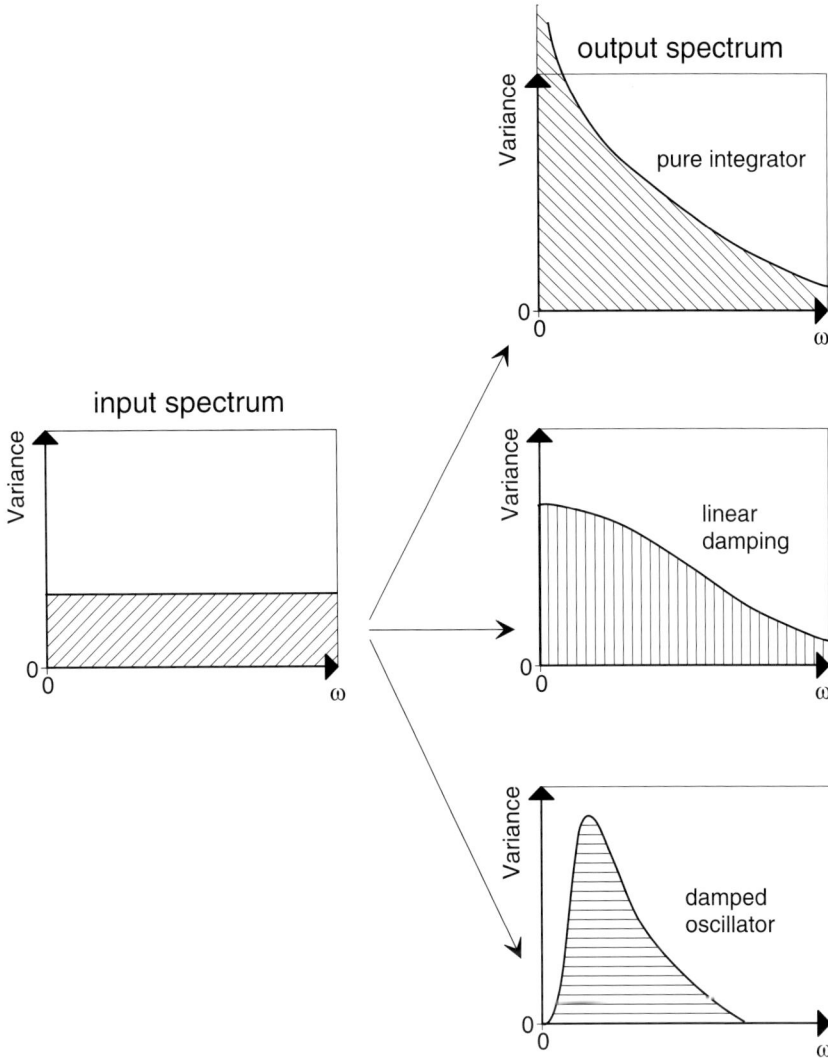

Fig. 4.5. Examples of climate response spectra (for explanations, see text)

longer be regarded as white[2]. The linear damping model corresponds to the Einstein-Uhlenbeck model of Brownian motion, which yields a variance spectrum $F_x \sim (\lambda^2 + \omega^2)^{-1}$, where λ is the damping factor; it provides a good representation of the observed variability for many climate variables, such as mid-latitude sea surface temperatures and sea-ice anomalies [4.8, 4.9, 4.19, 4.32, 4.45] or the water level of a lake [4.1, 4.3]. The damped oscillator model corresponds to a variance spectrum proportional to $\omega^2\{(\omega^2 - \omega_0^2)^2 + \lambda^2\omega^2\}^{-1}$, where ω_0 denotes the oscillator eigenfrequency. It reproduces the climate response in the neighborhood of an eigenmode of the system, for example in the simulation of the variability of the ocean circulation on century time scales [4.41].

---

[2] In real life, it also represents a model of the bank account of a gambler.

Climate Forecast Skill

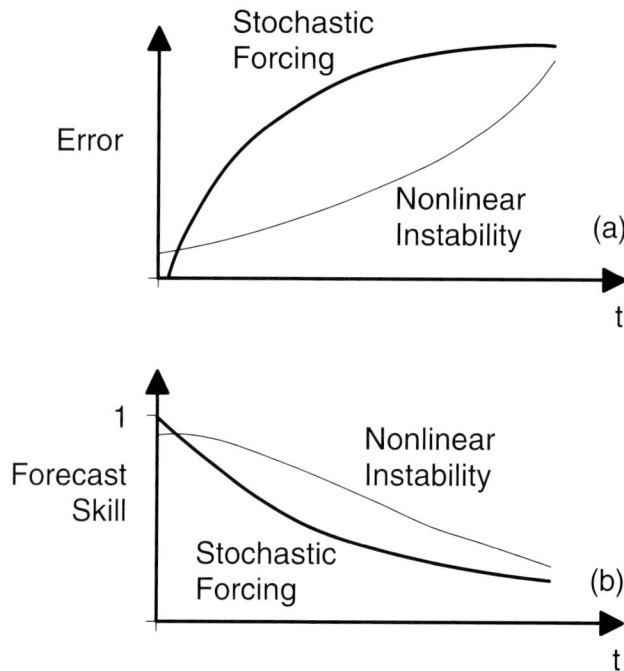

Fig. 4.6. Error growth (**a**) and forecast skill (**b**) for a climate system that is stable with respect to initial-state perturbations but stochastically forced and a nonforced climate system that is unstable with respect to initial conditions

It should be emphasized, however, that the stochastic forcing model is not limited to linear response systems, but applies generally for any nonlinear model of the slow system. In the simulation experiment of Mikolajewicz and Maier-Reimer [4.41], for example, the response of the ocean circulation to atmospheric weather forcing was simulated by a fully nonlinear OGCM. The stochastic forcing model (4.5) exhibits different error growth characteristics than the standard deterministic closure model (4.4). The realistic assessment of the forecast limits for climate predictions of the first kind therefore depends critically on the inclusion of stochastic forcing terms (cf. Fig. 4.6). Without stochastic forcing, the prediction limits are determined entirely by the instability of the climate model with respect to variations of the initial conditions. The errors grow exponentially and are proportional to the amplitude of the initial perturbation. If stochastic forcing terms are included, however, errors are generated independently of perturbations in the initial conditions. The root-mean square error growth is initially proportional to $\sqrt{t}$ (in accordance with the undamped-integrator stochastic model) and therefore faster than exponential. However, as the errors grow, exponential growth ultimately takes over. Whether the effective prediction limit is set by the stochastic forcing in the initial growth period or by errors in the determination of the initial state depends on the ratio of these two error sources. The analysis of the relative contributions of these error sources in limiting the range of ENSO predictions and other climate predictions of the first kind is an important area of current research.

4.5 Climate Predictions of the Second Kind: Global Warming

The widespread public and political concern with the prospect of major climate change through human activities has motivated extensive scientific studies on the response of climate to external influences. If the dependence on the initial climate state is ignored, this corresponds to a standard climate prediction problem of the second kind. However, an important question in this context is whether the predicted anthropogenic climate change can be detected already today, and, if so, whether the observed climate change is consistent with the climate change signal predicted by the models. This requires consideration of both the forced climate response and the natural climate variability, including the recent history of natural climate variability and its impact on the observed present climate. Thus brings elements of climate predictability of the first kind into the analysis. In this and the following section, however, we consider first the computation of anthropogenic climate change for fixed initial conditions, as a pure prediction problem of the second kind. Subsequently, we turn to the detection and attribution problem and consider finally some nonlinear aspects of the interaction between natural climate variability and forced climate change.

Human activities affect climate in many ways: destruction of stratospheric ozone through the emission of clorofluorocarbons (CFCs), increase of tropospheric ozone through the emission of nitrous oxides, carbon monoxide and other pollutants, modifications of the Earth's surface through land use changes, and the emission of greenhouse gases (Fig. 4.7). We consider in the following only the greenhouse problem, which represents the most serious as yet unresolved global climate problem, the destruction of stratospheric ozone hopefully being halted through the Montreal and subsequent agreements on the elimination of CFC emissions. The computation of climate change due to the emission of greenhouse gases is normally carried out in two steps: first, one applies a carbon cycle and other greenhouse gas models to compute the atmospheric concentrations of CO_2 and other greenhouse gases produced by the assumed emissions; second, one applies a AOGCM to compute the climate

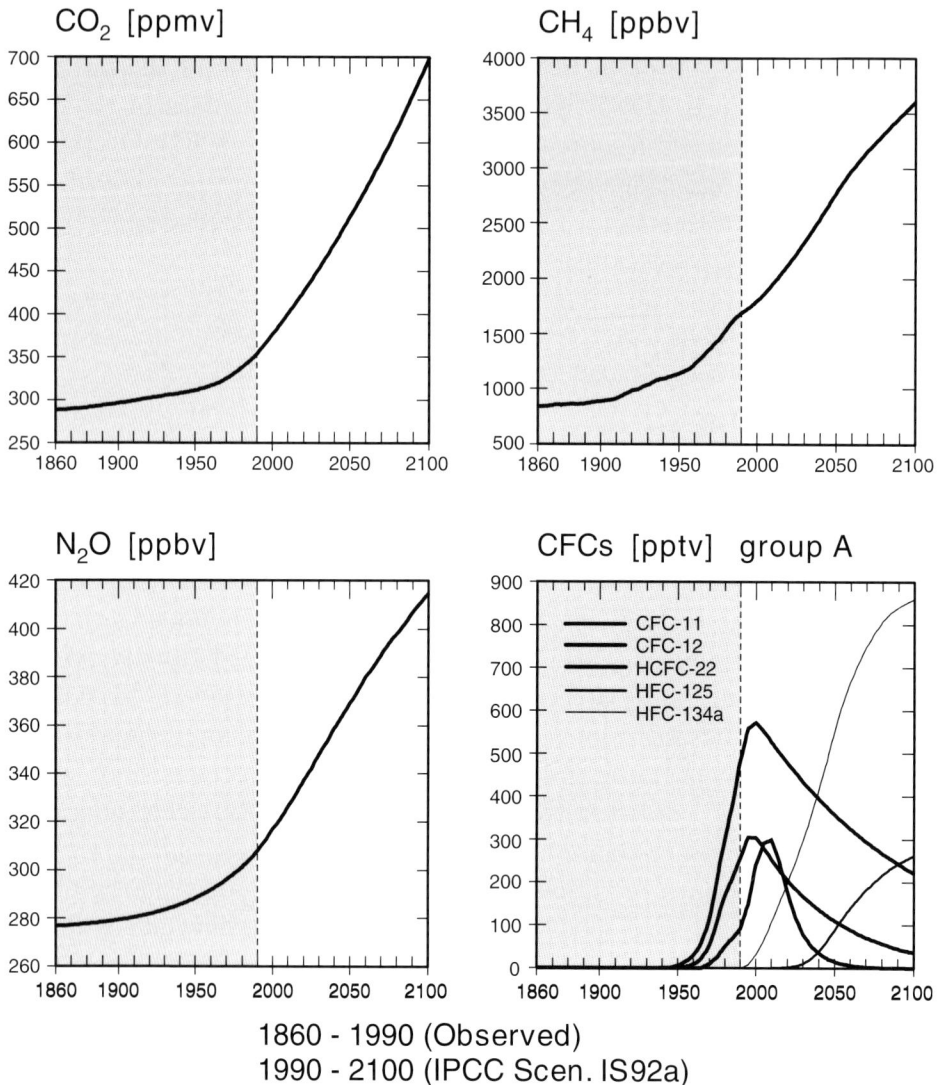

Fig. 4.7. Measured past greenhouse gas concentrations and projected future concentrations for the zero-regulation ('business-as-usual') case

change resulting from the computed greenhouse gas concentrations. To allow for the feedback of climate change on the carbon cycle, the two steps can also be combined in a single computation with a coupled carbon-cycle-AOGCM model [4.29,4.38,4.47,4.48]. Figure 4.8 shows an example of the CO_2 concentrations and resulting global mean temperature computed in this two-step manner by Cubasch et al. [4.5] for the 'business-as-usual' (BAU) scenario A and a reduced emissions scenario D originally proposed by the Intergovernmental Panel on Climate Change (IPCC) in its First Assessment Report [4.21]. Figures 4.9 and 4.10 show a set of similar computations of the global mean temperature and the resulting global warming patterns for a simpler BAU scenario representing a 1% increase in equivalent CO_2 emissions per year (in Figs. 4.8-4.10,

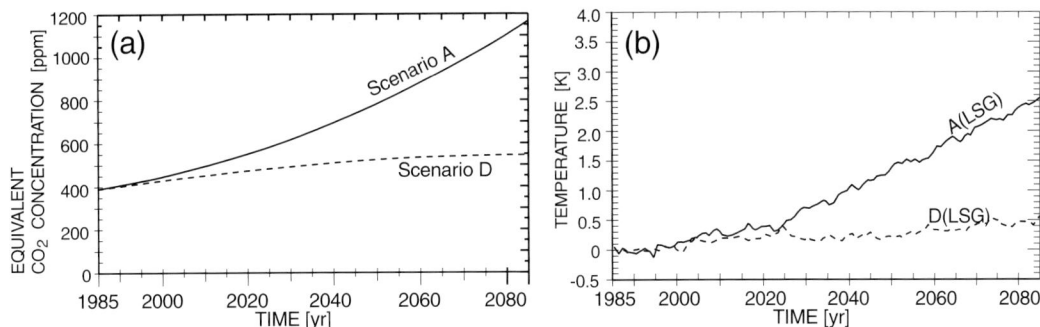

Fig. 4.8. Projected equivalent CO_2 concentrations (**a**) and global-mean near-surface temperatures (**b**) for the IPCC [4.21] nonregulated ('business-as-usual') Scenario A and strong reduction (Draconian measures) scenario D (from [4.5])

the cumulative radiative forcing of all greenhouse gases was translated for simplicity into an equivalent CO_2 concentration). The spread of predictions using different models in Figs. 4.9 and 4.10 provides some indication of the uncertainty of the predictions. These are normally estimated as $\pm 50\%$ for a 100 yr simulation. The model spread for the global mean temperature is somewhat smaller than for the global temperature distributions, which exhibit the largest deviations at high latitudes and over continents.

The differences between independent model simulations can be attributed primarily to differences in the model physics, in particular in the formulation of critical subscale parameterizations such as the formation of clouds or the amount of soil moisture. To a lesser extent they result also from the natural climate variability noise of the models, which is superimposed on the greenhouse warming signal. The model variability noise (in contrast to the observed climate variability) can be reduced, however, by averaging over an ensemble of simulations with different initial conditions (cf. [4.6]).

For prediction time scales of a hundred years, the differences between the typical AOGCM model simulations shown in Figs. 4.9 and 4.10 cannot be attributed to model instabilities with respect to predictions of the either the first or second kind. The models' natural variability levels grow more slowly than the anthropogenic signal, so that after a few decades, the model response to the greenhouse forcing dominates over the instabilities associated with variations in the initial conditions. The models also exhibit no instability of the second kind with respect to variations of the greenhouse forcing on these time scales.

The situation changes dramatically, however, if longer time scales or a broader class of models are considered. Most AOGCMs exhibit instabilities associated with a breakdown of the conveyor belt ocean circulation (cf. Chap. 6) at CO_2 concentrations higher than those attained in Figs. 4.9 and 4.10. Other instability mechanisms associated with processes not incorporated in standard AOGCMs have also been proposed, such as a break-off of the West-Antarctic

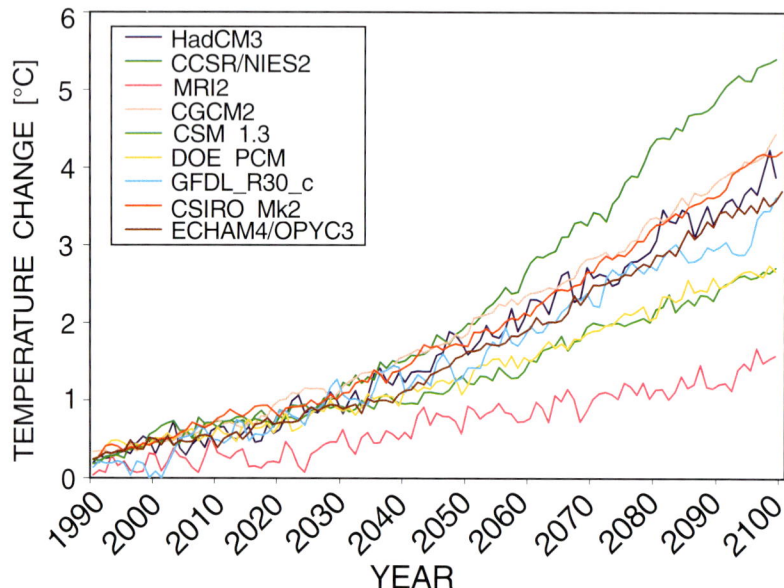

Fig. 4.9. Global mean temperature increase for the IPCC BAU scenario SRES-A2, computed by a number of different AOGCMs (for details, see [4.26])

ice shelf with rising sea level, the release of large quantities of methane currently trapped in permafrost regions through rising temperatures, leading to a run-away greenhouse effect, or phase-change type switches in circulation regimes such as the current El Niño/Monsoon system[3].

From the viewpoint of climate mitigation policy, the uncertainties associated with climate instabilities are at least as important as the impact of the major long-term warming predicted by models in the stable regime. For although the likelihood for the occurrence of instabilities on the century time scale is regarded as small, the impact of such instabilities, if they should occur, could well be catastrophic.

4.6 Linear Response Relations

Although the climate system is basically strongly nonlinear, the response δz of the climate state z to an external forcing F can nevertheless be approximated by a linear response relation if the forcing is sufficiently small. In this case the general forced-response equations for climate prediction of the second kind, given by (4.1) with the inclusion of an additional forcing term F,

$$\frac{dz}{dt} = P_z(z) + F, \qquad (4.6)$$

---

[3] Formally, bifurcation points of this kind represent instabilities with respect to predictions of both the first and second kind. However, they are usually regarded primarily as instabilities of the second kind that limit the predictability of the response to external forcing.

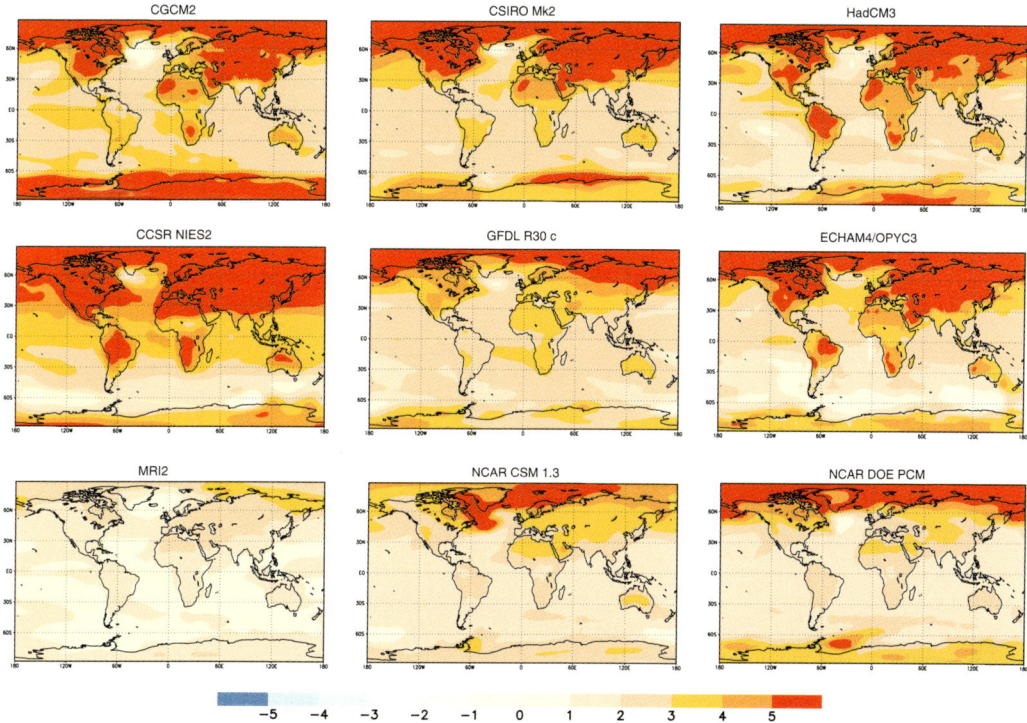

Fig. 4.10. Near-surface temperature distributions computed by the models shown in Fig. 4.9 at the time of a doubling of the CO_2 concentration (for details, see [4.26])

can be linearized in the form

$$\frac{d\delta z}{dt} = P'_z(z)(\delta z) + F, \qquad (4.7)$$

where $P'_z(z) = \partial P / \partial z$ represents the functional derivative of the propagator P. The integration of (4.7) yields

$$\delta z(t) = \int_0^t R(t - t')F(t')dt', \qquad (4.8)$$

where $R(t - t')$ denotes the impulse response (Green) function of the system[4].

Linearized response relations have been widely used for analyses of anthropogenic climate change [4.20, 4.28, 4.37, 4.40, 4.49]. They yield acceptable approximations for changes in the equivalent atmospheric CO_2 concentrations up to a factor of about two, corresponding to a global temperature increase up to about 3°C. Expressed in terms of absolute temperature appropriate for radiative forcing computations, this corresponds to a change in the global mean temperature (287 K) of about 1%. The approach breaks down, however, in the neighborhood of bifurcation points.

---

[4] It is assumed that the unperturbed reference climate state is statistically stationary, so that the general response function $R(t, t')$ for a time-dependent reference state can be replaced by $R(t - t')$.

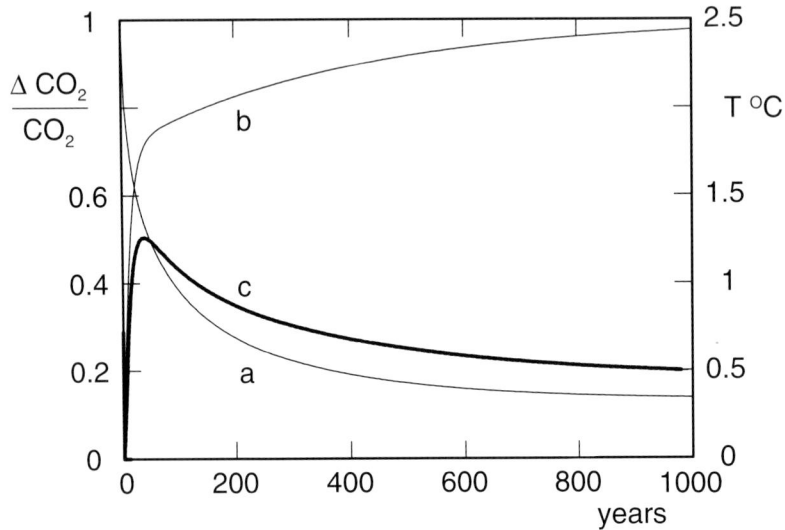

Fig. 4.11. Impulse response functions for the Hamburg carbon-cycle model (**a**), the Hamburg AOGCM climate model (**b**), and the complete coupled carbon-cycle AOGCM model (**c**). The curves (**a**), (**b**), and (**c**) show, respectively, the carbon-cycle response to a sudden doubling of the CO_2 concentration, the temperature-response to a CO_2 doubling which is then kept constant, and the net response of the coupled system to a CO_2 doubling which relaxes in accordance with curve (**a**) (from [4.20])

Impulse-response models can be constructed empirically by calibrating the response function $R(t - t')$ against simulations with high-resolution carbon-cycle or other greenhouse-gas models and AOGCMs. Once calibrated, the models can be run at computational costs far lower than the operational costs for regular three-dimensional carbon cycle models or AOGCMs. They imply nonetheless no loss of information relative to the complete models against which they are calibrated, for the climate state perturbation δz and the associated impulse response function R are of the same dimension as the climate state vector z. If the forcing function is a scalar (e.g. CO_2 emissions), the impulse response functions can be determined through a single reference simulation with the full climate model. However, if the forcing function exhibits higher dimensionality (for example, SO_2 emissions, which produce sulfate aerosols with relatively short half-lives and therefore spatially inhomogeneous distributions), a larger number of reference simulations will be needed to capture the principal distribution patterns. For this reason, linearizations of complex models have been limited so far to response models for CO_2 emissions.

The impulse response approach can be extended rather straightforwardly to include some of the more important nonlinearities of the climate system, such as the approximately logarithmic relation between greenhouse gas concentrations and radiative forcing and the decrease in the solubility of CO_2 in sea water with rising CO_2 concentrations [4.20]. However, the inclusion of all conceivable detailed nonlinearities of the parent models is, of course, not feasible within the conceptual constraint of the impulse response approach. Apart from their computational advantages, impulse response models are also useful tools for summarizing the temporal characteristics of the climate response to external forcing. Figure 4.11 shows as example the impulse response functions derived from the Hamburg carbon-cycle and AOGCM models, together with the resultant response inferred for the coupled carbon-cycle AOGCM system. The long memory of the climate system is determined by the slow uptake of the CO_2 injected into the atmosphere at time $t = 0$ by the oceans and the terrestrial as well as by the large thermal inertia of the oceans. The implications of these long time constants for the development of an effective climate policy, which has normally been focused on only the next decade or two, are not always fully appreciated.

4.7 Detection and Attribution of Climate Change

An important aspect of the prediction of anthropogenic climate change is verification (see also Chaps. 3 and 5).

Model computations of greenhouse warming predict a warming today of about 0.6°C since the end of the 19th century. This is comparable to the observed increase in the global mean temperature (Fig. 4.12). However, it is not immediately clear that the warming can indeed be attributed to human influence, as it could also be simply an expression of natural climate variability. In the conventional approach to detection and attribution (cf. [4.2,4.12,4.14,4.15,4.43,4.44]), the identification of the observed climate change as an anthropogenic signal requires, first, the demonstration that the observed signal exceeds the natural climate variability noise at some sufficiently high level of statistical significance (detection), and, second, that the observed signal is indeed consistent with the climate change signal pattern predicted by climate models (attribution). It is only rather recently that the observed climate change signal has grown sufficiently strong that detection and attribution have become feasible. While the first IPCC Science Assessment Report [4.21] stated that an anthropogenic climate signal could not yet be clearly detected, the Second Assessment Report [4.22] concluded cautiously that 'the balance of evidence suggest a discernible human influence on climate', a much debated statement that is succeeded nevertheless in the Third Science Assessment Report [4.26] by a clearly affirmative assessment. This rapid increase in confidence is partly also the fruit of a growing number of detection and attribution studies using more sophisticated multivariate (e.g. optimal fingerprint) analysis techniques (cf. [4.13], review by Barnett et al. [4.2] and discussion below).

However, the issue is still far from cut and dried and there remain many open questions. While most studies indicate that a climate change signal can

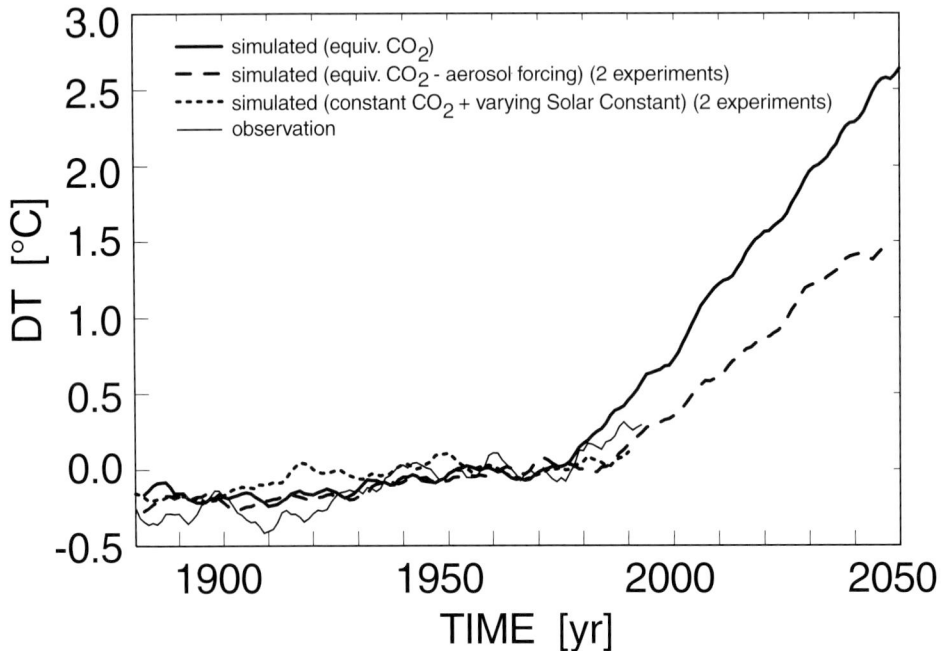

Fig. 4.12. Observed (from [4.27]) and simulated global warming since 1880 (from [4.17])

be clearly detected today above the natural variability noise at the 2σ (95% significance) level, the attribution of the signal to anthropogenic forcing remains marginal. The difficulty lies not only in uncertainties of the predicted anthropogenic signal, but also in the estimation of the natural variability noise against which the observed climate signal must be compared. The available observations are normally too sparse to determine the multivariate statistics of the natural climate variability with sufficient confidence for a reliable quantitative detection and attribution analysis. They must generally be augmented, therefore, by model simulations, which themselves are open to question.

In order to enhance the signal-to-noise ratio, the observed climate change signal is projected in the conventional detection and attribution analysis onto a lower dimensional space. This is obtained by representing the signal as a linear superposition of a small number of characteristic patterns (normally empirical orthogonal functions). If only a single anthropogenic forcing hypothesis is tested (for example, forcing by greenhouse gases only), the signal-to-noise ratio is maximized by projecting the signal onto only a single pattern. The optimal pattern (optimal fingerprint) is obtained by rotating the predicted signal in the low-dimensional pattern space away from directions of high natural variability towards directions of low noise. If more than one hypothesis is tested (for example, radiative forcing by greenhouse gases with and without inclusion of sulfate aerosols), the detection and attribution test is carried out in the two-

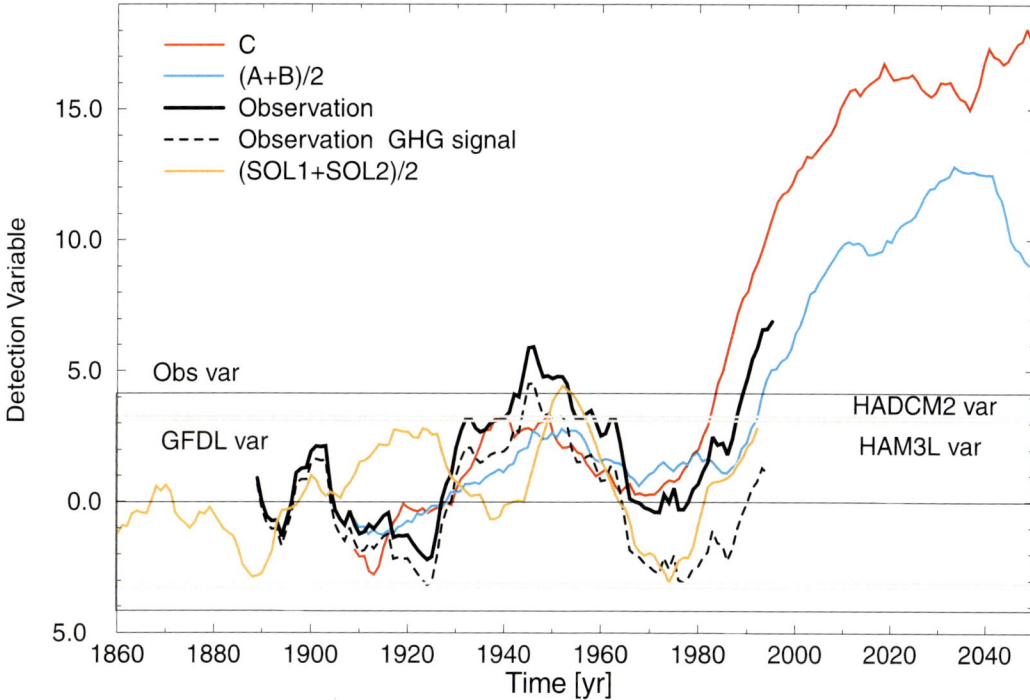

Fig. 4.13. Detection indices computed for near-surface temperatures for three externally forced climate change scenarios and two observation time series. Scenario C represents BAU forcing by greenhouse gases only, (A+B)/2 the average of two simulations for greenhouse-gas-plus-sulfate-aerosol forcing, (SOL1+SOL2)/2 the average of two simulations with variable solar forcing, and 'Observations - GHG signal' the observations minus an estimate of the anthropogenic climate change signal. Two independent estimates of the 95% detection significance level are given (from [4.2])

or three-dimensional pattern subspace spanned by the appropriately rotated predicted signals of the separate hypotheses[5].

Figure 4.13 shows as example the detection significance levels obtained by projection of the observed 30 yr trends of near-surface temperatures onto a single optimal fingerprint pattern for the case of forcing by greenhouse gases only. Similar positive results are obtained for the single-pattern test of a combined radiative forcing by greenhouse gases and sulfate aerosols. The warming trends in recent decades clearly exceed the 95% significance level estimated for the natural variability noise. The simulations indicate also that the warming cannot be explained by variations in the solar radiation [4.7], as has sometimes been suggested.

Qualitative support for this result is provided in Fig. 4.14 by a comparison of the warming observed in the twentieth century with the climate variations inferred from paleoclimatic data over the last thousand years [4.39]. Although a quantitative estimate of the detection significance level for these data was not attempted due to the uncertainties of the natural variability estimates

---

[5] In practice the natural variability noise normally limits the dimension of the pattern subspace to two, even when more than two hypotheses are tested, cf. [4.17, 4.18].

Surface temperatures of the last 1000 years

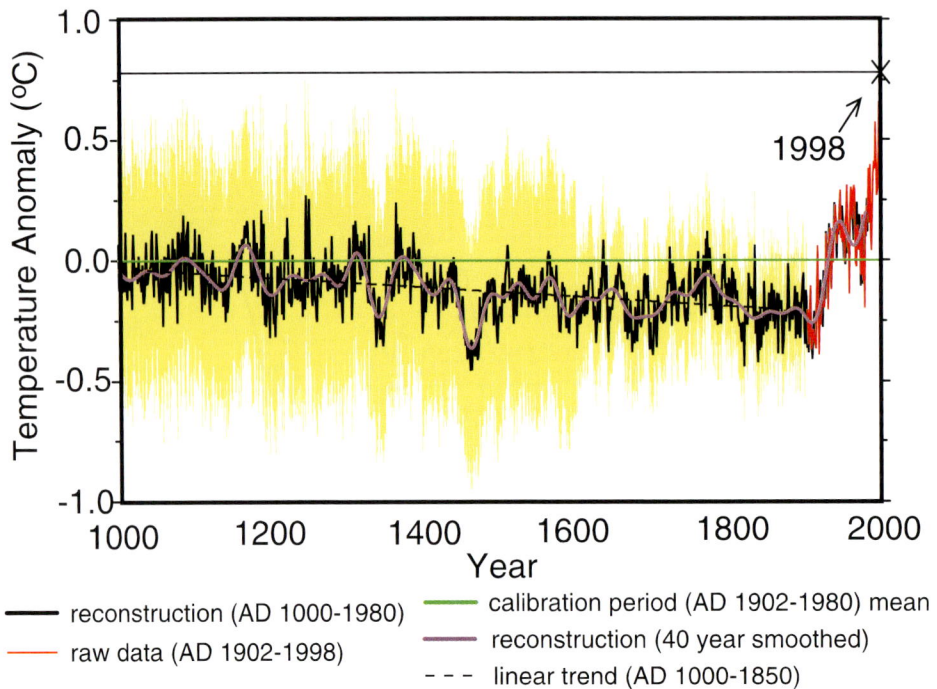

Fig. 4.14. Temperature variations since the last thousand years inferred from tree-ring and other proxy data (from [4.39]). The yellow shading represents two-standard-deviation limits

on these longer time scales, the rapid temperature increase in the twentieth century is striking. The parallel increase in the atmospheric CO_2 concentration shown in the front page of this article supports the causal relation between anthropogenic forcing and the observed global warming.

The corresponding results for a two-pattern test of the hypotheses of anthropogenic forcing by greenhouse gases only and by greenhouse gases together with sulfate aerosols is shown in Fig. 4.15. Most models yield significant detection and attribution for the second case of a superposition of radiative forcing by greenhouse gases and sulfate aerosols, but reject the hypothesis of forcing by greenhouse gases alone[6]. However, the attribution confidence levels are not large, and the large deviations between different models imply that the results should be treated with caution. Further work is clearly needed.

---

[6] This distinction could not be made by the one-dimensional test shown in Fig. 4.13, which accepted both hypotheses.

ECHAM3/LSG fingerprints:

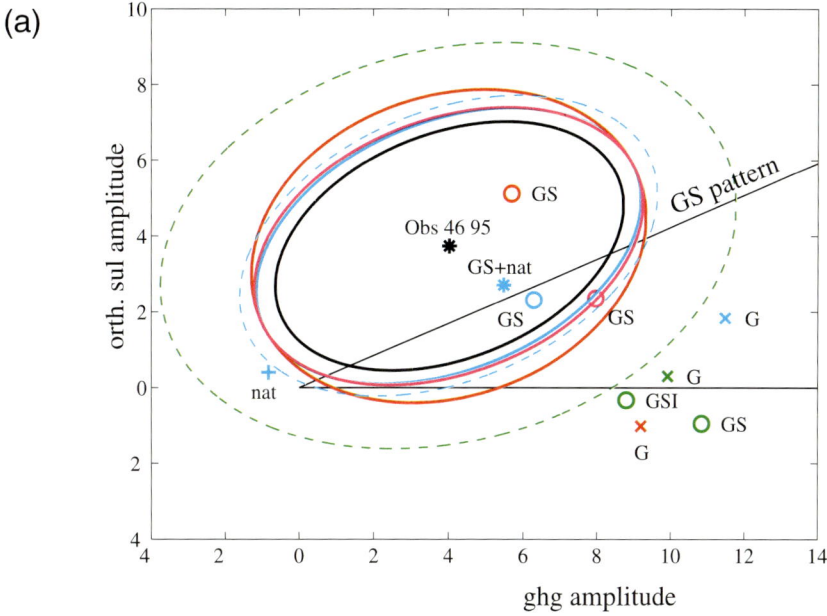

HadCM2 fingerprints

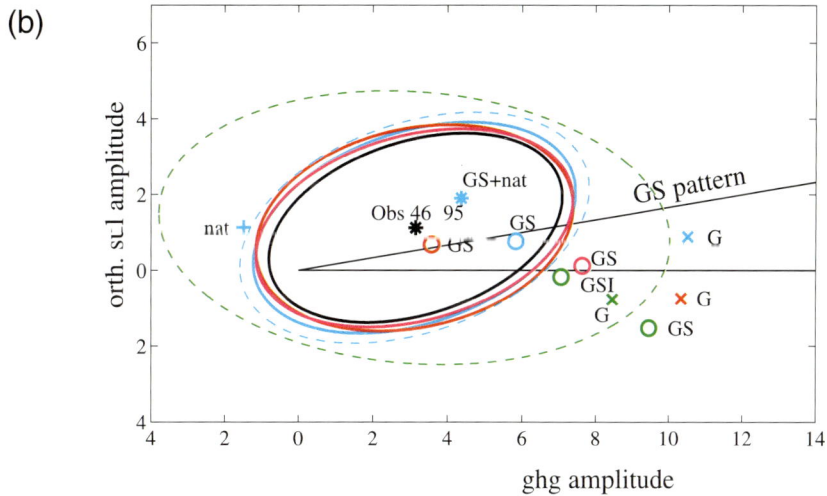

Fig. 4.15. Detection and attribution tests for two different anthropogenic forcing hypotheses: greenhouse gases only (signal along the x-axis), and greenhouse gases plus sulfate aerosols (signal along inclined line). Results are shown for the fingerprints derived from the Hamburg model (**a**) and the Hadley-Center model (**b**). Most models are consistent with the second hypothesis while rejecting the first hypothesis (for the descriptions of runs, e.g. GS, GSI, ... , see [4.17])

4.8 Nonlinear Signatures in Linear Response

In a recent paper, Corti et al. [4.4] questioned the application of linear response relations and the signal-plus-noise superposition principle underlying standard statistical detection and attribution analyses on the grounds that observed natural climate fluctuations exhibit a tendency to linger in selected states. This they explain as a nonlinear phenomenon, in analogy with the flipping between preferred states (quasi-attractors) in the nonlinear Lorenz system [4.33, 4.34]. As pointed out by Hasselmann [4.16], however, this interpretation is not in conflict with the linearization principle, since one is not in the neighborhood of a bifurcation point and the forcing can still be regarded as small. It points rather to an interesting interrelation between the inherent nonlinearity of the climate system, the linear system response to small perturbations, the impact of anthropogenic forcing on natural climate variability, and the stochastic forcing associated with short-term weather variability. These interrelations are particularly pertinent, since the principal impact of anthropogenic climate change will most likely be felt in the changes in the statistics of natural variability (storms, floods, droughts, heat waves, El Niño, and other extreme events) rather than in the changes in average properties such as the global mean temperature.

Corti et al. [4.4] base their findings on the observation that the probability distributions of climate variables are often found to be multimodal, a characteristic feature of nonlinear systems. A simple multimodular model that can be explained in terms of the climate-dynamic concepts discussed above is a stochastically forced multi-equilibrium system [4.16]. Consider a climate system for which the slow subsystem, without the coupling to the rapid weather variability of the atmospheric system, exhibits a number of separate equilibrium states. These can be envisaged, for example, as different circulation regimes of the coupled ocean-atmosphere system. In an analogous physical system characterized by potential forces, they would be represented by a number of potential wells. The introduction of the stochastic forcing by the weather variability of the atmosphere generates then climate-state trajectories that fluctuate around the potential well minima. If the potential wells are well separated, a given trajectory will be confined to the potential well in which the climate state happened to lie initially, and the resulting climate state probability distribution would be unimodal. However, if the ridges separating the different potential wells are sufficiently low compared to the energy level of the stochastic forcing, the trajectories will be able to cross over the ridges into neighboring potential wells, and the resulting probability distribution will be multimodal (Fig. 4.16). If an external force is applied to the system, the background potential surface becomes slanted. The trajectories will then tend to linger longer in the potential wells that lie lower on the slanted potential surface. The resulting probability distribution will retain its basic multimodal structure, but the weightings of the different probability peaks will be modified. Although the mean response of the climate system, integrated over the modified probability distribution, can still be described by a response pattern represented by a linear response relation of the form (4.8), the modifications in the probability distribution provide a more useful description of the impact of the external forcing on the climate system. For example, for the assessment of

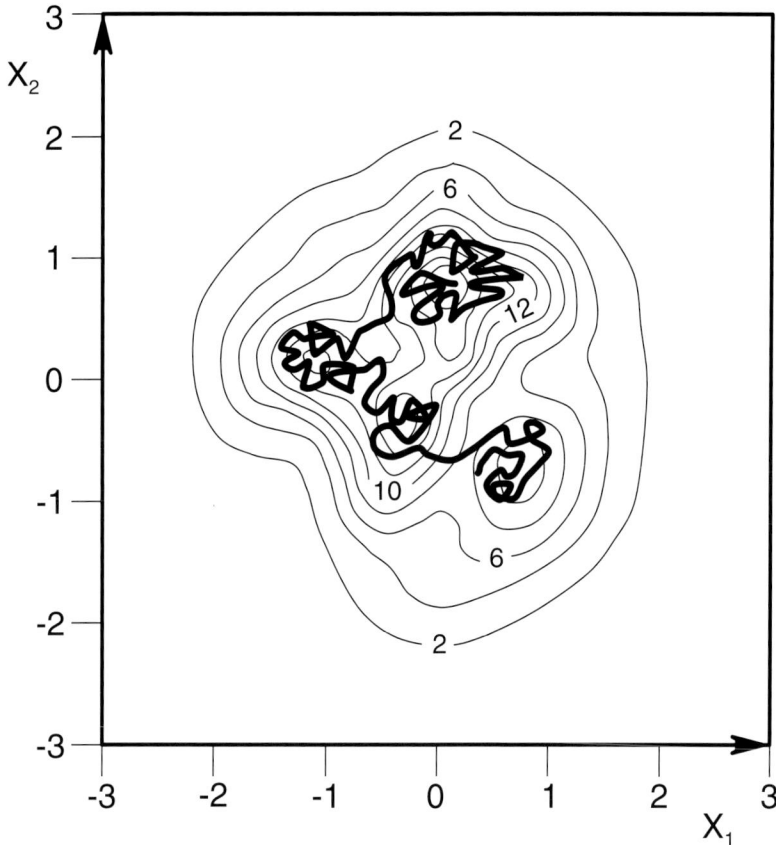

Fig. 4.16. Potential wells of a multi-equilibria nonlinear system, with random trajectories and tunneling between potential wells generated by stochastic forcing (from [4.16])

the impact of global warming on human living conditions and the environment, the changes in the frequency and magnitude of El Niño or the North Atlantic Oscillation, both of which have a strong influence on extreme weather events over large areas of the globe, are more relevant than the associated changes in the mean climate variables.

The impact of anthropogenic climate change on the probability distributions of climate variables and the occurrence of extreme weather events is an important current area of research that could reveal many interesting new features of the nonlinear dynamics of the stochastically forced climate system.

4.9 Conclusion

Our brief review demonstrates that climate is indeed predictable, both for predictions of the first and of the second kind. However, in both cases the predictability is limited. This is a general feature of chaotic systems, but for the climate system, the prediction limits have still to be more clearly delineated.

Successful climate predictions of the first kind, while encouraging and of
considerable economical value, have been achieved so far only for forecasts up
to about six months for the ENSO phenomenon. The predictability is largely
governed by upper-ocean equatorial wave dynamics, which are well understood
and relatively easily accessible to measurement. It remains to be investigated
whether the useful forecast time scale of ENSO can be significantly extended,
and whether longer period natural climate variations such as the North Atlantic Oscillation and similar fluctuations in the North Pacific region can be
predicted. Since these are presumably governed mainly by the longer time scale
properties of the ocean circulation, this should in principle be feasible, provided
the initial state of the ocean can be adequately established and sufficiently realistic ocean models developed. However, even if the initial ocean state could be
perfectly determined and used to initialize a perfect ocean model, the climate
predictability would still be limited ultimately by the basically unpredictable
stochastic forcing by short-term atmospheric weather fluctuations. The relative role of initial-state uncertainties and stochastic forcing in limiting the
predictability of the first kind of the ocean-atmosphere system have still to be
ascertained.

On millennium and still longer time scales (which were not discussed in this
brief overview) other components of the climate system, such as ice sheets and
biogeochemical cycles, need to be considered in addition to the ocean and atmosphere systems. Although many hypotheses have been proposed to explain
the observed natural climate variability on these time scales, a generally accepted satisfactory explanation of paleoclimatic climate variability, including,
for example, the ice-age cycles, has not emerged. An exception is the explanation of the variability observed in the discrete Milankovitch bands at periods
of 19, 22, and 40 kyr. This can be clearly attributed to the solar forcing due to
perturbations in the Earth's orbit (cf. [4.23]). However, although the forcing
is highly predictable, the variance in these bands constitutes only about 20%
of the total climatic variance on paleoclimatic time scales[7].

On the shorter time scales relevant for anthropogenic climate change, climate predictions of the second kind (or mixed first/second kind) face two
limitations: the superposition of natural climate variability noise on the computed mean anthropogenic response, and the possibility of the climate system
encountering an unstable bifurcation point.

The first limitation can be reduced in model simulations by averaging over
a number of independent predictions for modified initial conditions. However,
this option is not available for the empirical climate change signal against which
the predicted climate change signal needs to be compared. The uncertainty of

[7] Whether the large spectral peak observed at 100 kyr period can also be attributed to Milankowitch
forcing is still debated. The period corresponds to an orbital perturbation band, but the associated
solar forcing is too weak to explain the observed variance by straightforward linear response
arguments.

the empirical signal hinders the certain confirmation of an anthropogenic signal in the pattern of global warming observed during the twentieth century. Nevertheless, the application of multivariate optimal fingerprint methods indicates that the probability that the observed climate change signal can be attributed to natural variability noise is less than 5%. The observed signal pattern is furthermore found to be statistically consistent with the climate change signal predicted for a radiative forcing consisting of a superposition of greenhouse-gas and sulfate-aerosol forcing.

In addition to conventional detection and attribution tests based on estimated natural climate variability statistics, there exist also numerous indices of climate change (melting Arctic sea ice, increases in the occurrence of extreme events, re-treating glaciers, dying corals, etc.) which point collectively to anthropogenic global warming, but which cannot be tied down rigorously in the form of signal-to-noise ratios because of lacking information on the relevant natural variability levels. However, this information can be incorporated within the framework of a generalized Bayesian detection and attribution analysis (cf. [4.15, 4.46]).

The second limitation, the risk of encountering an unstable bifurcation point in the evolution of the climate change trajectory generated by anthropogenic forcing, is real but difficult to quantify. Essentially all coupled ocean-atmosphere climate models exhibit a breakdown of the present conveyor belt circulation of the ocean for sufficiently large warming or freshening of the North Atlantic surface waters that feed the deep ocean circulation and the associated Gulf Stream return flow system (cf. Chap. 6 and Fig. 6.0). The required forcing for most models lies within the range of anthropogenic forcings that can be anticipated for a 'business-as-usual' greenhouse warming scenario, but the onset of the instability differs widely from model to model. Similar uncertainties apply for other climate instabilities, such as a break-off of the West-Antarctic ice sheet or a runaway greenhouse effect triggered by a release of large quantities of methane currently trapped in Siberian permafrost regions. Although the probability for the occurrence of such an instability within the next one or two centuries is generally regarded as small, the impact could be unforeseeable. Thus, the possibility that greenhouse warming could lead to a climate instability must be regarded as one of the most serious concerns of anthropogenic climate change.

References

4.1 K. Arpe, L. Bengtsson, G.S. Golitsyn, I.I. Mokhov, V.A. Semenov, and P.V. Sporyshev, Geophys. Res. Abstr. **1**, 549 (1999).

4.2 T.P. Barnett, K. Hasselmann, M. Chelliah, T. Delworth, G. Hegerl, P. Jones, E. Rasmusson, E. Roeckner, C. Ropelewski, B. Santer, and S. Tett, Bull. Am. Meteorol. Soc. **12**, 2631 (1999).

4.3 E. Brückner, in *Geographische Abhandlungen*, edited by A. Penck (Hölzel, Vienna and Olmütz, 1890), p. 325.

4.4 S. Corti, F. Molteni, and T.N. Palmer, Nature **398**, 799 (1999).

4.5 U. Cubasch, K. Hasselmann, H. Hvck, E. Maier-Reimer, U. Mikolajewicz, B.D. Santer, and R. Sausen, Clim. Dyn. **2**, 55 (1992).

4.6 U. Cubasch, B.D. Santer, A. Hellbach, G.C. Hegerl, H. Höck, E. Maier-Reimer, U. Mikolajewicz, A. Stössel, and R. Voss, Clim. Dyn. **10**, 1 (1994).

4.7 U. Cubasch, G.C. Hegerl, R. Voss, and J. Waszkewitz, Clim. Dyn. **13**, 757 (1997).

4.8 C. Frankignoul, in *Analysis of Climate Variability: Applications of Statistical Techniques*, edited by H. v. Storch and A. Navarra (Springer, Berlin, 1995), p. 25.

4.9 C. Frankignoul and K. Hasselmann, Tellus **29**, 289 (1977).

4.10 GARP, *The Physical Basis of Climate and Climate Modelling, Global Atmospheric Prediction Programme* (World Meteorological Organization, Geneva, Rep.16, 1975).

4.11 K. Hasselmann, Tellus **28**, 473 (1976).

4.12 K. Hasselmann, in *Meteorology of Tropical Oceans*, Vol. 251 (R. Meteorol. Soc., Reading, 1979).

4.13 K. Hasselmann, Science **276**, 914 (1997).

4.14 K. Hasselmann, Clim. Dyn. **13**, 601 (1997).

4.15 K. Hasselmann, Q. J. R. Meteorol. Soc. **124**, 2541 (1998).

4.16 K. Hasselmann, Nature **398**, 755 (1999).

4.17 G.C. Hegerl, K. Hasselmann, U. Cubasch, J.F.B. Mitchell, E. Roeckner, R. Voss, and J. Waszkewitz, Clim. Dyn. **13**, 613 (1997).

4.18 G.C. Hegerl, P. Stott, M. Allen, J.F.B. Mitchell, S.F.B. Tett, and U. Cubasch, Clim. Dyn. **16**, 737 (2000)

4.19 K. Herterich and K. Hasselmann, J. Phys. Oceanogr. **12**, 2145 (1987).

4.20 G. Hooss, R. Voss, K. Hasselmann, E. Maier-Reimer, and F. Joos, Clim. Dyn. **18**(3/4), 189 (2001).

4.21 J.T. Houghton, G.L. Jenkins, and J.J. Ephraums, *Climate Change: The IPPC Scientific Assesment* (Cambridge University Press, Cambridge, 1990).

4.22 J.T. Houghton, L.G. Meira Filho, B.A. Callander, N. Harris, A. Kattenberg, and K. Maskell, *Climate Change 1995, The Science of Climate Change* (Cambridge University Press, Cambridge, 1996).

4.23 J. Imbrie and K.P. Imbrie, *The Ice Ages*, (Enslow, Short Hills, 1979).

4.24 P. Imkeller and J.-S. v. Storch, *Stochastic Climate Models* (Birkhäuser, Basel, 2001).

4.25 A. Indermühle, E. Monnin, B. Stauffer, T.F. Stocker, and M. Wahlen, Geophys. Res. Lett. **27**, 735 (2000).

4.26 J.T. Houghton, Y. Ding, D.J. Griggs, M. Noguer, P.J. van der Linden, and D. Xiaosu (eds.), *Climate Change 2001: The Scientific Basis, Contribution of Working Group I to the Third Assessment Report of the Intergovernmental Panel on Climate Change (IPCC)* (Cambridge University Press, Cambridge, 2001).

4.27 P.D. Jones and K.R. Briffa, The Holocene **2**, 165 (1992).

4.28 F. Joos, M. Bruno, R. Fink, U. Siegenthaler, T. Stocker, C. LeQuéré, and J. Sarmiento, Tellus **48B**, 397 (1996).

4.29 F. Joos, G.K. Plattner, T. Stocker, O. Marchall, and A. Schmittner, Science **284**, 464 (1999).

4.30 M. Latif and T.P. Barnett, Science **266**, 634 (1994).

4.31 P. Lemke, Tellus **29**, 385 (1977).

4.32 P. Lemke, E.W. Trinkl, and K. Hasselmann, J. Phys. Oceanogr. **12**, 2100 (1980).

4.33 E.N. Lorenz, J. Atmos. Sci. **20**, 130 (1963).

4.34 E.N. Lorcnz, J. Atmos. Sci. **20**, 448 (1963).

4.35 E.N. Lorenz, *The Nature and Theory of the General Circulation of the Atmosphere* (World Meteorological Organization, Geneva, 1967).

4.36 E.N Lorenz, *Climate Predictability, The Physical Basis of Climate and Climate Modelling, Global Atmospheric Prediction Programme* (World Meteorological Organization, Rep. 16, Geneva, 1975), p.132.

4.37 E. Maier-Reimer and K. Hasselmann, Clim. Dyn. **2**, 63 (1987).

4.38 E. Maier-Reimer, U. Mikolajewicz, and A. Winguth, Clim. Dyn. **12**, 711, (1996).

4.39 M.E. Mann, R.S. Bradley, and M.K. Hughes, Geophys. Res. Lett. **26**, 759 (1999).

4.40 R. Meyer, F. Joos, G. Esser, M. Heimann, G. Hooss, G. Kohlmaier, W. Sauf, R. Voss, and U. Wittenberg, Global Biogeochemi. Cycles **13**, 785 (1999).

4.41 U. Mikolajewicz and E. Maier-Reimer, Clim. Dyn. **4**, 145 (1990).

4.42 J.M. Mitchell Jr., Stochastic models of air sea interaction and climate fluctuation, *Symp. Arctic Heat Budget and Atmospheric Circulation, Lake Arrowhead, Calif.*, Mem. RM-5233-NSF, The Rand Corporation, Santa Monica (1966).

4.43 G.R. North, K.Y. Kim, S.P. Shen, and J.W. Hardin, J. Clim. **8**, 401 (1995).

4.44 G.R. North and K.Y. Kim, J. Clim. **8**, 409 (1995).

4.45 R.W. Reynolds, Tellus **30**, 97 (1978).

4.46 J. S. Risbey, M. Kandlikar, and D. Karoly, Clim. Res. **16**, 61 (2000).

4.47 J. Sarmiento and E. Sundquist, Nature **356**, 589 (1992).

4.48 J. Sarmiento, T. Hughes, R. Stauffer, and S. Manabe, Nature **393**, 245 (1998).

4.49 U. Siegenthaler and H. Oeschger, Science **199**, 388 (1978).

4.50 M.C. Wang and G.E. Uhlenbek, Rev. Mod. Phys. **17**, 323 (1945).

5. Atmospheric Persistence Analysis: Novel Approaches and Applications

*Armin Bunde, Shlomo Havlin, Eva Koscielny-Bunde,
and Hans Joachim Schellnhuber*

The persistence of short-term weather states is a well known phenomenon: a warm day is more likely to be followed by a warm day than by a cold one and vice versa. In a recent series of studies we have shown that this rule may well extend to months, years, and decades, and on these scales the decay of the persistence seems to follow a universal power-law. We review these studies and discuss how the law can be used to test the state-of-the-art climate models.

5.1 Introduction

The climate system, i.e. the intricately and strongly coupled complex of atmosphere, oceans, inland and sea ice, terrestrial and marine biota, pedosphere and lithosphere, is a perfect example for nonlinear dynamics (see e.g. [5.1] and the references therein). In particular, this system is capable of irregular ('singular') behavior in two qualitatively different ways:

First, the intrinsic processes of the climate system machinery tend to generate episodic excursions at all scales that wander far away from the long-term averages of pertinent variables such as the regional sea-surface temperature (see the famous ENSO phenomenon as reviewed in [5.2]).

Second, the system may operate in fundamentally distinct global modes as exemplified by the quasi-periodic alternation of glaciations and deglaciations, which seem to have affected the entire planet in the distant past [5.3]. What ultimately causes natural (or physiogenic) transitions between the possible 'eigenmodes' of climate dynamics still remains an enigma, but recent progress

---

◄ **Fig. 5.0.** Typical result of a weather persistence situation in Niger (Keita, Sahel): a long period of drought, which started in 1965/1966, reached its peak in 1984. This has led to a continous decline in agricultural production and rapid desertification in the Keita district. Courtesy of FAO Media Server/F. Paladini

in intermediate-complexity Earth system modeling [5.4, 5.5] may open up new ways of understanding. There is growing evidence, however, that civilizatory (or anthropogenic) forcing through incessant greenhouse gas emissions and land cover change is about to drive the climate system towards a qualitatively different state [5.6]. This journey may be accompanied by disruptive events like the shut-down of North Atlantic deep water formation (NADW) (cf. Chap. 6), or the occasional (if not long-term) suppression of continental monsoons [5.7].

Irrespective of the underlying causes, it is unfortunately very hard to realize (let alone prove) whether an observed variation of the atmospheric variable considered is either an ordinary excursion within a stable climate system regime, or the precursor/expression/repercussion of a regime flip, or already an episodical aberration from a new equilibrium state that has been established in an insidious way. Sophisticated methods are needed, in particular, to discriminate 'trends' from 'fluctuations' (regarding this discussion, see also Chap. 4 in this book).

The challenge of understanding weather and climate variability is motivated by several practical objectives as well: operational weather forecasting, the management of meteorological extremes [5.8], and - of course - climate protection at large. The last goal is an eminent one, and can only be reached if climate system science provides a number of insights fairly soon. Pertinent questions are the following ones: is the onset of global warming already discernible in the observational body of evidence? Is it really humankind that causes the observed excursions from centennial or even millennial average behavior? How far will anthropogenic interference probably drive the climate system away from its pre-industrial regime? How will global climate modification translate into new patterns of regional weather conditions and their fluctuations? The list can easily be extended.

The answers to most of these questions require, above all, a thorough examination of the patterns and laws of intrinsic atmospheric variability, and of the ways by which this variability can be affected in character by extrinsic (e.g. anthropogenic) perturbations. At present, the 'signal-to-noise-ratio' in identifying extrinsic and intrinsic forcing of weather climate excursions is certainly not satisfactory, and has to be improved considerably. The World Climate Research Programme (WCRP) has acknowledged this dilemma by launching a major initiative towards understanding the self-organized variability of the coupled atmosphere-ocean system [5.9].

Atmospheric and oceanographic research usually proceeds along two main avenues that we are going to describe next. (i) Inspection and processing of the available meteorological records by appropriate time series and spatial distribution analysis techniques; (ii) generation and analysis of observation-based, interpolated or purely simulated weather and climate records, respectively, through the operation of a hierarchy of models. Regarding the second avenue, we anticipate a crucial point here. Simulation models have to be calibrated (if not validated) in order to mimick the past and present climate including the associated fluctuations/persistence characteristics. If models fail to reproduce crucial traits of the recorded observations, then the confidence in their prognostic power is hard to justify. This is particularly true when it comes to the prediction of future regimes of extreme meteorological events. We will elaborate on this issue in the final part of our contribution.

5.2 Analysis of Meteorological Methods

Among the standard mathematical techniques that have been used are calculations of means, variations, and power spectra, decomposition into empirical orthogonal functions (EOF) and/or principal components (PC) (see, for example, Chap. 3). Very recently, more advanced techniques such as detrended fluctuation analysis (see Chaps. 7, 8) and wavelet analysis (see Chap. 2) have been used which are able to systematically separate trends from fluctuations at different time scales. In addition neural networks have been applied to study atmospheric time series [5.13]. The amount of relevant literature on data analysis is enormous (for reviews, see e.g. [5.14]), and we will refer here only to the work most relevant to our specific study.

A considerable amount of effort has been devoted to analyzing temporal correlations that characterize the persistence of weather and climate regimes. The persistence of short-term weather states is a well-known phenomenon: there is a strong tendency for subsequent days to remain similar, a warm day is more likely to be followed by a warm day than by a cold day and vice versa. The typical time scale for weather changes is about one week, a time period which corresponds to the average duration of so-called 'general weather regimes' or 'Grosswetterlagen' (see Chap. 3). This property of persistence is often used as a 'minimum skill' forecast for assessing the usefulness of short to medium range numerical weather forecasts. Longer term persistence of synoptic regimes up to time scales of several weeks is often related to circulation patterns associated with blocking [5.18]. A blocking situation occurs when a very stable high pressure system is established over a particular region and remains in place for several weeks, as opposed to the usual time scale of 3–5 days for synoptic systems. As a result the weather in the region of the high remains fairly persistent throughout the period. Furthermore, transient low pressure systems are deflected around the blocking high so that the region downstream of the high experiences a larger than usual number of storms.

There have also been indications that weather persistence exists over many months or seasons [5.19], between successive years, and even over several decades [5.20, 5.21]. Such persistence is usually associated with slowly varying external (boundary) forcing such as sea surface temperatures and anomaly patterns. On the scale of months to seasons, one of the most pronounced phenomena is the El Niño/Southern Oscillation (ENSO) event which occurs every 3-5 years and which strongly affects the weather over the tropical Pacific as well as over North America [5.22]. It has also been recently suggested that El Niño years are associated with increased rainfall over the Eastern Mediterranean [5.23]. Although the link between extratropical weather/climate and sea surface temperature has been more difficult to establish, several recent studies have successfully proven that a connection exists on multiyear to decadal time scales between (i) the climate of North America and the North Pacific Ocean [5.24], and (ii) the climate over Europe and the North Atlantic Ocean as expressed by the North-Atlantic-Oscillation (NAO) index (see Chap. 3) [5.20,5.21,5.25]. On the even longer multidecadal to century time scales, external forcing associated with anthropogenic effects (e.g. increasing greenhouse gases and changing land use) also appear to play an important role in addition to the natural variability of the climate system [5.26]. Clearly separating the anthropogenic forcing from the natural variability of the atmosphere may prove to be a major challenge since the anthropogenic signal may project onto and therefore be hidden in the modes of natural climate variability [5.27] (regarding this attribution and detection problem, see Chap. 4).

A number of power spectra studies of long-term meteorological research also support temporal correlations (see e.g. [5.28]). A typical temperature power spectrum consists of background noise and oscillatory maxima. It was early realized that the noise was not just a constant in the power spectra ('white' noise representing the absence of correlations), but decayed somehow with frequency (representing the presence of correlations). To distinguish it from 'white' noise, the term 'red' noise was coined. While usually an exponential decay of the persistence was anticipated, it has been argued [5.29, 5.30] that power-law decays are also possible.

In general, traditional methods of statistical analysis anticipate that the statistical properties of a signal remain the same throughout the signal. This assumption of stationarity, however, is not true for many systems in nature. Nonstationarities occur, for example, in heartbeat time series [5.32, 5.33] (see Chaps. 7 and 8) in DNA sequences (see Chap. 2), [5.34, 5.37], in economic records [5.39] (see Chap. 12), as well as in hydrologic time series [5.40]. Temperature records can be affected by several types of trends, among them urban and/or global warmings.

To avoid detection of spurious correlations arising from nonstationarities, new statistical-physics tools (wavelet techniques (WT) (see e.g. [5.37] and Chap. 2) and detrended fluctuation analysis (DFA)) have been developed recently (see e.g. [5.32–5.34]). DFA and WT can systematically eliminate trends in the data and thus reveal intrinsic dynamical properties such as distributions, scaling, and long-range correlations very often masked by nonstationarities. In a recent series of studies [5.41] we have used DFA and WT to study temperature correlations in different climatic zones on the globe. The results indicate that a universal long-range power-law correlation may exist which governs atmospheric variability at all spatio-temporal scales: the persistence, characterized by the autocorrelation $C(s)$ of temperature variations separated by s days, approximately decays as

$$C(s) \sim s^{-\gamma}, \tag{5.1}$$

with roughly the same exponent $\gamma \simeq 0.7$ for all stations considered. The range of this universal persistence law exceeds one decade, and is possibly even longer than the range of the temperature series considered. There are two major consequences: (a) conventional methods based on moving averages can no longer be used to separate trends from fluctuations; (b) conventional methods for the evaluation of the frequency of extreme low or extreme high temperature are based on the hypothesis that the temperature fluctuations are essentially uncorrelated. The appearance of long-range correlations sheds doubts on these methods.

We will describe the detrending methods in Sect. 5.4 and their application to temperature records of several meteorological stations around the globe in Sect. 5.5.

5.3 The Modeling Approach

Regarding the modeling approach towards simulating and explaining atmospheric variability on various time scales, major progress has been made during the last two decades. Today, the research community routinely and extensively makes use of atmospheric, oceanic, and coupled ocean-atmosphere circulation models where the major physical processes are included. Moreover, these models include representations of land-surface processes, sea-ice related processes, and many other complex forcing mechanisms within the climate system. All processes are represented by mathematical equations which are solved numerically using a three-dimensional grid with a domain, resolution, and complexity determined by the topic of interest. For very long time integrations of thousands of years it is impossible to apply full general circulation models (GCM) due to their extremely high computational demand. Consequently, intermediate complexity models are used in these cases.

For global climate simulations on time scales ranging from months to decades or centuries, GCMs are used with typical resolutions of 200–300 km in the horizontal and 1 km in the vertical [5.47, 5.48]. For regional climate simulations, similar models are used but the domain is limited to a few thousand kilometers and the horizontal resolution is typically increased to 50 km or less [5.49, 5.50]. In any case the grid resolution can never explicitly simulate all relevant scales of motion. This necessitates the parameterization of the smaller, subgrid scale processes such as cloud physics, radiative transfer, and macroscale turbulent mixing. Recently, it has also been found that previously neglected processes such as dust induced heating may also be important [5.51].

The state-of-the-art coupled atmosphere–ocean general circulation models (AOGCM) are able to simulate many of the important large-scale features of the climate system rather well. This includes seasonal, horizontal, and vertical variations. They also explain the response to greenhouse gases and aerosols in terms of physical processes. In addition to this, other less pronounced variations in climate are reproduced with reasonable accuracy (e.g. the relationship between El Niño and rainfall in Central America and the northern part of South America).

The systematic evaluation and intercomparison of climate model results has proven to be a useful and effective mechanism for identifying common model weaknesses. In general evaluations have been conducted for the atmospheric, oceanic, land-surface, and sea-ice components of the models, and for the sensitivity of the links among these components. Until now these validations have not addressed the question of whether such models can reproduce the long-term climate memory in an appropriate way. If the simulations of the model are valid, then the patterns and relationships discovered by analyzing real observations and data must also be identifiable in the virtual world as represented by the model outputs.

In Sect. 5.6 we show how the current GCMs can be tested by applying the detrending techniques of Sect. 5.4 to the model data. We show that we can judge the models by their ability of reproducing the proper type of trends and long-range correlations inherent in the real data.

5.4 Record Analysis: Detrending Techniques

Consider, for instance, a record T_i of maximum daily temperatures measured at a certain meteorological station. The index i counts the days in the record, $i = 1, 2, \ldots, N$. For eliminating the periodic seasonal trends, we concentrate on the departures of the T_i, $\Delta T_i = T_i - \overline{T}_i$, from the mean maximum daily temperature $\overline{T}_i$ for each calendar date i, say 1st of April, which has been obtained by averaging over all years in the temperature series.

We describe two detrending analysis methods: (a) detrended fluctuation analysis (DFA) and (b) wavelet methods (WT). The DFA was originally developed by Peng et al. [5.34] to investigate long-range correlations in DNA sequences and heart beat intervals, where nonstationarities similar to the nonstationarities in the temperature records [5.41] can occur. The wavelet methods in general are very convenient techniques to investigate fluctuating signals [5.44].

A very useful introduction to the wavelet technique with several applications is given in Chap. 2 of this book.

Both DFA and wavelet techniques have been used to analyze the correlation function $C(s)$. The correlation function describes how the persistence decays in time. $C(s)$ is defined by $C(s) = \langle \Delta T_i \Delta T_{i+s} \rangle$. The average $\langle ... \rangle$ is over all pairs with same time lag s. For reducing the level of noise present in the finite temperature series, we consider the 'temperature profile'

$$Y_n = \sum_{i=1}^{n} \Delta T_i, \qquad n = 1, 2, \ldots, N. \tag{5.2}$$

We can consider the profile Y_n as the position of a random walker on a linear chain after n steps. The random walker starts at the origin and performs, in the ith step, a jump of length ΔT_i to the right, if ΔT_i is positive, and to the left, if ΔT_i is negative. According to random walk theory, the fluctuations $F^2(s)$ of the profile, in a given time window of size s, are related to the correlation function $C(s)$. For the relevant case (5.1) of long-range power-law correlations, $C(s) \sim s^{-\gamma}$, $0 < \gamma < 1$, the mean-square fluctuations $\overline{F^2(s)}$, obtained by averaging over many time windows of size s (see below) increase by a power-law [5.52],

$$\overline{F^2(s)} \sim s^{2\alpha}, \quad \alpha = 1 - \gamma/2. \tag{5.3}$$

For uncorrelated data (as well as for correlations decaying faster than $1/s$), we have $\alpha = 1/2$.

To find how the square-fluctuations of the profile scale with s, we first divide each record of N elements into $K_s = \lfloor N/s \rfloor$ nonoverlapping subsequences of size s starting from the beginning and K_s nonoverlapping subsequences of size s starting from the end of the considered temperature series. We determine the square fluctuations $F_\nu^2(s)$ in each segment ν and obtain $\overline{F^2(s)}$ by averaging over all segments. When plotted in a double logarithmic way the fluctuation function

$$F(s) \equiv [\overline{F^2(s)}]^{\frac{1}{2}} \sim s^\alpha \tag{5.4}$$

is a straight line at large s values, with a slope $\alpha > \frac{1}{2}$ in the case of long-range correlations. The various methods differ in the way the fluctuation function is calculated.

5.4.1 Fluctuation Analysis (FA)

In the simplest type of analysis (where trends are not going to be eliminated), we obtain the fluctuation functions just from the values of the profile at both endpoints of the νth segment,

$$F_\nu^2(s) = [Y_{\nu s} - Y_{(\nu-1)s}]^2, \tag{5.5}$$

and average $F_\nu^2(s)$ over the $2K_s$ subsequences

$$\overline{F^2(s)} = (1/K_s) \sum_{\nu=1}^{K_s} F_\nu^2(s). \tag{5.6}$$

Here, $\overline{F^2(s)}$ can be viewed as mean square displacement of the random walker on the chain, after s steps. We obtain Fick's diffusion law $\overline{F^2(s)} \sim s$ for uncorrelated ΔT_i values.

We like to note that this fluctuation analysis corresponds to the R/S method introduced by Hurst (for a review, see e.g. [5.53]). Since both methods do not eliminate trends, they do not give a clear picture when used alone. In many cases they cannot distinguish between trends and long-range correlations when applied to time records without supplementary calculations.

5.4.2 Detrended Fluctuation Analysis (DFA)

There are different orders of DFA that are distinguished by the way the trends in the data are eliminated. In lowest order (DFA1) we determine, for each subsequence ν, the best *linear* fit of the profile, and identify the fluctuations by the standard deviation $F_\nu^2(s)$ of the profile from this straight line. This way, we eliminate the influence of possible linear trends on scales larger than the segment sizes. Note that linear trends in the profile correspond to patch-like trends in the original record. DFA1 has been proposed originally by Peng et al. [5.34] when analyzing correlations in DNA.

DFA1 can be generalized straightforwardly to eliminate higher order trends: in second order DFA (DFA2) one calculates the standard deviations $F_\nu^2(s)$ of the profile from best *quadratic* fits of the profile, this way eliminating the influence of possible linear and parabolic trends on scales larger than the segment considered. For an illustration, see Fig. 8.7 and 8.8 in Chap. 8. In general, in the nth-order DFA technique, we calculate the deviations of the profile from the best nth-order polynomial fit and can eliminate this way the influence of possible $(n-1)$th-order trends on scales larger than the segment size.

It is essential in the DFA-analysis that the results of several orders of DFA (e.g. DFA1-DFA5) are compared with each other. The results are only reliable when above a certain order of DFA they yield the same type of behavior. When compared with FA one can get additional insight into possible nonstationarities in the data.

To show how the trend elimination works, we have generated an artificial record T_i^{corr}, where the data are long-range correlated with an exponent $\gamma = 0.7$, and have added linear, quadratic, cubic, and quartic trends $T_i^{\mathrm{trend}} = 10^4(i/10^5)^p$, $p=1,2,3,4$ to the data. The variance of the T_i^{corr} values is one. The record consists of 100 000 data points, and is not much larger than the length-2 of the temperature records we have analyzed (between 30 000 and

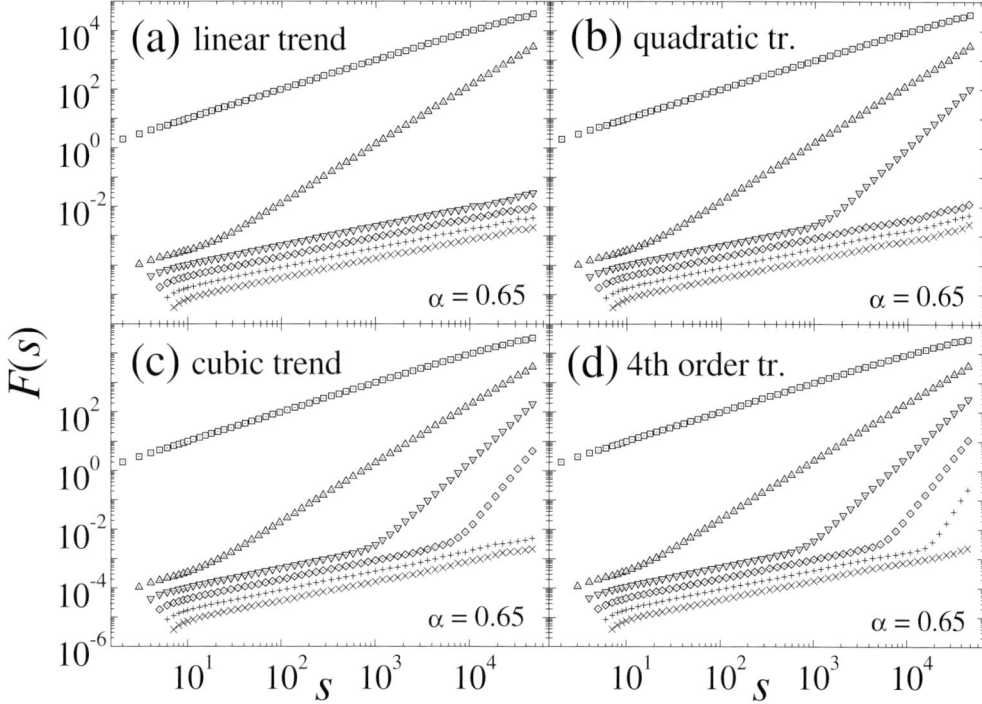

Fig. 5.1. Analysis of artificial long-range correlated data with (**a**) linear, (**b**) quadratic, (**c**) cubic, and (**d**) quartic trends. The correlation exponent is $\gamma = 0.65$. The four figures show the fluctuation functions obtained by FA ($\Box$), DFA1 ($\triangle$), DFA2 ($\triangledown$), DFA3 ($\Diamond$), DFA4 ($+$), and DFA5 ($\times$) for the four sets of data. The scale of the fluctuations is arbitrary (after [5.54])

$80\,000$). The results for $F(s) \equiv [\overline{F^2(s)}]^{\frac{1}{2}}$ from FA and DFA1-DFA5 are shown in Fig. 5.1a d. The fluctuation function $F(s)$ obtained by the FA method is nearly linear in s and fully dominated by the large trends. In DFA1-DFA5, when the trends are not completely eliminated, the fluctuation functions show a crossover from the predicted slope $\alpha = 1 - \gamma/2 = 0.65$ towards a larger slope. The crossover time increases with increasing order of DFA. From the trend elimination and the position of the crossovers one can estimate the trend. At short times an irrelevant DFA-typical crossover occurs, which can be eliminated by dividing by a proper normalization function [5.54]. In this chapter we shall focus on the application of DFA to temperature records. For completeness we describe also the wavelet technique that has been used in [5.41] and gives similar results.

5.4.3 Wavelet Techniques (WT)

The wavelet methods employed in [5.41] are based on the determination of the mean values $\overline{Y}_\nu(s)$ of the profile in each segment ν (of length s), and the calculation of the fluctuations between neighboring segments. First we divide, as above, the temperature profile into $2 \times K_s$ subsequences. Then we determine in each segment ν the mean values $\overline{Y}_\nu(s)$ of the profile. The various techniques we have used in analyzing temperature fluctuations differ in the way the fluctuations between the average profiles are treated and possible nonstationarities are eliminated.

(i) In the first-order wavelet method (WT1), we simply determine the fluctuations from the first derivative

$$F_\nu^2(s) = (\overline{Y}_\nu(s) - \overline{Y}_{\nu+1}(s))^2. \tag{5.7}$$

WT1 corresponds to FA where trends in the profile of a weather station originating, e.g. from trends consisting of different constant levels that might arise from the exchange of a thermometer, are eliminated, while linear increase of temperature, e.g. due to urban development around the station, are not eliminated.

(ii) In the second-order wavelet method (WT2), we determine the fluctuations from the second derivative

$$F_\nu^2(s) = (\overline{Y}_\nu(s) - 2\overline{Y}_{\nu+1}(s) + \overline{Y}_{\nu+2}(s))^2. \tag{5.8}$$

So, if the profile consists of a trend term linear in s and a fluctuating term, the trend term is eliminated. Regarding trend elimination, WT2 corresponds to DFA1.

(iii) In the third-order wavelet method (WT3), we determine the fluctuations from the third derivative

$$F_\nu^2(s) = (\overline{Y}_\nu(s) - 3\overline{Y}_{\nu+1}(s) + 3\overline{Y}_{\nu+2}(s) - \overline{Y}_{\nu+3}(s))^2. \tag{5.9}$$

By definition, WT3 eliminates linear and parabolic trend terms in the profile. In general, in 'WTn' we determine the fluctuations from the nth derivative, this way eliminating trends described by nth-order polynomials in the profile.

Finally, as for the DFA, we average in each case the quantity $F_\nu^2(s)$ over the $2K_s$ subsequences of the temperature series considered. Regarding the elimination of trends, the nth-order DFA corresponds to the $(n+1)$th-order wavelet method.

Methods (i)-(iii) are called wavelet methods, since they can be interpreted as transforming the profile by discrete wavelets representing first-, second-, and third-order cumulative derivatives of the profile (see Chap. 2). The first-order wavelets are known in the literature as Haar wavelets. In principle, one could also use different shapes of the wavelets (e.g. Gaussian wavelets with width s), which have been used by Arneodo et al. [5.37] to study long-range correlations in DNA.

5.4.4 Multifractal Analysis (MFA)

For a further characterization of meteorological records it is meaningful to extend (5.6) by considering the more general average

$$\overline{F^q(s)} = (1/K_s) \sum_{\nu=1}^{K_s} [F_\nu(s)]^q , \tag{5.10}$$

with q between $\pm\infty$. For $q \ll -1$ the small fluctuations will dominate the sum, while for $q \gg 1$ the large fluctuations are dominant. It is reasonable to assume that the q-dependent average scales with s as

$$\overline{F^q(s)} \sim s^{q\beta(q)}, \tag{5.11}$$

with $\beta(2) = \alpha$. Equation (5.11) generalizes (5.2). If $\beta(q)$ is independent of q, the generalized fluctuation function

$$F_q(s) \equiv (\overline{F_q(s)})^{1/q} \sim s^\alpha \tag{5.12}$$

is independent of q, and both large and small fluctuations scale the same. Accordingly, a single exponent is sufficient to characterize the time series, and we have a *monofractal*.

If $\beta(q)$ is not identical to α, we have a *multifractal*. In this case, the dependence of $\beta(q)$ on q characterizes the time series. Instead of $\beta(q)$ one considers frequently the spectrum $f(\omega)$ that one obtains from $\omega = d(q\beta(q))/dq$, $f(\omega) = q\omega - q\beta(q) + 1$. In the monofractal limit we have $f(\alpha) = 1$. Instead of the DFA method described here, the related wavelet transform modulus maxima method has been used often to study multifractal signatures of time series. For applications we refer to Chap. 2.

5.5 Analysis of Temperature Records

5.5.1 Correlation Analysis

Figures 5.2 and 5.3 show the results of the FA and DFA analysis of the maximum daily temperatures T_i of the following weather stations (the length of the records is written within the parentheses): Luling (USA, 90 yr), Kasan (Russia, 96 yr) (Fig. 5.2a, c), Tucson (USA, 97 yr), Melbourne (136 yr), Seoul (86 yr), Prague (218 yr) (Fig. 5.3a-d). The results are typical for a large number of records that we have analyzed so far (see [5.41]). In the log-log plots, all curves are (except at small s-values) approximately straight lines, with a slope $\alpha \cong 0.65$. There exists a natural crossover (above the DFA-crossover) that can be best estimated from FA and DFA1. As can be verified easily, the

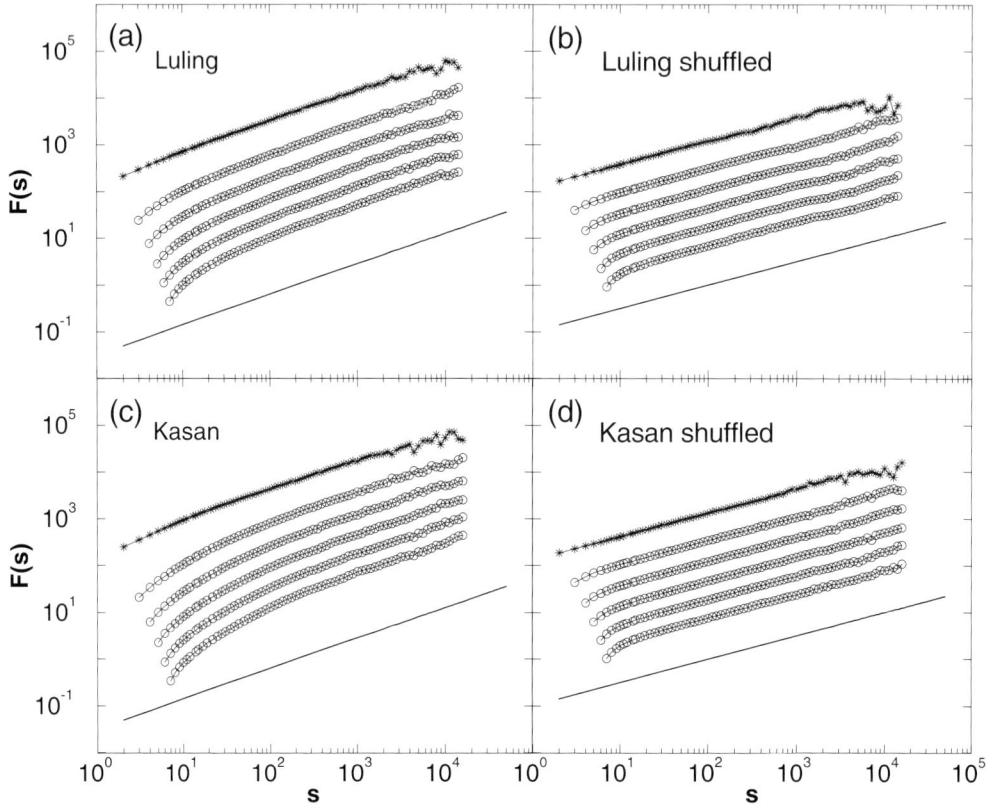

Fig. 5.2. Analysis of daily maximum temperature records of Luling and Kasan. The analysis of the real data shown in (**a**) and (**c**) is compared with the analysis of the corresponding shuffled data shown in (**b**) and (**d**). As in Fig. 5.1, the four panels show the fluctuation functions obtained by FA, DFA1, DFA2, DFA3, DFA4, and DFA5 (from top to bottom) for the four sets of data. The scale of the fluctuation is arbitrary

crossover occurs roughly at $t_c = 10d$, which is the order of magnitude for a typical Grosswetterlage. Above t_c, there exists long-range persistence expressed by the power-law decay of the correlation function with an exponent $\gamma = 2 - 2\alpha \cong 0.7$. The results seem to indicate that the exponent is universal, i.e. does not depend on the location and the climatic zone of the weather station. Below t_c, the fluctuation functions do not show universal behavior and reflect the different climatic zones.

To test our claim that the slope $\alpha \cong 0.65$ is due to long-range correlations, and does not result from a singular behavior of the probability distribution function of the ΔT_i we have eliminated the correlations by randomly shuffling the ΔT_i. By definition this shuffling has no effect on the probability distribution function of the ΔT_i, which we found to be approximately Gaussian at large temperature variations. The right hand sides of Fig. 5.2b, d show the effect of shuffling on the fluctuation functions. By comparing the left hand sides of the figures with the right hand sides we see the effect of correlations. The exponent

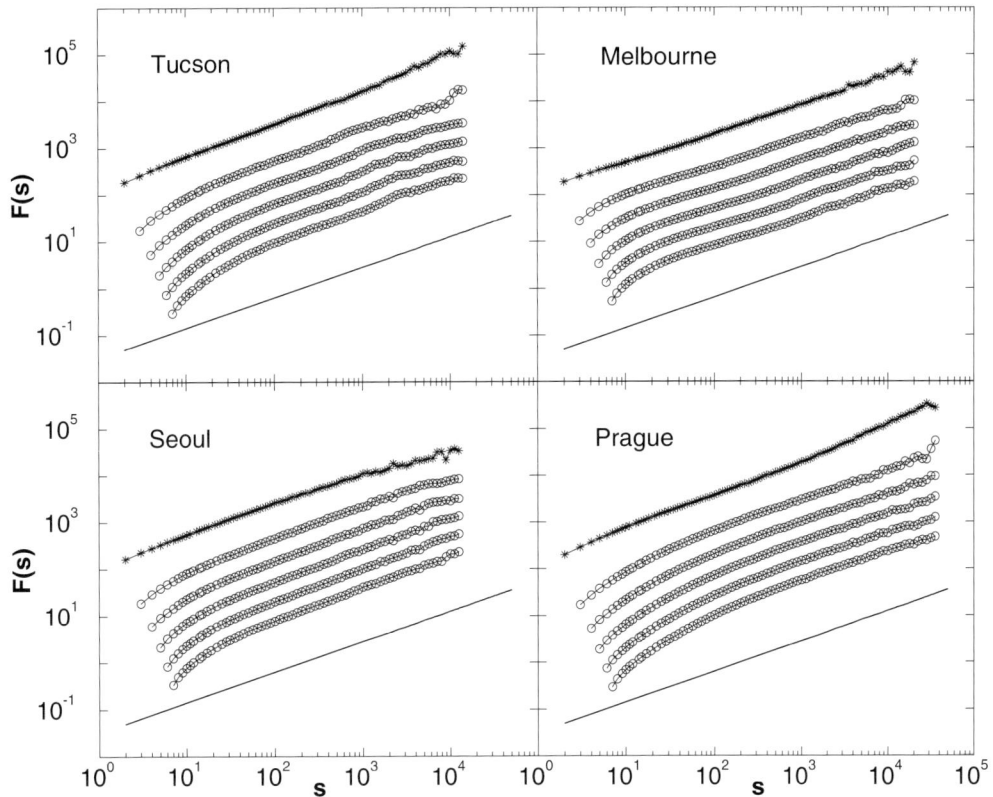

Fig. 5.3. Analysis of the daily maximum temperature records of Tucson, Melbourne, Seoul, and Prague (as Fig. 5.1)

α characterizing the fluctuations in the shuffled uncorrelated sequence is $1/2$, as expected.

Since the exponent does not depend on the location of the meteorological station and its local environment, the power-law behavior can serve as an ideal test for climate models where regional details cannot be incorporated and therefore regional phenomena like urban warming cannot be accounted for. The power-law behavior seems to be a global phenomenon and therefore should also show up in the simulated data of the GCMs. As mentioned earlier, the presence of long-range correlations has far-reaching consequences on the possible detection of trends directly from the record and on the evaluation of extreme events.

5.5.2 Multifractal Analysis

Figure 5.4 shows, for Melbourne and Luling, the generalized fluctuation functions $F_q(s)$ defined in (5.10) and (5.12). On the left hand sides, we show $F_q(s)$ for the real temperature record, while the right hand sides show $F_q(s)$ for the randomly shuffled data. The q-values range from -20 to $+20$ in all cases. Except from the large absolute values of q (where the accuracy is not so high),

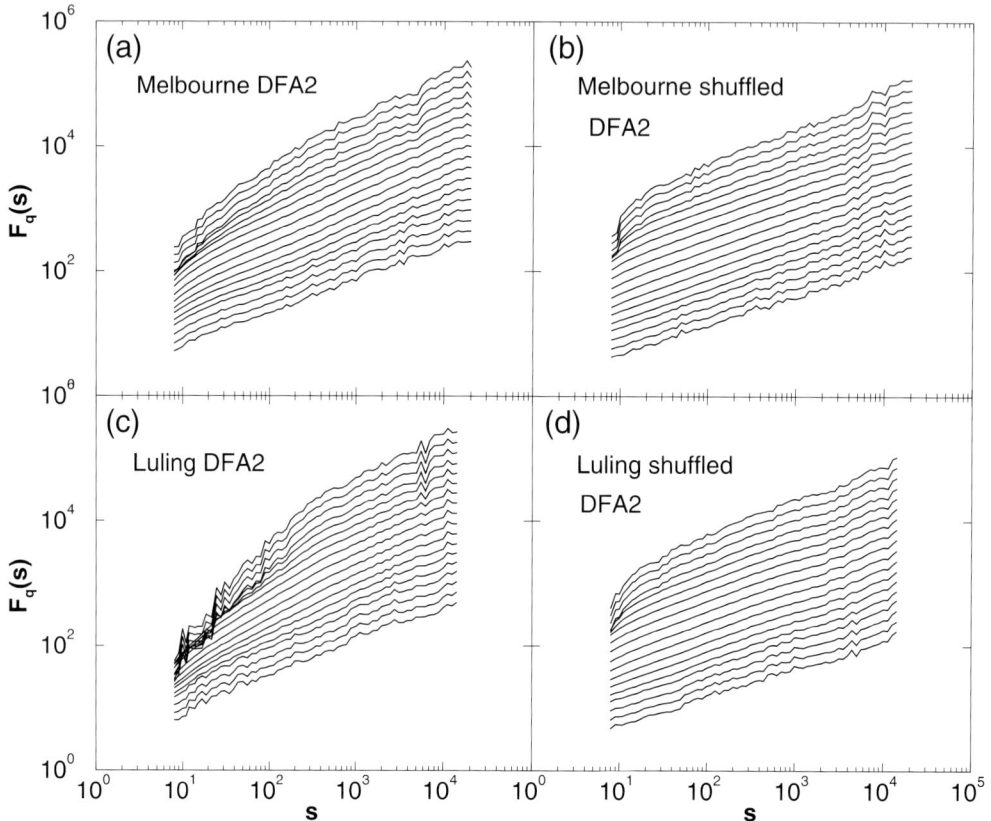

Fig. 5.4. Multifractal analysis of the daily maximum temperature records of Melbourne (**a**)-(**b**) and Luling (**c**)-(**d**) for both the real sequence of data and a shuffled sequence of data. The generalized fluctuation functions $F_q(s)$ are obtained by DFA2 for the q values $q = -20, -15, -12, -10, -6, -4, -2, -1, -0.1, 0.1, 1, 2, 4, 6, 10, 12, 15, 20$ (from top to bottom). The scale of the fluctuations is arbitrary

the functions $F_q(s)$ are approximately parallel lines in the double-logarithmic presentation. For the original temperature records in Fig. 5.4a, c, we have approximately $\beta(q) \simeq 0.65$ for large time scales s. For the shuffled data, all slopes are close to $\beta(q) = 0.5$, as expected.

These results are representative for all stations considered. The results suggest that, if there is a multifractality, it is weak. Since the spectra look the same for all stations, they may well (within the accuracy of the calculations) present universal features of the atmosphere and cannot be used as a 'fingerprint' of a meteorological station.

5.6 Analysis of Simulated Temperature Records

Next we turn to the analysis of simulated data that were obtained by four general circulation models around Prague. We have chosen Prague, since the Prague record is the longest record we could get. The models are:

1. GFDL-R15-a

This is the latest version of a coupled atmosphere-ocean model (AOGCM) that has been developed over many years [5.55, 5.56]. The atmospheric submodel is a spectral model with a horizontal truncation of rhomboidal 15 (R15), a transform grid of 48×40 longitude-latitude points (7.5×4.5 degrees), and nine vertical levels. The ocean submodel is a grid point model with a latitude-longitude grid spacing of 4.5×3.75 degrees and twelve vertical layers. To reduce model drift, flux corrections are applied to the heat and water fluxes at the surface. In the control run, the CO_2 concentration is kept fixed at the 1958 value while for the climate change greenhouse gases are represented by equivalent CO_2 concentrations which increase at a rate of roughly 1% per year according to the IPCC IS92a scenario (business-as-usual) [5.57].

2. CSIRO-Mk2

The CSIRO model is a coupled AOGCM which contains atmospheric, oceanic, sea-ice, and biospheric submodels. The atmospheric submodel is a spectral model with R21 truncation, a transform grid of 64×56 longitude-latitude points (5.6×3.2 degrees), and nine vertical layers. The ocean submodel is a grid point model that uses the same horizontal grid as the atmosphere and has 21 vertical levels. Flux correction is applied to the heat, fresh water, and momentum fluxes at the surface. All greenhouse gases are combined into an equivalent CO_2 concentration which follows observations from 1880 to 1989 and are then projected into the future according to the IS92a scenario [5.57]. This model was developed during the years 1994–1995 [5.58, 5.59].

3. ECHAM4/OPYC3

The coupled AOGCM ECHAM4/OPYC3 was developed as a cooperative effort between the Max-Planck-Institut für Meteorologie (MPI) and Deutsches Klimarechenzentrum (DKRZ) in Hamburg. The atmospheric model was derived from the European Centre for Medium Range Weather Forecasts (ECMWF) model. It is a spectral model with triangular truncation T42, a longitude-latitude transform grid of 128×64 points (2.8 degrees), and 19 vertical levels. The ocean model (OPYC3) is a grid point model with 11 isopycnal layers and it is run on the same grid as the atmosphere. Flux correction is applied to the heat, fresh water, and momentum fluxes at the surface [5.60–5.62]. Historic greenhouse gas concentrations are used from 1860–1989 and from 1990 onward they are projected according to the IS92a scenario.

4. HADCM3

HADCM3 is the latest version of the coupled AOGCM developed at the Hadley Centre [5.63]. Unlike the other models described above, here the atmospheric model is a grid point model with a longitude-latitude grid of 96×73 points (3.75×2.5 degree spacing) and it has 19 vertical levels. The ocean model has a horizontal resolution of 1.25 degrees in both latitude and longitude and 20 vertical levels. No flux correction is applied at the surface. Historic greenhouse gas concentrations are used during the period 1860-1989. From 1990 and onward they are increased according to the IS95a scenario (a slightly modified version of IS92a).

For each model, we obtained the temperature records (mean monthly data) of the four grid points closest to Prague from the Internet [5.64]. We interpolated the data at the location of Prague.

Figure 5.5a shows the results of the FA- and DFA-analysis for the real temperature record of Prague that starts in 1775 and ends in 1992.

Figure 5.5b, c, d show the results obtained from ECHAM4, CSIRO, and HADCM3, that end up at the same year as the real record. The available data of GFDL cover only 40 yr, so we do not present them here. We are interested in the way the models can reproduce the actual data regarding (a) trends and (b) long-range correlations [5.65]. Of course, we cannot expect the models to reproduce local trends like urban warming or short-term correlation structures. But the long-range correlations we discussed in the previous section show characteristic universal features that are actually independent of the local environment around a station. So we can expect that successful models with good prognostic features will be able to reproduce them.

As discussed already above (for the daily data), the FA- and DFA fluctuation functions for the real temperature record of Prague have approximately the same slope of $\alpha \simeq 0.65$ in the double-logarithmic plot (shown as a straight line in the figure). At large time scales there is a slight increase of the FA-function (which clearly indicates a weak trend). In contrast, the FA-results for the HADLEY and ECHAM4 data show a pronounced trend above 100 months represented by a large slope. For CSIRO, the FA-result is not so conclusive since they scatter considerably at large scales. It seems that two of the three models overestimate the trend in the past. Regarding scaling, the DFA curves for CSIRO show good straight lines in a double logarithmic presentation. The exponents are close to the exponents from the Prague record. In contrast, ECHAM4 and HADLEY show a crossover to an exponent $\alpha = 0.5$ after about 3 yr. The exponent $\alpha = 0.5$ indicates loss of persistence. Hence ECHAM4 and HADLEY reproduce datasets that show a linear trend at large time scales and simultaneously the lack of correlations exceeding 3 yr, in contrast to reality.

The scaling features remain very similar when the datasets from the models are extended into the 21st century (see Fig. 5.6). We consider this an important issue, since it shows an internal consistency of the models. However, the trends,

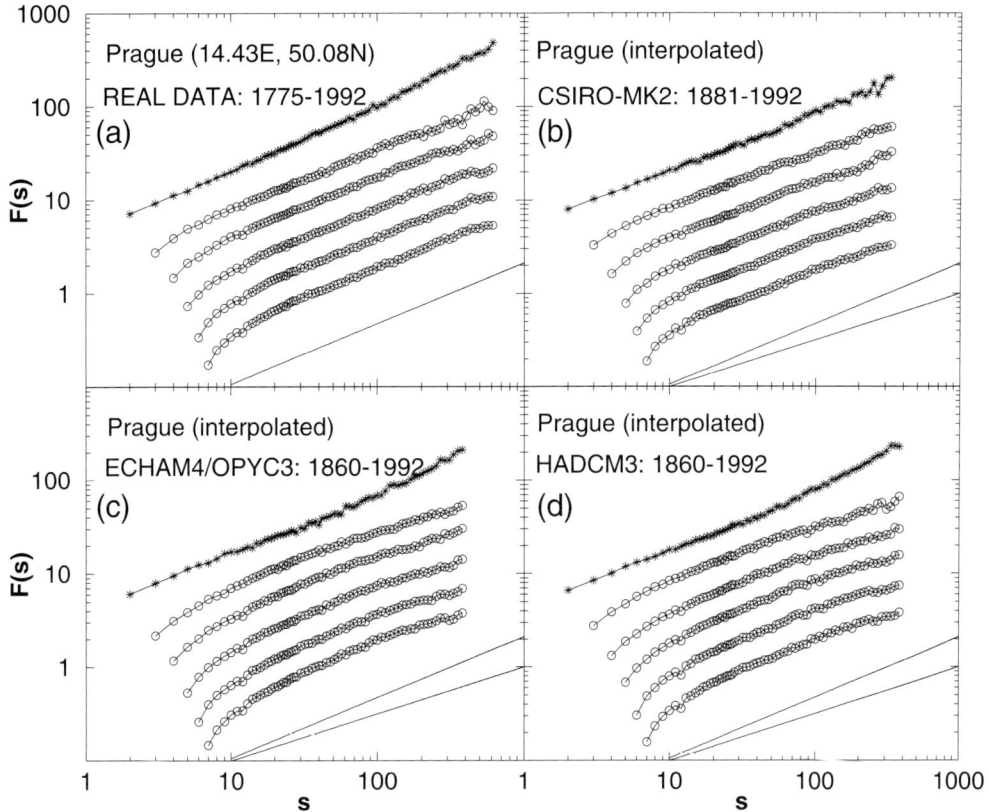

Fig. 5.5. FA- and DFA-analysis of (**a**) the monthly mean temperature record of Prague and simulated interpolated monthly mean temperature records at the geographical position of Prague, for three general circulation models: (**b**) CSIRO, (**c**) ECHAM4, and (**d**) HADCM3. The panels show the fluctuation functions obtained by FA, DFA1, DFA2, DFA3, DFA4, and DFA5 (from top to bottom) for the four sets of data. The scale of the fluctuations is arbitrary (after Govindan et al. [5.65])

which show up in FA and DFA1, are much more pronounced compared to the data from Fig. 5.5. We have obtained similar qualitative behavior also for other simulated temperature records. From the trends, one estimates the warming of the atmosphere in the future. Since the trends are almost not visible in the real data and overestimated by the models in the past, it seems possible that the trends are also overestimated by the models in the future. From this point of view it cannot be excluded that the global warming in the next 100 yr will be less pronounced than predicted by the models.

5.7 Conclusion

In this chapter we have discussed the presence of universal long-range correlations in the climate system and have shown how they can be effectively applied to test the state-of-the-art climate models. Long-range correlations occur in many other complex systems, too, for example in the heart rhythm (see

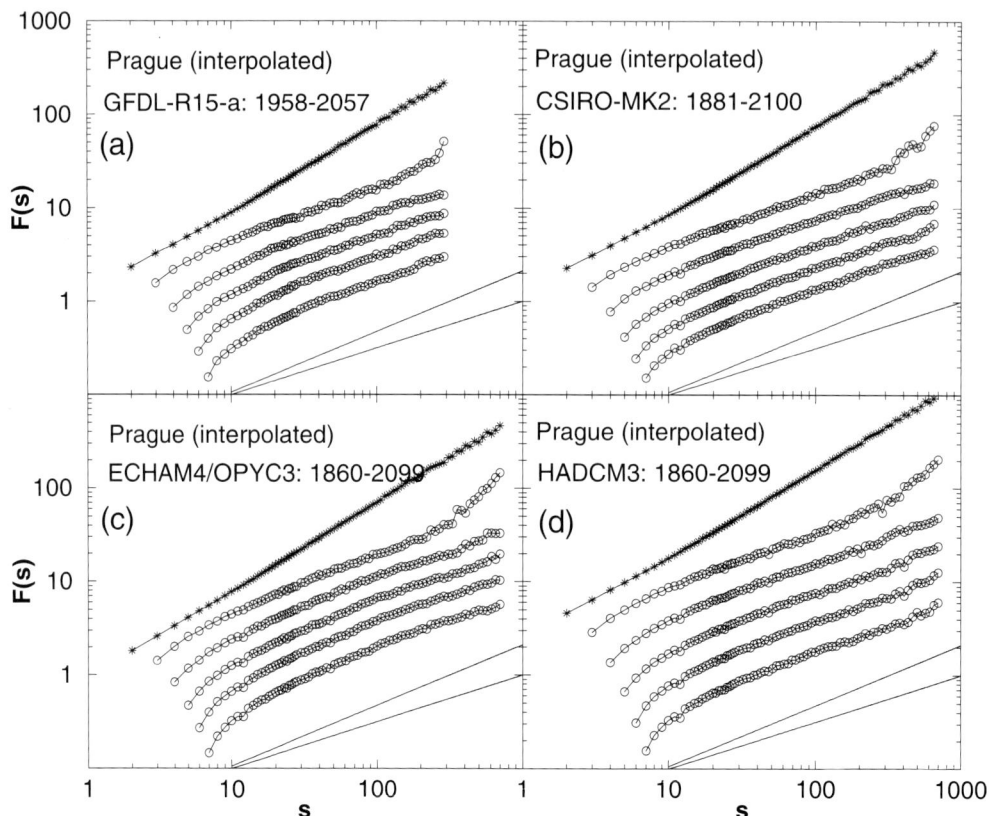

Fig. 5.6. FA- and DFA-analysis of the simulated interpolated monthly mean temperature records of the geographical position of Prague, for four general circulation models: (**a**) GFDL, (**b**) CSIRO, (**c**) ECHAM4, and (**d**) HADCM3. While Fig. 5.5 considered only data in the past, this figure considers the whole set of data (past and future). The data is available on the Internet [5.64] and the analysis is after Govindan et al. [5.65]

Chaps. 7 and 8 in this book), in DNA sequences (see Chaps. 1 and 2), as well as in the stock market (see, for example, Chaps. 12 and 14). But the implications of long-range correlations in the climate system (also river flows seem to be long-range correlated but with exponent α varying from place to place) are most severe:

(a) Trends can no longer be detected by linear fits to the data (a procedure that is still being used by many meteorologists and hydrologists).

(b) Extreme events (like very cold or very warm days or very high river flows) may be coupled over very large time periods. It is no longer sufficient to extract the information on extreme events just from the standard procedure, where the distribution function is just extrapolated toward large values.

In the presence of long-range correlations, new tools are needed to extract the relevant information on trends and extreme events from the data, which we consider as a great challenge for the future.

Acknowledgements. The authors would like to thank especially F.W. Gerstengarbe, P.C. Werner, and their colleagues from the Potsdam Institute for providing and preparing a huge mass of high-quality meteorological data from all over the world. We greatfully wish to thank the German Research Foundation (DFG) for financial support.

References

5.1 T.N. Palmer, Rep. Prog. Phys. **63**, 17 (2000).

5.2 R. Allan, J. Lindesay, and D. Parker (CSIRO, Collingword, 1997).

5.3 W. Hyde, T.J. Crowley, S.K. Baum, and W.R. Peltier, Nature **405**, 425 (2000).

5.4 H.J. Schellnhuber, Nature **402**, 6761 C19 Suppl. (1999).

5.5 A. Ganopolski and S. Rahmstorf, Nature **409**, 153 (2001).

5.6 J.T. Houghton, Y. Ding, D.J. Griggs, M. Noguer, P.J. van der Linden, and D. Xiaosu (eds.), *Climate Change 2001: The Scientific Basis, Contribution of Working Group I to the Third Assessment Report of the Intergovernmental Panel on Climate Change (IPCC)* (Cambridge University Press, Cambridge, 2001).

5.7 J.J. McCarthy, O.F. Canziani, N.A. Leary, D.J. Dokken and K.S. White (eds.), *Climate Change 2001: Impacts, Adaptation, and Vulnerability , Contribution of Working Group II to the Third Assessment Report of the Intergovernmental Panel on Climate Change (IPCC)* (Cambridge University Press, Cambridge, 2001).

5.8 P.M. Blaikie, T. Cannon, I. Davı, and B, Wisner, *At Risk: Natural Hazards, Vulnerability and Disasters* (Routledge, London, 1994).

5.9 WCRP, *Climate Variability & Predictability, CLIVAR Initial Implementation Plan*, WCRP No. 103 (1998).

5.10 IPCC, *The Regional Impacts of Climate Change. An Assessment of Vulnerability* (Cambridge University Press, Cambridge, 1998).

5.11 H. Grassl, Interdiscip. Sci. Rev. **24**(3), 185 (1999).

5.12 K. Hasselmann, Clim. Dyn. **13**, 601 (1997) and references therein.

5.13 C.D. Schönwiese, M. Denhard, J. Grieser, and A. Walter, Theor. Appl. Climatol. **57**(1-2), 119 (1997).

5.14 J.M. Craddock, Weather **34**, 332 (1979).

5.15 D. Bayer and C.D. Schönwiese, in Environment Canada (ed.), 5th Int. Meeting Statist. Climatol., 11 (1992).

5.16 R. Sneyers, *On the Statistical Analysis of Series of Observations* (WMO Publ. No. 415, Geneva, 1992).

5.17 C.D. Schönwiese, and J. Rapp, *Climate Trend Atlas of Europe Based on Observations 1891-1990* (Kluwer, Dordrecht, 1997).

5.18 J.G. Charney and J.G. Devore, J. Atmos. Sci. **36**, 1205 (1979).

5.19 J. Shukla, Science **282**, 728 (1998).

5.20 R.L. Molinari, D.A. Mayer, J.F. Festa, and H.F. Bezdek, J. Geophys. Res. **102**, 3267 (1997).

5.21 R.T. Sutton and M.R. Allen, Nature **388**, 563 (1997).

5.22 S.G. Philander, *El Niño, La Nina and the Southern Oscillation.* International Geophysics Series, Vol. 46 (Academic, New York, 1990).

5.23 C. Price, L. Stone, A. Huppert, B. Rajagopalan, and P. Alpert, Geophys. Res. Lett. **25**, 3963 (1998).

5.24 M. Latif and T.P. Barnett, Science **266**, 634 (1994).

5.25 M.J. Rodwell, D.P. Rodwell, and C.K. Folland, Nature **398**, 320 (1999).

5.26 K. Hasselmann, Science **276**, 914 (1997).

5.27 S. Corti, F. Molteni, and T.N. Palmer, Nature **398**, 799 (1999).

5.28 J. Malcher and C.-D. Schönwiese, Theor. Appl. Climatol. **38**, 157 (1987).

5.29 W. Ebeling, Physica D **109**, 42 (1997).

5.30 C. Nicolis, W. Ebeling, and C. Baraldi, Tellus **49A**, 10 (1997).

5.31 C.-K. Peng, J. Mietus, J.M. Hausdorff, S. Havlin, H.E. Stanley, and A.L. Goldberger, Phys. Rev. Lett. **70**, 1343 (1993).

5.32 P.Ch. Ivanov, A. Bunde, L.A.N. Amaral, S. Havlin, J. Fritsch-Yelle, R.M. Baevsky, H.E. Stanley, and A.L. Goldberger, Europhys. Lett. **48**, 59, 4 (1999).

5.33 A. Bunde, S. Havlin, J.W. Kantelhardt, T. Penzel, J.H. Peter, and K. Voigt, Phys. Rev. Lett. **85**, 3736 (2000).

5.34 C.-K. Peng, S.V. Buldyrev, A.L. Goldberger, S. Havlin, F. Sciortino, M. Simons, and H.E. Stanley, Nature **356**, 168 (1992).

5.35 S.V. Buldyrev, A.L. Goldberger, S. Havlin, C.K. Peng, M. Simons, and F. Sciortino, Phys. Rev. Lett. **71**, 1776 (1993).

5.36 S.V. Buldyrev, A.L. Goldberger, S. Havlin, R.N. Mantegna, M.E. Matsa, C.-K. Peng, M. Simons, and H.E. Stanley, Phys. Rev. E **51**, 5084 (1995).

5.37 A. Arneodo, E. Bacry, P.V. Graves, and J.-F. Muzy, Phys. Rev. Lett. **74**, 3293 (1995).

5.38 A. Arneodo, Y. d'Aubenton-Carafa, E. Bacry, P.V. Graves, J.-F. Muzy, and C. Thermes, Physica D **96**, 291 (1996).

5.39 R.N. Mantegna and H.E. Stanley, *An Introduction to Econophysics* (Cambridge University Press, Cambridge, 1999).

5.40 P. Braun, T. Molnar, and H.B. Kleeberg, Hydrol. Proc. **11**, 1219 (1997).

5.41 E. Koscielny-Bunde, A. Bunde, S. Havlin, H.E. Roman, Y. Goldreich, and H.J. Schellnhuber, Phys. Rev. Lett. **81**, 729 (1998).

5.42 E. Koscielny-Bunde, H.E. Roman, A. Bunde, S. Havlin, and H.J. Schellnhuber, Philos. Mag. B **77**, 1331 (1998).

5.43 E. Koscielny-Bunde, A. Bunde, S. Havlin, and Y. Goldreich, Physica A **231**, 393 (1996).

5.44 G.E.P. Box, G.M. Jenkins, and G.C. Reinsel, *Time Series Analysis* (Prentice Hall, Englewood Cliffs, 1994).

5.45 M. Vetterli and J. Kovacevic, *Wavelets and Subbane Coding* (Prentice Hall, Englewood Cliffs, 1995).

5.46 S. Brenner, J. Clim. **9**, 3337 (1996).

5.47 R. Voss and U. Mikolajewicz, *Long-term climate changes due to increased CO_2 concentration in the coupled atmosphere-ocean general circulation model ECHAM3/LSG.* Max-Planck-Institut fur Meteorologie, Report No. 298 (1999).

5.48 S. Brenner, J. Clim. 9, 3337 (1996).

5.49 F. Giorgi, J. Clim. **3**, 941 (1990).

5.50 E.S. Takle, W.J. Gutowski Jr., R.W. Arritt, Z. Pan, C. Anderson, R. Silva, D. Caya, S. Chen, J.H. Christensen, S.-Y. Hong, H.-M.H. Juang, H. Katzfey, W. Lapenta, R. Laprise, P. Lopez, J. McGregor, and J.R. Roads, J. Geophys. Res. **104**, 19 443 (1999).

5.51 P. Alpert, Y.J. Kaufman, and Y. Shay-El, Nature **395**, 367 (1998).

5.52 A. Bunde and S. Havlin (eds.), *Fractals in Science* (Springer, New York, 1995).

5.53 J. Feder, *Fractals* (Plenum, New York, 1989).

5.54 J.W. Kantelhardt, E. Koscielny-Bunde, H.A. Rego, A. Bunde, and S. Havlin, Physica A **295**, 441 (2001).

5.55 S. Manabe, R.J. Stouffer, M.J. Spelman, and K. Bryan, J. Clim. **4**, 785 (1991).

5.56 S. Manabe, M.J. Spelman, and R.J. Stouffer, J. Clim. **5**, 105 (1992).

5.57 J. Leggett, W.J. Pepper, and R.J. Swart, in *Climate Change 1992: The Supplementary Report to the IPCC Scientific Assessment*, edited by J.T. Houghton, B.A. Callander, and S.K. Varney (Cambridge University Press, Cambridge, 1992) p. 75.

5.58 H. B. Gordon and S. P. O'Farrell, Mon. Weather Rev. **125**, 875 (1997).

5.59 A.C. Hirst, H.B. Gordon, and S.P. O'Farrell, Geophys. Res. Lett. **23**, 3361 (1996).

5.60 L. Bengtsson, K. Arpe, E. Roeckner, et al., Clim. Dyn. **12**(4), 261 (1996).

5.61 ECHAM3: Atmospheric General Circulation Model, Editor: DKRZ-Model User Support Group (October 1996).

5.62 The OPYC Ocean General Circulation Model, J.M. Oberhuber (October 1992); Deutsches Klimarechenzentrum (DKRZ): Report No. 7 (1997).

5.63 C. Gordon, C. Cooper, C.A. Senior, et al., Clim. Dyn. **16**(2-3), 147 (2000).

5.64 The IPCC Data Distribution Centre: Access to the GCM Archive: [ipcc-ddc.cru.uea.ac.uk/dkrz/dkrz_index.html].

5.65 R.B. Govindan, D. Vjushin, S. Brenner, A. Bunde, S. Havlin, and H.J. Schellnhuber, Physica A **294**, 239 (2001).

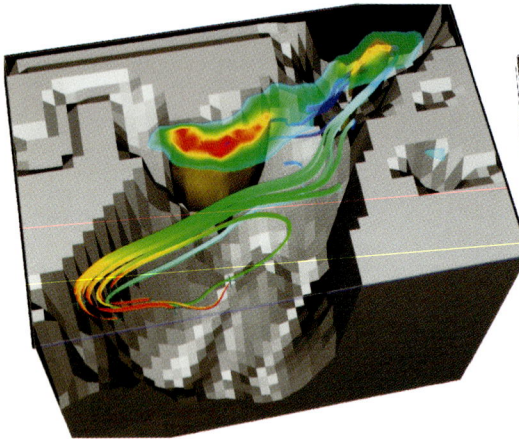

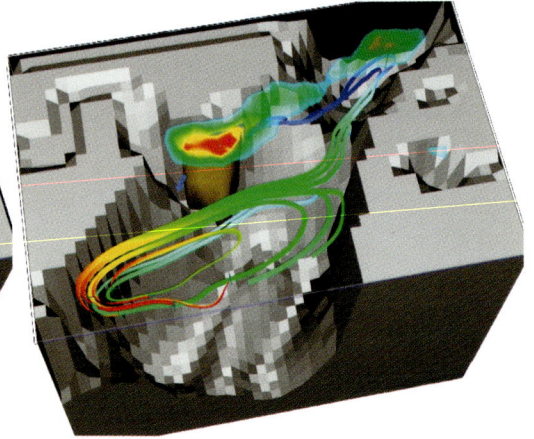

6. Assessment and Management of Critical Events: The Breakdown of Marine Fisheries and The North Atlantic Thermohaline Circulation

Jürgen Kropp, Kirsten Zickfeld, and Klaus Eisenack

In the coming decades, global environmental change is likely to have important repercussions on humanity itself, e.g. ways of living or nutritional possibilities. Simultaneously there exists considerable uncertainty regarding the general trends and rates of change. The scientific community has been instrumental in drawing attention to these dimensions in order to prevent/mitigate disastrous events. Although it is rather difficult to make accurate predictions and guarantee sensible and effective decisions under the conditions of nonlinearity and vague knowledge, new and smart techniques open some promising roads for scientific progress. In this chapter we address these issues by discussing novel model approaches with respect to the key features of two examples (i) the breakdown of a bioeconomic sector and (ii) the North Atlantic thermohaline circulation. We show what is already possible today and where we have to go in order to make substantial steps forward.

6.1 Introduction

Recent years have seen an increasing number of studies conceptionalizing research efforts regarding a global policy making which could support paths to sustainability. Examples are the Montreal Protocol combating ozone depletion

---

◄ **Fig. 6.0.** Composite cartoon of the main issues of this chapter: industrial fishing methods, e.g. as shown a single trawlers catches up to 400 t per trip, and unadopted subsidies policies (North Atlantic cod fishery) threaten the world's food security by inducing an ongoing overexploitation of renewable marine resources. In addition, in many cases the catch is not utilized for food, but used for the production of fishmeal (upper panel). The lower panel shows a 3-D image of the state of the Gulf Stream (left) under current climate conditions and under an anthropogenically induced global warming (business-as-usual, right). The deep water formation sites in the Labrador and Norwegian Seas (red and yellow, depth shown by dark green color) are substantially weakened if global warming occurs. Courtesy of [6.1] (lower panel)

in the stratosphere [6.2], or global expert and research networks, such as the Intergovernmental Panel of Climate Change (IPCC) (cf. [6.3]) or the International Geosphere-Biosphere Programme (IGBP) (cf. [6.4]). On one hand, these efforts have led to increased knowledge on the variety of processes driving the dynamics of the entire Earth. On the other hand, the endeavor to re-integrate this knowledge will be difficult, because modelers are often trapped in a tension between systems complexity and tractability.

It is a key characteristic of Earth system analysis that it often deals with exceptional dynamics involving the interaction of innumerable system parts. In common with a variety of important areas in this context, sustainability science aims to supply general strategies to manage singular or sets of subsystems of the entire Earth. Therefore, there exists an urgent need for improving our understanding of disaster dynamics, which might be triggered by human action as well as by other external and internal forcing factors, and could lead to abrupt changes in the geophysical, ecological, and socioeconomic inventory [6.5]. However, it is neither possible nor necessary to understand the Earth's complexity completely in order to steer it successfully. Consequently, we do not need the formal analysis of more and more details, but rather an analytical approach, which allows us to determine the general functionalities of a system. Such an approach should include the identification of potential precursor signals indicating critical events, and the relaxation of such events to non-critical states, or it should at least supply a weak prognosis of a system's potential development paths, for example, in terms of probabilities.

Our purpose here is to illustrate the general problem of disaster-related decision making under uncertainty by discussing two specific examples of global importance. It is shown that in spite of the different types of indeterminacy involved in these two cases it appears possible to take judicious measures to prevent or mitigate absolutely unacceptable outcomes. Additionally, we highlight the relevance of thresholds, because transgressions of these critical values could have tremendous effects, such as irreversible developments.

The first example discusses the impacts of industrial marine capture fisheries as a pattern of global change (cf. [6.6]). These impacts have important implications for the world's food security (Sect. 6.2). Several aspects of this system can - due to vague knowledge - only be expressed by qualitative properties, but nevertheless, even on the basis of structural (qualitative) generalizations, valuable hints for future policy actions and for prevention of a catastrophic systems failure can be derived. A second example concerns the potential instability of the North Atlantic thermohaline circulation (THC). Paleoclimatic reconstructions and model results show that in glacial times the system was prone to rapid transitions between different qualitative states triggered by events associated with changes in the freshwater budget of the North Atlantic such as, e.g. ice sheet surges. In addition, there also exists sufficient evidence that anthropogenically induced global warming might have similar effects, but because of the large uncertainties involved in climate projections, it is not yet possible to determine appropriate threshold values. Because the Earth cannot be used as an arbitrary laboratory, strategies guaranteeing a minimum of future safety standards should be investigated (cf. [6.8]). This is done along the lines of the guardrail approach (cf. [6.69]) which allows to operationalize the concept of 'emissions corridors' and define likelihood domains for tolerable systems behavior (Sect. 6.3).

Nevertheless, it should be borne in mind that a paradigm shift in 'political philosophy' might be needed since the present style of decision making is at least ostensibly based on explicit quantified findings. Given the complexity of the systems in question - such as the total planetary machinery - it must be accepted that one may have to do with weak and soft prognoses only.

6.2 The Role of Market Mechanisms in Marine Resource Exploitation

When we look at marine fish stocks, the 'human factor' has held sway over global marine biodiversity since historical times [6.9, 6.10]. Especially during the last decades marine fish stocks have been of particular interest, because approximately 20% of global protein consumption relies on fishery products, and thus they play an important role for the world's food security [6.11, 6.12]. Nevertheless, a considerable part of marine resources are overexploited or even fully exploited [6.10, 6.13] and policy actions guaranteeing precautionary management are not yet in sight. One reason for this situation is that policy suffers from a lack of long-term planning and uncoordinated regulatory frameworks [6.14, 6.15]. On the other hand, decision makers argue that only insufficient or too weakly defined knowledge exists to allow the set up of adequate intervention and/or management strategies. Despite these facts awareness needs to be raised that bioeconomic systems, such as commercial fisheries, exhibit per se an inherent uncertainty [6.16], e.g. decisions on investments of the market participants are kept secret. Thus, combining knowledge of different quality from the natural and/or socioeconomic realms in a model seems to be rather difficult, because rigorous and precise mathematical representations for such problems are often out of reach. A strategy to overcome these difficulties is opened by smart qualitative techniques which are being developed to an increasing extent in artificial intelligence research and which are very suitable tools for integrating a variety of knowledge, which may be exact, fragmentary, or subjective. Successful applications have been reported in various fields (cf. [6.17–6.22]).

6.2.1 From an Analytical to a Qualitative Model

Commercial fisheries represent a 'coevolutionary' system which consists of an economic and a natural systems part, and which is far from being fair and effective. In order to prevent collapsing fish populations and disastrous economic situations, however, it is a necessary precondition to know what are the general functionalities of this system, i.e. what are the essential driving forces and are points of 'no-return' observable in the potential systems development.

In a more general context, market mechanisms and current policy constraints have set up the optimization paradigm in industrial fisheries with its two peculiarities 'prosperity' and 'greed' (regarding the optimization and other sustainability paradigms cf. [6.23]). Consequently, economic pressure on marine renewable resources, e.g. capital accumulation, has been a major topic discussed in environmental economics over the last two decades (cf. [6.25–6.32]). Some key characteristics in this context are, e.g. a highly specialized capital stock which cannot readily be converted to other uses. But despite recent efforts, intrinsic system properties (uncertainty) have led to a situation in which current modeling approaches are still rudimentary, because vagueness seems to be an obstacle for improved models. In order to sail around the enigma of tractability, state-of-the-art models are mainly solved for equilibrium cases or include unsuitable simplifications, such as linear cost functions.

Here we surmount these difficulties by introducing a more general approach which contains rational choices of fishing firms and utilizes qualitative knowledge for modeling efforts. Let us point out that the model focuses explicitly on the system dynamics, because bioeconomic systems normally tend to stay far away from equilibrium. In a first step an analytic model based on more realistic assumptions is developed (for a model derivation cf. [6.33]), and in a second step it is abstracted in order to prepare it for qualitative modeling issues.

Focusing on the dynamics of the economic and the biological stock, one can consider a resource x with an associated recruitment function $R(x)$. The time development of the stock can be modeled by $\dot{x} = R(x) - (h + h')$, where h denote the harvest of the firm under consideration and h' those of all the others. Further, each firm's capital stock is described by $\dot{C} = I - \delta C$, where $I \geq 0$ represents (irreversible) investment and δ a constant depreciation rate. Assuming moreover that fishery firms apply an optimization strategy, it is the goal of each fishing company to select a plan for investment and harvest which maximizes the discounted (current-value) profit

$$\Pi := \int_J e^{-rt} \left(p(h + h')h - v(h, x, C) - c(I) \right) dt \qquad (6.1)$$

subject to the equations for stock and capital. Here the parameter r represents a constant discount rate and the demand for fish is expressed by a downward sloping inverted demand function $p(h + h')$ which assigns the obtained market price to the total harvest. The convex function $v(h, x, C)$ refers to the variable costs which increase in h and decrease in x and C. Investment costs are given by the strictly convex and increasing function $c(I)$, whereby the convexity reflects the inelastic supply of highly specialized equipment and rising adjustment costs. Finally, $J = [0, T]$ represents a planning interval. Actually, such a problem can be described as a noncooperative differential game introducing the Nash assumption that each rival rests its own decisions on given levels of harvest and investment of the other $N - 1$ market participants competing for the common property resource (for further examples regarding individual and collective behavior cf. Chap. 11).

Presuming also that all firms are characterized by the same technology and behave in the same way one obtains $h + h' = Nh$. Applying optimization principles [6.34] and, in particular, the theorem of Mangasarian [6.35], we derive a system of differential equations describing the optimal evolution of investment and harvest (for derivation see [6.33])[1]

$$\dot{x} = R(x) - Nh, \tag{6.2}$$

$$\dot{C} = I - \delta C, \tag{6.3}$$

$$\dot{I} = \frac{1}{c_{II}(I)}\big((r + \delta)c_I(I) + v_C\big), \tag{6.4}$$

$$v_h = \left(1 - \frac{\epsilon}{N}\right)p(Nh). \tag{6.5}$$

In (6.2) R refers to a concave recruitment function (cf. Fig. 6.1) of the resource and in (6.5), ϵ represents the inverse price elasticity of demand. Equation (6.5) denotes the usual equality between marginal variable costs and marginal revenue.

Even though a numerical analysis of (6.2)–(6.5) is conceivable, functions and parameterizations have to be chosen explicitly. This has, due to a variety of uncertainties, more the character of a 'rule of thumb', pursuing an ad-hoc avenue rather than a systematic approach. We thus make use of so-called qualitative differential equations (QDE) [6.36], which allow a system to be integrated in a sense of broader universality or robustness with respect to uncertainty and generalization of different fishery systems. It is the advantage of this approach that relations between variables are introduced in terms of monotonicity assumptions instead of exact parameterized numerical specifications.

For instance, it is rather difficult to determine the exact behavior of the recruitment of different target species, as well as the investment decisions of the distinct market competitors. Nevertheless, in the view of stringency several qualitative deductions can be drawn: we impose, for example, the assumption

---

[1] For sake of readability, X_z denotes the first and X_{zz} the second derivate of the function X with respect to z.

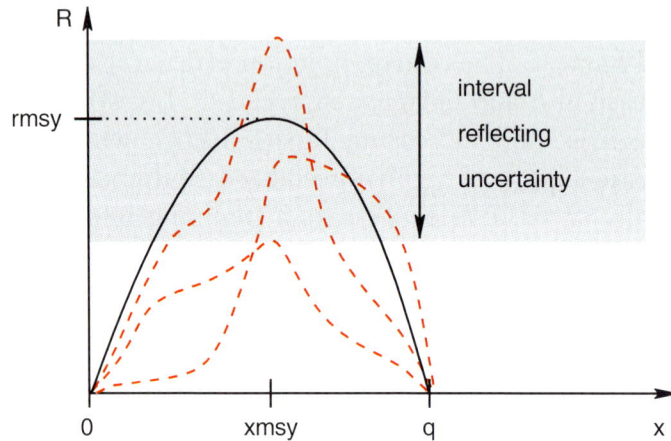

Fig. 6.1. Qualitative portrayal of a recruitment function in a qualitative model. The dashed red lines refer to possible behaviors of the fish stock x which are considered by the same qualitative model, because one can made only assumptions on the monotonicity properties, i.e. if $x < xmsy \Rightarrow R' = [+]$; $x > xmsy \Rightarrow R' = [-]$

that $R(0) = R(q) = 0$, where q denotes the carrying capacity of the biological system and $xmsy$ a stock size, where a maximal sustainable yield $rmsy$ is possible (Fig. 6.1). Qualitatively speaking $q, xmsy$, and $rmsy$ are values of particular interest and for each model variable a well-ordered set of these values, the so-called 'quantity space', is defined. Specifically, they are called 'landmarks' and they are points where the monotonicity properties of the model functions change. Moreover, relations between variables, e.g. between R and x, are expressed by 'constraints'. Considering the U-shaped downward behavior as shown in Fig. 6.1 one can write $R = U^-_{(xmsy,rmsy)}(x)[(0,0); (xmsy, rmsy); (q,0)]$ which requires that R is given by a function $f(x)$ which is increasing if $x < xmsy$, attains a maximum if $x = xmsy$, and decreases if $x > xmsy$. The tuples in the squared brackets indicate corresponding values, i.e. it holds that $f(0) = 0, f(xmsy) = rmsy$, and $f(q) = 0$. It is the result of this approach that the phase space of the system is segmented in a finite set of regions, which are bounded by landmarks of specific interest defined by the modeler.

In this framework the time evolution of each model variable, or qualitative trajectory, can be characterized by a sequence of landmarks or intervals between landmarks and the variable's direction of change, e.g. for the harvest (cf. Fig. 6.2)

$$h(t_i) : \langle (0, ymsy), \uparrow \rangle, \dots, \langle (ymsy, max) \circ \rangle, \dots, \langle (ymsy), \downarrow \rangle, \dots,$$
$$\langle (0, ymsy), \circ \rangle. \quad (6.6)$$

Here, for instance, the initial time interval indicates an increasing harvest ($\uparrow$) which resides in the open interval $(0, ymsy)$. In the following time steps it approaches a local maximum ($\circ$) in the landmark interval $(ymsy, max)$ and starts to decrease in the following one.

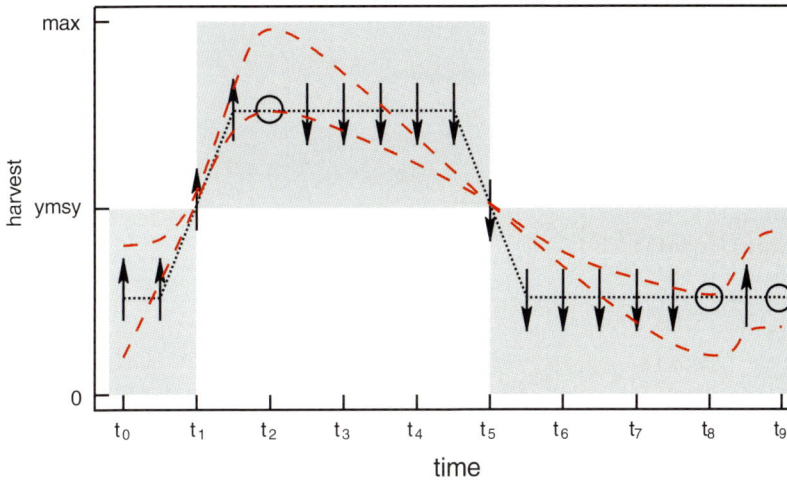

Fig. 6.2. An exemplary qualitative trajectory for the harvest h (dotted line). The symbols denote the direction of change of the function, i.e. ($\uparrow$): increasing, ($\downarrow$): decreasing, and ($\circ$): steady. Such a time development is representative for a class of (quantitative) time developments as indicated by the dashed red lines and the gray areas

The process of assigning qualitative assumptions to a given ordinary differential equation is called abstraction. For commercial fisheries the analytical model given by (6.2)–(6.5) can be abstracted as shown in Fig. 6.3. However, due to the uncertainties, the abstraction process introduces ambiguities to the model and thus one obtains not necessarily a single solution, but a set of solutions whose evaluation could become rather cumbersome for larger models. It is worth noting that these qualitative solutions contain all solutions of all ordinary differential equations which have the specified monotonicity properties in common. In order to display larger solution sets in a clear and concise manner several additional techniques can be used for simplification and visualization (cf. [6.37]). For example, some of these solutions only differ in the dynamics of auxiliary variables. Thus, they can be neglected and the solution set can be projected on the state space of variables of interest (focus variables). Moreover, behaviors which are equal with respect to these variables can be further aggregated which results finally in a digraph (Fig. 6.5). The digraph is a special kind of a state-transition-diagram, where each vertex represents a class of qualitative states (region in the state space) consistent with the underlying QDE and which are equivalent in the observables (for details see [6.38]). It displays the overall dynamics of the system and allows a system to be discussed under the perspective of general properties.

```
(quantity-spaces
 (stock   (0 xmsy q inf))                        'x'
 (capital (0 inf))                               'C'
 (invest  (0 inf))                               'I'
 (recruit (minf 0 rmsy))                         'R'
 (harvest (0 ymsy inf))                          'h'
 (mprofit (0 inf))                               'Π'
 (dstk    (minf 0 inf))                          'ẋ'
 (dcap    (minf 0 inf))                          'Ċ'
 (dinv    (minf 0 inf)))                         'İ'
(constraints
 ((D/DT stock   dstk))                           'D/DT: first qualitative
 ((D/DT capital dcap))                                  time derivative'
 ((D/DT invest  dinv))
 ((ADD  dcap    capital invest))                 'ADD:  qualitative addition'
 ((ADD  dstk    harvest recruit) (0 ymsy rmsy))
 ((ADD  mprofit dinv invest))
 ((U-   stock   recruit (xmsy rmsy)(0 0)(q 0))) 'U-:   parabola, bottom open'
 ((MULT stock   capital harvest))                'MULT: qualitative multiplication'
 (((M- -) stock capital mprofit)))               'M--:  monotonically decreasing
                                                        in two components'
```

Fig. 6.3. Qualitative variables and constraints of the bioeconomic model as LISP code. Instead of real numbers a variable is described qualitatively in terms of its quantity space, defined by a finite set of ordered symbols, e.g. `stock`: 0 < xmsy < q < inf. Constraints represent qualitative relations between variables, see e.g. (6.3): $\dot{C} = I - \delta C \Longleftrightarrow$ `dcap` + `capital` = `invest`. Corresponding values allow to make use of more definite assumptions on the interconnections between certain variables, e.g. $x(t) = xmsy$ if, and only if, $R(t) = rmsy$

6.2.2 Simulation Results, Critical Points, and Lessons to be Learnt

In contrast to previous approaches in bioeconomic modeling, the qualitative model can reconstruct the impact of capital accumulation in commercial fisheries, as is shown clearly by the case of blue whale hunting (Figs. 6.4b and 6.4c, equivalent to the development path #1 → #12 in Fig. 6.5). Discussing the results of a qualitative model in general and in the sense of political measures allows a variety of conclusions to be drawn on the systems behavior:

- Using the qualitative approach, previous analytical bioeconomic models can be fundamentally extended allowing a detailed focus on the dynamic behavior.
- Critical points can be determined, where the system either shifts to a more seriously damaged or to an ameliorated situation. In addition, there exist only limited feasibilities for a crossover from a sustainable to a nonsustainable systems state (Fig. 6.5).
- Overcapitalization in the fishery economy cannot be avoided in general (under current model assumptions) [6.33]. A situation where harvest decreases while the gross investment is still taking place (vertices: #3, #6, #8, #18,

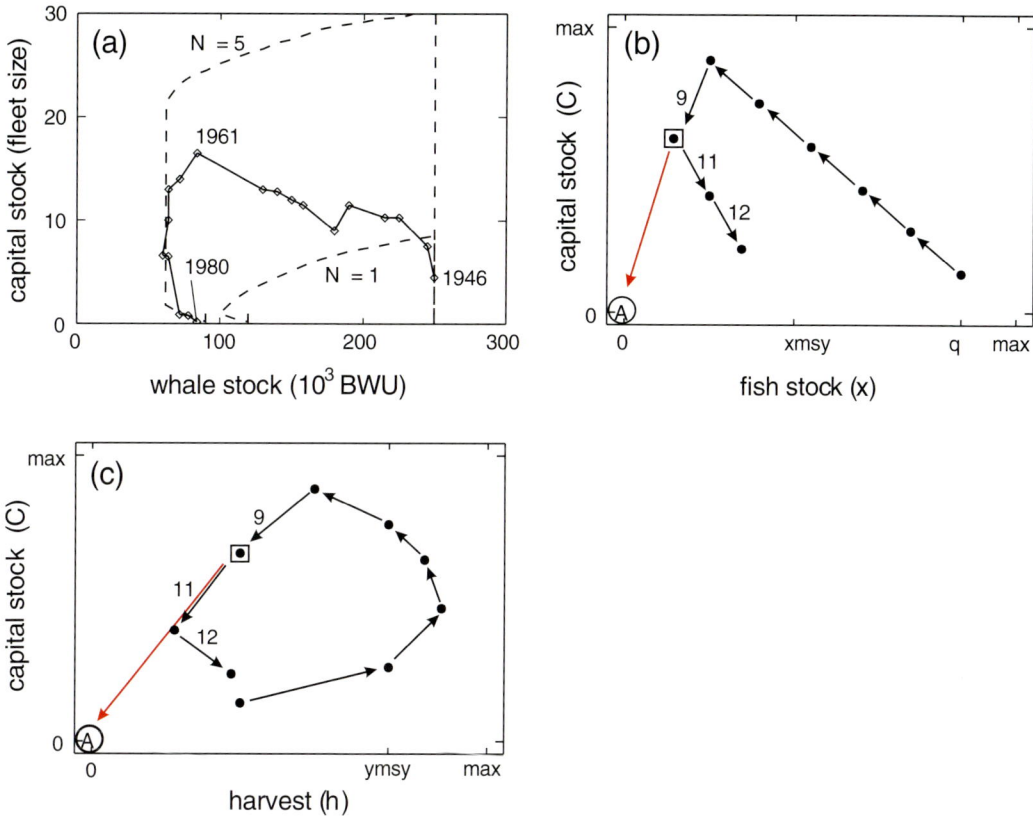

Fig. 6.4. Phase plots for the situation in blue whale hunting between 1946 and 1980. (**a**) Empirically estimated measures for the capital vs. blue whale stock, (the dashed lines refer to model outputs generated by the McKelvey model [6.30], N = number of firms). (**b,c**) Corresponding qualitative phase plots obtained by the QDE model, (the numbers comply with those of the numbered vertices in Fig. 6.5, red and blue trajectory). The boxed dot illustrates a critical point where the system either stabilizes at a low level or develops to a disaster scenario. It is worth mentioning that the black arrows of Fig. 6.4c form a quasi-hysteresis. Nevertheless the 'loop' is not completely closed, because a decline of an industry induces normally a structural change and not an instantaneous revitalization of the fisheries economy if the resource recovers

#11, #14, Fig. 6.5) can be clearly characterized as a build-up of overcapacities. There exists no development path that does not traverse one of these vertices.

The properties of the state-transition-diagram (Fig. 6.5) allow a detailed investigation of the systems behavior. Let us start, for instance, with fishing activities at vertex #1 (relatively undisturbed stock, industry at a low level; quadrant I, $x > xmsy, h < ymsy$) and refer to the trajectory indicated by the red edges, where the stock is decreasing and both the harvest and the capital stock are increasing. The first irreversible point the system attains is represented by the transition from vertex #1 to #2 (quadrant II). Here already the crucial situation occurs, that the catch reaches a nonsustainable level ($h > ymsy$), while the capital stock is still increasing. Focusing on the next step - the irreversible edge from #2 to #5 - harvest is still increasing and resides above

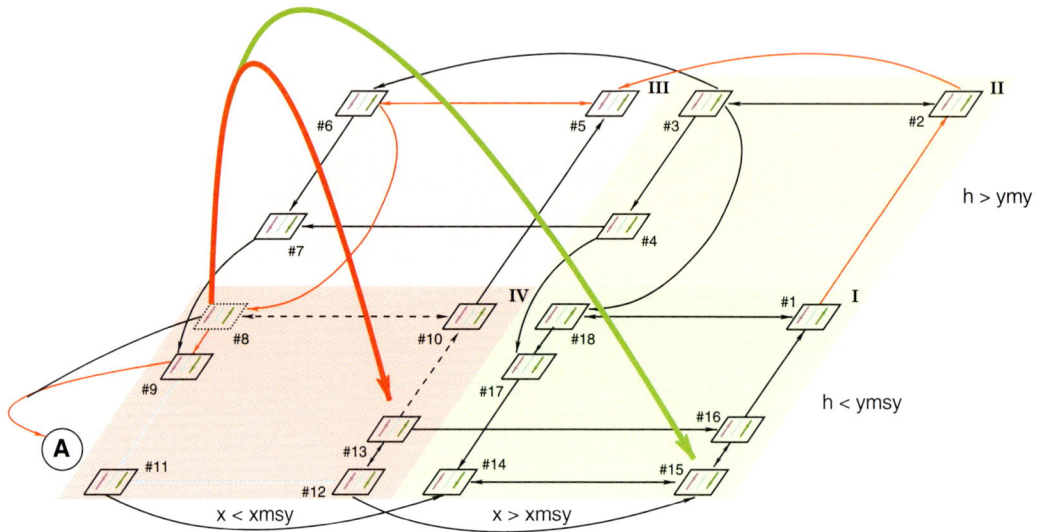

Fig. 6.5. Graph theoretic representation (state-transition-diagram) of possible behaviors of the qualitative model focusing on the target variables. They are sorted in the order of stock x (magenta), harvest h (cyan), and capital C (green) and indicated by arrows in the vertices labeling the direction of the focus variable's first derivative. (A) refers to the disaster case, where the stocks are fully exploited and a complete recession of industry occurs. The bold arrows (red and green) label the outcome of external interventions which start at the dashed box (for details see text)

the sustainable yield, although the resource is facing a continuing degradation (additionally now $x < xmsy$ holds). After this event, a direct reversal to quadrant II is impossible and a more threatening situation is unavoidable (#6, quadrant III). Here we observe a typical situation where the capital stock is further increased in an effort to offset the impact of a dwindling stock on the catch. But, as shown in Fig. 6.5, it is already far too late, because the downward development to #8 is inevitable, where the industry and the fish stock are likely to be ruined (A). Other possible paths through the state-transition-diagram that avoid vertex #8, reach vertex #9 at some time and face the same threat. It should be noticed that reaching vertex #8 or #9 is unavoidable, if the system leaves quadrants II. At these vertices the collapse of the fishery can be prevented if, due to a rapid comedown of the industry, the stock recovers again (indicated by the blue edges, Fig. 6.5). The latter illustrates the situation which we have already described for blue whale hunting and which is shown in Fig. 6.4a.

These different possible outcomes at vertices #8 and #9 qualifies them as critical branchings (qualitative bifurcations). Therefore, fishery management should take account of this general problem in two ways: (i) How can the likelihood of the system to recover at the critical branchings be strengthened? (ii) How can it be avoided that the fishery comes back to this 'bifurcation' after the resource has recovered?

To approach the first question, a detailed analytical investigation of the critical point is needed. It is the qualitative approach that has helped us to identify this important region of the phase space and it provides a starting point for further investigations (i.e. detection of 'precursor signals'). The outcome depends on how early net investment decreases (#8 to #9) and how early the stock recovers (#9 to #11). For the first step the reduction rate of investment and capital stock when the system enters vertex #8 is decisive. The outcome of the second step depends on the fish stock at the entry of vertex #9, its regeneration rate and the speed of harvest reduction. This shows that management actions focusing on a singular gain, e.g. industry as well as environmental protection, cannot be effective. In fact, a multicriteria evaluation is needed for viable management and it has preferably to be implemented at a very early stage, due to the indolence of both capital and stock.

The graph also throws light on the causalities hidden in the story of the Atlantic cod fishery from a systematic point of view. This example shows how unsuitable actions take effect (red and green arrows in Fig. 6.5) and is described in more detail in [6.39]. In the Canadian case (red arrow) allowable catch quotas were tardily introduced. In order to mitigate parallel economic losses, massive subsidies for technology and new ships were payed, leading to ever more efficient fish-killing and a tripling of licensed fishermen in Newfoundland alone (cf. Fig. 6.0 upper panel). The catch dropped from $\approx 1.9 \cdot 10^6$ t in 1969 to $\approx 10^3$ t in the early 1990s and the Grand Bank fishing grounds were closed. After the intervention, an additional path is traversed indicated by the vertices 13–10–8–9 (dashed arrows, Fig. 6.5) which finally leads to (A). This was the end of the story, because the cod stocks have been fully exploited up to the present day and more than 40 000 jobs were lost in the early 1990s (see also Fig. 6.0, inset of upper panel) [6.40, 6.41]. In Norway policy makers were facing a similar disaster, but they had to some extent learnt their lessons. Firstly, they introduced individual transferable quotas which halted the race for fish, secondly, they set up subsidies of about \$71 Mio. in order to remove ships from the fishery and to diversify the coastal economy [6.39], and thirdly, they banned fishing from spawning grounds. In the context of our model this is illustrated by a jump to vertex #15 (green bold arrow in Fig. 6.5). But a closer look at the Norwegian actions shows that their policy actions have been only half-hearted, because they paid subsidies for their large trawlers to leave Norwegian waters for other fishing grounds (e.g. New Zealand) [6.39]. Thus, the intrinsic problem has only been suspended and some serious developments in the near future are foreseeable (cf. [6.41]).

Discussing these examples in the context of control and policy actions is only one side of the coin. A more crucial point are the intrinsic system properties, i.e. unavoidable overcapitalization, making it rather difficult to steer the bioeconomic system to both a safe economic and ecological situation. Hence, even early interventions to create a controlled fishery are desirable, but under current boundary conditions this dynamic steering problem needs a massive perpetual adjustment. This looks more like a ride along the edge of an abyss than pursuing a sustainable strategy.

6.3 Could Europe's Heating System be Threatened by Human Interference?

Another hot topic debated under the perspective of precautionary policy actions is the potential breakdown of the thermohaline (i.e. density driven) component of the Atlantic ocean circulation which contributes a major part to the heat budget of the North Atlantic region [6.42]. Such a collapse might have far-reaching consequences for north-western Europe, because the oceanic heat transport (in the order of $1\,\mathrm{PW}=10^{15}\,\mathrm{W}$) has a significant warming effect for this region [6.43]. Also, it would be associated with major changes in the physical properties (i.e. salinity, temperature, boundary currents) of the North Atlantic waters, with possibly severe impacts on ecosystems and fisheries [6.44, 6.45]. The crucial point is that the thermohaline circulation system (sometimes dubbed 'conveyor belt') shows a typical nonlinear threshold dynamics switching into and out of certain modes, where either the thermohaline circulation shuts down completely or a shift in the locations of deep water formation (NADW) occurs [6.46]. Thus, the major questions which have to be discussed in this context are whether human interference may influence the natural dynamics and can induce transitions between the modes, and if so, when this may occur? However, we are just at the beginning of an understanding of the underlying mechanisms (see also Chap. 4), but facing potential evolutions like this, we have to ask how we can prevent potentially disastrous consequences and which mitigation strategies are on hand under the 'paradigm of uncertainty'.

6.3.1 Multiple Equilibria of the Thermohaline Circulation

Firstly, a brief illustration should highlight the main physical mechanisms underlying the Atlantic thermohaline circulation and elucidate the reasons for its nonlinearity. Figure 6.6 shows a highly simplified cartoon of the Atlantic THC: the Gulf Stream and its northern extension, the North Atlantic Drift, transport warm and saline water from the tropics towards the northern latitudes. On its way, the water releases heat to the atmosphere and gets cooler and denser. When in the high northern latitudes the water density is higher than that of the underlying water column, it sinks and spreads southward as North Atlantic Deep Water. The main regions of deep water formation are at present located in the Labrador and the Greenland Seas (cf. Fig. 6.0, lower panel).

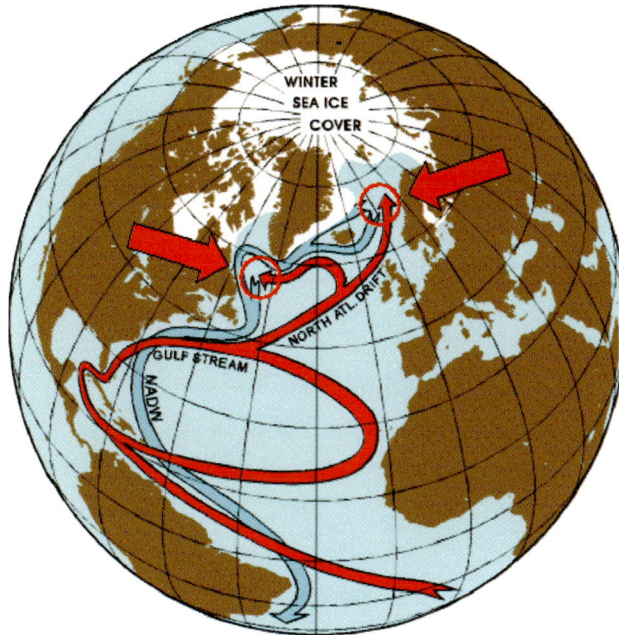

Fig. 6.6. Highly simplified cartoon of the Atlantic circulation in the present climate: surface currents are shown in red, the flow of North Atlantic Deep Water in blue. The red circles indicate the two main areas of deep water formation in the Labrador and Greenland Seas. Courtesy of S. Rahmstorf

Two feedback mechanisms have been identified as the primary reason for the nonlinearity of the THC: an advective and a convective feedback [6.47]. The advective feedback is linked to the northward transport of salt, which increases the salinity and thus the density of the North Atlantic waters. This, in turn, leads to an enhanced north-south density gradient and a stronger circulation. The convective feedback operates as follows: in the regions of net precipitation, convective mixing effectively transports freshwater downward, and thus prevents the formation of a fresh and light surface layer which would inhibit convection. Both feedbacks are positive, i.e. they tend to reinforce a given circulation pattern once it is active. This is the reason for the existence of multiple equilibrium states of the circulation. The most prominent and dramatic example is that as a consequence of the advective feedback climatic states with and without deep water formation in the North Atlantic (also referred to as conveyor *on* and *off* states) are both stable equilibria. This bistability was first demonstrated by Stommel with a simple box model [6.48] and later re-discovered by experiments with general circulation models [6.43]. As far as the convective feedback is concerned, some models indicate that it may lead to stable states with different convection patterns in the North Atlantic, e.g. with or without convection in the Labrador Sea [6.49].

The mode transition associated with the advective feedback can be illustrated with a simple box model of the Atlantic. It is a cross-hemispheric extension of Stommmel's model [6.48], i.e. it consists of four well-mixed boxes, representing the southern, tropical, northern, and deep Atlantic, respectively [6.50]. Assuming the water in the northern box is denser than that in the southern box, a pressure-driven circulation develops with northward flow near the surface and southward flow at depth. The associated meridional mass transport (or 'overturning') m is described by the quadratic equation (for a derivation see [6.50])

$$m^2 + k\alpha\left(T_2 - T_1\right)m + k\beta S_0 F_1 = 0. \tag{6.7}$$

Here $T_2 - T_1$ is the temperature difference between the northern and southern Atlantic boxes, k an empirical constant and α and β are thermal and haline expansion coefficients. F_1 is the freshwater transport (multiplied by a reference salinity S_0 for conversion into a salt flux) from the southern to the tropical box. Equation (6.7) yields m as a function of freshwater transport F_1

$$m = -\frac{1}{2}k\alpha\left(T_2 - T_1\right) \pm \sqrt{\frac{1}{4}\left[k\alpha\left(T_2 - T_1\right)\right]^2 - k\beta S_0 F_1}. \tag{6.8}$$

The solutions $m(F_1)$ are shown in Fig. 6.7 (dotted line). Since only solutions with the positive root are stable equilibria,(6.8) imposes a constraint on F_1. For the present mode of operation of the THC, i.e. for a thermally driven and freshwater inhibited circulation (see [6.50]) with $T_1 > T_2$ and $F_1 > 0$, this implies $F_1 < k\alpha^2\left(T_2 - T_1\right)^2/4\beta S_0$. The point $(F_1^{\mathrm{crit}} = k\alpha^2\left(T_2 - T_1\right)^2/4\beta S_0$, $m^{\mathrm{crit}} = k\alpha\left(T_2 - T_1\right)/2)$ is a saddle node bifurcation (point 'S' in Fig. 6.7). When the system hits this bifurcation point, it resides on an unstable branch and a transition to the *off* state occurs. The stability behavior illustrated with the simple box model agrees surprisingly well with that found in comprehensive climate models, such as general circulation models (GCMs). Figure 6.7 compares the quasi-equilibrium response of the THC as simulated in the box model with that of a GCM: the behavior of the THC in the two models is very similar. Note that the GCM locates the present climate in a region of the stability diagram, where two stable equilibria exist, THC *on* or *off* (to the left of the origin only the *on* mode is stable).

6.3.2 Were Mode Transitions of the THC Triggered in the Past?

The existence of multiple equilibria of the THC has raised concerns that anthropogenic climate change might lead to transitions from one mode to another. Indications on whether such mode transitions occurred in the past and on the forcing mechanisms that potentially triggered them, might give a hint on the behavior of the THC in the future.

On geological time scales Earth's climate is subject to regular alterations between ice ages and interglacials, which are induced by astronomical forcing factors ('Milankovitch cycles'). Over 1.2 million years the glaciations of the

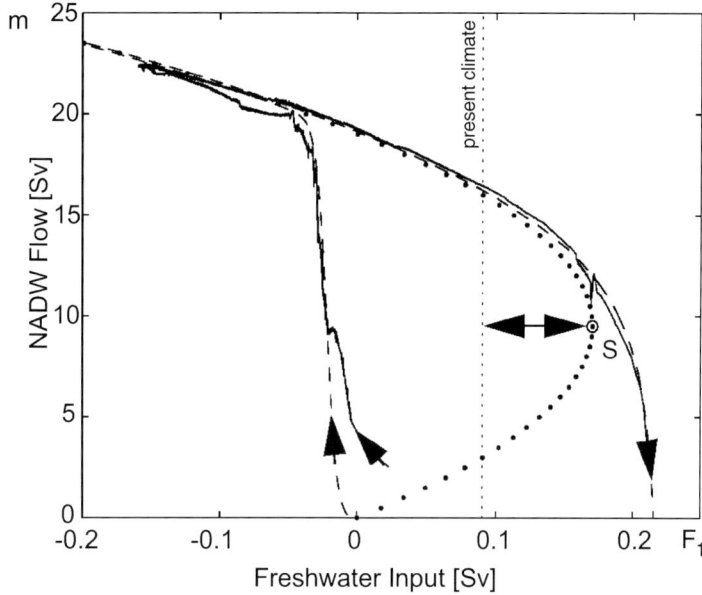

Fig. 6.7. Equilibrium solutions $m(F_1)$ of the box model (dotted line) and hysteresis curves of the box model (dashed line) as well as of a general circulation model (solid line). The latter two are traced by slowly increasing and decreasing the freshwater flux into the North Atlantic ($1\,\mathrm{Sv} = 10^6\,\mathrm{m}^3\,\mathrm{s}^{-1}$). The initial state of the GCM is marked as 'present climate'. Arrows show the distance of the present climate to the bifurcation point S. Courtesy of [6.47]

Earth have followed a 100 000 yr cycle which can be closely connected with the eccentricity of its orbit.

Paleoclimatic reconstructions [6.51] complemented by model simulations (cf. [6.52, 6.53]) have unraveled the fact that at the height of the last ice age (the 'Last Glacial Maximum' (LGM)), when temperatures were on average $\approx 6°\,\mathrm{C}$ cooler [6.54], the THC resided prominently in a state qualitatively different from today's: deep water formation occurred farther south, the outflow of deep water was shallower and bottom water of Antarctic origin pushed northward and filled the deep Atlantic. Besides this mode ('cold' mode), another qualitative state existed: an unstable 'warm' mode similar to today's with deep water formation in the Nordic Seas. This state was associated with temperatures in the North Atlantic sector that were 5–10° C warmer compared to the 'cold' mode. Ganopolski and Rahmstorf [6.52] showed that trigger events (of unknown origin) associated with freshwater export from the North Atlantic, could have induced mode transitions between these two qualitative states (transition at point B in Fig. 6.9b, d), leading to the abrupt warming events ('Dansgaard-Oeschger' events) evident in the Greenland ice core records [6.51] (cf. Fig. 6.8). Since this mode is not stable under glacial conditions, it decays after a waiting time of several hundred years (transition at point A in Fig. 6.9b, d). During glacial times also the *off* mode apparently existed. Transition to this circulation state could have been induced by large freshwater perturbations due to massive ice-sheet surges ('Heinrich events'; cf. [6.55],

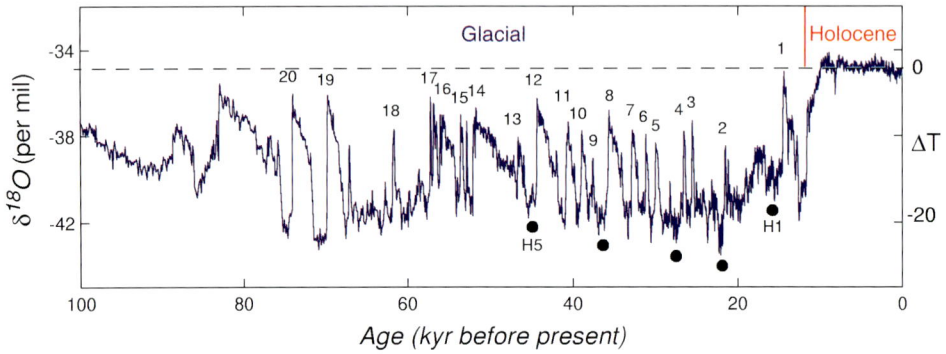

Fig. 6.8. Abrupt climate changes over Greenland recorded in ice-core data. Shown is the $\delta^{18}O$ record, a proxy for atmospheric temperature (the approximate relative temperature range (in $^\circ$ C) is given on the right). The glacial climate is punctuated by Dansgaard-Oeschger warming events (numbered). The black dots denote the timing of Heinrich events. Courtesy of [6.52]

Fig. 6.8). Such a transition would explain the unusual warming events seen in the Antarctic ice-core records and the synchronous cold events over Greenland (the so-called 'bipolar seesaw'; cf. [6.56]): indeed, a shutdown of the THC would impede the oceanic heat transport from the southern to the northern hemisphere, warming the first and cooling the latter. According to the model results of Ganopolski and Rahmstorf [6.52] the *off* mode was not stable under glacial conditions [6.52]. The reason is that in glacial times NADW formation is allowed to retreat gradually southward when forced by freshwater influx. In the modern climate NADW formation is confined to the Nordic Seas: it cannot retreat south because temperatures are too warm there. This leads to the smoother response of the glacial THC in comparison to the modern (compare the black hysteresis curves in Fig. 6.9c, d).

6.3.3 A Glimpse into the Future

Based on physical considerations and on the instability of the THC during glacial times when rapid transitions apparently occurred between the different modes (see Sect. 6.3.2), concerns have been raised that anthropogenic climate change might trigger a similar instability in the future. Model simulations indeed show that under global warming a transition to a state without deep water formation might occur. Manabe and Stouffer [6.57], for example, have simulated a complete shutdown of the THC under a quadrupling of atmospheric CO_2, while Rahmstorf and Ganopolski [6.58] obtained the same transition for a transient peak in CO_2 (see Fig. 6.10). It took several centuries until the circulation shut down completely in both studies. A regional shutdown of deep water formation in the Labrador Sea was simulated by Wood et al. [6.59] for a

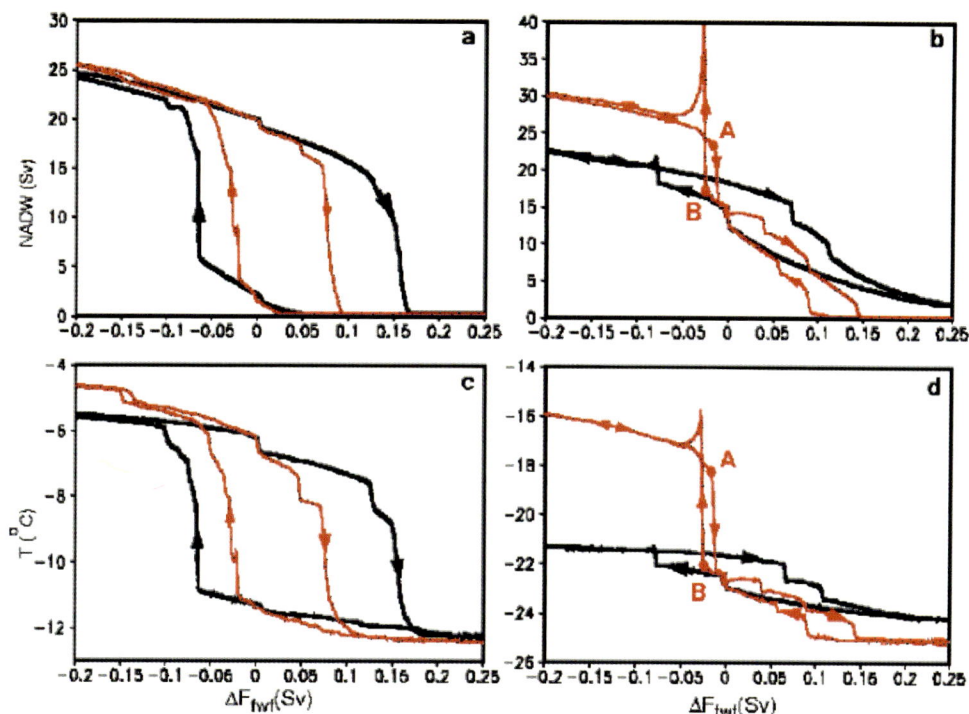

Fig. 6.9. Stability diagrams of the Atlantic thermohaline circulation for the glacial ((**b**) and (**d**)) and present climate conditions ((**a**) and (**c**)). The upper panels ((**a**) and (**b**)) show the ocean circulation response to slowly increasing and decreasing freshwater input, the lower ((**c**) and (**d**)) panels the associated air temperature in the North Atlantic sector (60–70° N). The black curves are obtained for the freshwater perturbation added outside the convection regions, while the red curves were computed by adding fresh water directly to the latitudes of the Nordic Seas (50–70° N). The hysteresis loops are narrower in the latter case because a shut-down or start-up of convection is triggered locally before the basin-scale advective stability limits are reached. Courtesy of [6.52]

'business-as-usual' scenario (IS92a [6.60]). This collapse occurred early in the 21st century on a time scale of less than a decade (cf. Fig. 6.0, lower panel).

These simulations clearly identify the potential for THC instability in a warming world. However, because of the large uncertainties involved in climate modeling, it is not yet possible to determine appropriate threshold values. A large uncertainty is associated, for example, with the changes in the net freshwater fluxes entering the North Atlantic, which play the role of bifurcation parameters in the system (see Fig. 6.7): estimates of evaporation and precipitation changes differ largely between models [6.61], as well as estimates of freshwater runoff from the Greenland ice sheet and other melting glaciers in the North Atlantic catchment [6.62]. Another uncertainty is related to the location of the pre-industrial climate on the stability diagram (Fig. 6.7): it determines the distance of the system from the bifurcation point, and hence the amount of freshwater influx that the system can sustain without being pushed beyond the critical threshold. Also the uncertainty in climate sensitivity (i.e. the response of global average surface air temperature to a doubling

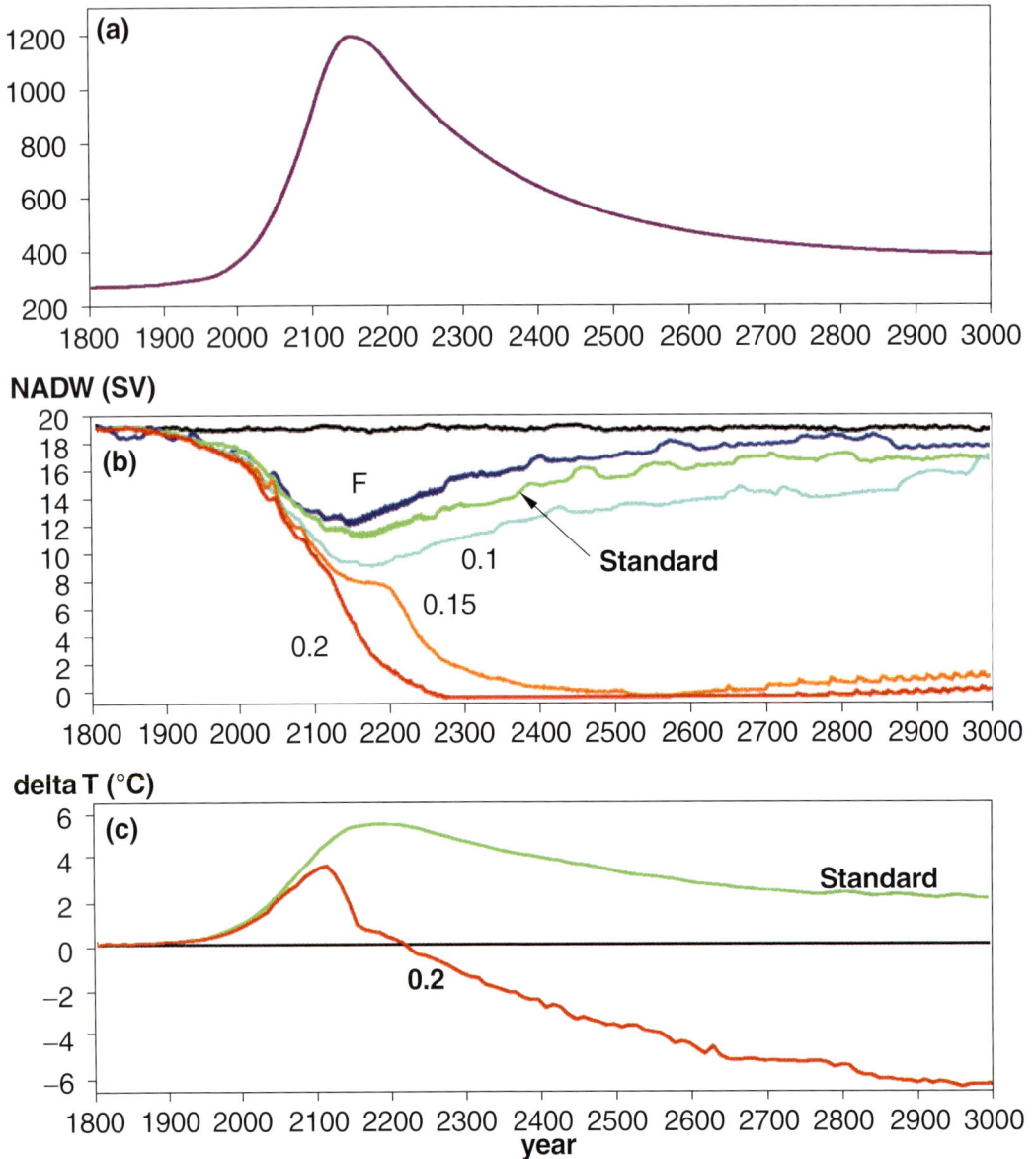

Fig. 6.10. Response of NADW flow and north Atlantic temperature to a global warming scenario as computed with the CLIMBER-2 model [6.63]. (**a**) CO_2 forcing scenario, (**b**) NADW circulation rate ($1\,Sv = 10^6\,m^3\,s^{-1}$), (**c**) winter surface air temperature change over the Atlantic at 55°N. The response of NADW flow to the forcing scenario in (**a**) differs because of different assumptions about changes in the freshwater budget of the North Atlantic. The experiment labeled 'F' is a run with fixed preindustrial freshwater forcing, 'Standard' is a run with standard model parameter settings. '0.1', '0.15', '0.2' are experiments with artificially increased freshwater forcing (i.e. giving a freshwater anomaly in the North Atlantic of 0.1, 0.15, 0.2 Sv at the height of warming). The NADW response shows that there is a threshold value in the freshwater forcing beyond which the circulation collapses. A complete shutdown of the circulation leads to an abrupt reversal of the warming trend and a cooling in the north Atlantic area (curve labeled '0.2' in (**b**) and (**c**)). Courtesy of [6.58]

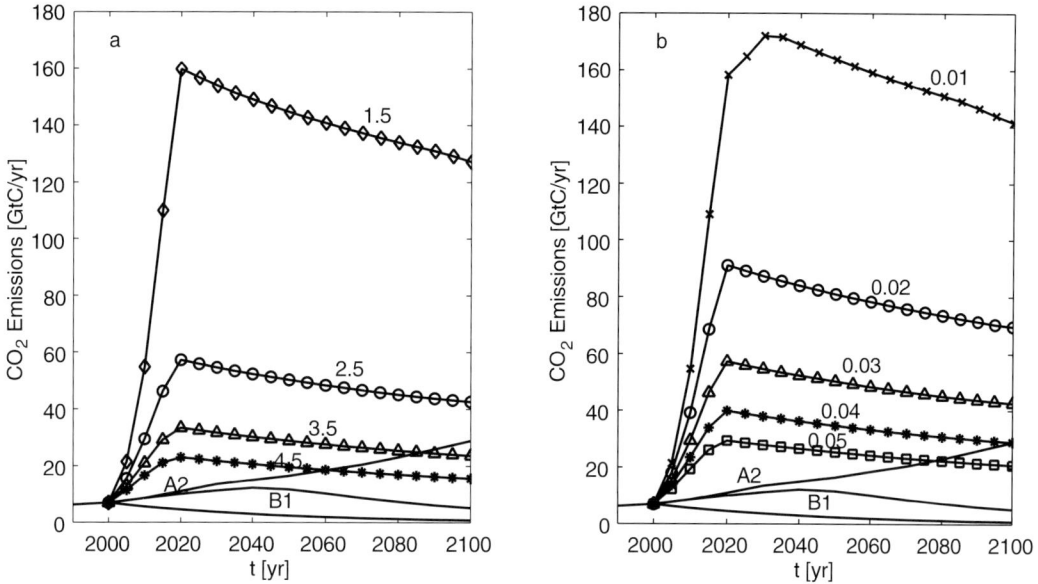

Fig. 6.11. Emisssions corridors for different assumptions about climate and hydrological sensitivities, which are among the main uncertain quantities in projecting the fate of the THC. An emissions corridor is given by the area delimited by the lower (solid line) and the upper boundary (marked lines). Note that the upper boundary depends upon the specific parameter values, while the lower boundary is the same for all values of climate and hydrological sensitivity. (**a**) Emissions corridors for a hydrological sensitivity of 0.03 Sv/°C and different values of the climate sensitivity [°C]. (**b**) Emisssions corridors for a climate sensitivity of 2.5° C and different values of the Atlantic hydrological sensitivity [Sv/°C]. For reference we also show a high and a low emissions scenario for the 21st century (SRES scenario A2 and B1 [6.64]). The analytical tool employed for the calculation of the emissions corridors consists of a four-box model of the THC coupled to a reduced-form climate model (cf. [6.65]). The corridor boundaries are determined by solving a sequence of dynamic optimization problems (cf. [6.66])

of atmospheric CO_2) plays a role: it determines the increase in the strength of the hydrological cycle as well as the heat flux to the ocean and hence water densities. The uncertainty in climate sensitivity is indicated by the IPCC as spanning the range 1.5–4.5°C [6.67].

The fact that the time scale on which climate change happens is similar to the response time of the ocean complicates matters further. Stocker and Schmittner [6.68], for example, have shown that the THC is sensitive not only to the final level of temperature increase, but also to the rate of change.

6.3.4 Guidelines for a Precautionary Policy

The review of our state of knowledge about the 'conveyor belt' clearly shows that even the use of quantitative models cannot instantaneously produce guidelines for strict policy advice. In comparison to the example discussed in Sect. 6.2, which focuses more on the mitigation of concrete situations, forward-looking strategies have to be implemented here in order to prevent disastrous outcomes. But how can we cope with uncertainty in the latter case? As in

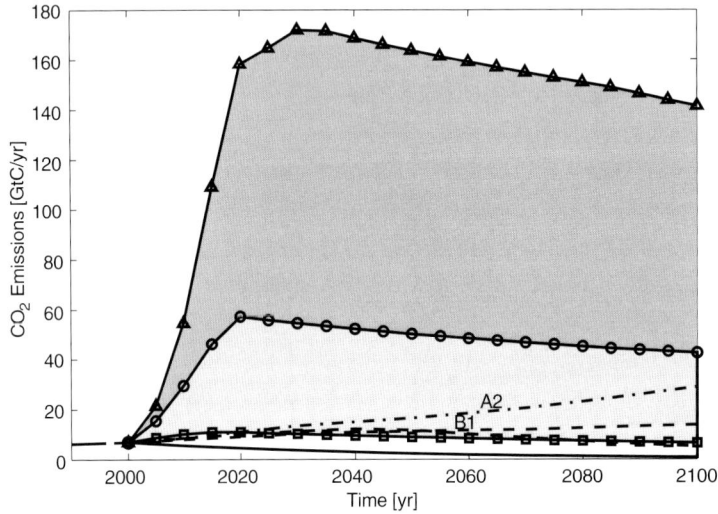

Fig. 6.12. Emissions corridors for the 'best case' (triangles), 'best guess' (circles), and 'worst case' (squares) combination of model parameter values (see text). The shaded areas between the corridors indicate likelihood domains for a shutdown of the THC: the darker the shading, the higher the probability that any given emissions paths entering that domain triggers a breakdown of the THC. For reference we show a high and a low emissions scenario for the 21st century (SRES scenario A2 and B1, respectively [6.64])

the fisheries example, vagueness is a system immanent property, but in the previous example a validation can be performed by a qualitative reconstruction of observable events, whereas this is impossible for the oceanic circulation. Thus, from the systems viewpoint, a 'simplification' with respect to this issue is unavoidable. We have to keep in mind that the idea of an unperturbed 'environment' is a human conception, which must be enhanced by socially and politically accepted guardrails. In addition, there should be a minimal consensus that these guardrails keep us away from catastrophic domains [6.69], and create a minimum of safety standards.

For the THC example this means that a set of potential strategies is conceivable that limit anthropogenic global warming to a magnitude at which a breakdown of the circulation is unlikely according to state-of-the-art scientific evidence. This conception is formalized in the notion of 'emissions corridors': it represents the range of long-term CO_2 greenhouse gas emissions which are allowed under a predefined set of normative climate policy goals, such as, for example, avoiding a collapse of the THC while assuring that future economic prosperity is not endangered (cf. [6.65]). Figure 6.11 displays emissions corridors achieving these goals for different assumptions made about two of the main uncertain quantities in climate change projections, i.e. climate and hydrological sensitivity. The latter is a measure for the enhancement of the hydrological cycle for a given amount of warming (see Sect. 6.3.3). Figure 6.11 indicates a very strong dependence of the width of the emissions corridor on the specific assumptions: for small values of climate and hydrological sensitivities the width of the corridor is much larger than the range of CO_2 emissions projected for the 21st century [6.64], implying that no immediate mitigation measure would be necessary to preserve the THC. For medium values of both quantities, however, the upper boundary is still out of reach, but the width of the emissions corridors is considerably reduced. For high values the corridors shrink further so that if emissions were to follow a high scenario (e.g. SRES emissions scenario A2 [6.64]), the corridor boundaries would be transgressed.

Given these large uncertainties, which emissions strategies are thus most likely to avert a collapse of the THC? This problem may be tackled by introducing the notion of 'worst case' emissions corridor (i.e. the corridor which is obtained for the most pessimistic assumptions about parameter values) and by drawing upon a fundamental strength of the guardrail approach: any emissions path complying with the guardrails under worst case assumptions would not loose this property whatever the 'true' parameter values turn out to be. This means that emissions strategies attaining the predefined goal of preserving the THC under worst case assumptions would in any case retain their validity. At best, i.e. if new scientific evidence allows the uncertainty ranges to be narrowed toward more optimistic values, existing greenhouse gas mitigation strategies could even be relaxed. Figure 6.12 illustrates these concepts by displaying emissions corridors for the 'best case', 'best guess', and 'worst case' assumptions about climate and hydrological sensitivity (i.e. the parameter combinations $2.5°$ C, 0.01 Sv/$°$C; $2.5°$ C, 0.03 Sv/$°$C; $4.5°$ C, 0.05 Sv/$°$C, respectively; cf. [6.65]). The areas delimited by the upper boundaries of these corridors (represented as shaded areas in Fig. 6.12) may be interpreted as likelihood domains for a collapse of the THC: the darker the shading, the higher the probability that any given emissions paths entering the respective domain triggers a complete and irreversible breakdown of the THC. Note that for the 'worst case' combination of parameters the corridor almost vanishes. This implies that the leeway for any climate policy committed to the precautionary principle could be tight. Indeed, even the nonintervention low emissions scenario B1 [6.64] leaves the area of the 'worst case' corridor, implying the necessity to turn away from 'business-as-usual' as soon as possible.

6.4 Conclusion

In our contribution we show that the tackling of large parts of the Earth system by unconventional modeling concepts even in a 'vague environment' could improve understanding of complex dynamics substantially, and in addition, supplies valuable hints for future decision making. The most important novel aspect of our analysis is to employ a set of dynamic patterns as primary units, i.e. typical trajectories of commercial fisheries as well as transitions between different qualitative states of the thermohaline circulation. All kinds of solutions for the concomitant problems - for the potential breakdown of marine fisheries as well as for the North Atlantic thermohaline circulation - are conceivable and implicate actions for mitigation/prevention of disastrous events which are based on soft-modeling techniques. We emphasize, however, that uncertainties in the model-based predictions presented here are large, but on the other hand, studying systems from hitherto novel methodological perspectives could pave the road towards an integrated modeling and assessment. Nevertheless, further work is clearly needed, in particular regarding the identification of potential precursor signals when approaching critical branchings.

Acknowledgement. We are grateful to T. Bruckner, M.K.B. Lüdeke, J. Scheffran, and G. Petschel-Held for many fruitful discussions. Especially, we wish to thank S. Rahmstorf and H. Welsch for their kind intellectual support during the preparation of this chapter. The authors thanks the German Federal Ministry of Science for financial support of their work (under grant no. 03F0205B9 and 01LD0016).

References

6.1 J. Biercamp, M. Boettinger, and K. Hasselmann, *Klimasimulation: Vorhersage des globalen Wandels* (Movie) (Deutsches Klimarechenzentrum (DKRZ), Hamburg, 1995).

6.2 Ozone Secretariat United Nations Environment Programme, *The Montreal Protocol on Substances that deplete the Ozone Layer* (UNON Printshop, Nairobi, 2000); [available from `www.unep.org/ozone/Montreal-Protocol/Montreal-Protocol2000.shtml/`].

6.3 IPCC, *Climate Change 2001: The Scientific Basis* (WGI); *Impacts, Adaption and Vulnerability* (WGII); *Mitigation* (WGIII); Intergovernmental Panel of Climate Change (IPCC) (Cambridge University Press, Cambridge, 2001); [available from `www.ipcc.ch`].

6.4 W. Steffen, J. Jäger, D.J. Carson, and C. Bradshaw (eds.), *Challenges of a Changing Earth* (Springer, Berlin, 2002).

6.5 J.B. Smith, H.J. Schellnhuber, and M.M.Q. Mirzu, in *Climate Change 2001: Impacts, Adaptation, and Vulnerability, Contribution of Working Group II to the Third Assessment Report of the Intergovernmental Panel on Climate Change (IPCC)*, edited by J.J. McCarthy, O.F. Canziani, N.A. Leary, D.J. Dokken, and K.S. White (Cambridge University Press, Cambridge, 2001) p. 913.

6.6 H.J. Schellnhuber, A. Block, M. Cassel-Gintz, J. Kropp, G. Lammel, W. Lass, R. Lienenkamp, C. Loose, M.K.B. Lüdeke, O. Moldenhauer, G. Petschel-Held, M. Plöchl, and F. Reusswig, GAIA **6**(1), 19 (1997).

6.7 G. Petschel-Held, A. Block, M. Cassel-Gintz, J. Kropp, M.K.B. Lüdeke, O. Moldenhauer, F. Reusswig, and H.J. Schellnhuber, Environ. Mod. Assess. **4**(4), 295 (1999).

6.8 H.J. Schellnhuber and G. Yohe, in *Achievements, Benefits, and Challenges* (World Meteorological Organization, Geneva, 1997) p. 179.

6.9 J.B.C. Jackson, M.X. Kirby, W.H. Berger, K.A. Bjørndal, L.W. Botsford, B.J Bourque, R.H. Bradbury, R. Cooke, J. Erlandson, J.A. Estes, T.P. Hughes, S. Kidwell, C.B. Lange H.S. Lenihan, J.M. Pandolfi, C. H. Peterson, R.S. Steneck, M.J. Tegner, and R.R. Warner, Science **293**(5530), 629 (2001).

6.10 FAO, *The State of World Fisheries and Aquaculture* (Food and Agriculture Organization of the United Nations, Rome, 2001).

6.11 FAO, *The State of World Fisheries and Aquaculture* (Food and Agriculture Organization of the United Nations, Rome, 1997).

6.12 M. Williams, *The Transition in the Contribution of Living Aquatic Resources to Food Security* (International Food Policy Research Institute, Washington, 1996).

6.13 S. Garcia and C. Newton, *Current Situation, Trends and Prospects in World Capture Fisheries* (American Fisheries Society, Bethesda, 1997).

6.14 O. Flaaten, A.G.V. Salvanes, T. Schweder, and O. Ulltang, Fish. Res. **37**, 1 (1998).

6.15 J.F. Caddy, Rev. Fish Biol. Fisher. **9**, 1 (1999).

6.16 K.I. Stergiou, Fish. Res. **55**, 1 (2002).

6.17 G. Guariso, A. Rizzoli, H. Werthner, IEEE Trans. Syst. Man Cyb. **22**(5), 1075 (1992).

6.18 U. Heller and P. Struss, in *Umweltinformatik Aktuell* (Metropolis, Marburg, 1996), p. 358.

6.19 G. Brajnik and M. Lines, J. Artif. Soc. Social Simul. **1**, 1 (1998).

6.20 C. Béné, L. Doyen, and D. Gabay, Ecol. Econ. **36**, 385 (2001).

6.21 K. Eisenack and J. Kropp, Mar. Pollut. Bull. **43**(7-12), 215 (2001).

6.22 G. Petschel-Held and M.K.B Lüdeke, Integr. Assess. **2**(3), 123 (2001).

6.23 H.J. Schellnhuber and J. Kropp, Naturwissenschaften **85**, 411 (1998).

6.24 H.J. Schellnhuber, in *Earth System Analysis*, edited by H.J. Schellnhuber and V. Wenzel (Springer, Berlin 1998).

6.25 C.W. Clark, F.H. Clarke, and G.R. Munro, Econometrica **47**, 25 (1979).

6.26 C.W. Clark, *Mathematical Bioeconomics: The Optimal Management of Renewable Resources* (Wiley, München, 1990).

6.27 J.R. Boyce, J. Environ. Econ. Manage. **28**, 324 (1995).

6.28 A. Hatcher, Mar. Pol. **24**(2), 129 (2000).

6.29 R. McKelvey, J. Environ. Econ. Mange. **12**(4), 287 (1985).

6.30 R. McKelvey, in *Applications of Control Theory in Ecology*, edited by M. Cohen (Springer, Berlin, 1986), p. 57.

6.31 Q. Weninger and K.E. McConnell, Can. J. Econ. **33**(2), 394 (2000).

6.32 S. Pascoe and S. Mardle, Eur. Rev. Agric. Econ. **28**(2), 161 (2001).

6.33 K. Eisenack, J. Kropp, and H. Welsch, The role of capital accumulation in open-access fishery: a qualitative modeling approach, submitted to Resource and Energy Economics (2002); [available from `de.arXiv.org/abs/cs.AI/0202004`].

6.34 B.D. Craven, *Control and Optimization* (Chapman & Hall, New York, 1995).

6.35 O.L. Mangasarian, J. Control **4**, 139 (1966).

6.36 B. Kuipers, *Qualitative Reasoning: Modeling and Simulation with Incomplete Knowledge* (MIT, Cambridge, 1994).

6.37 D.J. Clancy, *Solving Complexity and Ambiguity Problems with Qualitative Simulation* (PhD thesis, University of Texas, 1997).

6.38 J. Kropp and K. Eisenack, in *Working Paper of the 15th Workshop on Qualitative Reasoning*, edited by G. Biswas (University of Texas, San Antonio, 2001) p. 187.

6.39 M. Harris, *The Lament for an Ocean: The Collapse of the Atlantic Cod Fishery, a True Crime Story* (McClelland & Stewart, Toronto, 1998).

6.40 F. Pearce, New Scientist **153**(2068), 6 (1997).

6.41 Anonymous, Economist [February issue], 86 (2002).

6.42 D.H. Roemmich and C. Wunsch, Deep-Sea Res. **32**, 619 (1985).

6.43 S. Manabe and R.J. Stouffer, J. Clim. **1**, 841 (1988).

6.44 J.M. Napp, and G.L. Hunt Jr., Fish. Oceanogr. **10**, 61 (2001).

6.45 C.M. O'Brien, C.J. Fox, B. Planque, and J. Casey, Nature **404**, 142 (2000).

6.46 S. Rahmstorf, Climatic Change **46**, 247 (2000).

6.47 S. Rahmstorf, in *Reconstructing Ocean History: A Window into the Future*, edited by F. Abrantes and A.C. Mix (Kluwer, New York, 1999).

6.48 H. Stommel, Tellus **13**, 224 (1961).

6.49 S. Rahmstorf, Nature **372**, 82 (1994).

6.50 S. Rahmstorf, Clim. Dyn. **12**, 799 (1996).

6.51 W. Dansgaard, S.J. Johnsen, H.B. Clausen, N.S. Dahl-Jensen, N.S. Gundestrup, C.U. Hammer, C.S. Hvidberg, J.P. Steffensen, A.E. Sveinbjørnsdottir, J. Jouzel, and G. Bond, Nature **364**, 218 (1993).

6.52 A Ganopolski and S. Rahmstorf, Nature **409**, 153 (2001).

6.53 C.D. Hewitt, A.J. Broccoli, J.F.B. Mitchell, and R.J. Stouffer, Geophys. Res. Lett. **28**, 1571 (1998).

6.54 A. Ganopolski, S. Rahmstorf, V. Petoukhov, and M. Claussen, Nature **391**, 351 (1998).

6.55 H. Heinrich, Quat. Res. **29**, 143 (1988).

6.56 T.F. Stocker, Science **282**, 61 (1998).

6.57 S. Manabe and R.J. Stouffer, Nature **364**, 215 (1993).

6.58 S. Rahmstorf and A. Ganopolski, Climatic Change **43**, 353 (1999).

6.59 R.A. Wood, A.B. Keen, J.F.B. Mitchell, and J.M. Gregory, Nature **399**, 572 (1999).

6.60 J.T. Houghton, B.A. Callander, and S.K. Varney (eds.), *Climate Change 1992: The Supplementary Report to the IPCC Scientific Assessment* (Cambridge University Press, Cambridge, 1992).

6.61 U. Cubasch and G.A. Meehl, in *Climate Change 2001: The Scientific Basis - Contribution of Working Group I to the Third Assessment Report of the IPCC*, edited by J.T. Houghton, Y. Ding, D.G. Griggs, M. Noguer, P.J. van der Linden, X. Dai, K. Maskell, and C.A. Johnson (Cambridge University Press, Cambridge, 2001) p. 525.

6.62 J.A. Church and J.M. Gregory, in *Climate Change 2001: The Scientific Basis - Contribution of Working Group I to the Third Assessment Report of the IPCC*, edited by J.T. Houghton, Y. Ding, D.G. Griggs, M. Noguer, P.J. van der Linden, X. Dai, K. Maskell, and C.A. Johnson (Cambridge University Press, Cambridge, 2001) p. 639.

6.63 V. Petoukhov, A. Ganopolski, V. Brovkin, M. Claussen, A. Eliseev, C. Kubatzki, and S. Rahmstorf, Clim. Dyn. **16**, 1 (2000).

6.64 N. Nakićenović and R. Swart (eds.), *Emission Scenarios* (Cambridge University Press, Cambridge 2000).

6.65 K. Zickfeld and T. Bruckner, in *Integrated Assessment and Decision Support, Proc. 1st Biennial Meeting of the International Environmental Modelling and Software Society*, edited by A.E. Rizzoli and A.J. Jakeman (Servizi Editoriali Associati, Como, 2002).

6.66 M. Leimbach and T. Bruckner, Comput. Econ. **18**, 173 (2001).

6.67 A. Kattenberg, F. Giorgi, H. Grassl, G.A. Meehl, J.F.B. Mitchell, R.J. Stouffer, T. Tokioka, A.J. Weaver, and T.M.L. Wigley, in *Climate Change 1995: The Science of Climate Change - Contribution of Working Group I to the Second Assessment Report of the IPCC*, edited by J.T. Houghton, L.G. Meira Filho, B.A. Callander, N. Harris, A. Kattenberg, and K. Maskell (Cambridge University Press, Cambridge, 1995) p. 285.

6.68 T.F. Stocker and A. Schmittner, Nature **388**, 862 (1997).

6.69 T. Bruckner, G. Petschel-Held, F. Tóth, H.M. Füssel, C. Helm, M. Leimbach, and H.J. Schellnhuber, Environ. Mod. Assess. **4**, 217 (1999).

Part III

Biodynamics

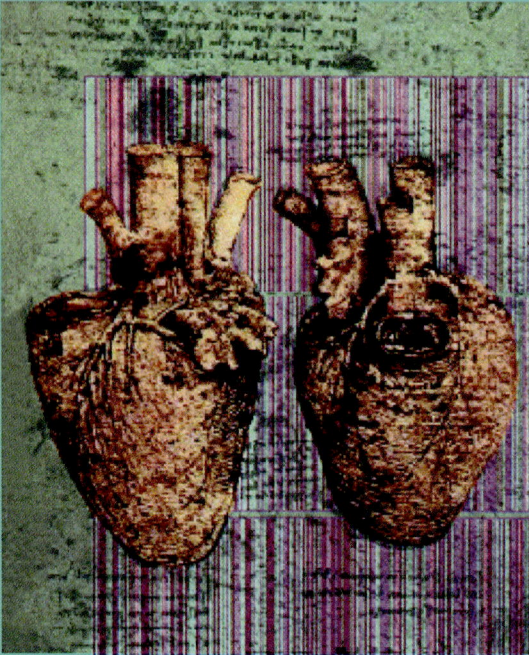

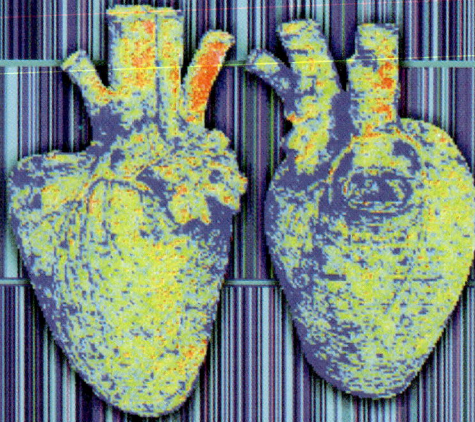

7. Fractal and Multifractal Approaches in Physiology

Plamen Ch. Ivanov, Ary L. Goldberger, and H. Eugene Stanley

We explore the degree to which concepts developed in statistical physics can be usefully applied to physiological signals. We first review recent progress using two analysis methods: (i) detrended fluctuation analysis to quantify homogeneous structures, termed monofractals, which are characterized with the same scaling properties throughout the entire signal, and (ii) wavelet-based multifractal analysis to quantify signals of higher complexity, termed multifractals, which require many exponents to characterize their scaling properties. We next illustrate the problems related to physiological signal analysis with representative examples of heartbeat dynamics under healthy and pathological conditions. We discuss the findings of fractal and multifractal properties in the human heartbeat and how they change with disease.

7.1 Introduction

Even under healthy, basal conditions, physiological systems show erratic fluctuations resembling those found in dynamical systems driven away from a single equilibrium state. Do such 'nonequilibrium' fluctuations simply reflect the fact that physiological systems are being constantly perturbed by external and intrinsic noise? Or, do these fluctuations actually contain useful, 'hidden' information about the underlying nonequilibrium control mechanisms? We report some recent attempts to understand the dynamics of complex physiological

◄ **Fig. 7.0.** Leonardo da Vinci's hand drawing of the human heart with panels on the background illustrating how the local Hurst exponent (vertical color bars) changes in time (horizontal axis). Healthy heart records (shorter panels, bottom) appear polychromatic indicating multifractality, while heart failure records (longer panels) are more monochromatic (blue color predominantly) indicating loss of multifractality. Courtesy of Z.R. Struzik (CWI, Amsterdam, The Netherlands) and Anna Ludwicka (Artgraph, Warszawa, Poland)

fluctuations by adapting and extending concepts and methods developed very recently in statistical physics.

The central task of statistical physics is to study macroscopic phenomena that result from microscopic interactions among many individual components. This problem is akin to many investigations undertaken in biology. In particular, physiological systems under neuroautonomic regulation, such as heart rate regulation, are good candidates for such an approach, since: (i) the systems often include multiple components, thus leading to very large numbers of degrees of freedom, and (ii) the systems usually are driven by competing forces. Therefore, it seems reasonable to consider the possibility that dynamical systems under neural regulation may exhibit temporal structures which are similar, under certain conditions, to those found in physical systems. Indeed, concepts and techniques originating in statistical physics are showing promise as useful tools for quantitative analysis of complicated physiological systems.

An unsolved problem in biology is the quantitative analysis of a nonstationary time series generated under *free-running* conditions [7.1–7.3]. The signals obtained under these constantly varying conditions raise serious challenges to both technical and theoretical aspects of time series analyses. A central question is whether such noisy fluctuating signals contain dynamical patterns essential for understanding underlying physiological mechanisms. There are three particularly vexing features of physiological time series:

(i) *Nonstationarity.* Traditional methods of statistical analysis assume that the statistical properties of a signal are the same throughout the signal. A time series is *stationary* if its statistical characteristics such as the mean and the variance are invariant under time shifts, i.e. if they remain the same when t is replaced by $t + \Delta$, where Δ is arbitrary. Then the probability densities, together with the moment and correlation functions, do not depend on the position of the points on the time axis, but only on their relative configuration [7.4]. Nonstationarity, an important feature of biological variability, can be associated with regimes of different drifts in the mean value of a given signal, or with changes in its variance or autocorrelation function, which may be gradual or abrupt. This is the case for many signals of interest in physiology, e.g. the statistical properties of the heart rate change when a subject rises to a standing position (Fig. 7.1a). Furthermore, nonstationarities are important features of the data under both healthy and perturbed conditions [7.5]. Such nonstationarity problems arise in other contexts in the discipline of statistical physics, and novel techniques such as *wavelets* (see also Chap. 2 in this book) have successfully been developed to study nonstationary signals.

(ii) *Nonlinearity.* Traditional methods of analysis also assume that to a large degree the system can be viewed as linear, so that departures from linearity can be treated perturbatively. This is not true for most physiological systems, which are intrinsically nonlinear. A salient feature of nonlinear

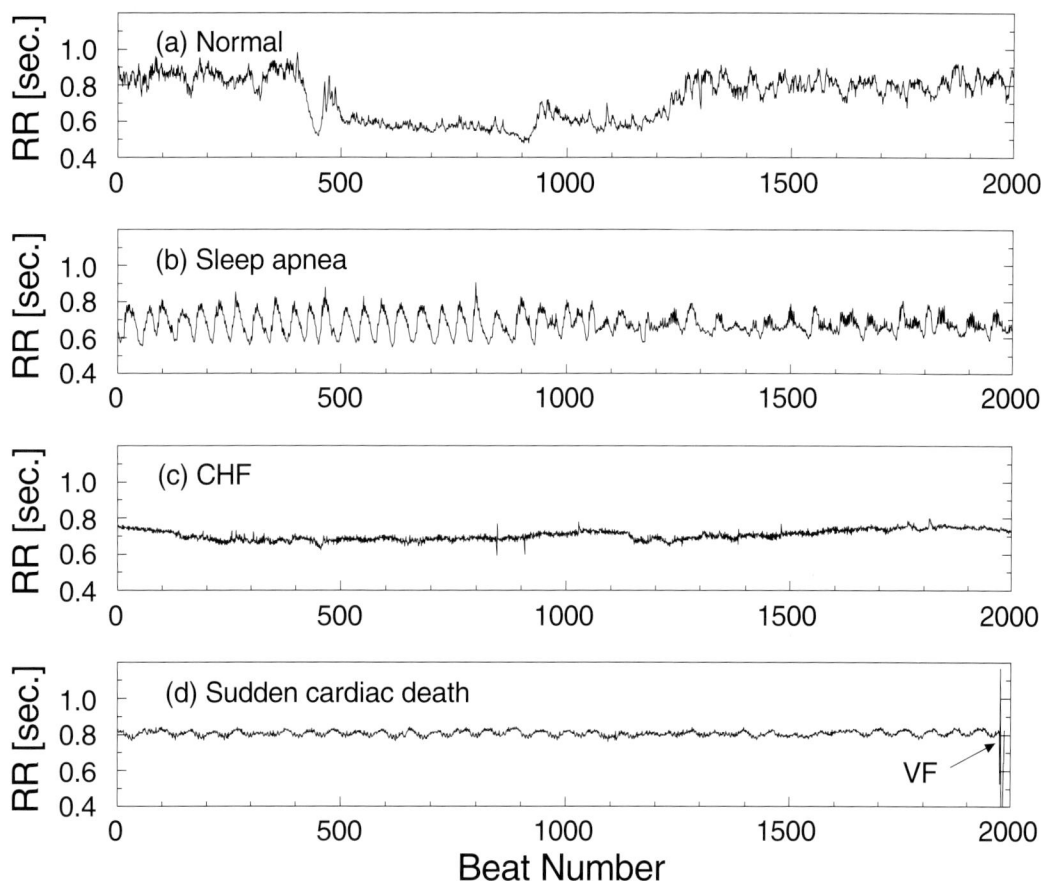

Fig. 7.1. Representative complex physiological fluctuations. Cardiac interbeat interval (normal sinus rhythm) time series of 2 000 beats from (**a**) a healthy subject, (**b**) a subject with obstructive sleep apnea, (**c**) a subject with congestive heart failure (CHF), and (**d**) a sudden cardiac death subject with ventricular fibrillation (VF). Note the nonstationarity of these time series (most apparent in (**a**) and (**b**)), which limits the applicability of traditional methods of analysis and modeling

systems is that their components interact with each other, and therefore their outputs are not proportional to the strength of the inputs. The field of statistical physics has in the past years focused on nonlinear systems, and has developed a conceptual framework within which a wide range of nonlinear phenomena can be usefully treated.

(iii) *Nonequilibrium Phenomena.* The fundamental principle of homeostasis asserts that physiological systems seek to maintain a constant output after perturbation. From the time of Claude Bernard and Walter Cannon [7.6–7.8], it has been assumed that physiological systems possess feedback and control mechanisms that serve to restore an equilibrium-like state when a system is perturbed away from some set point. Our research, however, indicates that physiological systems are inherently out-of-equilibrium systems. Nonequilibrium statistical mechanics has made advances in recent years that have yet to be applied in the physiological domain [7.9, 7.10].

Here we will explore the degree to which the concepts developed in statistical physics can be usefully applied to physiological signals. We illustrate the problems related to physiological signal analysis with representative examples of complex dynamical behavior under healthy and pathological conditions (see Fig. 7.1). Specifically, we focus on interbeat interval variability as an important quantity to help elucidate possibly nonhomeostatic physiological variability because (i) the heart rate is under direct neuroautonomic control, (ii) interbeat interval variability is readily measured by noninvasive means, and (iii) analysis of these heart rate dynamics may provide important practical diagnostic and prognostic information. Figure 7.1a shows a physiological cardiac interbeat time series, the output of a spatially and temporally integrated neuroautonomic control system. The time series shows 'erratic' fluctuations and 'patchiness'. These fluctuations are usually ignored in conventional studies which focus on averaged quantities. In fact, these fluctuations are often labeled as 'noise' to distinguish them from the true 'signal' of interest. Generally, in the conventional approach it is assumed that there is no meaningful structure in apparent noise, and, therefore, one does not expect to gain any understanding about the underlying system through the study of these fluctuations. However, by adapting and extending methods developed in modern statistical physics and nonlinear dynamics, we find that the physiological fluctuations shown in Fig. 7.1a exhibit unexpected hidden *scaling* structures. Furthermore, the dynamical patterns of these fluctuations and the scaling features associated with them *change* with pathological perturbations (see Fig. 7.1b–7.1d). These findings raise the possibility that understanding the origin of such temporal structures and their alterations may (a) elucidate certain basic features of heart rate control mechanisms, and (b) have practical value in clinical monitoring.

We first review recent progress using two analysis methods, i.e. detrended fluctuation analysis and wavelet techniques, appropriate for quantifying *monofractal* structures. We then describe very recent work that quantifies *multifractal* features of interbeat interval series, and the discovery that the multifractal structure of healthy subjects is different than that of diseased subjects. The analytic tools we discuss may be used on a wider range of physiological signals.

7.2 Limitations of Traditional Techniques

A technique widely used to analyze time series is the study of the moments of the distribution of measured values. Figure 7.2 shows two sequences of interbeat intervals, one for a normal individual and one for a subject with congestive heart failure. Visual inspection makes clear the existence of differences in the dynamics generating the two signals. However, the signals have the same averages and standard deviations. Hence additional methods which are sensitive

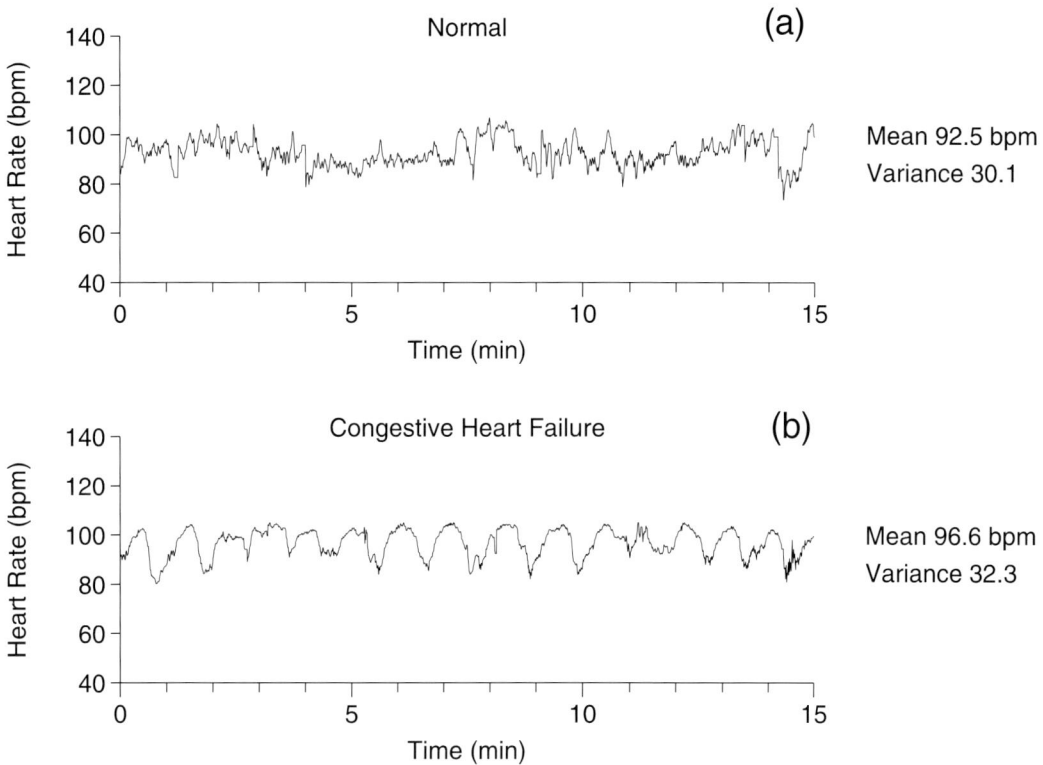

Fig. 7.2. Two heart rate time series with similar values in their mean and standard deviation. However, dramatic differences in their dynamics can be easily visualized. Note that the healthy subject (**a**) shows a complex type of variability while the subject with heart failure (**b**) shows a more periodic pattern. From [7.15]

to local patterns related to nonstationarities are required if these two signals are to be distinguished.

A quantity widely used to measure correlations in a time series is the power spectrum, which measures the relative frequency content of a signal. A power spectrum calculation assumes that the signal studied is stationary, and when applied to nonstationary time series can lead to misleading results. To illustrate this point, we analyze two signals: one is stationary (Fig. 7.3a), i.e. two different frequencies are present at all times. The other is nonstationary (Fig. 7.3b), i.e. one frequency is present in the first half of the signal and another frequency in the other half. The calculation of the power spectrum for these signals leads to almost *identical* results (Fig. 7.3c, d)! Similarly, the presence of linear or higher order polynomial trends can mask the frequency content of a signal. Moreover, the power spectrum carries no information on the nonlinear properties of the signal [7.11, 7.12]. Since the power spectrum is incapable of distinguishing between these types of behavior, it must not be used as the *only* form of analysis for nonstationary signals.

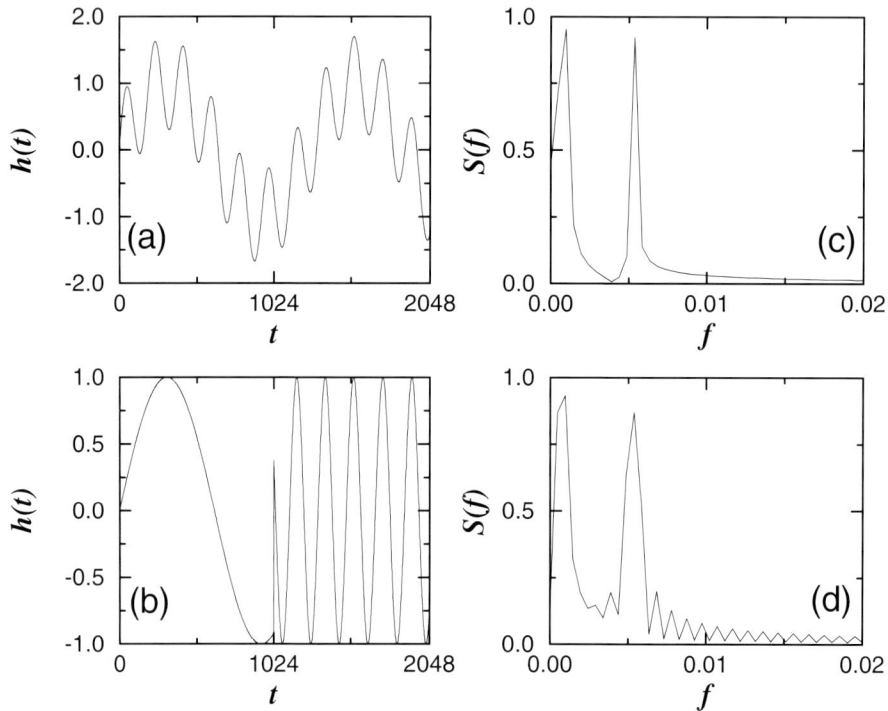

Fig. 7.3. (a) Stationary signal resulting from the sum of two sine waves with frequencies $1/(200\pi)$ and $1/(60\pi)$. (b) Nonstationary signal with a first regime consisting of a sine wave with frequency $1/(200\pi)$, and a second regime consisting of a sine wave with frequency $1/(60\pi)$, (c) and (d). Note how the power spectra of the two signals are almost identical. (They would be identical except for the small high frequency fluctuations due to the spurious singularity at $x = 1\,024$.) Thus, a power spectrum analysis cannot distinguish these signals, despite their obvious differences. Courtesy of L.A.N. Amaral

7.2.1 Power Spectra of Heartbeat Intervals and Heartbeat Interval Increments

These constraints of a power spectrum analysis of nonstationary signals become even more apparent when we consider heart data. The normal electrical activity of the heart is usually described as 'regular sinus rhythm' [7.6–7.8, 7.13]. However, time series of beat-to-beat (RR) heart rate intervals obtained from digitized electrocardiograms fluctuate in an irregular manner in healthy subjects (Fig. 7.1), even at rest [7.14, 7.15]. The mechanism underlying this complex heart rate variability is related to competing neuroautonomic inputs [7.16, 7.17]. Parasympathetic stimulation decreases the firing rate of pacemaker cells in the heart's sinus node. Sympathetic stimulation has the opposite effect. The nonlinear interaction (coupling) of the two branches of the nervous system is the postulated mechanism for the type of 'erratic' heart rate variability recorded in healthy subjects [7.18–7.20]. The complex behavior of the heartbeat manifests itself through the nonstationarity and nonlinearity of interbeat interval sequences [7.17].

In recent years the study of the statistical properties of interbeat interval sequences has attracted the attention of researchers from different fields [7.21–7.25]. Analysis of heartbeat fluctuations focused initially on short time oscillations associated with breathing, blood pressure and neuroautonomic control [7.26, 7.27]. Fourier and power spectrum analysis proved instrumental for recognizing the existence and role of characteristic frequencies (time scales) in cardiac dynamics. Studies of longer heartbeat records, however, revealed $1/f$-like scale-free behavior [7.28, 7.29]. Traditional approaches such as the power spectrum and correlation analysis [7.27, 7.30] are not suited for such nonstationary sequences. Because of that, researchers were faced with the task to consider only portions of the data and to test these portions for stationarity before performing power spectrum analysis. To illustrate the limitations of the power spectrum analysis for nonstationary time series, we consider 6 hr records ($n \approx 10^4$ beats) of interbeat intervals for a healthy subject during sleep and wake activity. We show that there is *no true* $1/f$ power spectrum for the interbeat intervals in the real heart. Instead, we find that the power spectrum of the interbeat intervals has different regimes with different scaling behavior and that the rounded crossover between the different regimes is the reason why it seems, to first approximation, to scale as $1/f$ (Fig. 7.4).

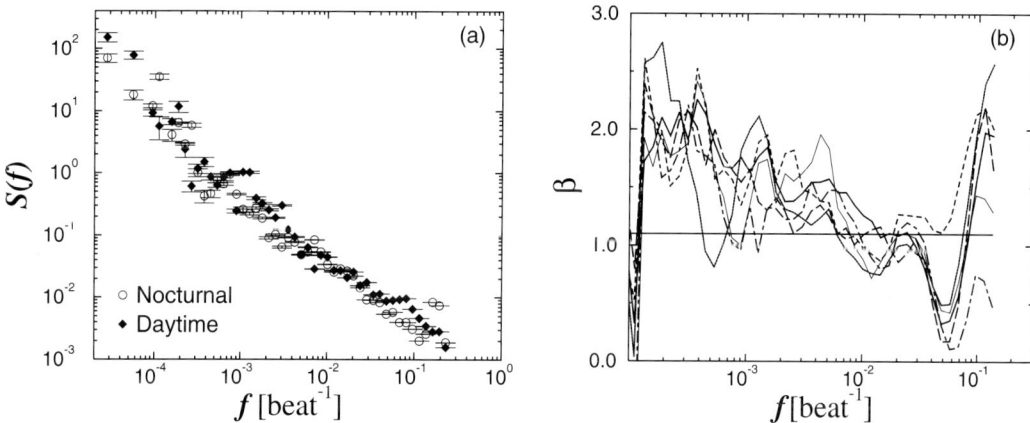

Fig. 7.4. (a) Power spectrum from 6 hr records of interbeat intervals for a healthy subject during day and night. (b) We plot the local exponent β calculated from the power spectrum for six healthy subjects. The local value of β shows a persistent drift, so *no true scaling exists*. This is not surprising, having in mind the nonstationarity of the signals. The horizontal line shows the value of the exponent obtained from a least square fit to the data. From [7.58]

Recent analyses of very long time series (up to 24 hr: $n \approx 10^5$ beats) show that under healthy conditions, interbeat interval increments $I(n)$ exhibit power-law anticorrelations [7.32]. Since $I(n)$ is stationary, we can apply standard spectral analysis techniques (Fig. 7.5) and we show that *true* scaling does exist. The fact that the log-log plot of the power spectrum $S_I(f)$ vs. f is linear implies:

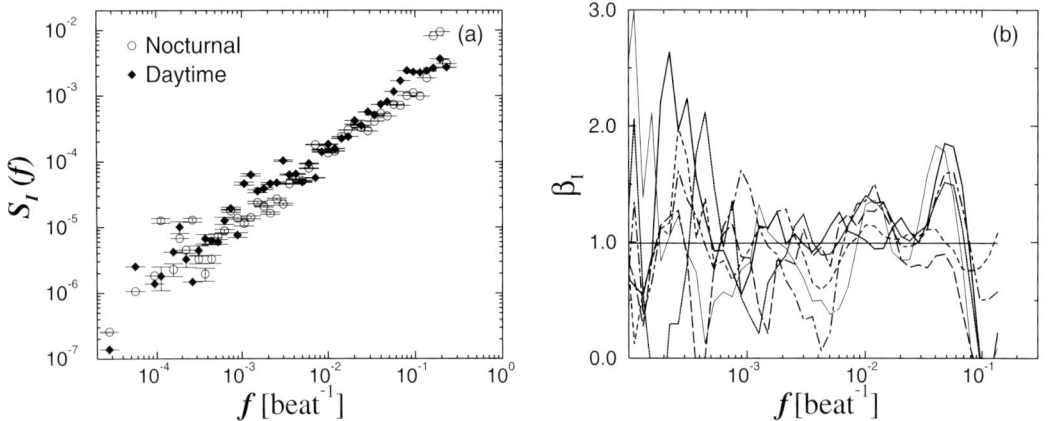

Fig. 7.5. (a) Power spectrum of the interbeat interval increments from 6 hr record for the same healthy subject as in Fig. 7.4. (b) The local exponent β_I for the power spectrum of the increments for the same six healthy subjects as in Fig. 7.4. Note that the exponent β_I fluctuates around an average value close to one, so *true scaling does exist*. The horizontal line shows the value of β_I obtained from a least square fit. Note however, that the difference between wake and sleep dynamics cannot be observed from the power spectra. Error bars are calculated as the standard deviation of the power spectrum values for frequencies within the binning interval. From [7.58]

$$S_I(f) \sim f^{-\beta}. \tag{7.1}$$

The exponent β is related to the mean fluctuation function exponent α by $\beta = 2\alpha - 1$ [7.33] and can serve as an indicator of the presence and type of correlations: (i) if $\beta = 0$, there is no correlation in the time series $I(n)$ ('white noise'); (ii) if $0 < \beta < 1$, then $I(n)$ is correlated such that positive values of I are likely to be close (in time) to each other, and the same is true for negative I values; (iii) if $-1 < \beta < 0$, then $I(n)$ is also correlated. However, the values of I are organized such that positive and negative values are more likely to alternate in time ('anticorrelation') [7.33].

For interbeat interval records from healthy subjects we obtain $\beta \simeq -1$, suggesting *nontrivial* power-law long-range correlations in the heartbeat. Furthermore, the anticorrelation properties of I indicated by the negative β are consistent with a nonlinear feedback system that 'kicks' the heart rate away from extremes. This tendency, however, does not only operate locally on a beat-to-beat basis but on a wide range of time scales up to thousands of beats (Fig. 7.5). The emergence of such scale-invariant properties in the seemingly 'noisy' heartbeat fluctuations is believed to be a result of highly complex, nonlinear mechanisms of physiological control [7.34].

Extracting increments from a time series is only a first step in effectively treating problems related to nonstationarities. Note that the power spectrum of the increments in the heartbeat intervals (Fig. 7.5) *does not* distinguish between wake and sleep dynamics (Fig. 7.9, Sect. 7.3.4). One needs to do better, e.g. by taking into account the presence of polynomial trends in the time series. We discuss such an approach in the following section.

7.3 Monofractal Analysis

An important question is whether the 'heterogeneous' structure of physiological time series arises trivially from external and intrinsic perturbations which push the system away from a homeostatic set point. An important alternative hypothesis is that the fluctuations are, at least in part, due to the underlying dynamics of the system. The key problem is how to decompose subtle fluctuations (due to intrinsic physiological control) from other nonstationary trends associated with external stimuli.

7.3.1 Detrended Fluctuation Analysis (DFA)

Recently the *detrended fluctuation analysis* (DFA) method [7.35] was introduced to detect long-range correlations in physiological fluctuations when these are embedded in a seemingly nonstationary time series (cf. Chap. 8). The advantage of the DFA method over conventional methods, such as power spectrum analysis, is that it avoids the spurious detection of apparent long-range correlations that are an artifact of nonstationarity related to linear and higher order polynomial trends in the data. The DFA method has been tested on control time series that consist of long-range correlations with superposition of a nonstationary external trend. It has also been successfully applied to detect long-range correlations in highly heterogeneous DNA sequences [7.35–7.38]. Of note is a recent independent review of fractal fluctuation analysis methods which determined that DFA was one of the most robust methods [7.39].

Briefly, the DFA method involves the following steps:

1. The interbeat interval time series (of total length N) is integrated, $y(k) \equiv \sum_{i=1}^{k}[RR(i) - RR_{\text{ave}}]$, where $RR(i)$ is the ith interbeat interval and RR_{ave} is the average interbeat interval.
2. The integrated time series is divided into boxes of equal length, n.
3. In each box of length n, a least squares line is fitted to the data (representing the *trend* in that box). The y coordinate of the straight line segments is denoted by $y_n(k)$ (see Fig. 7.6).
4. The integrated time series, $y(k)$, is detrended by subtracting the local trend, $y_n(k)$, in each box.
5. For a given box size n, the characteristic size of fluctuation for this integrated and detrended time series is calculated:
$$F(n) \equiv \sqrt{\frac{1}{N}\sum_{k=1}^{N}[y(k) - y_n(k)]^2}.$$
6. The above computation is repeated over all time scales (box sizes n) to provide a relationship between $F(n)$ and the box size n (i.e. the number of beats in a box or the size of the window of observation).

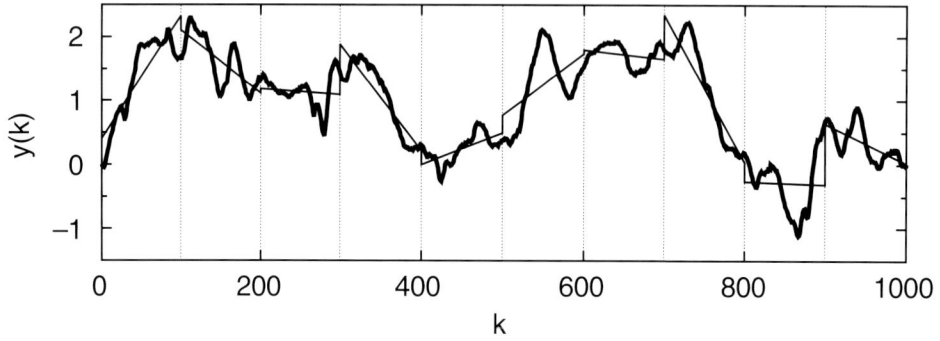

Fig. 7.6. The integrated time series: $y(k) = \sum_{i=1}^{k}[RR(i) - RR_{\text{ave}}]$, where $RR(i)$ is the interbeat interval shown in Fig. 7.2a. The vertical dotted lines indicate boxes of size $n = 100$, the solid straight line segments are the estimated 'trend' in each box by least-squares fit

The power-law relation between the average root-mean-square fluctuation function $F(n)$ and the number of beats n in a box indicates the presence of scaling: the fluctuations can be characterized by a scaling exponent α, a self-similarity parameter, defined as $F(n) \sim n^{\alpha}$.

7.3.2 Long-Range Anticorrelations in Physiological Heartbeat Dynamics

Assessing correlations under pathological conditions is likely to be particularly informative for patients with congestive heart failure due to severe left ventricular dysfunction since these individuals have abnormalities in both the sympathetic and parasympathetic control mechanisms [7.40] that regulate beat-to-beat variability. Previous studies have demonstrated marked changes in short-range heart rate dynamics in heart failure compared to healthy function, including the emergence of intermittent relatively low frequency (~ 1 cycle/minute) heart rate oscillations associated with the well-recognized syndrome of periodic (Cheyne-Stokes) respiration, an abnormal breathing pattern often associated with low cardiac output [7.40]. Of note is the fact that patients with congestive heart failure are at very high risk of sudden cardiac death.

Figure 7.7 compares the DFA analysis of representative 24 hr interbeat interval time series of a healthy subject and a patient with congestive heart failure. Notice that for large time scales (asymptotic behavior), the healthy subject shows almost perfect power-law scaling over more than two decades ($20 \leq n \leq 10\,000$) with $\alpha = 1$ (i.e. $1/f$ noise) indicating long-range power-law anticorrelations in the heartbeat fluctuations. For the pathological dataset $\alpha \approx 1.3$ (closer to Brownian noise). This difference in the scaling exponents (fractal properties) could be utilized as a marker of pathological deviation from healthy cardiac dynamics. We also note that the difference ('gap') in the values of $\log F(n)$ for healthy and heart failure at scale $n \approx 20$ beats could also be used for diagnosis (a scale-dependent measure) [7.41, 7.42].

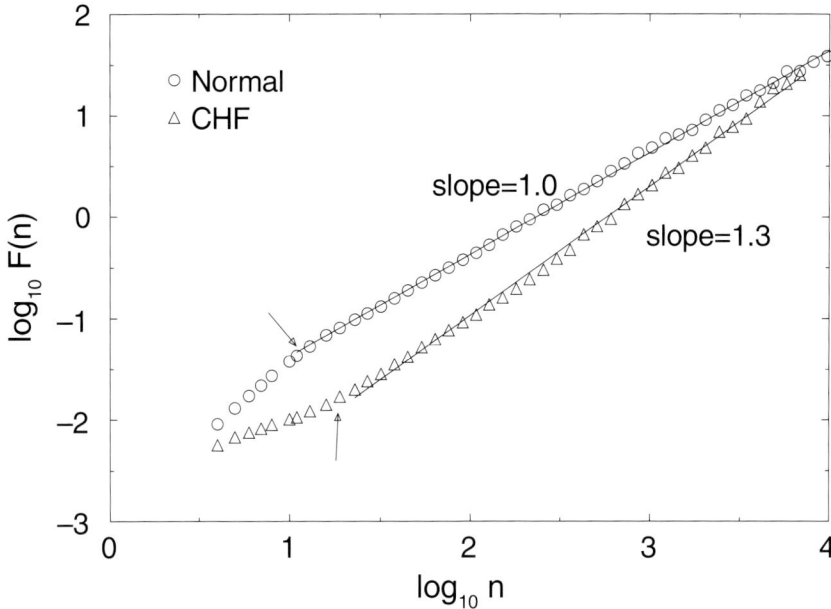

Fig. 7.7. Plot of $\log F(n)$ vs. $\log n$ for two long interbeat interval time series ($\sim 24\,\mathrm{hr}$). The circles are for a representative healthy subject while the triangles are from a subject with congestive heart failure. Arrows indicate 'crossover' points in scaling. Note altered scaling with heart failure, suggesting apparent perturbations of both short and long-range correlation mechanisms. Note also the 'gap' in the values of $\log F(n)$ for healthy and heart failure at scale $n \approx 20$ beats. From [7.43]

To study the alteration of long-range correlations with pathology, cardiac interbeat data from three different groups of subjects were analyzed: (i) 29 adults (17 male and 12 female) without clinical evidence of heart disease (age range: 20–64 yr, mean 41), (ii) 10 subjects with fatal or near-fatal sudden cardiac death syndrome (age range: 35–82 yr) and (iii) 15 adults with severe heart failure (age range: 22–71 yr; mean 56). Data from each subject contains approximately 24 hr of ECG recording encompassing $\sim 10^5$ heartbeats [7.44].

For the normal control group, we observed $\alpha = 1.00 \pm 0.10$ (mean value $\pm$ S.D.) [7.43]. These results indicate that healthy heart rate fluctuations exhibit long-range power-law (fractal) correlation behavior, similar to that observed in many dynamical systems far from equilibrium [7.45, 7.46]. Furthermore, both pathological groups show significant deviation of the long-range correlation exponent α from the normal value, $\alpha = 1$. For the group of heart failure subjects, we find that $\alpha = 1.24 \pm 0.22$, while for the group of sudden cardiac death syndrome subjects, we find that $\alpha = 1.22 \pm 0.25$. Of particular note, we obtained similar results when we divided the time series into three consecutive subsets (of ~ 8 hr each) and repeated the above analysis [7.43]. Therefore our findings are not simply attributable to different levels of daily activities. Recently these findings have been independently verified [7.47].

7.3.3 Clinical Utility

Recently, our fractal scaling analysis has been applied to three retrospective clinical studies [7.48–7.50]. Results from all three studies indicate that additional information can be extracted from heart rate time series with the DFA method. Furthermore, this information can be used for prognostic purposes. We have also shown [7.42] that the scale-free parameters we introduce have a greater potential for more accurate diagnosis than recently suggested scale-specific measures [7.41].

Forecasting Clinical Outcomes: Framingham Heart Study: A major question regarding this new fractal long-range measurement is the following: *does DFA have clinically predictive value, independent of conventional time- and frequency-domain indices?* To answer this question, we have studied the predictive power of the DFA exponent in comparison with 10 other conventional indices. Ho et al. analyzed two hour ambulatory ECG recordings of 69 participants (mean age 71.7 ± 8 yr) in the Framingham Heart Study, a prospective, population-based study [7.48]. Importantly, they found that this fractal measurement carries prognostic information about mortality not extractable from traditional methods of heart rate variability analysis.

Heart Rate Dynamics in Patients at High Risk of Sudden Death After Myocardial Infarction: Mäkikallio and co-workers compared short-term (< 11 beats, α_1) and long-term (> 11 beats, α_2) correlation properties of RR interval data in three groups: (i) 45 postinfarction patients with a recent history of ventricular tachyarrhythmia and inducible ventricular tachyarrhythmia by programmed electrical stimulation, (ii) 45 postinfarction patients without clinical ventricular tachyarrhythmia events or inducible ventricular tachyarrhythmia, and (iii) 45 healthy control subjects. The short-term scaling exponent (α_1) was significantly lower in the ventricular tachyarrhythmia group than in the postinfarction control group ($p < 0.001$) or the healthy controls ($p < 0.001$) [7.49]. In stepwise multiple regression analysis, the short-term exponent was the strongest independent predictor of vulnerability to ventricular tachyarrhythmia. The data suggest that short-term correlation properties of RR interval dynamics are altered in postinfarction patients with vulnerability to ventricular tachyarrhythmia, and that abnormal beat-to-beat heart rate dynamics may be related to vulnerability to ventricular tachyarrhythmia in postinfarction patients.

Heart Rate Dynamics in Patients with Stable Angina Pectoris:
Recently, Mäkikallio and co-workers [7.50] also compared conventional measures of heart rate variability with short-term (≤ 11 beats, α_1) and long-term(> 11 beats, α_2) fractal correlation properties and with approximate entropy of RR interval data in 38 patients with stable angina pectoris without previous myocardial infarction or cardiac medication at the time of the study and 38 age-matched healthy controls. The short- and long-term fractal scaling exponents (α_1, α_2) were significantly higher in the coronary patients than in the healthy controls (1.34 ± 0.15 vs. 1.11 ± 0.12 ($p < 0.001$) and 1.10 ± 0.08 vs. 1.04 ± 0.06 ($p < 0.01$), respectively), and they also had lower approximate entropy ($p < 0.05$), standard deviation of all RR intervals ($p < 0.01$), and high-frequency spectral component of heart rate variability ($p < 0.05$). The short-term fractal scaling exponent performed better than other heart rate variability parameters in differentiating patients with coronary artery disease from healthy subjects, but it was not related to the clinical or angiographic severity of coronary artery disease or any single nonspectral or spectral measure of heart rate variability in this retrospective study.

7.3.4 Sleep-Wake Differences in Scaling Behavior of the Heartbeat

It is known that circadian rhythms are associated with periodic changes in key physiological processes [7.13,7.17,7.51]. Typically the differences in the cardiac dynamics during sleep and wake phases are reflected in the average (higher at sleep) and standard deviation (lower at sleep) of the interbeat intervals [7.51]. Such differences can be systematically observed from plots of the interbeat intervals recorded from subjects during sleep and wake hours (Fig. 7.8). In recent studies we have reported on sleep-wake differences in the distributions of the amplitudes of the fluctuations in the interbeat intervals, a surprising finding indicating higher probability for larger amplitudes during sleep [7.31,7.52,7.53]. Next, we ask the question if there are characteristic differences in the scaling behavior between sleep and wake cardiac dynamics. We hypothesize that sleep and wake changes in cardiac control may occur on all time scales and thus could lead to systematic changes in the scaling properties of the heartbeat dynamics. Elucidating the nature of these sleep-wake rhythms could lead to a better understanding of the neuroautonomic mechanisms of cardiac regulation.

To answer this question we apply the detrended fluctuation analysis (DFA) method [7.35] (for further applications see Chaps. 5 and 8) to quantify long-range correlations embedded in nonstationary heartbeat time series. We analyze 30 datasets, each with 24 hr of interbeat intervals, from 18 healthy subjects and 12 patients with congestive heart failure [7.44]. We analyze the nocturnal and diurnal fractions of the dataset of each subject, which correspond to the 6 hr ($n \approx 22\,000$ beats) from midnight to 6 am and noon to 6 pm.

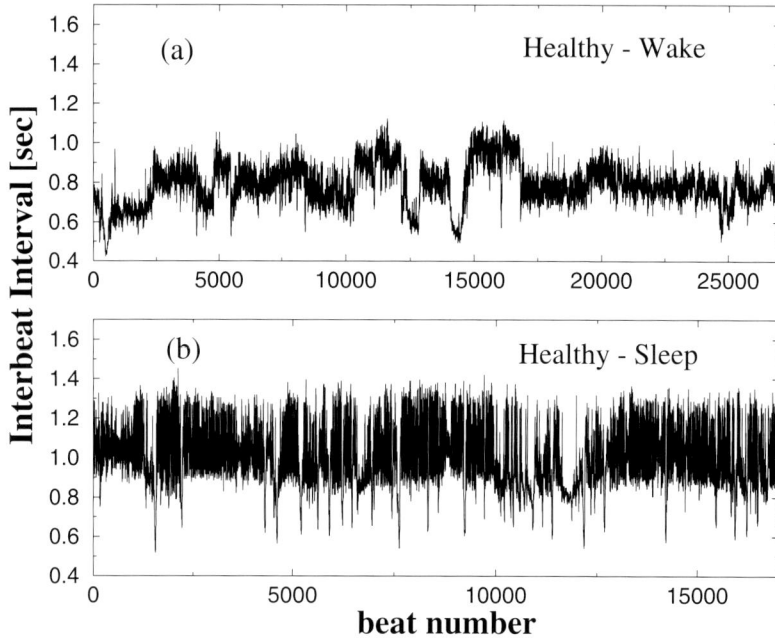

Fig. 7.8. Consecutive heartbeat intervals are plotted vs. beat number for 6 hr recorded from the same healthy subject during: (**a**) wake period: 12 pm to 6 pm and (**b**) sleep period: 12 am to 6 am. (Note that there are fewer interbeat intervals during sleep due to the larger average of the interbeat intervals, i.e. slower heart rate). From [7.54]

We find that at scales above ≈ 1 min ($n > 60$) the data during wake hours display long-range correlations over two decades with average exponents $\alpha_W \approx 1.05$ for the healthy group and $\alpha_W \approx 1.2$ for the heart failure patients. For the sleep data we find a systematic crossover at scale $n \approx 60$ beats followed by a scaling regime extending over two decades characterized by a smaller exponent: $\alpha_S \approx 0.85$ for the healthy and $\alpha_S \approx 0.95$ for the heart failure group (Fig. 7.9a, c) [7.54]. Although the values of the sleep and wake exponents vary from subject to subject, we find that for all individuals studied, the heartbeat dynamics during sleep are characterized by a smaller exponent (Table 7.1 and Fig. 7.10).

Table 7.1. Comparison of the scaling exponents from the three groups in our database. Here N is the number of datasets in each group, α is the corresponding group average value, and σ is the standard deviation of the exponent values for each group. The differences between the average sleep and wake exponents for all three groups are statistically significant ($p < 10^{-5}$ by Student's t-test)

Group	N	α	σ
Healthy wake	18	1.05	0.07
Healthy sleep	18	0.85	0.10
Cosmonaut wake	17	1.04	0.12
Cosmonaut sleep	17	0.82	0.07
Heart failure wake	12	1.20	0.09
Heart failure sleep	12	0.95	0.15

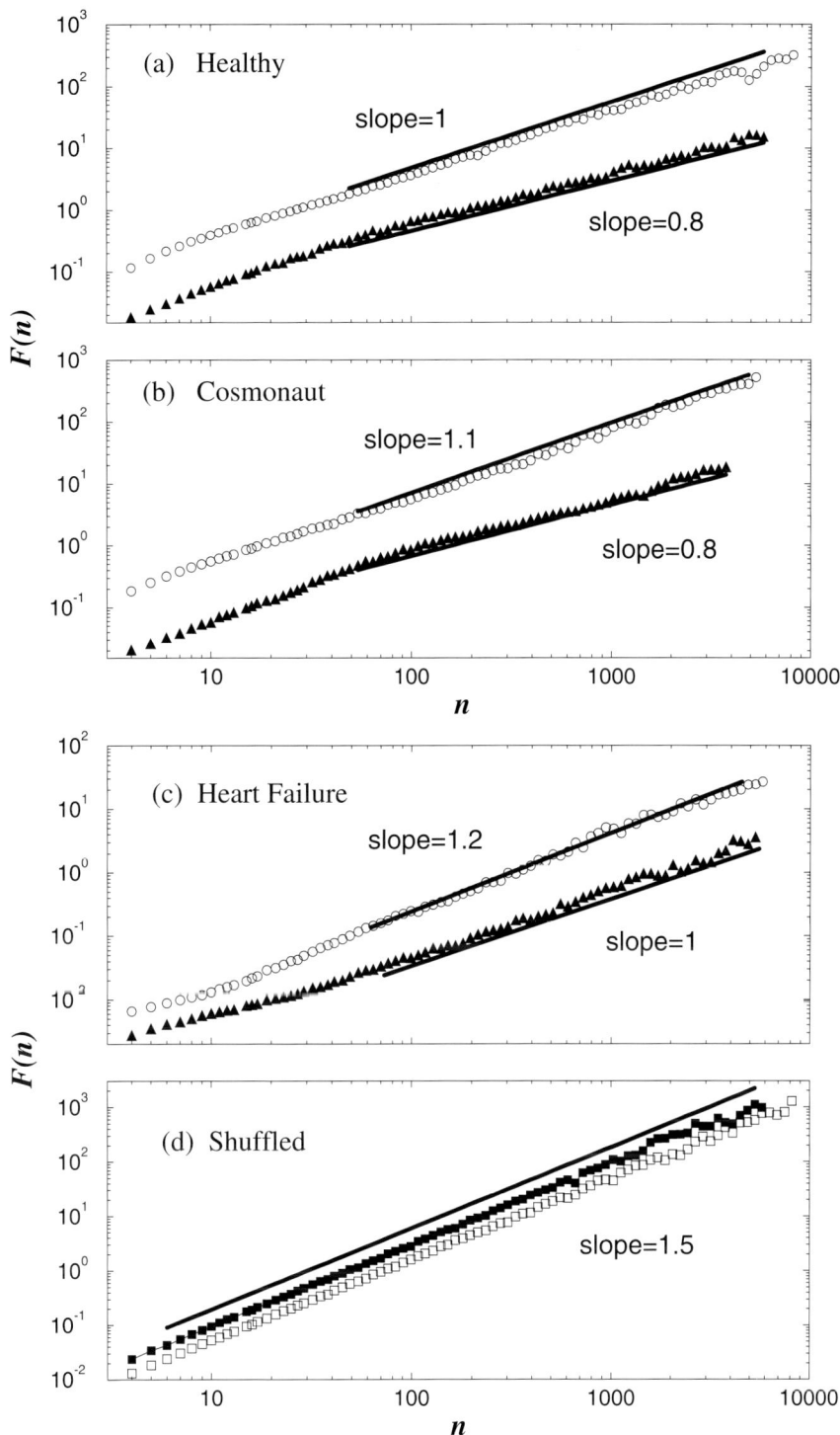

Fig. 7.9. Plots of $\log F(n)$ vs. $\log n$ for 6 hr wake (open circles) and sleep records (filled triangles) of (**a**) one typical healthy subject; (**b**) one cosmonaut (during orbital flight); and (**c**) one patient with congestive heart failure. Note the systematic lower exponent for the sleep phase (filled triangles), indicating stronger anticorrelations. (**d**) As a control, we reshuffle and integrate the interbeat increments from the wake and sleep data of the healthy subject presented in (**a**). We find a Brownian noise scaling over all time scales for both wake and sleep phases with an exponent $\alpha = 1.5$, as one expects for random walk-like fluctuations. From [7.54]

As a control, we also perform an identical analysis on two surrogate datasets obtained by reshuffling and integrating the increments in the interbeat intervals of the sleep and wake records from the same healthy subject presented in Fig. 7.9a. As expected, both surrogate sets display uncorrelated random walk fluctuations with a scaling exponent of 1.5 (Brownian noise) (Fig. 7.9d). A scaling exponent larger than 1.5 would indicate persistent correlated behavior, while exponents with values smaller than 1.5 characterize anticorrelations (a perfectly anticorrelated signal would have an exponent close to zero). Our results therefore suggest that the interbeat fluctuations during sleep and wake phases are long-range anticorrelated but with a significantly greater degree of anticorrelation (smaller exponent) during sleep.

An important question is whether the observed scaling differences between sleep and wake cardiac dynamics arise trivially from changes in the environmental conditions (different daily activities are reflected in the strong nonstationarity of the heartbeat time series). Environmental 'noise', however, can be treated as a 'trend' and distinguished from the more subtle fluctuations that may reveal intrinsic correlation properties of the dynamics. Alternatively, the interbeat fluctuations may arise from nonlinear dynamical control of the neuroautonomic system rather than being an epiphenomenon of environmental stimuli, in which case only the fluctuations arising from the intrinsic dynamics of the neuroautonomic system should show long-range scaling behavior.

Our analysis suggests that the observed sleep-wake scaling differences are due to intrinsic changes in the cardiac control mechanisms for the following reasons: (i) the DFA method removes the 'noise' due to activity by detrending the nonstationarities in the interbeat interval signal and analyzing the fluctuations along the trends. (ii) Responses to external stimuli should give rise to a different type of fluctuations having characteristic time scales, i.e. frequencies related to the stimuli. However, fluctuations in both diurnal and nocturnal cardiac dynamics exhibit scale-free behavior. (iii) The weaker anticorrelated behavior observed for all wake phase records cannot be simply explained as a superposition of stronger anticorrelated sleep dynamics and random noise of day activity. Such noise would dominate at large scales and should lead to a crossover with an exponent of 1.5. However, such crossover behavior is not observed in any of the wake phase datasets (Fig. 7.9). Rather, the wake dynamics are typically characterized by a stable scaling regime up to $n = 5 \times 10^3$ beats.

To test the robustness of our results, we analyze 17 datasets from six cosmonauts during long-term orbital flight on the Mir space station [7.55]. Each dataset contains continuous periods of 6 hr data under both sleep and wake conditions. We find that for all cosmonauts the heartbeat fluctuations exhibit an anticorrelated behavior with average scaling exponents consistent with those found for the healthy terrestrial group: $\alpha_W \approx 1.04$ for the wake phase and $\alpha_S \approx 0.82$ for the sleep phase (Table 7.1). This sleep-wake scaling difference is observed not only for the group averaged exponents but for each individual

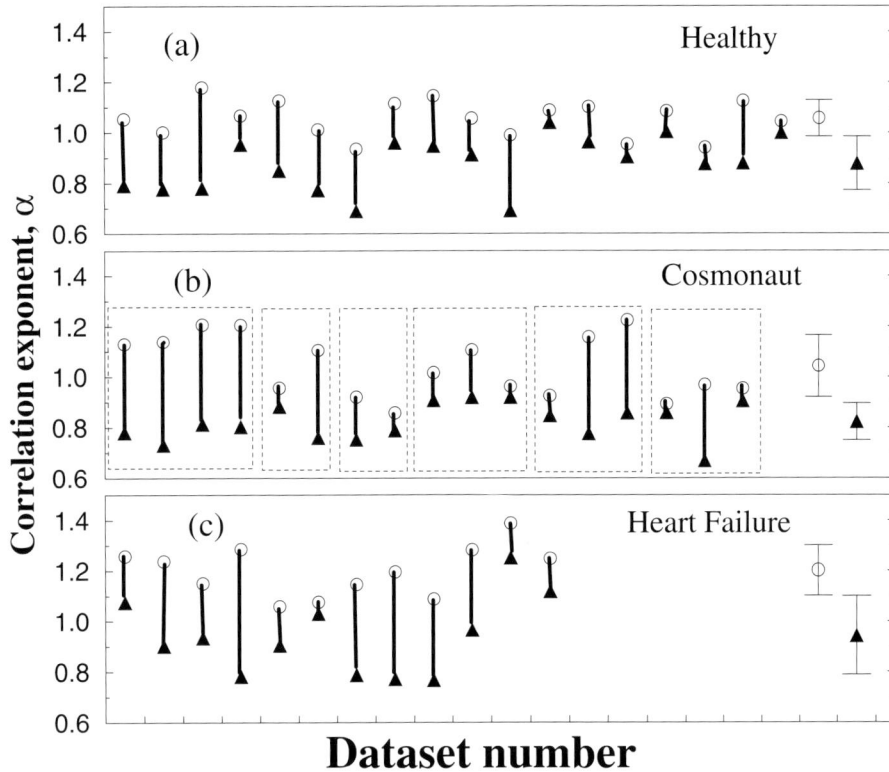

Fig. 7.10. Values for the sleep (filled triangles) and wake activity (open circles) exponents for all individual records of (**a**) the healthy, (**b**) the cosmonaut, and (**c**) the heart failure groups. For the healthy and heart failure groups each dataset corresponds to a different individual. Data from the six cosmonauts are grouped in six blocks, where each block contains data from the same individual, recorded on different days during orbital flight and ordered from early to late flight, ranging from the 3rd to the 158th day in orbit. For all individuals in all groups, the day exponent exhibits systematically a higher value than the sleep exponent. The sleep and wake group averages and standard deviations are presented on the right of each panel. From [7.54]

cosmonaut dataset (Fig. 7.9b and Fig. 7.10). Moreover, the scaling differences are persistent in time, since records of the same cosmonaut taken on different days (ranging from the 3rd to the 158th day in orbit) exhibit a higher degree of anticorrelation in sleep.

We find that even under the extreme conditions of zero gravity and high stress activity, the sleep and wake scaling exponents for the the cosmonauts are statistically consistent ($p = 0.7$ by Student's t-test) with those of the terrestrial healthy group. Thus, the larger values for the wake phase scaling exponents cannot be a trivial artifact of activity. Furthermore, the larger value of the average wake exponent for the heart failure group compared to the other two groups (Table 7.1) cannot be attributed to external stimuli either, since patients with severe cardiac disease are strongly restricted in their physical activity. Instead, our results suggest that the observed scaling characteristics in the heartbeat fluctuations during sleep and wake phases are related to intrinsic mechanisms of neuroautonomic control.

The mechanism underlying heartbeat fluctuations may be related to countervailing neuroautonomic inputs. Parasympathetic stimulation decreases the heart rate, while sympathetic stimulation has the opposite effect. The nonlinear interaction between the two branches of the nervous system is the postulated mechanism for the type of complex heart rate variability recorded in healthy subjects [7.13, 7.17]. The fact that during sleep the scaling exponents differ more from the value $\alpha = 1.5$ (indicating 'stronger' anticorrelations) may be interpreted as a result of stronger neuroautonomic control. Conversely, values of the scaling exponents closer to 1.5 (indicating 'weaker' anticorrelations) for both sleep and wake activity for the heart failure group are consistent with previously reported pathological changes in cardiac dynamics [7.32]. We note, however, that the average sleep-wake scaling difference remains the same (≈ 0.2) for all three groups. These sleep-wake changes in the scaling characteristics may indicate different regimes of intrinsic neuroautonomic regulation of the cardiac dynamics, which may 'switch' on and off associated with circadian rhythms. Surprisingly, we find that for the regime of large time scales ($n > 60$) the average sleep-wake scaling difference is *comparable* to the scaling difference between health and disease; cf. Table 7.1. At small time scales ($n < 60$) we do not observe systematic sleep-wake difference; however, the scaling difference between the healthy and pathological dynamics grows even stronger [7.43].

We also note that the scaling exponents for the heart failure group during sleep are close to the exponents observed for the healthy group (Table 7.1). Since heart failure occurs when the cardiac output is not adequate to meet the metabolic demands of the body, one would anticipate that the manifestations of heart failure would be most severe during physical stress when metabolic demands are greatest, and least severe when metabolic demands are minimal, i.e. during rest or sleep. The scaling results we obtain are consistent with these physiological considerations: the heart failure subjects should be closer to normal during minimal activity. Of related interest, recent studies indicate that sudden death in individuals with underlying heart disease is most likely to occur in the hours just after awakening and before sleep [7.56, 7.57]. Our findings raise the intriguing possibility that the transition between the sleep and wake phases is a period of potentially increased neuroautonomic instability because it requires a transition from strongly to weakly anticorrelated regulation of the heart.

The findings of *stronger* anticorrelations [7.54], as well as higher probability for larger heartbeat fluctuations during sleep [7.31, 7.52, 7.53] are of interest from a physiological viewpoint, since they support a re-assessment of the sleep as a surprisingly *active dynamical* state and may motivate new modeling approaches [7.58]. Perhaps the restorative function of sleep may relate to an increased reflexive responsiveness of neuroautonomic control not just at one characteristic frequency, but over a broad range of time scales.

Very recent work has focused on the heartbeat dynamics in the different sleep stages - deep, light, and rapid eye movement (REM) sleep - that reflect different brain activity and neuroautonomic heart rate regulation [7.59]. By means of the DFA method the authors find different types of correlations in the heartbeat fluctuations associated with the different sleep stages: from long-range anticorrelations in the 'dream'-REM stage to random behavior over certain range of time scales during deep sleep. These new findings further underline the complexity of physiological systems and the need of appropriate tools for their analysis.

7.3.5 Self-Similar Cascades in the Heartbeat Fluctuations

Many simple systems in nature have correlation functions that decay with time in an exponential way. For systems comprising many interacting subsystems, physicists discovered that such exponential decays do not occur. Rather, correlation functions were found to decay with a power-law form. The implication of this discovery is that in complex systems, there is no single characteristic time [7.45, 7.46, 7.60]. If correlations decay with a power-law form, we say the system is 'scale free' since there is no characteristic scale associated with a power-law. Since at large time scales a power-law is always larger than an exponential function, correlations described by power-laws are termed 'long-range' correlations, i.e. they are of longer range than exponentially-decaying correlations.

The findings of long-range power-law correlations [7.43, 7.54] and the recently reported scaling in the distributions of heartbeat fluctuations [7.31, 7.52] (i.e. 'data collapse' of the distributions for different time scales) suggest absence of a characteristic scale and indicate that the underlying dynamical mechanisms regulating the healthy heartbeat have statistical properties which are *similar* on different time scales. Such statistical self-similarity is an important characteristic of *fractal* objects [7.24, 7.61]. However, how do we see this supposed fractal structure in the seemingly 'erratic' and noisy heartbeat fluctuations? The wavelet decomposition of beat-to-beat heart rate signals can be used to provide a visual representation of this fractal structure (Fig. 7.11). The brighter colors indicate larger values of the wavelet amplitudes (corresponding to large heartbeat fluctuations) and white tracks represent the wavelet transform maxima lines. The structure of these maxima lines shows the evolution of the heartbeat fluctuations with scale and time. The wavelet analysis performed with the second derivative of the Gaussian (the Mexican hat) as an analyzing wavelet uncovers a hierarchical scale invariance (Fig. 7.11 (top panel)), which is characterized by the stability of the scaling form observed for the distributions and the power-law correlations [7.43, 7.52, 7.54]. The plots reveal a self-affine cascade formed by the maxima lines: a magnification of the central portion of the top panel shows identical branching patterns (Fig. 7.11 (lower panel)).

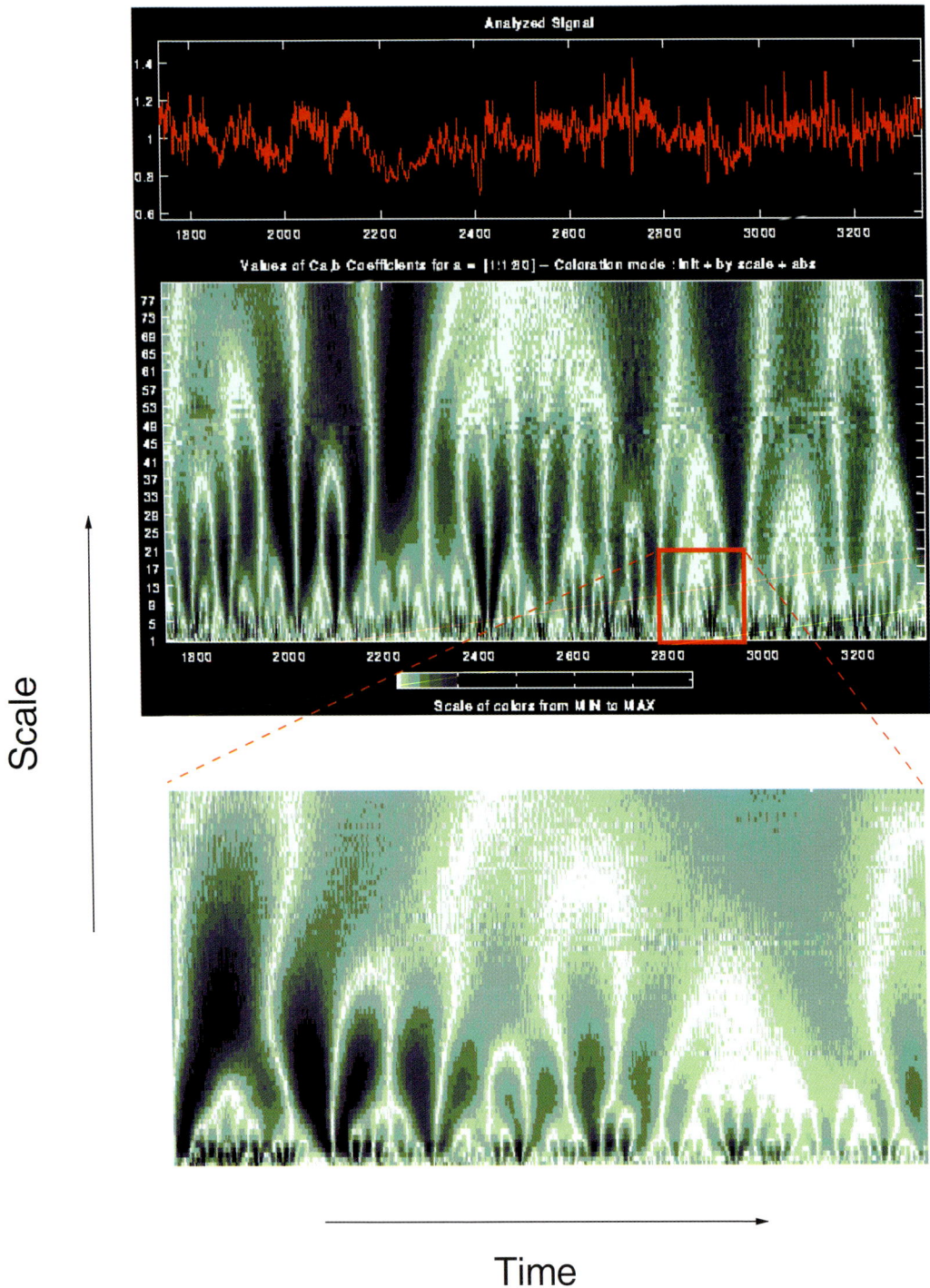

Fig. 7.11. Color coded wavelet analysis of a heartbeat interval signal. The x-axis represents time ($\approx 1\,700$ beats) and the y-axis indicates the scale of the wavelet used ($a = 1, 2, \ldots, 80$; i.e. $\approx$ from 5 s to 5 min) with large scales at the top. This wavelet decomposition reveals a self-similar fractal structure in the healthy cardiac dynamics: a magnification of the central portion of the top panel with 200 beats on the x-axis and wavelet scale $a = 1, 2, \ldots, 20$ on the y-axis shows identical branching patterns (lower panel). Adapted from [7.31]

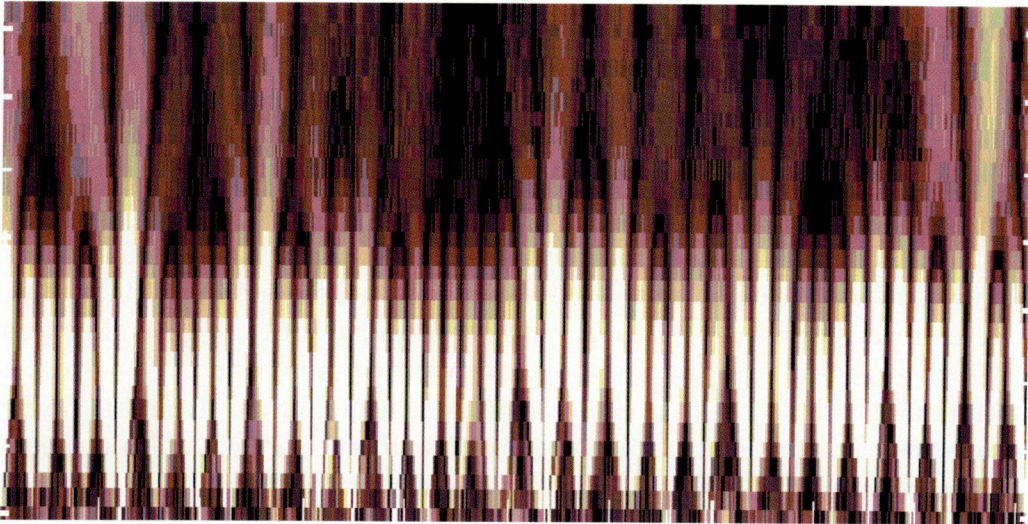

Fig. 7.12. Wavelet decomposition (cf. Fig. 7.11) for a heartbeat interval record from a subject with sleep apnea. The self-similar branching pattern observed for the healthy heartbeat fluctuations is lost. A repetitive periodic structure is observed over a range of time scales indicating pathological mode locking. From [7.31]

Such fractal cascade results from the interaction of many nonlinearly coupled physiological components, operating on different scales (polynomial trends due to daily activity are filtered out).

Thus the wavelet transform, with its ability to remove local trends and to extract interbeat variations on different time scales, enables us to identify fractal patterns (arches) in the heartbeat fluctuations even when the signals change as a result of background interference. Data from a pathologic state (e.g. sleep apnea) lack these patterns (Fig. 7.12). Fractal characteristics of the cardiac dynamics and other biological signals can be successfully studied with the generalized multifractal formalism based on the wavelet transform modulus maxima method which we discuss in the next section (see also Chap. 2 in this book).

Analysis of the variance of the distributions for healthy cardiac dynamics at different time scales shows a power-law behavior with an exponent close to zero [7.52]. This relates to previous studies reporting long-range anticorrelations in the heartbeat variations [7.32]. The findings that correlation functions and distributions describing physiological systems are not characterized by a single time scale become more plausible if we consider the survival advantage conferred upon organisms that evolved with an infinite hierarchy of time scales compared to organisms that evolved with a single characteristic time scale. Organisms with a physiological control system generated by a single time scale are analogous, formally, to the famous Tacoma Narrows bridge, which survived until by chance a wind storm occurred that happened to correspond to the characteristic frequency (inverse of the characteristic time scale). Organisms that have survived millions of years have plausibly evolved some feature to render them immune from the analog of the Tacoma bridge disaster, and this feature would seem to be the absence of any characteristic time scales (compare Figs. 7.1a and 7.11, which show scale-invariant complex fluctuations, with Figs. 7.1b, d and 7.12, which show pathological mode locking).

What are the possible adaptive advantages of the apparently far from equilibrium scale-free behavior that appears to characterize the free-running dynamics of certain neural control systems? First, we note that complex erratic fluctuations shown in Fig. 7.1a are *not* inconsistent with the general concept that physiological systems must operate with certain bounds. However, an intriguing possibility is that these complex nonequilibrium dynamics, rather than classical homeostatic *constancy* [7.6–7.8], may be a mechanism for maintaining physiological stability. Such complex multiscale variability keeps the system from becoming 'locked' to a dominant frequency (mode locking), a common manifestation of pathological dynamics (Fig. 7.1b). At the same time, long-range fractal correlations underlying these complex fluctuations may provide an important organizational mechanism for systems that lack a characteristic spatial or temporal scale. Finally, the intrinsic 'noisiness' of far-from-equilibrium dynamics may facilitate coping with unpredictable environmental stimuli.

7.4 Multifractal Analysis

Multifractal structures have been uncovered in a number of classical physical problems such as voltage drops across a random resistor network [7.62], spatial distribution of the dissipation field of fully developed turbulence [7.63, 7.64], viscous fingering [7.65, 7.66], and diffusion limited aggregation [7.67, 7.68]. However, in physics and other applied sciences, fractals appear not only as singular objects (measures) [7.69, 7.70] but also as singular functions generated by dynamical systems. There have been only a few attempts to extend the concept of multifractality to singular functions: for velocity in turbulence [7.71] and for rough surfaces [7.72].

Physiological signals are generated by complex self-regulating systems that process inputs with a broad range of characteristics. Monofractal signals are homogeneous and have 'linear' properties. Many physiological time series, e.g. interbeat interval sequences, are in fact inhomogeneous, suggesting that different parts of the signal have different scaling properties. In addition, there is evidence that heartbeat dynamics exhibits nonlinear properties [7.73–7.77]. Up to now, robust demonstration of multifractality for nonstationary time series has been hampered by problems related to a drastic bias in the estimate of the singularity spectrum due to diverging negative moments. Moreover, the classical approaches based on the box-counting technique and structure function formalism fail when a fractal function is composed of a multifractal singular part embedded in regular polynomial behavior [7.78].

7.4.1 Multifractality: Nonstationarity in Local Scaling

Monofractal signals are homogeneous in the sense that they have the same scaling properties, characterized locally by a single singularity exponent h_0, throughout the entire signal [7.60, 7.79–7.83]. Therefore monofractal signals can be indexed by a single *global* exponent, the Hurst exponent $H \equiv h_0$ [7.84], which suggests that they are *stationary* from the viewpoint of their local scaling properties. On the other hand, multifractal signals can be decomposed into many subsets - possibly infinitely many - characterized by different *local* Hurst exponents h, which quantify the local singular behavior and thus relate to the local scaling of the time series. Thus multifractal signals require many exponents to fully characterize their scaling properties [7.69, 7.79, 7.83] and are intrinsically more complex, and *inhomogeneous*, than monofractals.

The statistical properties of the different subsets characterized by these different exponents h can be quantified by the function $D(h)$, where $D(h_0)$ is the fractal dimension of the subset of the time series characterized by the local Hurst exponent h_0 [7.69, 7.79, 7.83, 7.85–7.87]. Thus, the multifractal approach for signals, a concept introduced in the context of multi-affine functions [7.88, 7.89], has the potential to describe a wide class of signals that are more complex then those characterized by a single fractal dimension (such as classical $1/f$ noise).

Biomedical signals are generated by complex self regulating systems that process inputs with a broad range of characteristics [7.90, 7.91]. Many physiological time series are extremely inhomogeneous and nonstationary, fluctuating in an irregular and complex manner. The analysis of the fractal properties of such fluctuations has been restricted to second order linear characteristics such as the power spectrum and the two-point autocorrelation function. These analyses reveal that the fractal behavior of healthy, free-running physiological systems is often characterized by $1/f$-like scaling of the power spectra [7.27, 7.28, 7.43, 7.92].

In a recent study we establish the relevance of the multifractal formalism for the description of a physiological signal, e.g. the human heartbeat [7.93]. The motivation for our work is not merely looking for yet another example of multifractality, this time in the biological sciences. In fact, if we consider the neuroautonomic control mechanisms responsible for the generation of heartbeats, it is natural to expect the need for multifractal concepts for their description, since the heartbeats are a result of the interaction of many physiological components operating on different time scales. These interactions are nonlinear and self-regulating (through feedback control), leading to the *nonlinear* char-

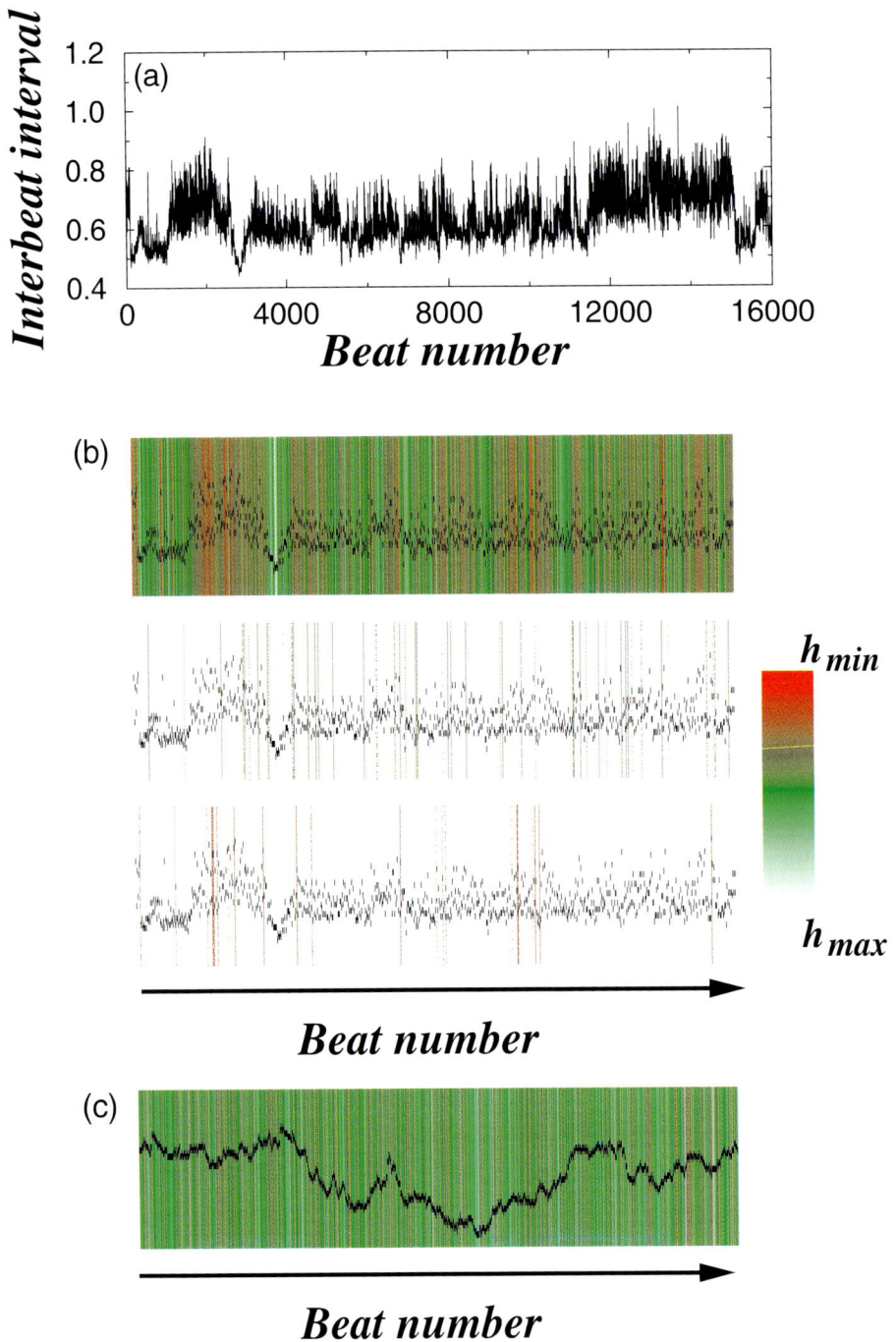

Fig. 7.13. (**a**) Consecutive heartbeat intervals measured in seconds are plotted vs. beat number from an approximately 3 hr record of a representative healthy subject. The time series exhibits very irregular and nonstationary behavior. (**b**) The top panel displays in color the local Hurst exponents calculated for the same 3 hr record shown in (**a**). (**c**) The panel displays in color the local Hurst exponents calculated for a *monofractal* signal, i.e. fractional Brownian motion with $H = 0.6$. The homogeneity of the signal is represented by the nearly monochromatic appearance of the signal which indicates that the local Hurst exponent h is the same throughout the signal and identical to the global Hurst exponent H. From [7.93]

acter of the output signal and to the heterogeneous features of heartbeat time series.

In contrast, the assumption of heartbeat monofractality, which has been the scope of studies in the field so far, is unrealistic because the monofractal hypothesis assumes that the scaling properties of the signal are the same throughout time, and are characterized by the *same* local Hurst exponent h. However, by looking at the heart signals it is clear that they are heterogeneous and therefore they might require more exponents for their description. Since the power spectrum and the correlation analysis (DFA method) can measure only one exponent characterizing a given signal, this implies that these methods are more appropriate for the study of monofractal signals. Moreover, the power spectrum and the correlation analysis reflect only the linear characteristics, while the heartbeat dynamics exhibits nonlinear properties. Thus the multifractal analysis can reveal new information on the nature of the nonlinearity, and we show (see Fig. 7.17) that the multifractal character and nonlinear properties of the signal are encoded in the Fourier phases.

We test whether a large number of exponents is required to characterize heterogeneous heartbeat interval time series (Fig. 7.13) by undertaking multifractal analysis. The first problem is to extract the local value of h. To this end we use methods derived from wavelet theory [7.94]. The properties of the wavelet transform make wavelet methods attractive for the analysis of complex nonstationary time series such as one encounters in physiology [7.52]. In particular, wavelets can remove polynomial trends that could lead box-counting techniques to fail to quantify the local scaling of the signal [7.64]. Additionally, the time-frequency localization properties of the wavelets makes them particularly useful for the task of revealing the underlying hierarchy that governs the temporal distribution of the local Hurst exponents [7.95]. Hence, the wavelet transform enables a reliable multifractal analysis [7.64, 7.95]. As the analyzing wavelet, we use derivatives of the Gaussian function, which allows us to estimate the singular behavior and the corresponding exponent h at a given location in the time series. The higher the order n of the derivative, the higher the order of the polynomial trends removed and the better the detection of the temporal structure of the local scaling exponents in the signal.

The concept of multifractality is exemplified in Fig. 7.13a, b for a heartbeat intervals record from a healthy subject. The heterogeneity of the healthy heartbeat is represented by the broad range of local Hurst exponents h (colors) present and the complex temporal organization of the different exponents. The middle and bottom panels illustrate the different fractal structure of two subsets of the time series characterized by different local Hurst exponents. The value of the local Hurst exponent for each subset is represented with a shade of green and red respectively. The two subsets display different temporal structure which can be quantified by different fractal dimension $D(h)$. While the healthy signal is represented by a *multicolor* plot, reflecting *multifractal* behavior through the variety of values for the local Hurst exponent, the fractional Brownian motion (a *monofractal* signal) (Fig. 7.13c) is essentially *monochromatic* indicating that the local Hurst exponent h is the same throughout the signal.

7.4.2 Multifractality in Heartbeat Dynamics

We evaluate the local exponent h through the modulus of the maxima values of the wavelet transform at each point in the time series, i.e. by the wavelet transform modulus maxima method [7.95] (for a more detailed description of this technique see Chap. 2). However, heartbeat time series contain densely packed, *nonisolated* singularities which unavoidably affect each other in the time-frequency decomposition. Therefore, rather than evaluating the distribution of the inherently unstable local singularity exponents (as shown in color in Fig. 7.13), we estimate the scaling of an appropriately chosen global measure - a partition function $Z_q(a)$, which is defined as the sum of the qth powers of the local maxima of the modulus of the wavelet transform coefficients at scale a. For each scale a these local maxima values are traced along the maxima lines obtained after the wavelet decomposition of the heartbeat signal (Fig. 7.11). As analyzing wavelet we use the third derivative of the Gaussian function. For small scales, we expect

$$Z_q(a) \sim a^{\tau(q)} . \tag{7.2}$$

For certain values of q, the exponents $\tau(q)$ have familiar meanings. In particular, $\tau(2)$ is related to the scaling exponent of the Fourier power spectra, $S(f) \sim 1/f^\beta$, as $\beta = 2 + \tau(2)$. For positive q, $Z_q(a)$ reflects the scaling of the large fluctuations and strong singularities, while for negative q, $Z_q(a)$ reflects the scaling of the small fluctuations and weak singularities [7.79, 7.83]. Thus, the scaling exponents $\tau(q)$ can reveal different aspects of cardiac dynamics (Fig. 7.14). Monofractal signals display a linear $\tau(q)$ spectrum, $\tau(q) = qH - 1$, where H is the global Hurst exponent. For multifractal signals, $\tau(q)$ is a nonlinear function: $\tau(q) = qh(q) - 1$, where $h(q) \equiv d\tau(q)/dq$ is not constant.

A previous obstacle to the determination of the multifractal spectrum of a time series has been the calculation of the negative moments. Until the application of the wavelet modulus maxima method, it was not possible to estimate $Z_q(a)$ for $q < 0$. We calculate $\tau(q)$ for moments $q = -5, 4, \ldots, 0, \ldots, 5$ and scales $a = 2 \times 1.15^i$, $i = 0, \ldots, 41$ from 6 hr records obtained from a healthy subject and a subject with congestive heart failure. In Fig. 7.14a, b we display the calculated values of $Z_q(a)$ for scales $a > 8$. The top curve corresponds to $q = -5$, the middle curve (shown heavy) to $q = 0$, and the bottom curve to $q = 5$. The exponents $\tau(q)$ are obtained from the slope of the $Z_q(a)$ curves in the region $16 < a < 700$, thus eliminating the influence of any residual

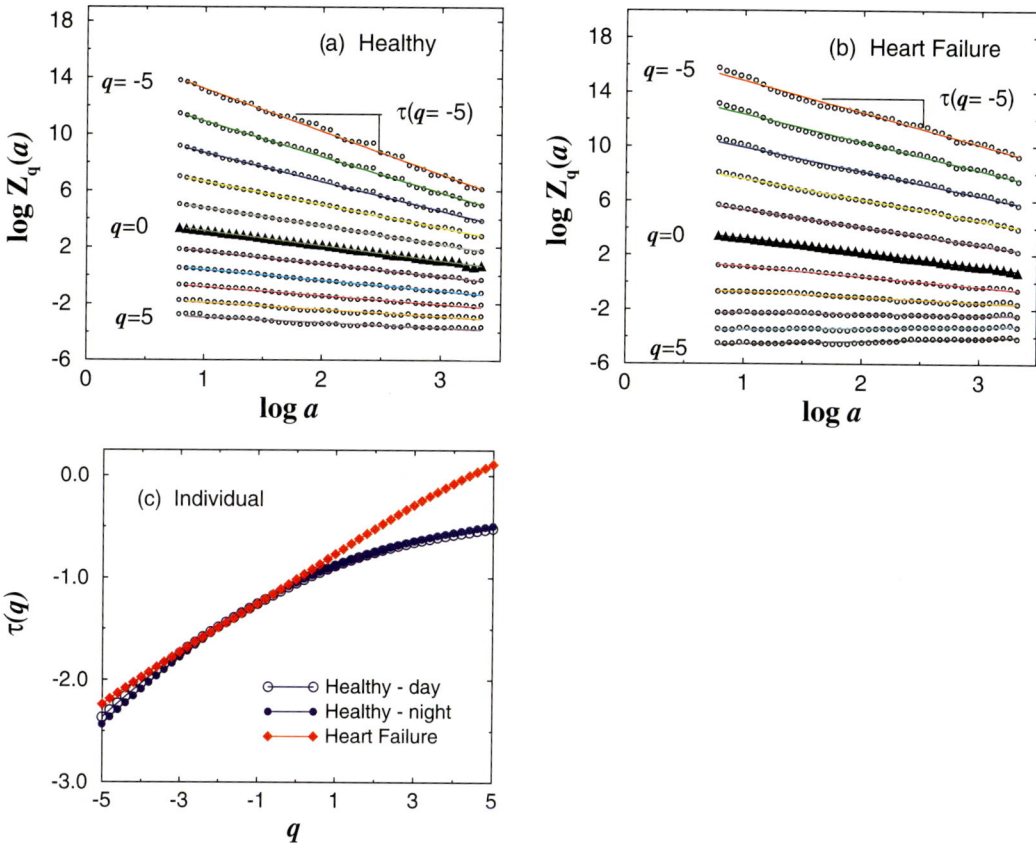

Fig. 7.14. Scaling of the partition function $Z_q(a)$ with scale a obtained from daytime records consisting of $\approx 25\,000$ beats for (**a**) a healthy subject and (**b**) a subject with congestive heart failure. (**c**) Multifractal spectrum $\tau(q)$ for the individual records in (**a**) and (**b**). From [7.93]

small scale random noise due to ECG signal pre-processing as well as extreme, large scale fluctuations of the signal. A monofractal signal would correspond to a straight line for $\tau(q)$, while for a multifractal signal $\tau(q)$ is nonlinear. Note the clear differences between the $\tau(q)$ curves for healthy and heart failure records (Fig. 7.14c). The constantly changing curvature of the $\tau(q)$ curves for the healthy records suggests multifractality. In contrast, $\tau(q)$ is linear for the congestive heart failure subject, indicating monofractality.

We analyze both daytime (12:00 to 18:00) and nighttime (0:00 to 6:00) heartbeat time series records of healthy subjects, and the daytime records of patients with congestive heart failure. These data were obtained by Holter monitoring [7.96]. Our database includes 18 healthy subjects (13 female and 5 male, with ages between 20 and 50, average 34.3 yr), and 12 congestive heart failure subjects (three female and nine male, with ages between 22 and 71, average 60.8 years) in sinus rhythm [7.44].

For all subjects, we find that for a broad range of positive and negative q the partition function $Z_q(a)$ scales as a power-law (Figs. 7.14a,b). In Fig. 7.15, we show $Z_q(a)$ for $q = -2$ and $q = 2$ for all 18 healthy subjects in our database.

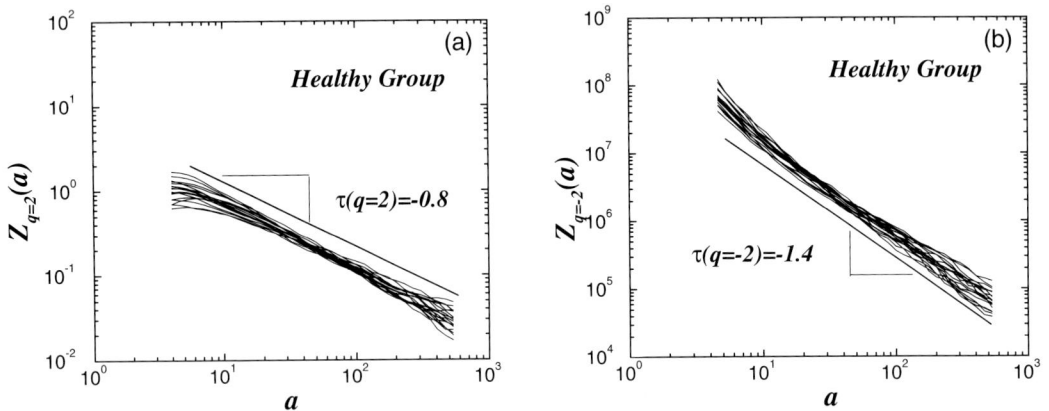

Fig. 7.15. $Z_q(a)$ for $q = 2$ (**a**) and $q = -2$ (**b**) moments from 6 hr long records for all 18 healthy subjects in our database

The figure very convincingly shows that there is good power-law scaling - that is, the data points fall on a straight line in a log-log plot - for all subjects and across nearly two orders of magnitude in a. Also, it is clear that the data have nearly the same slope for all subjects. Since the slope gives the value of the exponent $\tau(q)$, we can conclude that the representative group of healthy subjects are characterized by the *same* set of exponents.

Next, we obtain the fractal dimension $D(h)$. It is related to $\tau(q)$ through a Legendre transform,

$$D(h) = q\frac{d\tau(q)}{dq} - \tau(q) \,. \tag{7.3}$$

For all healthy subjects, we find that $\tau(q)$ is a nonlinear function (Fig. 7.14c and Fig. 7.16a), which indicates that the heart rate of healthy humans is a multifractal signal. Figure 7.16b shows that for healthy subjects, $D(h)$ has nonzero values for a broad range of local Hurst exponents h. The multifractality of healthy heartbeat dynamics cannot be explained by activity, as we analyze data from subjects during nocturnal hours. Furthermore, this multifractal behavior cannot be attributed to sleep-stage transitions, as we find multifractal features during daytime hours as well. The range of scaling exponents - $0 < h < 0.3$ - with nonzero fractal dimension $D(h)$ suggests that the fluctuations in the healthy heartbeat dynamics exhibit anticorrelated behavior ($h = 1/2$ corresponds to uncorrelated behavior while $h > 1/2$ corresponds to correlated behavior).

In contrast, we find that heart rate data from subjects with a pathological condition, i.e. congestive heart failure, show a clear *loss of multifractality* (Fig. 7.16a, b). For the heart failure subjects, $\tau(q)$ is close to linear and $D(h)$ is nonzero only over a very narrow range of exponents h indicating monofractal behavior (Fig. 7.16).

Our results show that, for healthy subjects, local Hurst exponents in the range $0.07 < h < 0.17$ are associated with fractal dimensions close to one. This

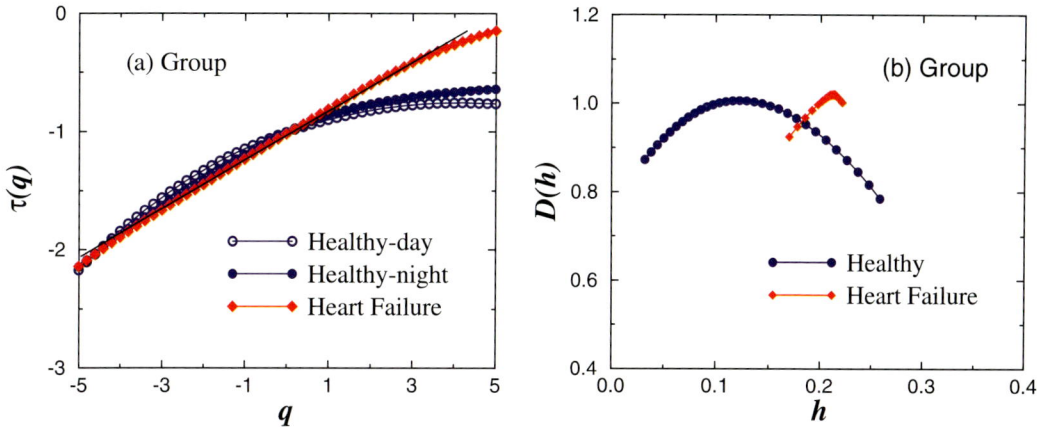

Fig. 7.16. (a) Multifractal spectrum $\tau(q)$ of the group averages for daytime and nighttime records for 18 healthy subjects and for 12 patients with congestive heart failure. The results show multifractal behavior for the healthy group and distinct change in this behavior for the heart failure group. (b) Fractal dimensions $D(h)$ obtained through a Legendre transform from the group averaged $\tau(q)$ spectra of (a). The shape of $D(h)$ for the individual records and for the group average is broad, indicating multifractal behavior. On the other hand, $D(h)$ for the heart failure group is very narrow, indicating monofractality. The different form of $D(h)$ for the heart failure group may reflect perturbation of the cardiac neuroautonomic control mechanisms associated with this pathology. From [7.93]

means that the subsets characterized by these local exponents are statistically dominant. On the other hand, for the heart failure subjects, we find that the statistically dominant exponents are confined to a narrow range of local Hurst exponents centered at $h \approx 0.22$. These results suggest that for heart failure the fluctuations are less anticorrelated than for healthy dynamics since the dominant scaling exponents h are closer to $1/2$. Our findings support the previous report of long-range anticorrelations of heartbeat fluctuations [7.43].

7.4.3 Multifractality and Nonlinearity

The multifractality of heartbeat time series also enables us to quantify the greater complexity of the healthy dynamics compared to pathological conditions. Power spectrum and detrended fluctuation analysis define the complexity of heartbeat dynamics through its scale-free behavior, identifying a *single* scaling exponent as an index of healthy or pathological behavior. Hence, the power spectrum is not able to quantify the greater level of complexity of the healthy dynamics, reflected in the heterogeneity of the signal. On the other hand, the multifractal analysis reveals this new level of complexity by the *broad* range of exponents necessary to characterize the healthy dynamics (Fig. 7.16). Moreover, the change in shape of the $D(h)$ curve for the heart failure group may provide insights into the alteration of the cardiac control mechanisms due to this pathology.

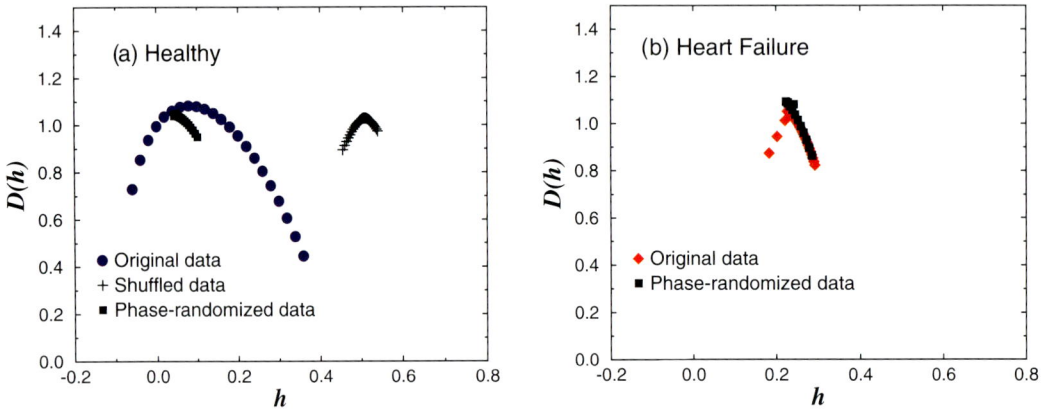

Fig. 7.17. (a) The fractal dimensions $D(h)$ for a 6 hr daytime record of a healthy subject. After re-shuffling and integrating the increments in this interbeat interval time series, so that all correlations are lost but the distribution is preserved, we obtain monofractal behavior: a very narrow point-like spectrum centered at $h \equiv H = 1/2$. Such behavior corresponds to a simple random walk. A different test, in which the $1/f$-scaling of the heartbeat signal is preserved but the Fourier phases are randomized (i.e. nonlinearities are eliminated) leads again to a monofractal spectrum centered at $h \approx 0.07$, since the linear correlations were preserved. These tests indicate that the observed multifractality is related to nonlinear features of the healthy heartbeat dynamics rather than to the ordering or the distribution of the interbeat intervals in the time series. (**b**) The fractal dimensions $D(h)$ for a 6 hr daytime record of a heart failure subject. The narrow multifractal spectrum indicates loss of multifractal complexity and reduction of nonlinearities with pathology. From [7.93]

To further study the complexity of the healthy dynamics, we perform two tests with surrogate time series. First, we generate a surrogate time series by shuffling the interbeat interval increments of a record from a healthy subject. The new signal preserves the distribution of interbeat interval increments but destroys the long-range correlations among them. Hence, the signal is a simple random walk, which is characterized by a single Hurst exponent $H = 1/2$ and exhibits monofractal behavior (Fig. 7.17a). Second, we generate a surrogate time series by performing a Fourier transform on a record from a healthy subject, preserving the amplitudes of the Fourier transform but randomizing the phases, and then performing an inverse Fourier transform. This procedure eliminates nonlinearities, preserving only the linear features of the original time series. The new surrogate signal has the *same* $1/f$ behavior in the power spectrum as the original heartbeat time series; however, it exhibits monofractal behavior (Fig. 7.17a). We repeat this test on a record of a heart failure subject. In this case, we find a smaller change in the multifractal spectrum (Fig. 7.17b). The results suggest that the healthy heartbeat time series contains important phase correlations canceled in the surrogate signal by the randomization of the Fourier phases, and that these correlations are weaker in heart failure subjects. Furthermore, the tests indicate that the observed multifractality is related to nonlinear features of the healthy heartbeat dynamics. A number of recent studies have tested for nonlinear and deterministic properties in recordings of interbeat intervals [7.73–7.77]. Our results suggest an explicit relation between

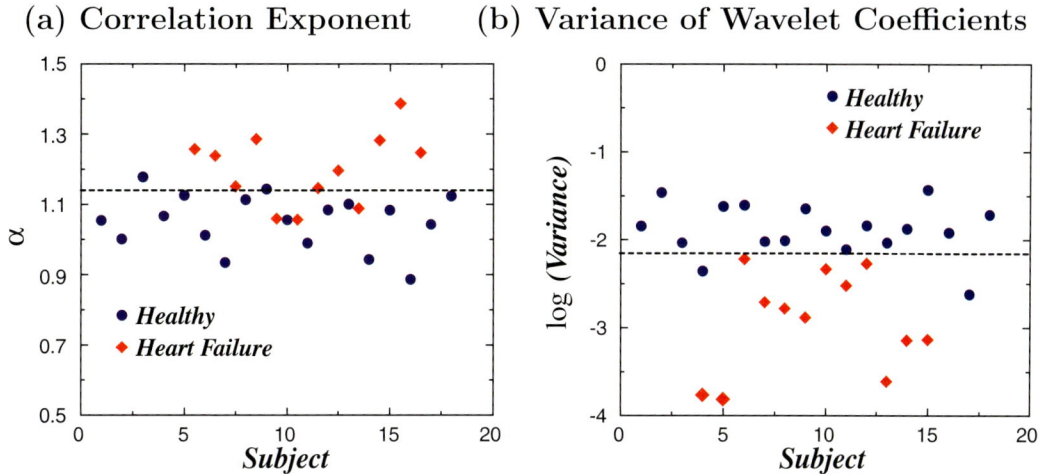

Fig. 7.18. Comparison of the discrimination power of (**a**) scale-invariant measure (the correlation exponent α), and (**b**) scale-dependent measure (the variance of the wavelet transform coefficients at scale $a = 32$)

the nonlinear features (represented by the Fourier phase interactions) and the multifractality of healthy cardiac dynamics (Fig. 7.17).

7.4.4 Clinical Applicability

The multifractal spectrum does in fact show good promise as a diagnostic tool. In particular, we find that the multifractal method outperforms other methods, both standard and fractal-related methods. We systematically compare our method with other widely used methods of heart rate time series analysis [7.41–7.43]. Several of these methods do not result in a fully consistent assignment of healthy vs. diseased subjects. As an example, we show in Fig. 7.18 two well-established methods.

The first is based on the measurement of long-range correlations on the fluctuations in heartbeat intervals [7.32]. These correlations have been quantified with both the power spectrum and the detrended fluctuation analysis (DFA) method. Figure 7.18a shows the values of the correlation exponent α measured through the detrended fluctuation analysis. The dashed line represents an in-sample threshold for discrimination of the healthy and heart failure groups. A second method [7.41] measures the variance of the coefficients of the wavelet transform of the heartbeat signal at a wavelet scale of $a = 32$, shown in Fig. 7.18b. These two methods do not result in a fully consistent separation between the healthy and heart failure groups.

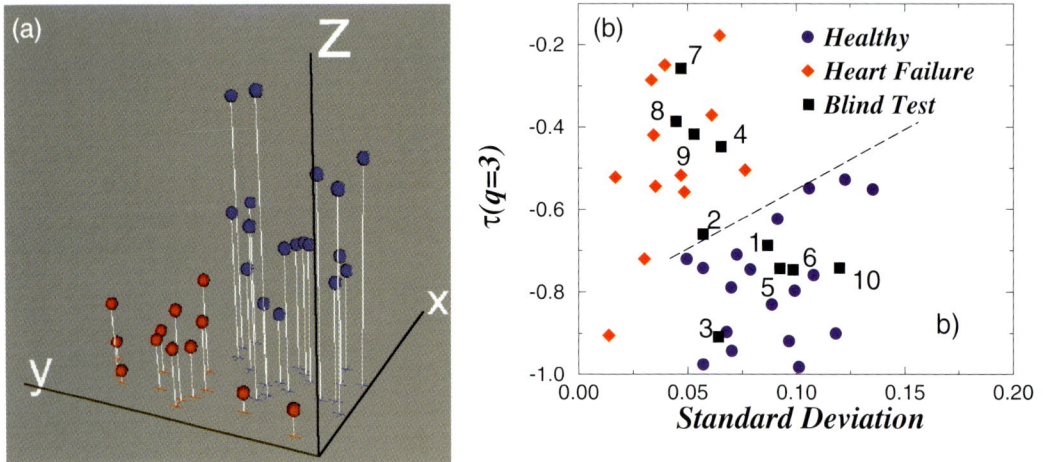

Fig. 7.19. (a) Discrimination method based on the multifractal formalism. Each subject in the database is characterized by three quantities. The first quantity (z-axis) is the degree of multifractality, which is the difference between the maximum and minimum values of local Hurst exponent h for each individual. The second quantity (y-axis) is the exponent value $\tau(q = 3)$ characterizing the scaling of the third moment $Z_3(a)$. The third quantity (x-axis) is the standard deviation of the interbeat intervals. The healthy subjects are represented by blue spheres and the heart failure subjects by red spheres. (b) Discrimination method based on multifractal formalism for the 'blind' datasets. The y-axis is the exponent value $\tau(q = 3)$, and the x-axis is the standard deviation of the time series of interbeat intervals. Marked in black, we show the results for the datasets of the blind test. From [7.93]

In Fig. 7.19a we illustrate the results of our method based on the multifractal formalism. Each subject's dataset is characterized by three quantities: (1) the standard deviation of the interbeat intervals; (2) the exponent value $\tau(q = 3)$ obtained from the scaling of the third moment $Z_3(a)$, and (3) the degree of multifractality, defined as the difference between the maximum and minimum values of local Hurst exponent h for each individual (Fig. 7.17). Note that the degree of multifractality takes value zero for a monofractal. We find that the multifractal approach robustly discriminates the healthy from heart failure subjects.

We next blindly analyze a separate database containing 10 records, five from healthy individuals and five from patients with congestive heart failure. The time series in the new database are shorter than the ones in our original database [7.44]; on average they are only 2 hr long, that is, less than 8 000 beats. Figure 7.19b shows the projection on the x-y plane of our data presented in Fig. 7.19a. Marked in black, we show the results for the blind test. Our approach clearly separates the blind test subjects into two groups: 1, 3, 5, 6, and 10 fall in the healthy group and 2, 4, 7, 8, and 9 in the heart failure group. Unblinding the test code reveals that indeed subjects 1, 3, 5, 6, and 10 are healthy, while 2, 4, 7, 8, and 9 are heart failure patients. We note also that two of the heart failure subjects (closest to the origin) are ambiguous; however, they are clearly identified as unhealthy in Fig. 7.19a because their degree of multifractality (z-axis) is close to zero (these two subjects are the red spheres in Fig. 7.19a closest to the origin). This case demonstrates our point that a single exponent is *not* sufficient to describe the complexity of heartbeat time series. We conclude that an analysis incorporating the multifractal method may add diagnostic power to contemporary analytic methods of heartbeat (and other physiological) time series.

As a second 'blind test', we show color panels with the local Hurst h exponent for six healthy individuals (Fig. 7.20) and six subjects with congestive heart failure (Fig. 7.21). Each panel represents 6 hr long record. The color code for these panels is the following: with increasing value of h, the spectrum goes from red to green to blue. A wider range of colors indicates a higher degree of multifractality. For this reason, records from healthy individuals should be more polychromatic. On the other hand, records from heart failure patients should be more monochromatic (with a single color predominating) indicating loss of multifractality. In addition, the color spectrum for the healthy individuals should be shifted to the red and for the heart failure patients shifted to the blue (as observed from the results in Fig. 7.16 where the peak of the multifractal spectrum $D(h)$ is centered at smaller values of h for the healthy group and at larger values of h for the heart failure group).

7.5 Conclusion

The discovery of multifractality in a physiological time series and its breakdown with pathology is significant from a number of perspectives.

First, contemporary analysis of heartbeats, and the study of physiological time series in general, have emphasized two important, but apparently unconnected properties: (i) the presence of nonlinearities and (ii) $1/f$-behavior (monofractality). The monofractal hypothesis assumes that the scaling properties of the signal are the same throughout. Yet the heterogeneous nature of the heartbeat interval time series clearly indicates nonlinear features. The finding of a multifractal mechanism for heart rate control provides a unifying connection between nonlinear and fractal properties, and, indeed indicates that they are aspects of a more fundamental type of mechanism. In particular, we show that both the multifractal character and the nonlinear properties of the signal are encoded in the Fourier phases (see Fig. 7.4).

Second, our analysis indicates that the healthy heartbeat is described by a broad range of scaling exponents h with a well-defined set of bounding parameters, $h_{\min}$ and $h_{\max}$. Furthermore, certain exponents appear to be 'forbidden' ($h < h_{\min}$ and $h > h_{\max}$) and the exponents present occur with a given structure characterized by the function $D(h)$.

Third, our findings may lead to new diagnostic applications.

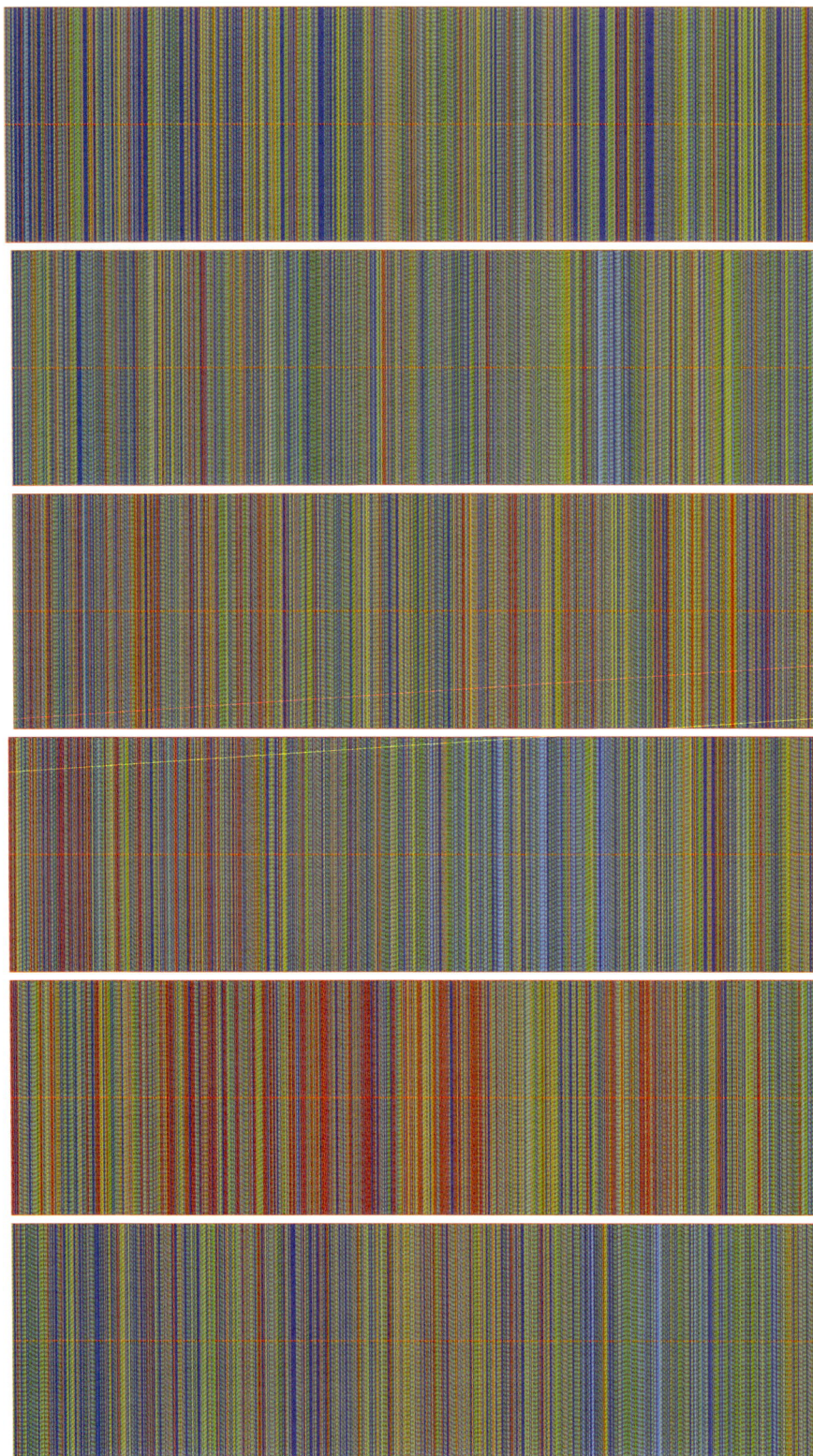

Fig. 7.20. Panels obtained from healthy individuals illustrating how the local Hurst exponent h (vertical color bars) changes with time (x-axis) (cf. Fig. 7.0). Each panel represents a 6 hr record. A broad range of colors indicates a broad multifractal spectrum. From [7.97]

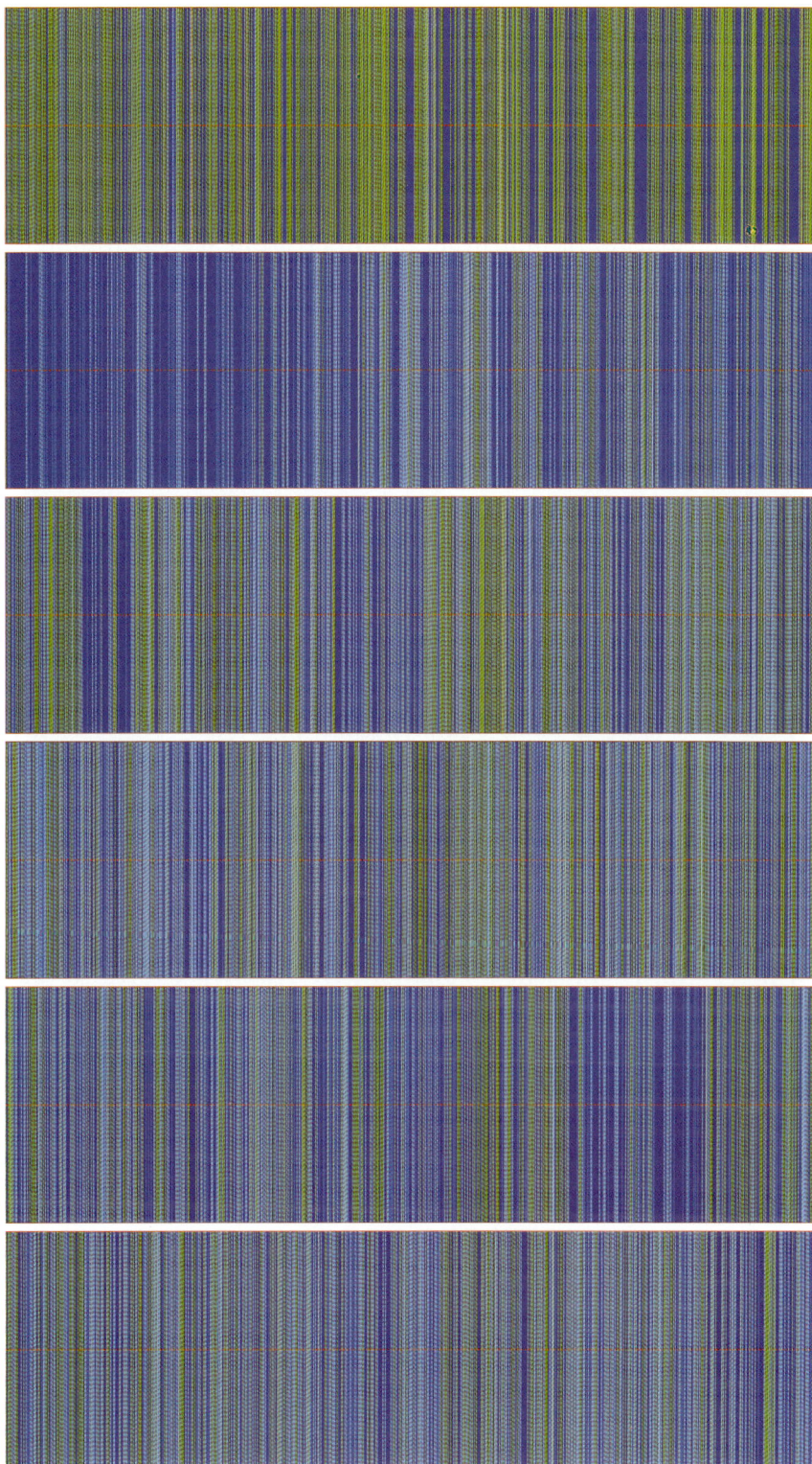

Fig. 7.21. Panels obtained from subjects with congestive heart failure illustrating how the local Hurst exponent h (vertical color bars) changes with time (x-axis) (cf. Fig. 7.0). Each panel represents a 6 hr record. An almost monochromatic appearance indicates a narrow multifractal spectrum, i.e. loss of multifractality. From [7.97]

Fourth, our analysis is based on a 'microscopic' approach which can identify the statistical properties of the self-affine cascade of heartbeat fluctuations at different scales (Fig. 7.11). Our finding of multifractality quantifies the complex dynamics of this cascade and suggests that a *multiplicative* mechanism might be the origin of this phenomenon. This finding will impact ongoing efforts to develop realistic models for better understanding mechanisms of cardiac control. The elucidation of multifractal properties of neuroautonomic control poses a major new challenge to physicists and physiologists, and any realistic model should be able to reproduce the empirical findings we uncover.

Finally, on a more general level, our approach provides a way of testing a broad range of $1/f$-type signals to see if they represent multifractal or monofractal processes. As such, these findings should be of interest to a very wide audience given the historic interest in elucidating the nature of different types of $1/f$ noise.

From a physiological perspective, the detection of robust multifractal scaling in the heart rate dynamics is of interest because our findings raise the intriguing possibility that the control mechanisms regulating the heartbeat interact as part of a coupled cascade of feedback loops in a system operating far from equilibrium [7.31, 7.71]. Furthermore, the present results indicate that the healthy heartbeat is even more complex than previously suspected, posing a challenge to ongoing efforts to develop realistic models of the control of heart rate and other processes under neuroautonomic regulation.

Acknowledgements. We are greatful to many individuals, including L.A.N. Amaral, A. Bunde, R.M. Baevsky, J. Fritsch-Yelle, S. Havlin, J. Mietus, C.-K. Peng, M.G. Rosenblum, and Z.R. Struzik, for major contributions to the results reviewed here, which represent a collaborative research effort. We also thank A. Arneodo, Y. Ashkenazy, F. Family, U. Frisch, J.M. Hausdorff, I. Grosse, H. Herzel, V. Horváth, H. Kallabis, J.W. Kantelhardt, J. Kurths, Y. Lee, T. Lopez-Ciudad, H. Makse, B. Rosenow, K.R. Sreenivasan, V. Vicsek, and B.J. West for valuable discussions. This work was supported by NIH/National Center for Research Resources (P41 RR13622), NSF, and The G. Harold and Leila Y. Mathers Charitable Foundation.

References

7.1 G.E.P. Box, G.M. Jenkins and G.C. Reinsel, *Time Series Analysis: Forecasting and Control* (Prentice-Hall, Englewood Cliffs, 1994).

7.2 M.F. Shlesinger, Ann. N. Y. Acad. Sci. **504**, 214 (1987).

7.3 L.S. Liebovitch, Biophys. J. **55**, 373 (1989).

7.4 R.L. Stratonovich, *Topics in the Theory of Random Noise*, Vol. I (Gordon and Breach, New York, 1981).

7.5 H. Kantz and T. Schreiber, *Nonlinear Time Series Analysis* (Cambridge University Press, Cambridge, 1997).

7.6   C. Bernard, *Les Phénoménes de la Vie* (Baillière, Paris, 1878).

7.7   B. van der Pol and J. van der Mark, Philos. Mag. **6**, 763 (1928).

7.8   W.B. Cannon, Physiol. Rev. **9**, 399 (1929).

7.9   E.W. Montroll and M.F. Shlesinger, in *Nonequilibrium Phenomena II: from Stochastics to Hydrodynamics*, edited by L.J. Lebowitz and E.W. Montroll (North-Holland, Amsterdam, 1984), p. 1.

7.10 C.-K. Peng, S.V. Buldyrev, J.H. Hausforff, S. Havlin, J.E. Mietus, M. Simons, H.E. Stanley, and A.L. Goldberger, Integr. Physiol. Behav. Sci. **29**, 283 (1994).

7.11 J. Theiler, S. Eubank, A. Longtin, B. Galdrikian, and J.D. Farmer, Physica D **58**, 77 (1992).

7.12 A.R. Osborne and A. Provenzale, Physica D **35**, 357 (1989).

7.13 R.M. Berne and M.N. Levy, *Cardiovascular Physiology*, 6th edn. (Mosby, St. Louis, 1996).

7.14 R.I. Kitney, D. Linkens, A.C. Selman, and A.A. McDonald, Automedica **4**, 141 (1982).

7.15 A.L. Goldberger, Lancet **347**, 1312 (1996).

7.16 M.N. Levy, Circ. Res. **29**, 437 (1971).

7.17 M. Malik and A.J. Camm (eds.), *Heart Rate Variability* (Futura, Armonk, 1995).

7.18 D.C. Michaels, E.P. Matyas and J. Jalife, Circ. Res. **55**, 89 (1984).

7.19 J.T. Bigger, Jr., C.A. Hoover, R.C. Steinman, L.M. Rolnitzky, and J.L. Fleiss, Am. J. Cardiol. **21**, 729 (1993).

7.20 G. Sugihara, W. Allan, D. Sobel, and K.D. Allan. Proc. Natl. Acad. Sci. USA **93**, 2608 (1996).

7.21 M. Mackey and L. Glass, Science **197**, 287 (1977).

7.22 M.M. Wolf, G.A. Varigos, D. Hunt, and J.G. Sloman, Med. J. Aust. **2**, 52 (1978).

7.23 L. Glass, P. Hunter, and A. McCulloch (eds.), *Theory of Heart* (Springer, New York, 1991).

7.24 J.B. Bassingthwaighte, L.S. Liebovitch, and B.J. West, *Fractal Physiology* (Oxford University Press, New York, 1994).

7.25 J. Kurths, A. Voss, P. Saparin, A. Witf, H.J. Kilner, and N. Wessel, Chaos **5**, 88 (1995).

7.26 R.I. Kitney and O. Rompelman, *The Study of Heart-Rate Variability* (Oxford University Press, London, 1980).

7.27 S. Akselrod, D. Gordon, F.A. Ubel, D.C. Shannon, A.C. Barger, and R.J. Cohen, Science **213**, 220 (1981).

7.28 M. Kobayashi and T. Musha, IEEE Trans. Biomed. Eng. **29**, 456 (1982).

7.29 J.P. Saul, P. Albrecht, D. Berger, and R.J. Cohen, *Computers in Cardiology* (IEEE Computer Society, Washington, 1987).

7.30 D. Panter, *Modulation, Noise and Spectral Analysis* (McGraw Hill, New York, 1965).

7.31 P.Ch. Ivanov, M.G. Rosenblum, C.-K. Peng, J. Mietus, S. Havlin, H.E. Stanley, and A.L. Goldberger, Physica A **249**, 587 (1998).

7.32 C.-K. Peng, J. Mietus, J.M. Hausdorff, S. Havlin, H.E. Stanley, and A.L. Goldberger, Phys. Rev. Lett. **70**, 1343 (1993).

7.33 S. Havlin et al., Phys. Rev. Lett. **61**, 1438 (1988).

7.34 M.F. Shlesinger and B.J. West, *Random Fluctuations and Pattern Growth: Experiments and Models* (Kluwer, Boston, 1988).

7.35 C.-K. Peng, S.V. Buldyrev, S. Havlin, M. Simons, H.E. Stanley, and A.L. Goldberger, Phys. Rev. E **49**, 1691 (1994).

7.36 S.V. Buldyrev, A.L. Goldberger, S. Havlin, R.N. Mantegna, M.E. Matsa, C.-K. Peng, M. Simons, and H.E. Stanley, Phys. Rev. E **51**, 5084 (1995).

7.37 S.V. Buldyrev, A.L. Goldberger, S. Havlin, C.-K. Peng, H.E. Stanley, and M. Simons, Biophys. J. **65**, 2673 (1993).

7.38 S.M. Ossadnik, S.V. Buldyrev, A.L. Goldberger, S. Havlin, R.N. Mantegna, C.-K. Peng, M. Simons, and H.E. Stanley, Biophys. J. **67**, 64 (1994).

7.39 M.S. Taqqu, V. Teverovksy, and W. Willinger, Fractals **3**, 785 (1996).

7.40 A.L. Goldberger, D.R. Rigney, J. Mietus, E.M. Antman, and M. Greenwald, Experientia **44**, 983 (1988).

7.41 S. Thurner, M.C. Feurstein, and M.C. Teich, Phys. Rev. Lett. **80**, 1544 (1998).

7.42 L.A.N. Amaral, A.L. Goldberger, P.Ch. Ivanov, and H.E. Stanley, Phys. Rev. Lett. **81**, 2388 (1998).

7.43 C.-K. Peng, S. Havlin, H.E. Stanley, and A.L. Goldberger, in *Proc. NATO dynamical disease conference*, edited by L. Glass, Chaos **5**, 82 (1995).

7.44 *Heart Failure Database* (Beth Israel Deaconess Medical Center, Boston, MA). The database now includes 18 healthy subjects (13 female and five male, with ages between 20 and 50, average 34.3 yr), and 12 congestive heart failure subjects (three female and nine male, with ages between 22 and 71, average 60.8 yr) in sinus rhythm. [Data are freely available at the following URL: `http://www.physionet.org`.]

7.45 F. Mallamace and H.E. Stanley (eds.), *Physics of Complex Systems: Proc. Enrico Fermi School Phys., Course CXXXIV* (IOS, Amsterdam, 1997).

7.46 P. Meakin, *Fractals, Scaling and Growth Far from Equilibrium* (Cambridge University Press, Cambridge, 1997).

7.47 R.G. Turcott and M.C. Teich, Ann. Biomed. Eng. **24**, 269 (1996).

7.48 K.K.L. Ho, G.B. Moody, C.-K. Peng, J.E. Mietus, M.G. Larson, D. Levy, and A.L. Goldberger, Circulation **96**, 842 (1997).

7.49 T.H. Mäkikallio, T. Seppänen, K.E.J. Airaksinen, J. Koistinen, M.P. Tulppo, C.-K. Peng, A.L. Goldberger, and H.V. Huikuri, Am. J. Cardiol. **80**, 779 (1997).

7.50 T.H. Mäkikallio, T. Ristimäe, K.E.J. Airaksinen, C.-K. Peng, A.L. Goldberger, and H.V. Huikuri, Am. J. Cardiol. **81**, 27 (1998).

7.51 H. Moelgaard, K.E. Soerensen, and P. Bjerregaard, Am. J. Cardiol. **68**, 777 (1991).

7.52 P.Ch. Ivanov, M.G. Rosenblum, C.-K. Peng, J. Mietus, S. Havlin, H.E. Stanley, and A.L. Goldberger, Nature **383**, 323 (1996).

7.53 P.Ch. Ivanov, M.G. Rosenblum, C.-K. Peng, S. Havlin, H.E. Stanley, and A.L. Goldberger, in *Wavelets in Physics*, edited by H. van der Berg (Cambridge University Press, Cambridge, 1998).

7.54 P.Ch. Ivanov, A. Bunde, L.A.N. Amaral, S. Havlin, J. Fritsch-Yelle, R.M. Baevsky, H.E. Stanley, and A.L. Goldberger, Europhys. Lett. **48**, 594 (1999).

7.55 A.L. Goldberger, M.W. Bungo, R.M. Baevsky, B.S. Bennett, D.R. Rigney, J.E. Mietus, G.A. Nikulina, and J.B. Charles, Am. Heart J. **128**, 202 (1994).

7.56 R.W. Peters et al., J. Am. Coll. Cardiol. **23**, 283 (1994).

7.57 Behrens S. et al., Am. J. Cardiol. **80**, 45 (1997).

7.58 P.Ch. Ivanov, L.A.N. Amaral, A.L. Goldberger, and H.E. Stanley, Europhys. Lett. **43**, 363 (1998).

7.59 A. Bunde, S. Havlin, J.W. Kantelhardt, T. Penzel, J.-H. Peter, and K. Voigt, Phys. Rev. Lett. **85**, 3736 (2000).

7.60 H.E. Stanley, Nature **378**, 554 (1995).

7.61 A. Bunde and S. Havlin, *Fractals in Science* (Springer, Berlin, 1994).

7.62 L. de Arcangelis, S. Redner, and A. Coniglio, Phys. Rev. B **31**, 4725 (1985).

7.63 U. Frisch, *Turbulence* (Cambridge University Press, Cambridge, 1995).

7.64 J.F. Muzy, E. Bacry, and A. Arneodo, Phys. Rev. Lett. **67**, 3515 (1991).

7.65 J. Nittmann, G. Daccord, and H.E. Stanley, Nature **314**, 141 (1985).

7.66 J. Nittmann and H.E. Stanley, Nature **321**, 663 (1986).

7.67 P. Meakin, H.E. Stanley, A. Coniglio, and T.A. Witten, Phys. Rev. A **32**, 2364 (1985).

7.68 H.E. Stanley and P. Meakin, Nature **335**, 405 (1988).

7.69 H.E. Stanley, in *Fractals and Disordered Systems*, 2nd edn., edited by A. Bunde and S. Havlin (Springer, Berlin, 1996), p. 1.

7.70 H.E. Stanley and P. Meakin, Nature **335**, 405 (1988).

7.71 C. Meneveau and K.R. Sreenivasan, Phys. Rev. Lett. **59**, 1424 (1987).

7.72 A.-L. Barabási, P. Szépfalusy, and T. Vicsek, Physica A **178**, 17 (1991).

7.73 J. Lefebvre et al., Chaos **3**, 267 (1993).

7.74 Y. Yamamoto et al., Biol. Cybern. **69**, 205 (1993).

7.75 J.K. Kanters, N.H. Holstein-Rathlou, and E. Agner, J. Cardiovasc. Electrophysiol. **5**, 128 (1994).

7.76 G. Sugihara, W. Allan, D. Sobel, and K.D. Allan, Proc. Natl. Acad. Sci. USA **93**, 2608 (1996).

7.77 C.-S. Poon and C.K. Merrill, Nature **389**, 492 (1997).

7.78 J F. Muzy, E. Bacry, and A. Arneodo, Phys. Rev. E **47**, 875 (1993).

7.79 T. Vicsek, *Fractal Growth Phenomena*, 2nd edn. (World Scientific, Singapore, 1993).

7.80 A. Bunde and S. Havlin (eds.), *Fractals in Science* (Springer, Berlin, 1994).

7.81 A. Bunde and S. Havlin (eds.), *Fractals and Disordered Systems*, 2 edn., (Springer-Verlag, Berlin, 1996).

7.82 A.-L. Barabási and H.E. Stanley, *Fractal Concepts in Surface Growth* (Cambridge University Press, Cambridge, 1995).

7.83 H. Takayasu, *Fractals in the Physical Sciences* (Manchester University Press, Manchester, 1997).

7.84 H.E. Hurst, Trans. Am. Soc. Civ. Eng. **116**, 770 (1951).

7.85 T.G. Dewey, *Fractals in Molecular Biophysics* (Oxford University Press, Oxford, 1997).

7.86 Z.R. Struzik, Fractals, **8**(2), 163 (2000).

7.87 Z.R. Struzik, Fractals, **9**(1), 77 (2001).

7.88 T. Vicsek and A.-L. Barabási, J. Phys. A: Math. Gen. **24**, L845 (1991).

7.89 A.-L. Barabasi and H.E. Stanley, *Fractal Concepts in Surface Growth* (Cambridge University Press, Cambridge, 1995), Chap. 24.

7.90 M.F. Shlesinger, Ann. N. Y. Acad. Sci. **504**, 214 (1987).

7.91 M. Malik and A.J. Camm (eds.), *Heart Rate Variability* (Futura, Armonk, 1995).

7.92 J.M. Hausdorff et al., J. Appl. Physiol. **80**, 1448 (1996).

7.93 P.Ch. Ivanov, L.A.N. Amaral, A.L. Goldberger, S. Havlin, M.G. Rosenblum, Z. Struzik, and H.E. Stanley, Nature **399**, 461 (1999).

7.94 I. Daubechies, *Ten Lectures on Wavelets* (S.I.A.M., Philadelphia, 1992).

7.95 J.F. Muzy, E. Bacry, and A. Arneodo, Int. J. Bifurcat. Chaos **4**, 245 (1994).

7.96 C.-K. Peng et al., J. Electrocardiol. **28**, 59 (1996).

7.97 P.Ch. Ivanov, L.A.N. Amaral, A.L. Goldberger, S. Havlin, M.G. Rosenblum, H.E. Stanley, and Z.R. Struzik, Chaos **11**, 641 (2001).

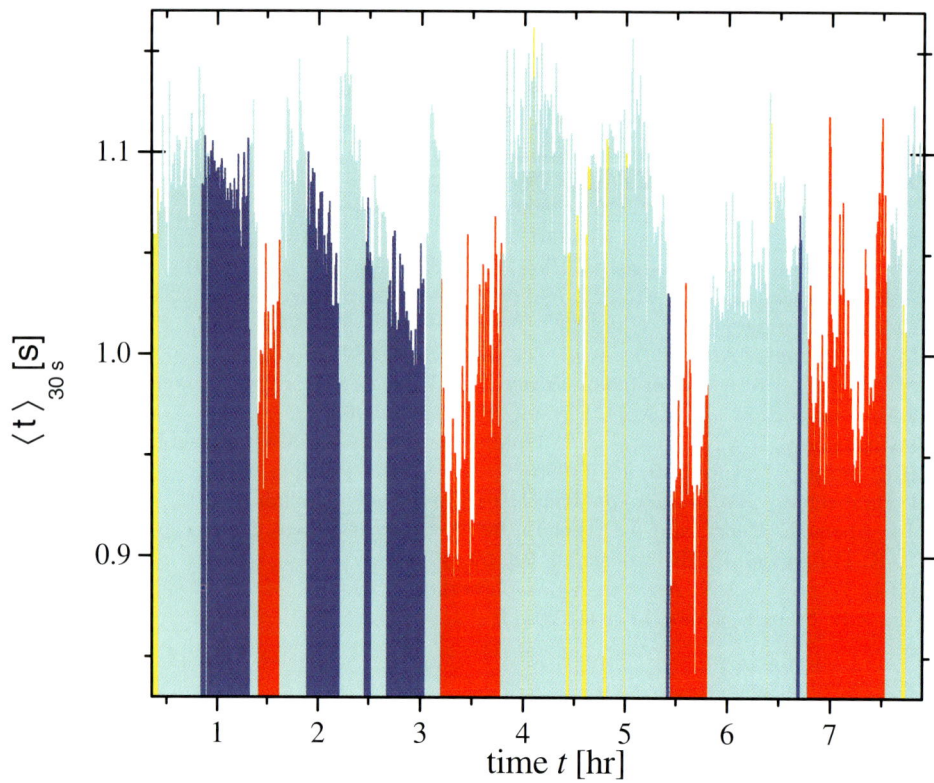

8. Physiological Relevance of Scaling of Heart Phenomena

Larry S. Liebovitch, Thomas Penzel, and Jan W. Kantelhardt

Most methods used to analyze experimental data are based on the assumption that the data is from a Gaussian distribution and uncorrelated. We describe methods to analyze data from scaling phenomena that do not have a Gaussian distribution [8.1–8.3] or involve long-range correlations [8.4–8.6]. We then show how those scaling methods have proved useful in characterizing the heart rate data from people who are healthy, from people who have a specific sleep disorder, and from people who have irregular heart rhythms.

8.1 Introduction

We are used to thinking that a single number is the best way to characterize our experimental data. For example, we typically determine the average value of a set of experimental measurements. This basically assumes that most of the values of the data have about this value, with perhaps some a bit larger and others a bit smaller. More specifically, we assume that the data has a Gaussian distribution. When this is the case, as we collect more data, the means of those samples of data approach a finite limiting value that we identify as the characteristic mean of the population, called the population mean. Our best estimate of the population mean is then how we characterize the data.

However, many systems have the universal feature that they extend over many scales. Thus, there is no single value, such as an average value, that can adequately characterize such systems. Such systems are fractal. For example, a tree has an ever larger number of ever smaller branches. There is no single, average value that properly characterizes the diameter of the branches of the

◄ **Fig. 8.0.** A one-night record of heartbeat intervals from a healthy subject with color-coded sleep stages: REM sleep (red), light sleep (light blue), deep sleep (dark blue), and intermediate wake stages (yellow)

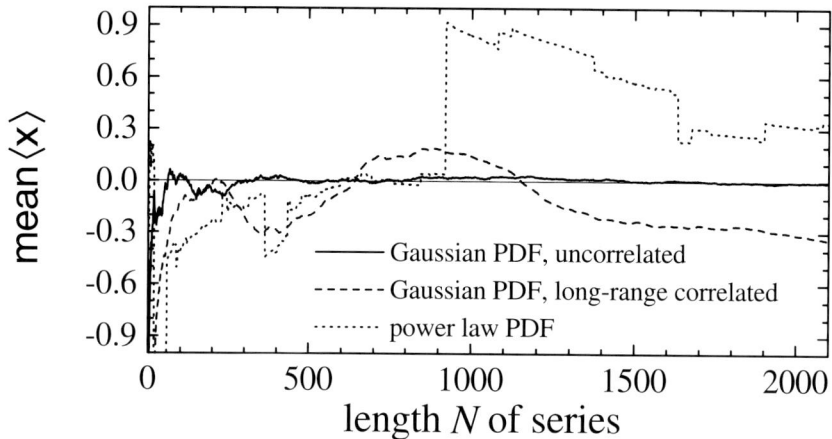

Fig. 8.1. The means $\langle x \rangle$ for three series of random numbers $(x_i), i = 1, \ldots, N$ are shown vs. the length N of the series. While the mean quickly converges to zero for the uncorrelated random data from a Gaussian probability density function (PDF), the means do not converge in the case of random numbers with a power-law PDF or long-range correlated Gaussian distributed numbers. In these cases the data is not described by a mean value but by the Hurst exponent $H = 1.1$, which corresponds to $\beta \approx 1.9$ in the case of the power-law PDF. All PDFs are symmetrical to $x = 0$

tree. A fractal object in space or a fractal process in time has a distribution of values that consist of a few large values, many medium values, and a huge number of small values. More specifically, the data has a distribution that is a power-law, that is, the plot of the logarithm of the relative frequency vs. the logarithm of the values is a straight line. When this is the case, as we collect more data, the means of those samples of data, continue to increase or decrease. The same happens when the data is not power-law distributed but involves long-range correlations, for example, when the probability density for the value at a given time depends on the previous values. The sample means do not approach a finite, limiting value that we can identify as the population mean. This behavior is illustrated in Fig. 8.1. It makes sense because there is no single value that characterizes the many scales spanned by the values of the data. For this case, the meaningful way to characterize the data is to determine how the sample means measured from the data depend on the resolution of the measurement. This can be done by plotting the logarithm of the average fluctuations of the sample means vs. the logarithm of the resolution at which they are determined. If the data is a simple fractal then this plot will be a straight line. If the data is multifractal, that is, consists of many different fractal dimensions, the slope of the straight lines will depend on the moment considered [8.9]. The fractal dimension is related to the slope of that line. The relationship between the slope and the fractal dimension depends on the type of measurement and on the definition of the fluctuations of the sample means. In the case of a power-law distribution of the values, the fractal dimension characterizes the relative frequency of the small values compared to the large values. For such fractal systems, it is the fractal dimension, rather than the mean, that is the meaningful way to characterize the data. In the case of long-range correlations the fractal dimension characterizes the scaling behavior of the mean, for example, as the Hurst exponent characterizes the fractional Brownian motion.

The properties of the higher moments are also different for experimental data that has Gaussian and fractal distributions. Consider the second moment, the variance, which is a measure of the dispersion in the data. For data with a Gaussian distribution, the variance measured from increasingly large samples of data approaches a finite, nonzero, limiting value. For data with a fractal distribution, the variance measured from increasingly large samples of data continues to increase. Again, the same holds if the data is not power-law distributed but involves long-range correlations. For example, in a fractal time series there are increasingly larger fluctuations over longer time scales. Thus, as the variance is measured over longer time windows, these ever larger fluctuations are included, and the variance measured increases. The properties of uncorrelated data from Gaussian distributions can therefore be properly characterized by the limiting values of their moments, particularly, the mean and variance. These are not meaningful measures for data from fractal distributions or for correlated data. However, data from fractal distributions can be properly characterized by determining how those moments depend on the resolution at which they are measured, which determines their fractal dimension.

8.2 Methods of Scaling Analysis

8.2.1 Probability Density Function

The probability density function (PDF) can be used to determine whether the distribution of experimental data is Gaussian or fractal. $\mathrm{PDF}(x)dx$ is the probability that a measurement has a value between x and $x + dx$. The PDF can be determined from a histogram of how often each range of values is found in the data. This method is limited by the fact that if the bins of the histograms are chosen to be small, then there is too little data at large x, and if the bins of the histogram are chosen to be too large, then the resolution is limited at small x. A better method is to determine the PDF by combining histograms of different binsize [8.7]. The PDF of data that is Gaussian has the form $\mathrm{PDF}(x) \propto (1/2\pi s^2)^{\frac{1}{2}} \exp[-(x-m)^2/2s^2]$ where m is the population mean and s is the population standard deviation. The PDF of data that is fractal has the power-law form

$$\mathrm{PDF}(x) \propto x^{-\beta} \tag{8.1}$$

which is a straight line with a slope β on a plot of $\log[\mathrm{PDF}(x)]$ vs. $\log(x)$.

8.2.2 Autocorrelation Function

In addition to the distribution of the values in the time series (x_i), we are interested in the correlation of the deviations of the values x_i and x_{i+T} from their mean $\langle x \rangle$ with different time lags T. Quantitatively, the correlation in the x_i can be determined by the autocorrelation function

$$C(T) = \langle \bar{x}_i \cdot \bar{x}_{i+T} \rangle = \frac{1}{N-T} \sum_{i=1}^{N-T} \bar{x}_i \cdot \bar{x}_{i+T} \qquad (8.2)$$

for a given time series with length N and $\bar{x}_i \equiv x_i - \langle x \rangle$. If the x_i are uncorrelated, $C(T)$ is zero for $T > 0$. For short-range correlations of the x_i, $C(T)$ declines exponentially and for long-range correlations it declines as a power-law

$$C(T) \propto T^{-\gamma} \qquad (8.3)$$

with an exponent $0 < \gamma < 1$. A direct calculation of $C(T)$ is not appropriate due to noise superimposed on the collected data x_i and due to underlying trends of unknown origin. For example, $\langle x \rangle$ need not be constant (as discussed above), which makes the definition of $C(T)$ problematic. Thus, we have to determine the correlation exponent γ indirectly using fractal scaling (dispersion) analysis.

8.2.3 Dispersion Analysis

For values from a time series of data, the moments of samples of the data can be determined as more data is included or as the time resolution of the measurement is changed. For uncorrelated data from a Gaussian distribution, the moments such as the mean and variance, approach finite, limiting values that we identify as the mean and variance of the population. For data from a fractal distribution and long-range correlated data, the moments, m, depend on the time window, T, that is used to measure them. Typically, the average fluctuations of the moments display a scaling relationship of a power-law form, AT^H, which is a straight line on a log-log plot. Several methods of scaling analysis are based on the scaling properties of the second moment, the dispersion. The dispersion can be measured by the mean squared deviation, the relative dispersion (standard deviation divided by the mean), the Fano factor (variance divided by the mean), or the Hurst rescaled range (maximum minus the minimum value of the running sum of the deviations from the mean divided by the standard deviation). In these methods, the parameter H is determined from the slope on a log-log plot of the dispersion measure vs. the time window over which it is measured.

The parameter H characterizes how the fluctuations in the measured values depend on the length of the data window analyzed. This is related to the fractal dimension, D. The relationship between H and D depends on the method used to measure the dispersion. It also characterizes the fractal correlations in the data. For example, for the Hurst rescaled range method, H is determined as the Hurst exponent. When the increments of a time series are uncorrelated and not taken from a power-law distribution, then $H = 0.5$. When the time series has positive correlations at all time scales, that is, when an increase at any time t is more likely to be followed by an increase at some times $t + T$ later, or when the values are power-law distributed, then $H > 0.5$. This is called persistence. When the values in the record are power-law distributed according to (8.1), $D = 2 - H$, and $\beta = 1 + 1/H$, while in the case of long-range correlations according to (8.3), $\gamma = 2 - 2H$. The same relations hold for the detrended fluctuation analysis (DFA). They allow the determination of the scaling exponents β or γ through dispersion analysis. When the time series has negative correlations, that is, when an increase at any time t is more likely to be followed by a decrease at some times $t + T$ later, then $H < 0.5$. This is called antipersistence.

8.2.4 Hurst Rescaled Range (R/S) Analysis

This is the oldest method of scaling analysis. It has been originally introduced by Hurst et al. in 1965 [8.8], see also [8.9]. As a measure of the dispersion it uses the range, R, which is the difference between the maximum and minimum of the deviation from the mean of the running sum of the values of the time series over a given number of values, T. The rescaled range R/S, is the range, R, divided by the standard deviation, S. The dependence of R/S on T is then found and the Hurst parameter H is then determined as the slope of the plot of $\log(R/S)$ vs. $\log(T)$. This is done by partitioning the total record of N values into $[N/T]$ consecutive segments of T values. The mean, $\langle x \rangle_{n,T}$, and standard deviation, $S_{n,T}$, of the values x_i, in the nth segment are determined from

$$\langle x \rangle_{n,T} = \frac{1}{T} \sum_{i=(n-1)T+1}^{nT} x_i \tag{8.4}$$

and

$$S_{n,T} = \left[\frac{1}{T} \sum_{i=(n-1)T+1}^{nT} (x_i - \langle x \rangle_{n,T})^2 \right]^{1/2}. \tag{8.5}$$

For i in the range $(n-1)T+1 \leq i \leq nT$ then

$$Y_{n,T}(i) = \sum_{k=(n-1)T+1}^{i} x_k - \langle x \rangle_{n,T}, \tag{8.6}$$

the range $R_{n,T}$ in the nth segment can be found from

$$R_{n,T} = \max\left[Y_{n,T}(i)\right] - \min\left[Y_{n,T}(i)\right] , \qquad (8.7)$$

and the rescaled range $(R/S)_{n,T}$ can be found from

$$(R/S)_{n,T} = R_{n,T}/S_{n,T} . \qquad (8.8)$$

The average rescaled range over all the windows is then given by

$$(R/S)_T = \frac{1}{[N/T]} \sum_{n=1}^{[N/T]} (R/S)_{n,T} . \qquad (8.9)$$

The average rescaled range $(R/S)_T$ is calculated for all window sizes T, and the scaling law

$$(R/S)_T \propto T^H \quad \Leftrightarrow \quad \log(R/S)_T \propto H \log T \qquad (8.10)$$

allows to determine the Hurst exponent H. As explained above, $H > 0.5$ indicates persistence, while $H < 0.5$ indicates antipersistence.

8.2.5 Detrended Fluctuation Analysis (DFA)

Often experimental data is affected by trends, e.g. slow temperature drifts, that modify the investigated system (for other examples see Chaps. 7 and 5). The human heartbeat is also affected by trends, which are caused by external stimuli and lead to drifts in the mean heart rate. Even rather sudden changes can occur, e.g. when the sleep stage changes. The effects of trends have to be well separated from the intrinsic fluctuations of the system. This task in not easy, since as discussed above the average value of the recorded quantity might already be unstable due to the intrinsic fluctuations. Hurst (R/S) analysis works well if we have long records without interruptions or trends. But if trends are present in the data, it might give wrong results. Very often we do not know the reasons for underlying trends in collected data and even worse we do not know the scales of the underlying trends. The underlying trends certainly follow their own laws which can be linear or can have different properties. Detrended fluctuation analysis (DFA) is a well-established method used to determine the scaling behavior for noisy data in the presence of trends without knowing their origin and shape [8.10–8.13].

In the procedure of DFA, we study the profile

$$Y(i) = \sum_{k=1}^{i} x_k - \langle x \rangle \qquad (8.11)$$

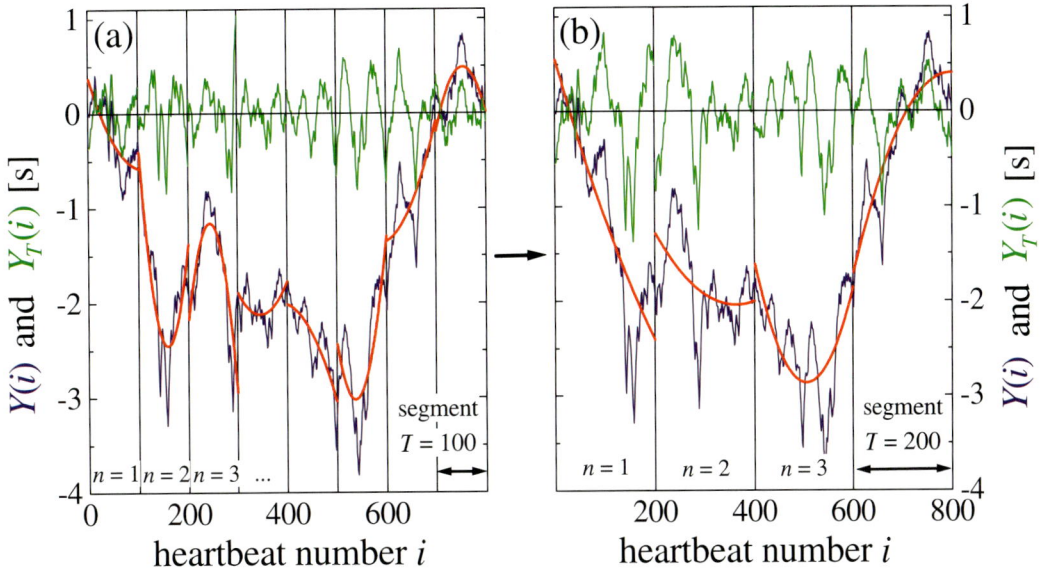

Fig. 8.2. Illustration of the detrending procedure in the detrended fluctuation analysis (DFA). For two segment durations (time scales) $T = 100$ (**a**) and 200 (**b**), the profiles $Y(i)$ (blue lines; defined in (8.11)), least square quadratic fits to the profiles (red lines), and the detrended profiles $Y_T(i)$ (green lines) are shown vs. the heartbeat number i for one record of interbeat intervals for a healthy subject

where $\langle x \rangle$ is the mean of the whole record (x_i) of length T. The second step is performed by cutting the profile $Y(i)$ into $[N/T]$ nonoverlapping segments of equal duration T (see Fig. 8.2). Note that this procedure is rather similar to the (R/S) analysis (cf. (8.6)), except that the splitting of the record is done after the integration. Since the record length N need not be a multiple of the considered time scale T, a short part at the end of the profile will remain. In order not to disregard this part, the same procedure is repeated starting from the other end of the record, giving $2[N/T]$ segments altogether. In the third step, we calculate the local trend for each segment n by a least-square fit of the data. Then we define the detrended time series for segment duration T, denoted by $Y_T(i)$, as the difference between the original time series and the fits.

Figure 8.2 illustrates this procedure for $T = 100$ and 200. In the example quadratic polynomials are used in the fitting procedure for each segment n, which is characteristic of quadratic DFA (DFA2). Linear, cubic, fourth order, or higher order polynomials can also be used in the fitting procedure (DFA1, DFA3, DFA4, and higher order DFA). Since the detrending of the time series is done by the subtraction of the fits from the profile, these methods differ in their capability to eliminate trends in the data. In qth order DFA, trends of order q in the profile and of order $q - 1$ in the original record are eliminated. Thus a comparison of the results for different orders of DFA allows to estimate the strength of the trends in the time series. In the fourth step, we calculate for each of the $2[N/T]$ segments the variance $F_T^2(n) = \langle Y_T^2(i) \rangle$ of the detrended time series $Y_T(i)$ by averaging over all data points i in the nth segment. At last we average over all segments and take the square root to obtain the DFA fluctuation function $F(T)$

$$F(T) = \left[\frac{1}{2[N/T]} \sum_{n=1}^{2[N/T]} F_T^2(n) \right]^{1/2} . \qquad (8.12)$$

The fluctuation function is calculated for all possible segment durations $T = 2$ to N. It is apparent, that the variance will increase with increasing duration T of the segments. For long-range power-law correlations in the data $F(T)$ increases by a power-law [8.13]

$$F(T) \propto T^H . \qquad (8.13)$$

We can plot $F(T)$ as a function of T with double logarithmic scales to measure H. Hence by measuring the exponent H we can detect the correlation exponent γ from (8.3) or the PDF exponent β from (8.1). For uncorrelated data and short-range correlations, we have $H = 0.5$, while $H > 0.5$ indicates long-range correlations or power-law distributed values. In this way H can be used to investigate correlation properties within a time series.

It is useful to determine if the value of H found from the experimental data is statistically significantly different from that when there are no autocorrelations in the time series. As explained above, a nonsimple scaling behavior (leading to $H \neq 0.5$ in the dispersion analysis) can be either due to a power-law PDF of the data, to long-range (auto)correlations, or to both. Randomizing the order of the values in the time series (shuffling) will remove the correlations in the data. The time series of the data can be randomized a number of different times and the mean H_m and standard deviation H_s of those time series determined. It is then possible to test the null hypothesis that the H from the time series is the same as the H_m from the uncorrelated surrogate time series. The probability that this is the case is given by probability that $z > |(H - H_m)|/s$ where z is the deviate of the normalized Gaussian distribution. If this probability is large, the nonsimple scaling is not significantly modified by the shuffling. In that case, the deviations from simple scaling with $H = 0.5$ must be due to power-law distributed data. On the contrary, if H is significantly different from $H_m \approx 0.5$, we know that long-range correlations present in the data are responsible for the nonsimple scaling. If H_m is neither close to 0.5 nor to H, both, long-range correlations and a power-law PDF are relevant.

8.3 Heart Rate During Sleep

8.3.1 Heart Rate Regulation

The heartbeat might be thought as a very regular activity solely following the instantaneous needs of the body. However, the measured interbeat intervals do actually fluctuate spontaneously according to the body's needs which themselves are not always very obvious and not completely observable. Not only diseases of the heart itself, but diseases of different origins as well, affect heart rate regulation both in terms of amplitude and variance. The human heartbeat is an essential vital signal which can be assessed over long periods of time using long-term ECG recorders, called a Holter recorder. This type of signal monitoring is now not very expensive and it is now widely available in medicine for clinical investigations and in physiology for research. Thus, this signal can be easily used to investigate the scaling properties and the correlational properties of human biology in healthy subjects and in patient with various diseases.

The heartbeat is regulated by the 'autonomous nervous system'. The autonomous nervous system has two branches, the sympathetic nervous activity which is responsible for heart rate accelerations and the parasympathetic nervous activity which is responsible for heart rate decelerations. These two opposing activities form an autonomic balance which determines the actual heart rate [8.14]. It is thought that these two activities influence the heart rate at different time scales, leading to spontaneous fluctuations of the heart rate. Actually, other influences and reflexes, such as respiration, body temperature regulation, humoral factors and the baroreceptor reflex are also involved in heart rate regulation. When disease is present additional factors can also influence the heart rate regulation. Therefore the model of heart rate regulation driven by sympathetic and parasympathetic activities, as accelerating and decelerating forces, has to be regarded as a simplified concept which helps to understand some, but not all, aspects of heart rate regulation.

The central nervous control of heart rate and blood pressure is located in the cardiovascular center in the brainstem. This center influences sympathetic and parasympathetic activities which control the heart rate and the blood pressure. The respiratory control system, that is also located in the brainstem, controls breathing. The respiratory neurons in the formatio reticularis receive feedback from mechanoreceptors of the lung and the chemoreceptors of the carotid bodies. Both the cardiovascular and the respiratory control centers interact when regulating heart rate, blood pressure and respiration.

Sleep is primarily a condition of rest and has a recreative function for the body and it produces a lower mean value of heart rate. It is assumed that this change is due to the fact that the autonomic balance is shifted towards parasympathetic predominance. Sleep, and its different states of brain activity, affects the cardiovascular and respiratory physiological control systems [8.15].

8.3.2 Sleep Physiology

Sleep is investigated in sleep laboratories where it is possible to continuously record many different physiological measures. These include brain activity (electroencephalogram, EEG), eye movements (electrooculogram, EOG) muscle activity of the chin (electromyogram, EMG submentalis), and the electrical activity of the heart (electrocardiogram, ECG) . In our sleep laboratory we also record respiration with inductive plethysmography, which uses a belt around the chest to monitor thoracic movements and a belt around the abdomen to monitor abdominal movements. These physiological measures can be used to determine the different sleep states. Figure 8.3 gives an example for some of the recorded signals.

The respiratory movements result in successful inspiration only if the upper airways are maintained open by activating the upper airway muscles located in the throat. The upper airways are collapsible because they also serve the function of swallowing and they play an important role in vocalization. To measure breathing as the result of respiratory movements we record airflow at the nose and the mouth. In addition we record the oxygen content of the blood, called oxygen saturation with a sensor at the finger tip. The muscle activity of legs (EMG tibialis) is recorded to monitor leg movements, which are normal when falling asleep and during changes of body position. A periodic occurrence of leg movements at other times of the night may also disturb sleep. This total set of signals is required for a polysomnography in a sleep laboratory in order to diagnose sleep disorders according to medical recommendations [8.16]. In many sleep laboratories all these parameters mentioned are recorded as a digital time series with digital sampling rates ranging from 1 to 250 samples per second dependent of the signal. The polysomnographic recordings last for 8 to 10 hr

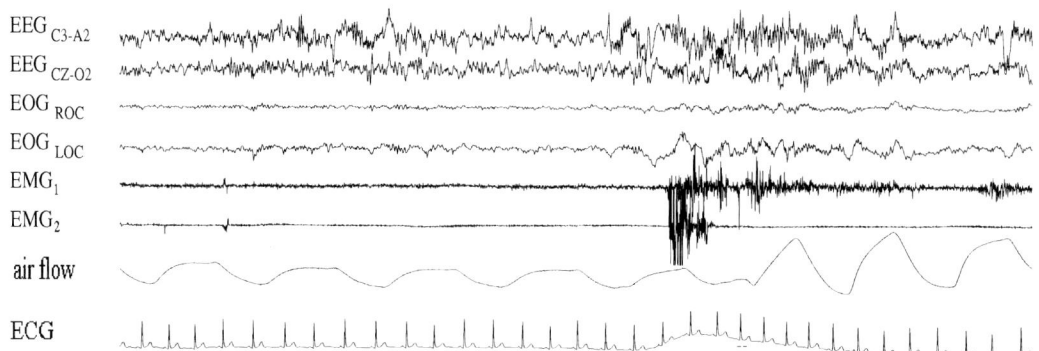

Fig. 8.3. Samples of the recordings taken in the sleep laboratory. The EEG, EOG and EMG records are used to determine the sleep stages, while air flow and ECG records are analyzed

and produce large amount of data which must be reviewed by the physician in charge in order to produce a diagnosis and a report.

Based on the recording of EEG, EOG and EMG sleep experts classify sleep states according to rules compiled by committee chaired by Rechtschaffen and Kales in 1968 [8.17]. In order to classify the sleep states the time series are displayed in 30 s segments to the expert. The expert looks at the pattern of the time series visualizing the brain waves, and according to the rules, distinguishes six categories which are awake, sleep stages 1 and 2 (also called light sleep), sleep stages 3 and 4 (also called deep sleep or slow-wave sleep), and REM (rapid eye movement) sleep. In contrast to the other five stages, the sleep expert recognizes REM sleep by reading the rapid eye movements in addition to the brain waves which have a similar pattern to wakefulness. REM sleep is also called paradoxical sleep because the investigated person is still asleep and difficult to arouse even if the brain waves resemble wakefulness. To emphasis the completely different pattern of REM sleep, the other four sleep stages (1 to 4) are summarized as non-REM sleep.

Deep sleep has a physical recreative function. This theory is supported by the finding that all the muscles are relaxed, energy consumption of the body is lowered and many humoral secretions such as growth hormone and stress hormone (cortisol) are activated during non-REM sleep. When awakened during REM sleep, persons report dreaming in 80% of all cases. Even today, there is still controversy as to the function of dreaming and REM sleep. One theory says that memory consolidation occurs during REM sleep and thus it has a mental recreative function, while another theory challenges that result [8.18].

In a normal night, sleep stages follow a specific temporal sequence. A few minutes of light sleep follow immediately after falling asleep. These roughly 15 min are only a transition to deep sleep. Deep sleep is maintained for the next 20 to 30 min. Then a few minutes of light sleep are passed as a transition to the first short period of REM sleep which lasts for 5 to 10 min. The entire sequence lasts 90 to 110 min and is called one sleep cycle. Four to six such sleep cycles occur in a normal night of sleep. In the beginning of the night the deep sleep is longer and dominates the sleep cycle. However, in the sleep cycles later in the night REM sleep dominates the cycle and may last 20 to 30 min. No more deep sleep is found in the last third of the night. The duration of light sleep also increases as sleep progresses. Figure 8.4 shows a typical example of the sleep stage pattern for a healthy subject.

8.3.3 Sleep Related Breathing Disorders

One important sleep disorder with a high prevalence of 4% in males and 2% in females within a range of 30 to 60 yr is 'obstructive sleep apnea' [8.19]. This disorder is characterized by cessations of respiratory airflow called apnea, each

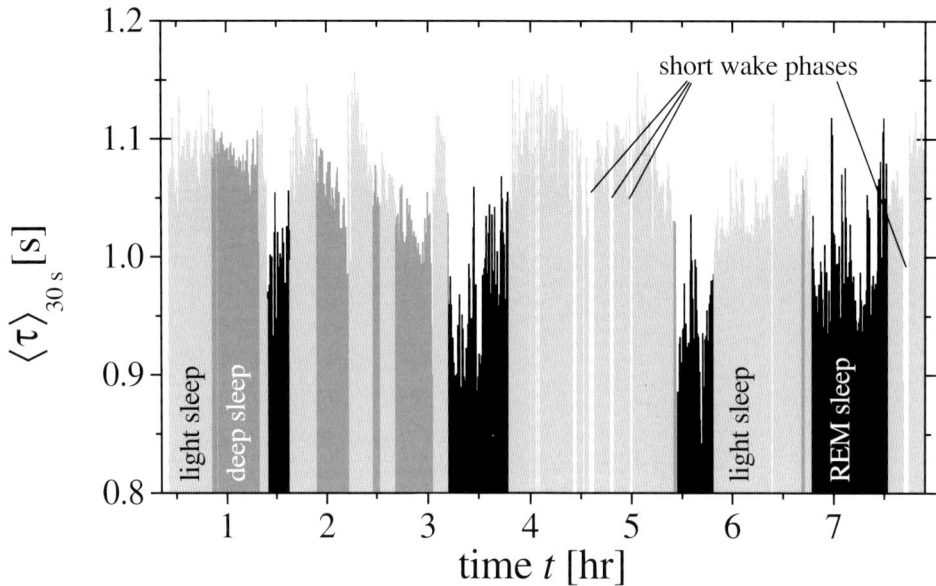

Fig. 8.4. A one-night record of heartbeat intervals from a healthy subject with sleep stages coded underneath the curve. Here RR interval values are averaged over 30 s corresponding to the time resolution of sleep stages. The areas in black mark the REM sleep stages, while light and dark gray have been chosen for light and deep sleep, respectively. It can be observed that the mean of RR intervals is lower in REM sleep than in non-REM sleep, and they fluctuate more strongly

of which is at least 10 s long [8.16]. These apneas episodes can occur up to 600 times per night [8.16]. During these individual apneas, the respiratory neurons in the brainstem continue to fire and activate the respiratory intercostal muscles and the diaphragm. Therefore, respiratory movements continue during these apneas. But since the upper airways and the pharynx collapse, the respiratory movements are not able to produce a successful ventilation. Because there is no gas exchange in the lung, the oxygen content of the blood decreases during the apnea and this finally causes a central nervous alarm reaction. This causes the brain to wake up and to re-establish respiration. But the patient seldomly becomes aware of these apneas and arousal reactions because the wake up is incomplete and below the level of consciousness.

The changes of respiration during sleep caused by obstructive sleep apnea, influence the autonomous nervous system and the control of heart rate during sleep. This has been established by the direct recording of sympathetic nerve activity with microneurography, a complicated technique which only allows undisturbed recording of nerve activity for a couple of minutes. This method demonstrated the influence of sleep apnea on sympathetic tone. The recordings proved, that during the course of each single apnea, the sympathetic tone increases and after the re-established respiration, the sympathetic tone decreases rapidly [8.20]. The normal reduction of sympathetic tone during sleep is overdriven.

With a much more simple and convenient diagnostic tool, as is the Holter recorder, a typical pattern of heart rate fluctuations was found to be associated with sleep apnea. This pattern is called 'cyclical variation of heart rate' [8.21]. Heart decelerates during the 20 to 60 s of apnea and accelerates during the following 10 to 20 s of breathing following each apnea. This repetitive cyclic pattern is used as an indirect indication for the detection of sleep apnea based on long-term recordings of heart rate, snoring sounds, and oxygen saturation [8.22].

8.3.4 Results for the PDFs

First we present an analysis of the time series of the interbeat intervals τ_i and a calculation of the PDFs of these interbeat intervals in the different sleep stages with their corresponding statistical moments $\langle \tau \rangle$ and $\langle \tau^2 \rangle - \langle \tau \rangle^2$. Figure 8.5 shows the PDFs for one healthy subject and one sleep apnea patient. All these distributions have approximately a Gaussian shape. The first and second moments of the PDFs in healthy subjects already allows one to distinguish the recordings made during the daytime and during sleep. The mean heart rate is lower and the variance is smaller during non-REM sleep than during wakefulness, as can also be seen in the histograms in Fig. 8.5. Only during REM sleep, does the heart rate increase and the variability is high. In patients with severe obstructive sleep apnea this relation between daytime states and non-REM sleep is reversed. This is caused by the cyclical variation of heart

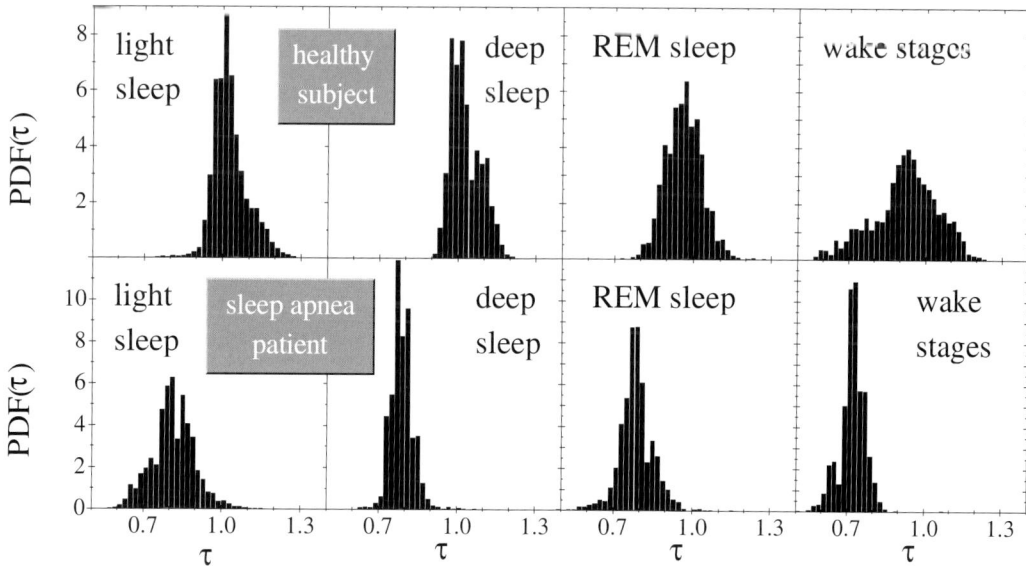

Fig. 8.5. Probability density functions (PDFs) for the heartbeat RR intervals τ_i in the different sleep stages. The first row shows the representative histograms for a one-night recording for a healthy subject, while the second row shows representative data from a sleep apnea patient (with RDI = 33 apneas per hr)

rate associated with the apneas. For the sleep apnea patient in Fig. 8.5 the variability (indicated by the width of the histograms) is largest in light sleep. Unfortunately this effect is not so obvious in patients with moderate or mild sleep apnea and it cannot be found in patients with additional heart diseases that also affect heart rate regulation.

8.3.5 Results from Power Spectrum Analysis and Implications

The power spectrum analysis of heart rate fluctuations provides another way to quantitatively assess the mechanisms of heart rate control [8.23]. Pharmacological blockage of sympathetic and parasympathetic activity and tone is related to specific frequency bands in the power spectrum. Parasympathetic blockage abolishes the mid- ($\approx 0.15\,\mathrm{Hz}$) and high frequency ($\approx 0.4\,\mathrm{Hz}$) peaks in the power spectrum, whereas sympathetic blockage reduces the low frequency ($\approx 0.04\,\mathrm{Hz}$) components of the power spectrum. This finding and other studies have led to recommendations for the interpretation of the spectral analysis of heart rate variability by the European Society of Cardiology and the American Society of Pacing and Electrophysiology [8.24]. The high frequency range is set as 0.15 to 0.4 Hz, the low frequency range is set as 0.04 Hz to 0.15 Hz, the very low frequency range is set as 0.003 to 0.04 Hz and the ultra low frequency range is below 0.003 Hz. It also has been proposed, that these spectral components of heart rate variability provide a measure of the degree of autonomic modulations rather than of the level of autonomic tone.

We also further investigated the textbook description of the progressive decreases of heart rate with sleep stages by using spectral analysis and the frequency bands defined above. It was found that the low frequency components of heart rate variability progressively decrease from wakefulness over sleep stage 1 and sleep stage 2 and reach their lowest values during slow wave sleep. During REM sleep the low frequency components are as high as those during wakefulness. However, the high frequency components of heart rate variability behave just the opposite. They increase from wakefulness over stage 1 and stage 2 to slow-wave sleep. During REM sleep these components are as low as during wakefulness [8.25]. This confirms the theory that high frequency components increase with an increase in parasympathetic activity and low frequency components decrease with a decrease in sympathetic activity. In recent studies heart rate variability was investigated using these frequency bands with the ultimate goal of detecting sleep stage changes based on heart rate changes alone. This ambitious goal has not been reached until now.

Patients suffering from sleep apnea have a specific pattern of heart rate variability associated with their disorder which is called cyclic variation of heart rate [8.21]. Spectral analysis of heart rate variability in patients with sleep apnea was used to identify the periodic changes associated with respiration and with the occurrence of sleep apnea [8.26]. We used nonoverlapping consecutive

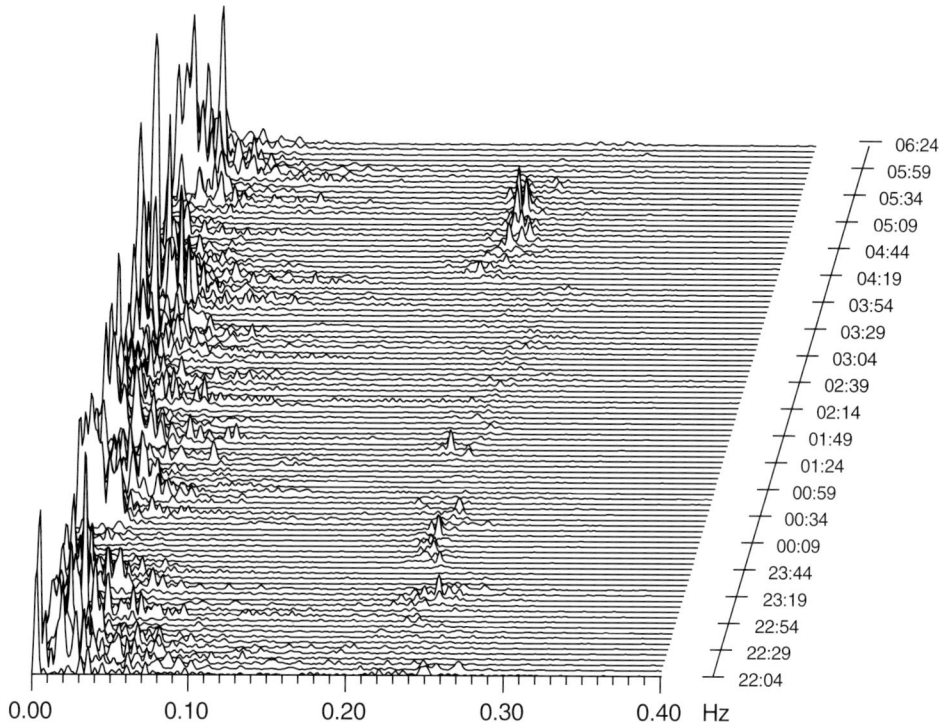

Fig. 8.6. Spectral analysis of heart rate variability in a patient with sleep apnea. The power spectra are shown vs. frequency ν for 20 segments of 5 min duration. The numbers on the right indicate the time of the recordings. The peaks around $\nu \approx 0.25$ Hz show the modulating effect of breathing on the heart rate. The low frequency peaks are caused by apneas. Both effects are modified in intensity and frequency by the sleep stages (not shown)

segments of heart rate recording with a duration of 5 min each. Heart rate was derived from ECG recordings as a part of the polysomnography recordings. The recording of EEG, EOG, EMG, and respiration was used to verify the sleep stages and sleep apnea for each 5 min segment. One peak in the power spectrum, at 0.25 Hz, was identified as the frequency of breathing. A second, more pronounced peak in the power spectrum, was found at 0.015 Hz. This frequency corresponds to the periodicity of apneas which is directly reflected in the cyclical variation of the heart rate. An example for a sleep apnea patient is shown in Fig. 8.6. Thus periods of sleep apnea can be identified using a spectral analysis of heart rate. Having completed this spectral analysis for all segments during the night, then the changes in the frequency of breathing could be observed by the variation of the peak corresponding to the respiratory rhythm. Moreover, the occurrence of sleep apnea could also be followed by the presence or absence of the peak related to the periodicity of sleep apnea. This peak also changed its centre frequency with sleep stages in some patients. We could confirm that the periodicity of apneas did change accordingly in these patients. We can conclude, that spectral analysis of heart rate in patients with sleep apnea allows us to visualize the periodicities in heart rate related to respiration and to sleep apnea. This analysis is less applicable in patients with additional arrhythmic events and with additional heart diseases which cause a reduced heart rate variability.

8.3.6 Results from Dispersion Analysis and Implications

The investigation of the PDFs of the heartbeat intervals (in Fig. 8.5) has shown that they are approximately Gaussian distributed in all sleep and wake stages. Since there are no power-laws involved in the distribution one would not expect any nonsimple scaling behavior. Nevertheless the record of mean interbeat intervals shown in Fig. 8.4 indicates that the mean RR interval is not constant and that the fluctuations might be different in the different sleep stages. In order to investigate the heartbeat fluctuations in the different sleep stages we employed the detrended fluctuation analysis (DFA), since (unknown) trends might be superimposed on the fluctuations (cf. Chaps. 5 and 7). The separation of the sleep stages also eliminated the rather sudden shifts in the heart rate that are due to a change in the sleep stage and not intrinsic to the heartbeat regulation (or the autonomous nervous system) itself.

First we investigated the correlation behavior in the normal volunteers in the different sleep stages [8.6, 8.27]. For many years it had been believed that the fluctuations of the heartbeat intervals are characterized by $1/f$ noise in all sleep and wake stages [8.5, 8.11, 8.28–8.32], and only recently some differences between day and night had been discovered [8.33]. $1/f$ noise corresponds to strong (positive) long-range correlations in the interbeat intervals τ_i. It can equivalently be described by strong long-range anticorrelations in the interbeat interval increments $\delta\tau_i = \tau_i - \tau_{i-1}$. We found that $1/f$ noise is present only for wakefulness and REM sleep which spans 20% of the whole sleep period. In contrast, in non-REM sleep the memory of the heart rates vanishes after a small number of heartbeats corresponding to the typical breathing cycle time, i.e. interbeat intervals separated by more than five heartbeats are actually uncorrelated in deep sleep and light sleep [8.6].

Figure 8.7 shows our results for a representative healthy subject. In light and deep sleep there are only short-range correlations, and we found uncorrelated behavior ($H \approx 0.5$) for larger T, while pronounced long-range correlations ($H \approx 0.85$) were observed in REM sleep [8.6]. For the higher order DFAs the scaling ranges reach up to the longest accessible time scales, since all trends are removed (see e.g. the $T > 500$ region in (a) for the trend removal). All correlations in the data are perfectly removed by shuffling the data, which indicates that that they are real correlations and not due to some special PDF. The strength of the short-range correlations in light and deep sleep can be estimated by a comparison of the results for the shuffled data.

When we applied detrended fluctuation analysis to patients with sleep apnea, we were surprised to find roughly the same correlation behavior for the

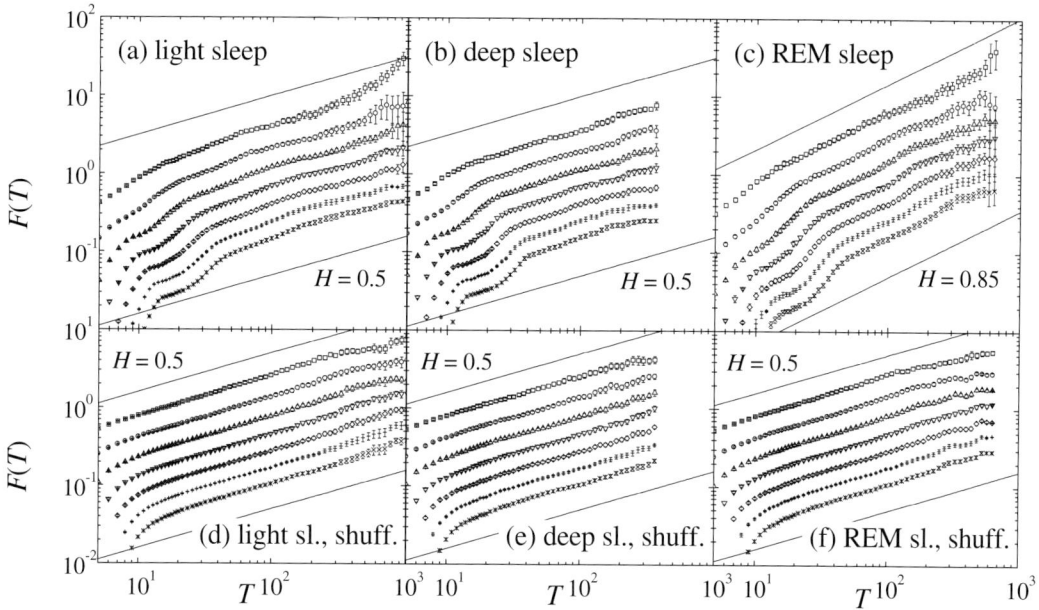

Fig. 8.7. Results of the DFA for a representative healthy subject. The DFA fluctuation functions $F(T)$ for DFA1 (top curves, squares) to DFA7 (bottom curves, crosses) have been plotted vs. the segment duration (time scale) T in log-log plots for (**a**) light sleep, (**b**) deep sleep, and (**c**) REM sleep. Parts (**d**) through (**f**) show the same for surrogate data obtained by shuffling the values of the corresponding records randomly

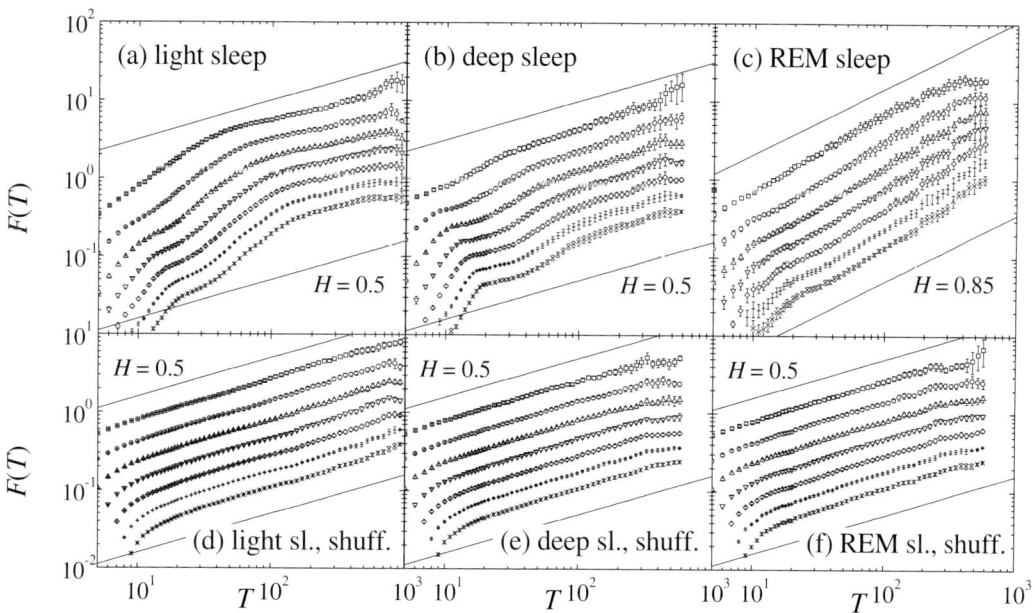

Fig. 8.8. Results of the DFA for a representative sleep apnea patient. For explanation see Fig. 8.7

sleep stages [8.27], see Fig. 8.8 with the results for a representative sleep apnea patient. Therefore we investigated this in more detail. We compared the correlation behavior in our subjects over two consecutive nights, which is called intraindividual variability and between different healthy subjects, which is called interindividual variability. Still we found the same laws of correlation behavior. The only difference is the range of the short-range correlations, which persist slightly longer for the sleep apnea patients than for healthy subjects (see Figs. 8.7 and 8.8). Also we repeated this comparison for patients with obstructive sleep apnea and again we confirmed the correlation behavior. We made a statistical analysis of the exponent H and found that the differences between normals and sleep apnea patients were smaller than the differences between sleep stages. We conclude that sleep apnea does represent a shift of the heart rate regulation but not a general change of the regulation of the autonomous nervous system. It may be possible that we will be able to distinguish those patients which are likely to die from cardiovascular consequences of sleep apnea and those which do not die by using this advanced analysis to assess their heart rate regulation. This has to be investigated in long-term and outcome studies which have access to mortality data in these patients.

The results shown in Figs. 8.7 and 8.8 are representative for the 30 interbeat records from 15 healthy individuals and 47 records from 26 individuals suffering from moderate sleep apnea with less than 22 apneas per hour that we analyzed [8.6]. The results are summarized in Fig. 8.9, where the histograms for the exponents H in the three sleep stages are shown. For light and deep sleep, the histograms are centered around $H \approx 0.5$ and show a large overlap. Both histograms are well separated from the histogram of REM sleep, that is centred around $H \approx 0.85$. We believe that our finding of the significant differences of the heartbeat correlations in the different sleep stages will lead to a better understanding of the different regulatory processes governing heart

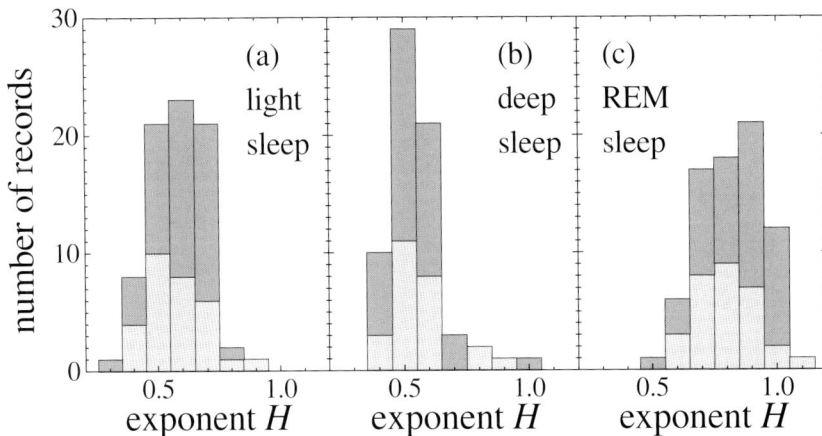

Fig. 8.9. Histograms of the fluctuation exponents H obtained from linear fits to log-log plots of $F(T)$ vs. T in the regime $70 < T < 300$ for (**a**) light sleep, (**b**) deep sleep, (**c**) REM sleep. The fitting range has been chosen to be above the regime of short-range correlations related to breathing and below the T values where the statistical errors become too large due to the finite length of the sleep stages. The data in (**a**) are based on all 30 records from healthy subjects (light gray) and on all 47 records from patients with moderate sleep apnea (dark gray). In (**b**) 10 records have been dropped since they where too short while in (**c**) only one record was too short and has been dropped

rate variability, which is an important diagnostic tool for pathophysiology. We also believe that the results will be useful to develop a sleep phase finder, that is based on the different heart rhythm in the different sleep stages, supplementing the quite tedious evaluation of the sleep stages by the standard electrophysiological procedures.

8.4 Timing Between Arrhythmic Events

8.4.1 Ventricular Tachycardia

Most of the fractal analysis of heart data has focused on analyzing the time between heartbeats. For example, it has been found that there are fractal scalings in the time between heartbeats and that these scalings are different between normal people and those with heart disease [8.29, 8.34]. There has been some previous fractal analysis on the times between events that disrupt the normal pattern of the heartbeat [8.35]. Here we describe how we used fractal methods to analyze the times between two different types of events that disrupt the heart rhythm.

The first type of events that we studied were episodes of rapid heart rate. These rapid heart rates are dangerous because they can lead to disorder in the heart contractions (ventricular fibrillation) so that the heart no longer pumps blood and death follows in a few minutes. A small computer, called a 'cardioverter defibrillator' can be implanted in the chest of patients at risk for these events. If the heart beats too fast this device produces an electrical shock to kick the heart back into a slower, safer rhythm. It can record the times when it was triggered and this can be played back with a radio transceiver when the patient returns to the hospital. We used this capability to determine the times between these events.

8.4.2 Premature Ventricular Contractions

The second type of events that we studied were episodes of additional heart beats. Both normal people and those with heart disease occasionally have additional heartbeats (premature ventricular contractions). Heartbeats can be detected by electrodes taped to the skin on the chest and recorded over a 24 hr period by a small monitor called a Holter recorder. We used a Rozzin Holter PC system to detect these additional heartbeats and determine the times between these events.

8.4.3 Probability Density Functions

As shown in Fig. 8.10, we found that the PDF of the time, t, between events of rapid heart rate is a straight line on a plot of $\log[\text{PDF}(t)]$ vs. $\log(t)$, indicating that it has a fractal, power-law form over a scaling regime from 20 seconds to 10 days. At times greater than 10 days, the data seemed to shift from a power-

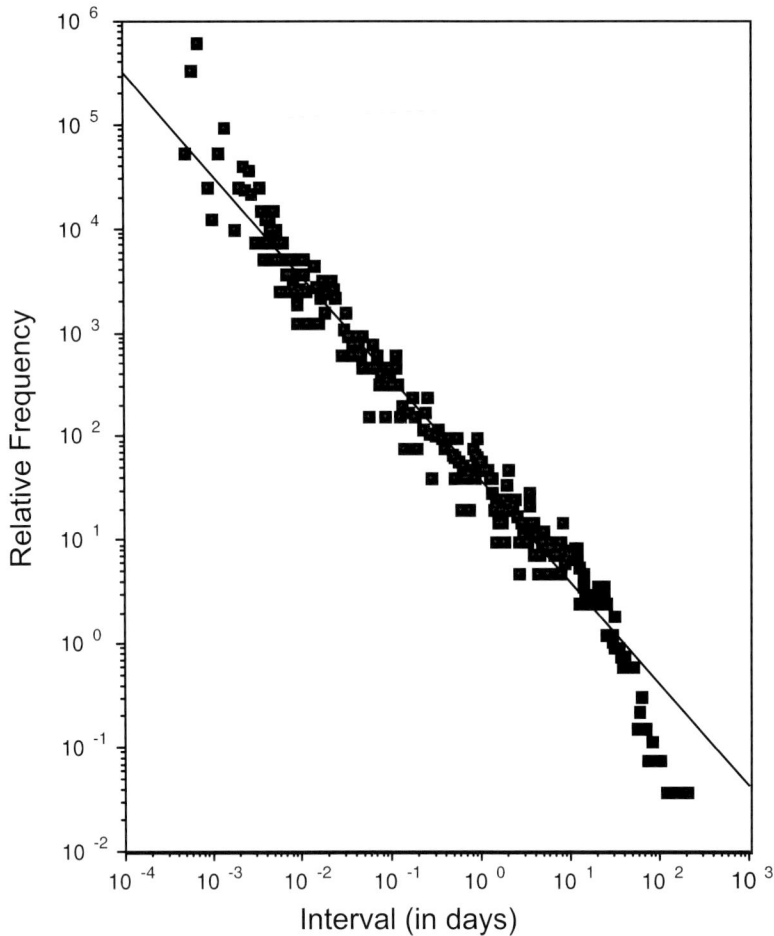

Fig. 8.10. The PDF (probability density function), the relative probability of the time, t, between episodes of rapid heart rate measured in 28 patients by implanted cardioverter defibrillators [8.7]. The straight line on this log-log plot over $20\,\mathrm{s} < t < 10\,\mathrm{days}$ indicates that the timing between these events has a fractal scaling over that time range. This means that no single, 'average' value can be used to characterize the time between these episodes. Such data can however, be meaningfully characterized by using the slope of the line on this plot, which is related to the fractal dimension

law fractal form to one with a characteristic time scale. The fractal scaling regime means that over that large range of times there was no single average value for the time between these events. Most of the time it was a short time between these events. Less often it was longer. Very infrequently it was very long. This means that the rate of these events, that is, the number of events per day will be different if it is measured over one day, or one week, or one month, or one year. However, the meaningful way to characterize this data is by the slope of the plots of $\log[\mathrm{PDF}(t)]$ vs. $\log(t)$. This is related to the fractal dimension. We are presently studying how the fractal dimension depends on different disease conditions and medical therapies.

The short times between many of these events means that the local average rate of events is sometimes quite high. This observation had led to the conclusion that there were 'storms' of these events [8.36]. However, the fractal scaling analysis shows that all the different times between these events are actually all part of the same PDF. This implies that the same physiological mechanisms produce both the short and long times between the events. It means that the high local average rate of events is therefore not due to a sudden physiological change [8.37]. The fractal form of this PDF raises important practical questions about how to evaluate the status of patients. A patient could be over-medicated in response to a high local average rate of events or under-medicated in response to a low local average rate of events. We are currently developing a measure, based on the slope and intercept of these PDF plots, to measure the relative risk of these events to the patient.

The PDF of the time, t, between the events of additional heartbeats was, to first order, also a straight line on a plot of $\log[\text{PDF}(t)]$ vs. $\log(t)$. However, these plots had additional curvature and patient to patient variability that was not found in the PDF plots of the times between events of rapid heart rates [8.7].

8.4.4 Hurst Rescaled Range Analysis

We also used the Hurst rescaled range analysis to determine if there are fractal correlations in time between the episodes of rapid heart rate [8.7]. This analysis requires more data than determining the PDF. There were only enough events recorded from two patients with rapid heart rate to reliable perform this analysis. For both patients we found that $H = 0.6$, namely, that there were persistent correlations between the times between these events at all time scales. However, that result was statistically significantly different from the uncorrelated $H = 0.5$ at the $p = 0.05$ level for only one of those patients. It is a surprising result that the times between these events are so weakly correlated. It is as if the heart almost completely resets itself after each event so that the timing of the previous event has only a little effect on the timing of the next event.

We also used the Hurst rescaled range analysis to determine if there are fractal correlations in the time between additional heartbeats. We found that $0.7 < H < 0.8$ and that these persistent correlations were statistically significant ($p < 0.05$) [8.7]. This implies that the timing of the previous event has a strong effect on the timing of the next event.

8.5 Conclusion

Distributions of experimental data that have fractal rather than Gaussian distributions or that are long-range correlated cannot be meaningfully characterized by their moments, such as the mean and variance. We have shown here how fractal measures, such as the power-law form of the PDF and the dispersion measured as a function of window size, by using the detrended fluctuations method (DFA) and the Hurst rescaled range method (R/S), are valid measures to characterize such fractal data.

In sleep physiology, these fractal measures have shown to be useful in the description of the heartbeat variability. While the PDF of the interbeat intervals has approximately Gaussian shape, long-range time autocorrelations in the records lead to fractal scaling of the dispersion function. Fourier transform has been employed to investigate the effect of obstructive sleep apnea on the heartbeat. Studying the heart rhythm in the different sleep stages (deep, light and REM sleep) that reflect different brain activities, we have shown that contrary to the common belief long-range correlations are present only in the 'dream'-REM stage, when the brain is very active as it is in the waking state. In contrast, in deep sleep the memory of the heart rates vanishes after a small number of beats that is below the order of the breathing cycle time. In light sleep finally, the heart rates seem to become uncorrelated as well, but the crossover occurs slightly above the breathing time. We believe that these findings will lead to a better understanding of the different regulatory processes governing heart rate variability and that they will be useful to develop a sleep phase finder.

In measuring heart arrhythmias, these fractal measures have shown that there is a fractal, power-law distribution in the times between episodes of rapid heart rate (ventricular tachycardia) and between episodes of additional heartbeats (premature ventricular contractions). It has also shown that the times between episodes of rapid heart rate are only weakly correlated, while the times between episodes of additional heartbeats are strongly correlated. Since this data is fractal it cannot be meaningfully characterized by its moments such as the mean and variance. That is, there is no single average rate per day of episodes of rapid heart rate or additional heartbeats. It is hoped that using fractal measures, such as the fractal dimension, to characterize this type of fractal data will lead to better diagnostic indicators and better ways to assess the effectiveness of medical therapies.

Acknowledgements. We would like to thank the National Institutes of Health (EY06234), the German Israeli Foundation, the Minerva Center for Mesoscopics, Fractals and Neural Networks, the Minerva Foundation, and the WE-Heraeus Foundation for financial support. The healthy volunteers were recorded as part of the SIESTA project funded by the European Union grant no. Biomed-2-BMH4-CT97-2040.

References

8.1 L.S. Liebovitch, *Fractals and Chaos Simplified for the Life Sciences* (Oxford University Press, New York, 1998).

8.2 L.S. Liebovitch and D. Scheurle, Complexity **5**, 34 (2000).

8.3 L.S. Liebovitch, A.T. Todorov, M.A. Wood, and K.A. Ellenbogen, in *Handbook of Research Design in Mathematics and Science Education*, edited by A.E. Kelly and R.A. Lesh, (Lawrence Erlbaum, Mahwah, 1999).

8.4 S. Havlin, R.B. Selinger, M. Schwartz, H.E. Stanley, and A. Bunde, Phys. Rev. Lett. **61**, 1438 (1998).

8.5 S.V. Buldyrev, A.L. Goldberger, S. Havlin, C.-K. Peng, and H.E. Stanley, in *Fractals and Science*, edited by A. Bunde and S. Havlin (Springer, Berlin, 1994).

8.6 A. Bunde, S. Havlin, J.W. Kantelhardt, T. Penzel, J.-H. Peter, and K. Voigt, Phys. Rev. Lett. **85**, 3736 (2000).

8.7 L.S. Liebovitch, A.T. Todorov, M. Zochowski, M. Scheurle, L. Colgin, M.A. Wood, K.A. Ellenbogen, J.M. Herre, and R.C. Bernstein, Phys. Rev. E **59**, 3312 (1999).

8.8 H.E. Hurst, R.P. Black, and Y.M. Simaika, *Long-term storage. An Experimental Study* (Constable, London, 1965).

8.9 J. Feder, *Fractals* (Plenum, New York, 1988).

8.10 C.-K. Peng, S.V. Buldyrev, S. Havlin, M. Simons, H.E. Stanley, and A.L. Goldberger, Phys. Rev. E **49**, 1685 (1994).

8.11 C.-K. Peng, S. Havlin, H.E. Stanley, and A.L. Goldberger, Chaos **5**, 82 (1995).

8.12 S.V. Buldyrev, A.L. Goldberger, S. Havlin, R.N. Mantegna, M.E. Matsa, C.-K. Peng, M. Simons, and H.E. Stanley, Phys. Rev. E **51**, 5084 (1995).

8.13 M.S. Taqqu, V. Teverovsky, and W. Willinger, Fractals **3**, 785 (1995).

8.14 F. Lombardi, A. Malliani, M. Pagani, and S. Cerutti, Cardiovasc. Res. **32**, 208 (1996).

8.15 R.L. Verrier, J.E. Muller, and J.A. Hobson, Cardiovasc. Res. **31**, 181 (1996).

8.16 American Academy of Sleep Medicine Task Force, Sleep **22**, 667 (1999).

8.17 A. Rechtschaffen and A. Kales, *A Manual of Standardized Terminology, Techniques, and Scoring System for Sleep Stages of Human Subjects*, (BIS/BRI, University of California, Los Angeles, 1968).

8.18 R.P. Vertes and K. Eastman, Behavior. Brain Sci. **23**, 867 (2000).

8.19 T. Young, M. Palta, J. Dempsey, J. Skatrud, S. Weber, and S. Badr, New Engl. J. Med. **328**, 1230 (1993).

8.20 V.K. Somers, M.E. Dyken, M.P. Clary, and F.M. Abboud, J. Clin. Invest. **96**, 1897 (1995).

8.21 C. Guilleminault, S.J. Connolly, R. Winkle, K. Melvin, and A. Tilkian, Lancet **I**, 126 (1984).

8.22 T. Penzel, G. Amend, K. Meinzer, J.H. Peter, and P. von Wichert, Sleep **13**, 175 (1990).

8.23 S. Akselrod, D. Gordon, F.A. Ubel, D.C. Shannon, A.C. Barger, and R.J. Cohen, Science **213**, 220 (1981),

8.24 Task Force of the European Society of Cardiology and the North American Society of Pacing and Electrophysiology, Circulation **93**, 1043 (1996).

8.25 M.H. Bonnet and D.L. Arand, Electrocncoph. Clinical Neurophysiol. **102**, 390 (1997).

8.26 T. Penzel, J.H. Peter, and P. von Wichert, in *Colloque INSERM*, edited by C. Gaultier, P. Escourrou, and L. Curzi-Dascalova (John Libbey Eurotext, London, 1991), p. 79.

8.27 T. Penzel, A. Bunde, J. Heitmann, J.W. Kantelhardt, J.H. Peter, and K. Voigt, IEEE Comp. Cardiol. **26**, 249 (1999).

8.28 C.-K. Peng, J. Mietus, J.M. Hausdorff, S. Havlin, H.E. Stanley, and A.L. Goldberger, Phys. Rev. Lett. **70**, 1343 (1993).

8.29 S. Thurner, M.C. Feurstein, and M.C. Teich, Phys. Rev. Lett. **80**, 1544 (1998).

8.30 L.A. Amaral, A.L. Goldberger, P.Ch. Ivanov, and H.E. Stanley, Phys. Rev. Lett. **81**, 2388 (1998).

8.31 C.-K. Peng, J. Mietus, J.M. Hausdorff, S. Havlin, H.E. Stanley, and A.L. Goldberger, Physica A **249**, 491 (1998).

8.32 P.C. Ivanov, M.G. Rosenblum, L.A.N. Amaral, Z. Struzik, S. Havlin, A.L. Goldberger, and H.E. Stanley, Nature **399**, 461 (1999).

8.33 P.C. Ivanov, A. Bunde, L.A.N. Amaral, S. Havlin, J. Fritsch-Yelle, R.M. Baevsky, H.E. Stanley, and A.L. Goldberger, Europhys. Lett. **48**, 594 (1999).

8.34 A. Goldberger, C.-K, Peng, J. Hausdorff, J. Mietus, S. Havlin, and H.E. Stanley, in *Fractal Geometry in Biological Systems*, edited by P.M. Iannaccone and M. Khokha (CRC, Boca Raton, 1995).

8.35 K.M. Stein, L.A. Karagounis, J.L. Anderson, P. Kligfield, and B.B. Lerman, Circulation **91**, 722 (1995).

8.36 S.C. Credner, T. Klingenheben, O. Mauss, C. Sticherling, and S.H. Hohnloser, J. Am. Coll. Cardiol. **32**, 1909 (1998).

8.37 M.A. Wood, K.A. Ellenbogen, and L.S. Liebovitch, J. Am. Coll. Cardiol. **34**, 950 (1999).

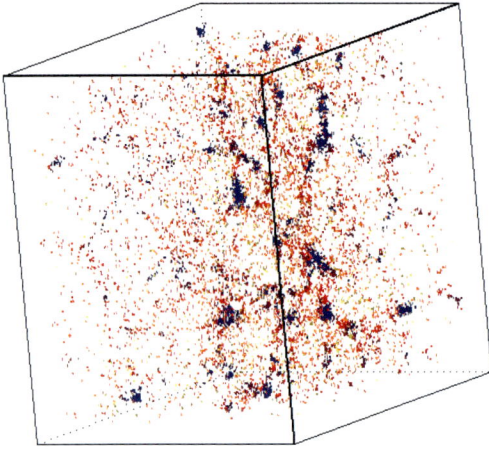

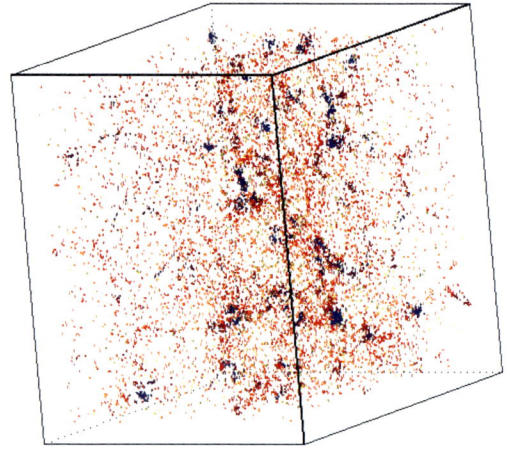

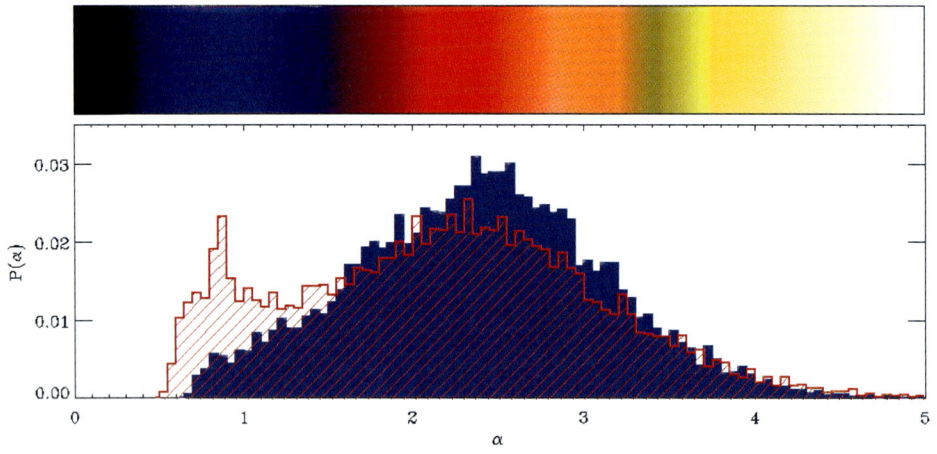

9. Local Scaling Properties
for Diagnostic Purposes

Wolfram Bunk, Ferdinand Jamitzky, René Pompl,
Christoph Räth, and Gregor E. Morfill

Two major sources of information in medicine and biology are time dependent recordings of biosignals and different forms of imaging. Whilst technology has made rapid progress, quantitative analysis techniques for supporting diagnostics and therapy control have lagged behind. A major reason lies in the requirement to characterize complex structures - either temporal patterns or image structures - reliably, reproducibly and with sufficient precision that the analysis technique may qualify as an accepted tool for the experts to use. New techniques, based on fundamental nonlinear approaches, which have originally been developed for the analysis of astronomical images and time series were transferred into the analysis of medical and biological systems. In this chapter, first the theoretical approach is outlined with some easily understood examples. Then selected applications in the areas of atomic force imaging, skin cancer detection, tumor diagnostics, and heart rate variability are discussed in greater detail.

9.1 Introduction

In almost all areas of everyday life we are faced with continuously growing levels of information: huge data masses that need to be understood in order to make correct decisions or to improve the modeling of the underlying dynamics. One reason for this trend is that some systems, especially man-made ones in finance, economics, and technology, are becoming more and more complex,

---

◄ **Fig. 9.0.** Large scale structure of the universe: the cubes show two simulated cold dark matter distributions (OCDM and τCDM) where the color coding corresponds to the respective scaling index of the points. The point distributions are not random but show many different structural features. Based on the feature content of point distributions, which is reflected in different $P(\alpha)$ spectra (for a definition see text), a discrimination between different cosmological models is possible

for example through globalization or technological advances. Frequently, the increasing flood of information is a result of the enhanced capabilities of measuring instruments: new sensor or detector technologies, for instance, provide multiparametric monitoring of technical or biological systems at high sampling rate and with high spectral or spatial resolution. At the same time experience has shown that very often the accompanying analysis techniques lag behind the technical progress of the hardware. This applies in particular to the field of medical diagnostics, where an added complication lies in the stringent requirements (imposed e.g. by the controlling authorities) on the software reliability and verification procedures.

The introduction of powerful new analysis techniques for the quantitative characterization of images and time series in a variety of medical and biological applications is, of course, necessary if the available information is to be exploited properly. In this chapter some recent developments will be described.

9.2 Reductionism

In order to grasp the 'essentials' of a given system under study, i.e. those aspects that have the greatest influence, it is necessary to reduce the available data to a few 'characteristics' of the system. These characteristics can be 'handled' by the investigator and, should the need arise, it becomes an acceptable risk to base decisions on these 'reduced measures'.

In reducing many different measurements obtained from a system to a few 'characteristics' we must not cut across the inherent interdependences. Otherwise some essential aspects of the system cannot be monitored, let alone be understood. This is easily demonstrated with the well known 'double pendulum'. In the 'linear' regime, where the amplitudes of oscillation are very small, the system can be described approximately by the superposition of two independently moving pendulums. However, as soon as the amplitudes are 'finite' - which is practically always the case - this simple description fails. Thus, reducing a coupled to two independent pendulums clearly leaves out some of the essential physics; a fact that becomes very clear, of course, from simple visual inspection of the motion of the coupled system.

It is easy to see, therefore, that with increasing system complexity a much more sophisticated approach is required to reduce the system to its essential components and to derive the corresponding characteristics from measurements of the system properties. This situation becomes progressively worse if certain parameters are not measurable or if we are dealing with a (partially) unknown system and wish to infer its components, including their interactions, from the available measurements. In such situations - which are the rule, rather than the exception - a new data analysis strategy has to be developed that goes

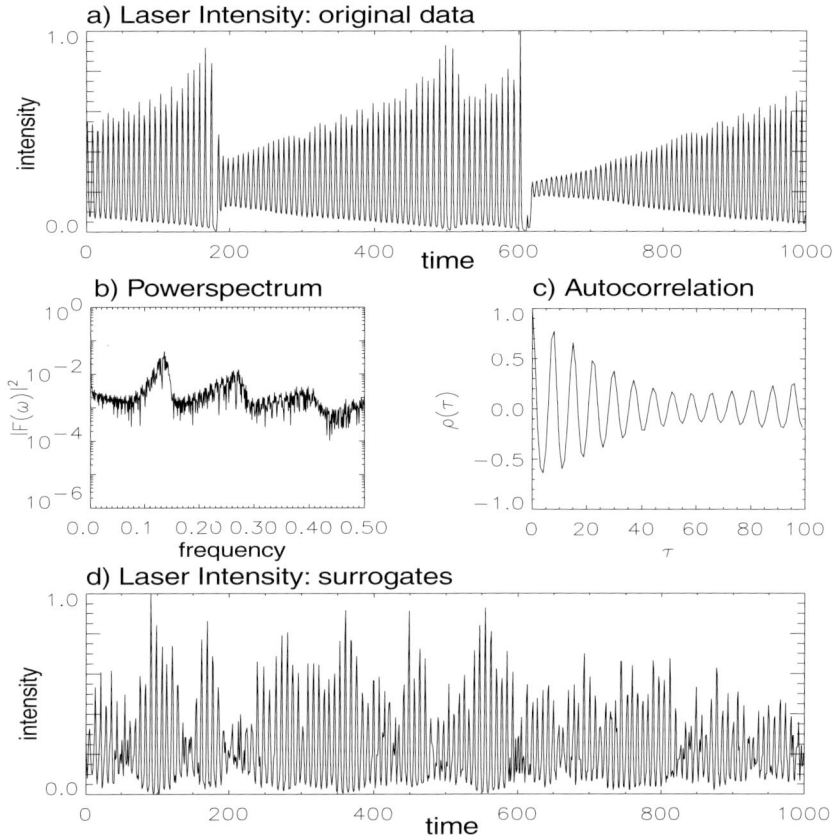

Fig. 9.1. Example of limitations of linear analysis techniques. The top panel (**a**) shows the time dependent variation of a laser intensity signal which was analyzed by Fourier and autocorrelation techniques - shown in the two middle panels (**b**), (**c**). The original data are then modified in such a way that the linear properties (e.g. power spectrum and autocorrelation function) remain but the phase information was randomized. This surrogate dataset is shown in the lowest panel (**d**). This example shows that two very different datasets may contain the same linear information content and illustrates the limitations of linear techniques

far beyond the classical techniques. In short, one has to characterize complex, nonlinear behavior.

One of the most widely and very successfully used classical reduction techniques is Fourier analysis. In such an analysis, a series of measurements made at regular time intervals is broken down into its different frequency components. Often, the dominant spectral features are then used as indicators or 'measures' of the system. As with all reduction techniques, Fourier analysis clearly involves removing some of the information; a fact that may be fatal as shown in Fig. 9.1. The problem in many cases lies in the fact that Fourier analysis as well as several other established techniques such as power-spectral analysis and autocorrelation analyses are 'linear' techniques. As a result, they cannot describe the nonlinear aspects of the measurements obtained from complex systems.

In order to do just that, new techniques and methods have been developed in recent years, taking nonlinearities and nonlinear correlations between different measurements into account.

These include neural networks, measures based on information theory like entropies or 'transinformation' [9.1,9.2] (for some applications based on entropy and transinformation concepts, see Chap. 1), algorithms characterizing the flow in state space like nearest-neighbor algorithms [9.3, 9.4] or the Lyapunov exponent [9.5], fractal dimensions [9.6, 9.7] and local scaling properties [9.8, 9.9] (cf. also Chaps. 2 and 7). In principle, they all can provide 'measures' for certain aspects of the nonlinearities contained in measurements, thus allowing additional insights into the system under study.

9.3 Scaling Index Method

At the Max-Planck-Institut für extraterrestrische Physik in Garching, near Munich, we have developed different nonlinear analysis methods - with one of the aims being the quantitative study of the irregular and highly structured distribution of the galaxies in the universe, in order to learn more about the physics at its earliest stages, to differentiate quantitatively between various theoretical models, and to compare these with measurements of large-scale galactic surveys (see Fig. 9.0). More recently, however, we have used the same techniques in a variety of industrial and medical applications, in particular for the early detection of skin cancer and better risk assessment of heart attacks.

The most powerful techniques which we have developed are based on the so-called scaling index method (SIM), which in turn is based on the following considerations: the 'state' of a given system can be thought of as a single point localized in an appropriate 'state space'. This state space is made up of the different 'state variables'. For example, each pixel in a static gray-scale image has three state variables - the x, y coordinates and gray-scale - and hence occupies a point in a three-dimensional state space. A pixel in a color image has five state variables: x, y and the three color components red, green, and blue. A tomographic image of a solid body has four state variables: x, y, z, and gray-scale.

A dynamic system may have many such variables that change with time. In this case, monitoring the trajectory of a system in state space can be an important tool for, say, risk analysis. A simple example of a dynamic system is the ideal pendulum without energy dissipation, which is described in phase space (coordinates x and dx/dt) by a stable ellipsoidal trajectory. The phase portrait of a damped pendulum is similar to a spiral showing a central focus, whereas the aforementioned coupled pendulum shows a variety of complex structures depending on the parameters of the system. The latter cannot be

represented adequately in a two-dimensional phase plane, and requires a more sophisticated approach for characterization.

Different regions in state space are generally not occupied uniformly by complex systems. Some regions will be favored, while others are less favored. In an image, for example, a local region with the same color will lead to a sheet-like structure in state space, while a line will remain a line. Hence the complexity of a system may be measured by the structures found in its state space.

The question then is how to quantify 'structure'? The answer comes from 'fractals' - geometric shapes that look similar at different scales. Many structures in nature can be quantified by their fractal geometry, including coastlines, mountains, and clouds.

The scaling index method makes use of this property to characterize the system and is a realization of computing the distribution of pointwise dimensions. It is briefly explained as follows. In order to calculate the local 'fractal dimension' or the 'scaling index', $\alpha(\boldsymbol{x})$, a sphere of radius r around the position $\boldsymbol{x}$ of a given system state in the state space is considered. The number of system states located inside this sphere, $N(\boldsymbol{x}, r)$, is counted. This is repeated for a range of sphere radii, all centered at $\boldsymbol{x}$, resulting in:

$$N(\boldsymbol{x}, r) = \Sigma_j \Theta(r - ||\boldsymbol{x}_j - \boldsymbol{x}||). \tag{9.1}$$

In this equation Θ is the Heaviside function with $\Theta(s) = 1$ if $s \geq 0$ and 0 if $s < 0$. The distribution function $N(\boldsymbol{x}, r)$ is computed within a specified range $r \in [r_1, r_2]$, the so-called scaling range. The value of the scaling index, $\alpha(\boldsymbol{x})$, can then be extracted from a log-log diagram of $N(\boldsymbol{x}, r)$ vs. r using the scaling assumption:

$$N(\boldsymbol{x}, r) \propto r^{\alpha(\boldsymbol{x})}.$$

We assume that this scaling law holds in the range of r chosen. (This is usually the case, but not always. From a pragmatic point of view this is not a problem in most applications.) The scaling index $\alpha_i \equiv \alpha(\boldsymbol{x}_i)$ is then determined by a power-law fit to $N(\boldsymbol{x}, r)$ for each point x_i. The first order approximation is given by the difference ratio:

$$\alpha_i = \frac{\log N(\boldsymbol{x}_i, r_2) - \log N(\boldsymbol{x}_i, r_1)}{\log r_2 - \log r_1}. \tag{9.2}$$

r_1 and r_2 are specified by the lower and upper limits of the scaling range. This scaling index α_i describes the structure of the point distribution in the neighborhood of the point i at location $\boldsymbol{x}_i$ and is a property of that point and of the location $\boldsymbol{x}_i$ in phase space, as well. This procedure is repeated for every system state located in state space. The frequency distribution (i.e. the

histogram) of the α-values is called the $N(\alpha)$ spectrum and the respective probability density $P(\alpha)$, where $\int_0^\infty P(\alpha)d\alpha = 1$.

We can determine the type of structure from the computed value of α_i. For example, a point-like structure will have $\alpha \approx 0$; a line-like structure has $\alpha \approx 1$; and a flat sheet has $\alpha \approx 2$. Curvy lines or sheets will have $1 \leq \alpha \leq 2$ and $2 \leq \alpha \leq 3$ respectively.

One of the most interesting properties is that for point distributions filling the n-dimensional space homogeneously, α-values close to the embedding dimension $n = d_E$ are assigned. The frequency distribution $N(\alpha)$ for a given point distribution is therefore in some sense a statistical measure of the distribution of elementary structural components. In particular, the indices located in a given range $n_1 < \alpha < n_2$ identify all system states located in a region of a certain structural order and can easily be extracted from the entire distribution. In addition, if the scaling law holds as r approaches zero, then we can regard the scaling index $\alpha(\boldsymbol{x})$ as a local property of the system at position $\boldsymbol{x}$. We can thus define a nonlinear property for each observed system state.

More important still, it turns out (after some algebra) that $\alpha(\boldsymbol{x})$ is directly proportional to the specific information gain I_i ($I_i = -c \log P_i$, where P_i is the occurrence probability for the system state i and c is a constant.), which we obtain with each new measurement i of a system state. For instance, if the system is static, all measured states will occur at the same location in state space. A new measurement will then not provide new information about the system, unless a new state is observed, i.e. at a new location in state space. Thus the scaling index is a very fundamental and independent property of each observed state of the system, which retains information about the nonlinearity of the system and complexity. This makes it a prime candidate for the quantitative analysis of complex systems. For specific applications there are a number of additional considerations, including the finite resolution, the choice of scaling lengths, and different sampling rates for different measurements. After these remarks on the connection between α, structural properties, and information gain, it is obvious that the choice of the scaling region $[r_1, r_2]$ is a crucial element in employing this technique in practice. From a purely theoretical point of view, the specification of the scaling region is irrelevant for the results, as long as the function $N(\boldsymbol{x}_i, r)$ is scale free (i.e. a true power-law).

9.4 Applications

The basic reason why time series and images - which are intrinsically of very different nature - can be analyzed in a similar way and with similar techniques, is that the underlying rules controlling the complex system manifest themselves

in geometric structures, which can be made visible by an appropriate data representation (e.g. in state space) and which can then be quantified by the definition of a metric.

In the following we illustrate this with a few examples, where we focus on the use of the scaling index method. We demonstrate some basic tasks of image analysis, which can be performed considering the local scaling properties in point distributions, and give an example of time series analysis by means of this technique.

The first example deals with the decomposition of structural components in atomic force microscopy (AFM) images of biological specimens. Scaling indices are used for image enhancement and noise reduction. Special emphasis is placed on the detection of line-like and point-like structures. The line-like structures correspond to single strands of DNA, so-called plasmid rings, and the point-like structures represent small nodes in the DNA and coiled regions, where two strands are twisted around each other.

The second example concerns the early detection of skin cancer. A number of measures are used for a computer assisted recognition of malignant melanomas; hence this is an example of digital image analysis for classificational purposes. Entropy measures, fractal dimensions, and local scaling laws for point distributions are used to define an unambiguous, objective and quantitative 'score' or 'measure' of the widely used dermatoscopic ABCD-rule (for a definition, see later). This empirically developed rule is employed by the dermatologists as a guide or procedure for the early recognition of malignant melanomas. The main objective of our image analysis is a kind of structural decomposition of the lesion in order to quantify the structural content in terms of the above mentioned 'rule'.

Related to this problem is a third example: the digital segmentation of tumorous tissues in computer tomographic (CT) images. Scaling indices in combination with other image quantities may allow a fast and accurate tumour volumetry for e.g. quantitative therapy control. Obtaining the diagnostically very important tumour volume 'by hand', may take over an hour of an expert's time, hence a reliable, fast and semi-automatic analysis system has a high clinical value in tumour therapy.

In the section dealing with time series we present some examples for analyzing beat-to-beat sequences extracted from electrocardiograms (ECG). One can easily show that specific types of complex arrhythmias and pathological heart rhythms can be found at distinct positions in a simply constructed artificial phase space. The $P(\alpha)$ spectrum is helpful for the characterization of cardiac dynamics and for a patient's risk assessment with respect to sudden cardiac death.

9.4.1 Pattern Recognition in AFM Images

Atomic force microscopy (AFM, [9.10]) has now become a standard tool in surface analysis and it has found wide application in biological research. Despite the widespread use of this microscopic technique, usually a very limited range of image analysis algorithms is used. The most common procedures, which are realized in nearly all commercially available instruments, involve the removal of nonlinearities induced by the scanning piezo from the topographic AFM images, the determination of surface roughness (e.g. [9.11]), and averaging together with symmetrization in combination with power spectral analysis for crystallographic applications (e.g. [9.12, 9.13]). There are also some attempts to include the nonlinear character of the so called tip sample 'convolution' in order to perform a 'blind' reconstruction of the specimen surface [9.14].

For many applications an automatic identification and localization of surface features is desirable. For example, a practical application to biochemical research could be the statistics of binding events of a DNA-binding protein on double stranded DNA. Image decomposition into features of different dimensionalities like spots (zero-dimensional, e.g. crossing-points of DNA strands), line like structures (one-dimensional, e.g. double stranded DNA), and areas (two-dimensional, e.g. proteins) offers a direct approach. However, there is often a significant amount of noise in the images, which can obscure minute surface features. Thus, automatic image processing must also be able to reduce noise in the images reliably without removing small features. In this section we discuss the application of the scaling index method for the analysis of topographic AFM images.

Decomposition of an Image: Since the scaling index is a measure of the dimensionality of the structure it can be used as a tool for the segmentation of an AFM image. As mentioned above, scaling indices order the basic structural elements of an image from low dimensionality (dots) to high dimensionality (random background noise). It is therefore obvious that the method can be used as a structural filter or as a tool for image decomposition.

One can choose, for example, the range from 0.5 to 1.5 in order to discriminate for line like structures, as the scaling index of a straight line is ideally equal to one. By using the scaling index in this way as a 'spectral filter' for different types of structures that occur in the image, we can produce different 'component images' that emphasize and select certain structures. Color coding and superposing such component images then yield a reproducible, objective decomposition of the original image, which allows the identification of preselected structures and their quantification. For the component images generated in such a way one can again compute the scaling indices. This iterative process can lead to an even better decomposition. One can also use a fuzzy decomposition where a pixel can belong to more than one component image.

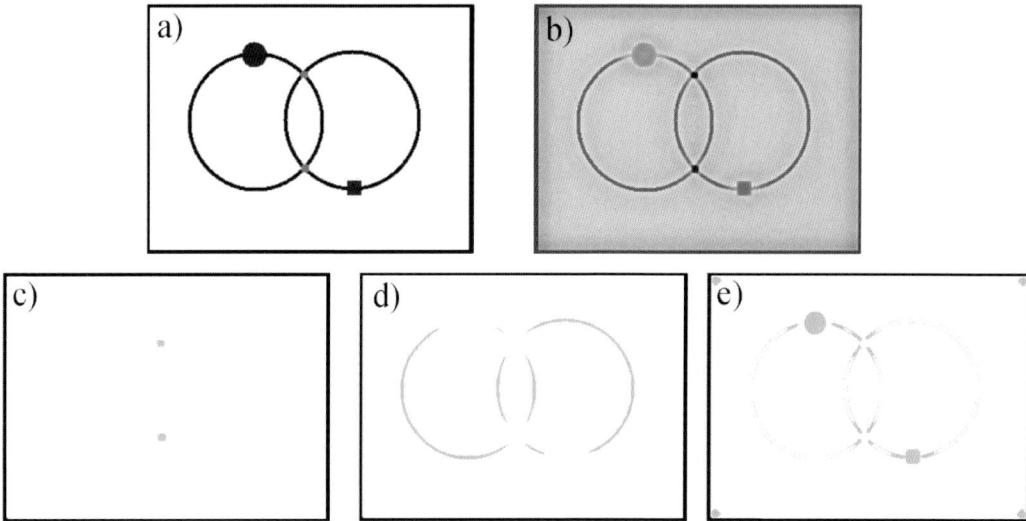

Fig. 9.2. Synthetic test image for the decomposition algorithm representing two intersecting plasmid rings with specifically bound proteins (filled rectangle and circle). (**a**) Artificial test image, (**b**) scaling index (scaling range 1 to 3 pixels), (**c**) image with scaling index $\alpha < 1.0$, (**d**) scaling index $1.0 < \alpha < 1.2$, and (**e**) scaling index $1.2 < \alpha < 1.6$. For larger scaling indices the background is recognized

In order to demonstrate the working of the algorithm, a synthetic test image was analyzed (see Fig. 9.2a, image size: 330×244 pixels, 8 bit gray-scale). This image consists of two circles which represent the stretched DNA in the practical application. At the points where both circles intersect, an additional brighter point was added. This models the crossing of two DNA double strands, which, due to geometric effects, results in a topographically higher (brighter) and broader structure. The square and the filled circle could be taken to represent e.g. plasmid-ring proteins with different shapes. Based on the scaling index (Fig. 9.2b), the image is decomposed into constituents of different dimensionality as discussed above. Both points, which originate from the intersection of the (fictive) DNA sequences are found in the image at scaling indices $\alpha < 1.0$ (Fig. 9.2c). Thread-like structures are stored in the image for scaling indices in the range $1.0 < \alpha < 1.2$ (Fig. 9.2d). Due to the decomposition process, both circles are interrupted in the regions where structures lead to a different scaling index. For a scaling index $1.2 < \alpha < 1.6$ area-like features are found (Fig. 9.3e). The regions in the image corners are distorted due to boundary effects. Additionally to the simulated 'proteins', also parts of the 'DNA-strand' are recognized by the algorithm. Thus, for practical applications, a subsequent image processing step has to be performed, which imposes a minimum-threshold to the size of recognized structures in order to recognize and reproduce the larger structures in the AFM.

Decomposition of an AFM Image of Plasmids: AFM images have been recorded with a commercial atomic force microscope (for the experimental

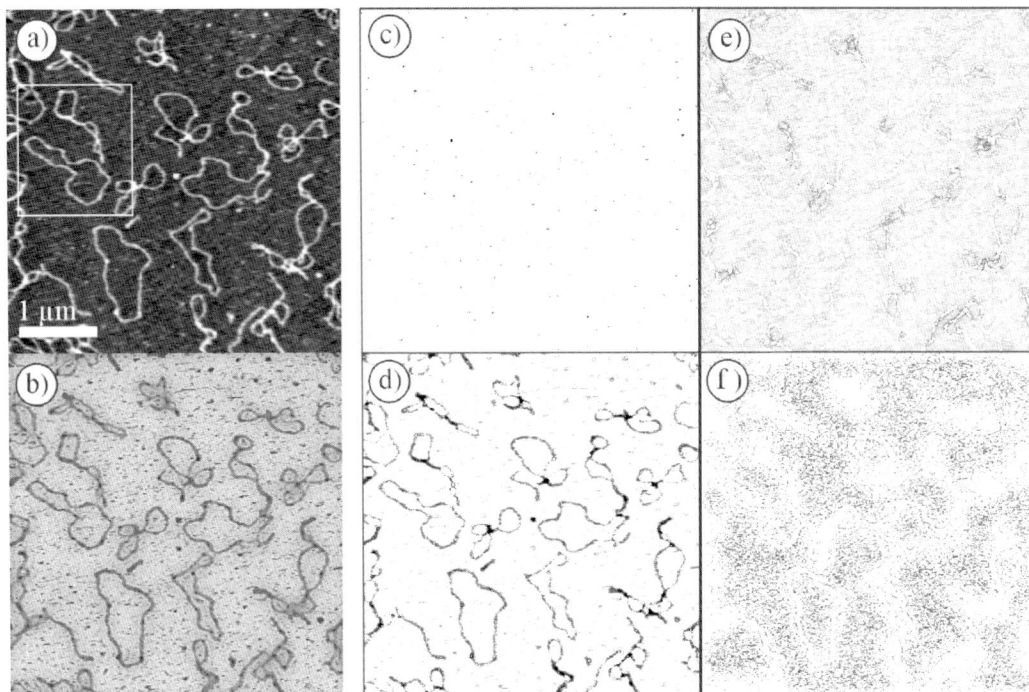

Fig. 9.3. (**a**) Topographic AFM image of plasmids PUC 8 on mica. The rectangle refers to the software zoom in Fig. 9.4. (**b**) Scaling index (scaling range two to five pixels) and decomposition: (**c**) $\alpha < 0.5$, i.e. point-like, (**d**) $0.5 < \alpha < 1.5$, i.e. line-like, (**e**) $1.5 < \alpha < 2.5$, i.e. area-like, (**f**) $2.5 < \alpha$, i.e. noise

setup, see [9.15]). The atomic force microscopic images were first processed with the image analysis software supplied with the microscope (SPMLab 3.06.06) in order to remove scan-induced artifacts (third-order plane fit, adjustment of color-scale for optimal 8 bit export). After export to a 8-bit gray-scale bitmap, further processing was done by the scaling index algorithm.

In order to test the decomposition algorithm in practice, AFM images of purified DNA-plasmids were investigated, without additional proteins. In Fig. 9.3a, the topographic image is displayed (image size: 500×500 pixels, 8 bit gray-scale). There are well spread rings as well as coiled DNA. The corresponding scaling index image, shown in Fig. 9.3b, was used for decomposition. For small scaling indices ($\alpha < 0.5$) point-like structures are detected (Fig. 9.3c). These points correspond to intersections of the plasmid rings, as well as to coiled areas and noise peaks on the substrate. The DNA-strands are found at scaling indices $0.5 < \alpha < 1.5$ (Fig. 9.3d). For larger scaling indices ($1.5 < \alpha < 2.5$) signals for area-like structures could be detected, for the substrate and for areas of coiled or crossing DNA (Fig. 9.3e). This effect is similar to the result obtained for the synthetic image (cf. Fig. 9.2). The noise in the image can be separated for $2.5 < \alpha$ (Fig. 9.3f).

Figure 9.4 shows a software zoom of the subimage indicated in Fig. 9.3a. This image was composed by overlaying the scaling index images Fig. 9.3c (in

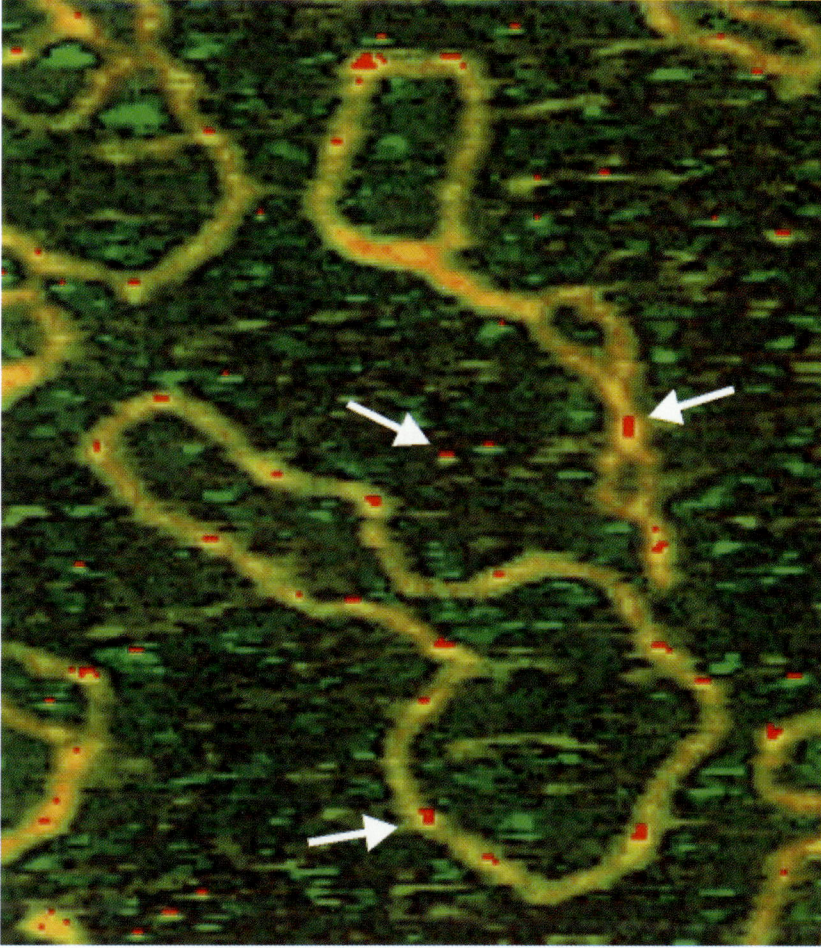

Fig. 9.4. Composite color coded image, based on the topographic images shown before. Figure 9.3a was color coded in the green channel of the RGB image. The scaling index images (Fig. 9.3c, red and Fig. 9.3d, yellow) were overlaid. The arrows point to three different features which were identified as points by the algorithm

red) and Fig. 9.3d (in yellow) with the topographic image (Fig. 9.3a, green). Area-like structures and noise were not included into the composite image, because for this purpose they do not carry information. Using this method, the different structures become prominent. The arrows point to some one-dimensional structures (points). Three different features are illustrated: shot noise, small nodes in the DNA (due to local coiling or ill defined adhesion to the substrate), and coiled regions, where two strands are twisted around each other. The DNA sequences can be easily separated from the background.

In a different study [9.16] it is shown that the method can be used to extract structural information of chromosomal data thus producing a 'fingerprint' of the chromosomes which helps in the interpretation and classification of the measured data.

9.4.2 Diagnosing the Malignant Melanoma

The malignant melanoma is one of the most dangerous forms of cancer in humans. Moreover, many studies have proven that the incidence has more than doubled within the last ten years in Western countries [9.25, 9.26] and the mortality is increasing almost just as much. A malignant melanoma occurs as a new lesion or may develop from an initially benign nevus. It is known that minor surgery is almost always successful if the malignancy is detected at a sufficiently early stage. On the other hand, if the discovery is too late, metastases may have formed (the malignant melanoma is particularly 'aggressive' in this respect), and if so then patient's prognosis is poor and the course of the disease is fatal. An early detection of malignant melanoma is therefore crucial since best prognosis requires that both the lesion and the depth of the tumour are very small. If the tumour thickness exceeds only a few millimeters, the chance of a complete cure decreases dramatically.

The diagnostic accuracy in the assessment of the dignity of a lesion by the naked eye hardly exceeds the 70% level, even for a dermatologist [9.21]. For this

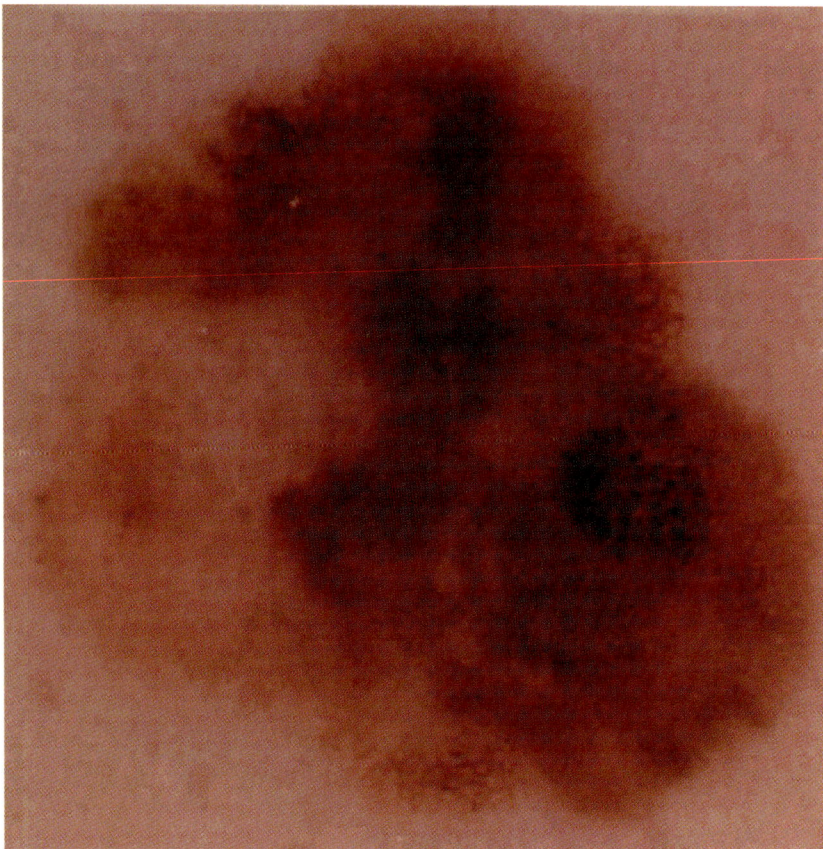

Fig. 9.5. Dermatoscopic image of a malignant melanoma at ten times magnification

reasons dermatologists investigate pigment changes using a magnifying glass of about ten times magnification, the so-called dermatoscope. This instrument allows an expert to identify substructures with a reasonable resolution, facilitates a better view of the border region between the lesion and the surrounding healthy skin, and helps to distinguish diagnostically relevant colors. Figure 9.5 shows an image of a malignant melanoma obtained with a 3-chip CCD camera at about 10 times magnification. The real size of the image area is about 11.8 mm × 11.8 mm [9.23].

In the late 1980s, some dermatologists developed a system, the so-called dermatoscopic ABCD-rule [9.18], which provides a recipe for characterizing pigmented skin lesions. The parameters are the asymmetry of the skin lesion (A), the border characteristics (B), the color variation across it (C), and the differential structure inside it (D). This systematic approach is very useful for clinicians, dermatologists, and medical practitioners in the early detection of skin cancer using dermatoscopy [9.20]. Without question, the diagnostic accuracy could be significantly improved by using the dermatoscope and following the ABCD-rule. Experts in dermatoscopy can now detect up to about 90% of early skin cancer signatures [9.19, 9.22].

The system of rules is well established, in particular in Europe, but the success depends strongly on the experience of the physician. Even a dermatologist usually does not see many malignant melanomas during the daily routine to build up a sufficient level of experience. As a result, to be on the safe side medical doctors tend to excise many more lesions than necessary. From histopathological results one knows that only about one out of 20 excised moles is actually positive (i.e. malignant). Ignoring economical arguments, it is evident that even this minor surgery, i.e. the excision of the supposed tumour with a collar of surrounding healthy tissue, is not without risk for the patient.

Therefore, accurate diagnostic support should prove invaluable in screening tests or routine inspections, provided it does not take up too much of the dermatologist's time. Moreover, such a need is likely to grow in the future as the trend towards 'tele-medicine' increases. In these cases, digital images could be evaluated on site while data from borderline candidates could be sent to an expert dermatology center for further scrutiny and advice. This centre could double up as a data archive that may be used for epidemiological studies and to develop new analysis techniques.

In cooperation with expert dermatologists we have developed quantitative measures for some of the most predictive features for malignant melanoma, according to the classification algorithm of the ABCD-rule. For illustration purposes, we focus on the quantification of complex patterns [9.17] in the images using our scaling index method, namely the border characteristics and the content of differential structures. For these color images, the 'state variables' are x, y, r, g, b as mentioned earlier.

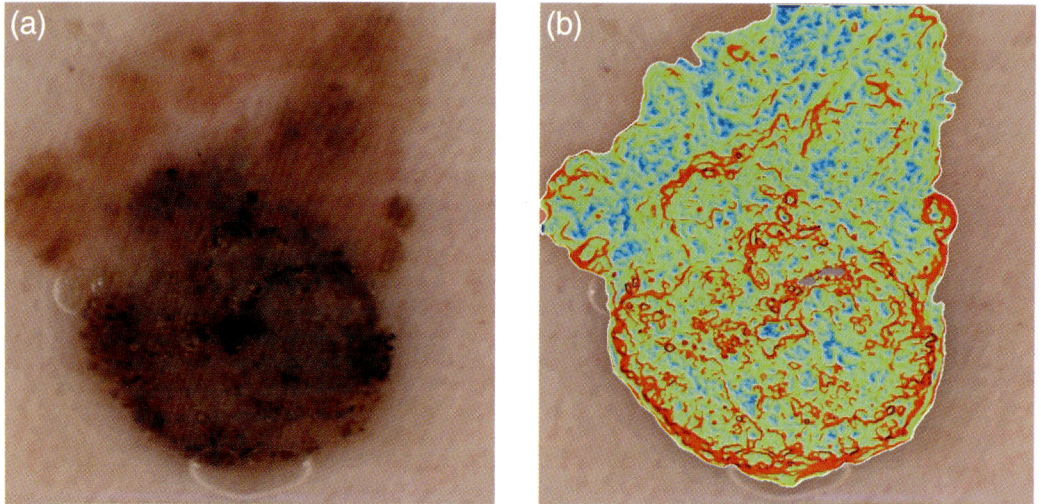

Fig. 9.6. Quantification of differential structures of a malignant melanoma (**a**) using the scaling index method. The color coded image (**b**) indicates the scaling indices of the lesion pixels. In addition to the quantitative characterization of the structural content, this technique provides a visualization of the analysis result

Border Characteristics: In order to characterize the border between the lesion and the healthy skin, segmentation of the lesion and removal of artefacts in it (e.g. hairs) is necessary. The basic idea for the quantitative description of an irregular border is related to the well-known estimation of the fractal dimension of a boundary line, in our case not only the contour line is used but also a more extended transition zone. This border region is automatically defined by means of an optimized region growing algorithm, based on the intensity of a selected color channel. For the investigation of the border properties the color content is neglected, the border region is therefore represented by a binary mask, only. The scaling index method is then applied to the resulting mask. In this way a scaling index α is assigned to each of the border pixels. The result is visualized using color coding (see Fig. 9.7, middle panel), where green colors correspond to a sharply defined transition ($\alpha < 2$) while red colors ($\alpha > 2$) indicate a more diffuse boundary. We find that the border between pigmented and normal skin tends to be much more diffuse in malignant skin lesions than in benign ones (Fig. 9.7).

This visualization is important for an easy-to-understand optical representation of the border features. The actually computed quantitative 'score' for the computer assisted diagnosis is derived from the resulting $P(\alpha)$ spectrum (the normalized histogram of the α-distribution) which characterizes the border property as a whole. From many previous validation studies and from the empirically derived weights of the individual components in the ABCD-rule, it is known that the border feature is of minor importance for distinguishing between benign and malignant melanomas. This is confirmed in our quantitative investigation but as it can be seen from the receiver operating characteristic curve (ROC curve [9.24]; Fig. 9.8), even a single quantitatively computed feature already yields a significant contribution for the classification problem.

Differential Structures: Malignant growth of nevi often leads to the development of characteristic patterns. Besides the commonly found 'netlike structures', the dermatologists also look for so-called 'dots', 'branched streaks', 'networks', 'globules', and some other features. Experience has shown that it is the diversity of the structural content rather than the occurrence or frequency of specific elements, which points to the pernicious nature of the lesion.

Applying the scaling index method on the segmented image of the lesion yields a continuous structural measure of the pigmented area. The resulting $P(\alpha_i)$ spectrum assigns the appropriate structure in state space to each pixel i. As a result of this procedure we obtain not only a global measure, characterizing the structural content of the image as a whole, but also a complete set of local measures, which allows the visualization of the image properties pixel by pixel.

In the right image of Fig. 9.6 the natural color of each pixel $c(x, y)$ is replaced by its scaling index α. An appropriate color coding of the values of α easily reveals relevant features of the lesion, which are of diagnostic importance. In this figure blue colors indicate sheet like areas, red colors emphasize line-like or net-like components, while green colors coincide with noisy areas. From an analysis of a sample of more than 700 melanocytic skin lesions it was found that a large area fraction in blue color (i.e. sheet-like structures) indicates a harmless lesion.

A quantitative characterization requires further reduction of the structural analysis to a few measures that can be used for the initial classification and then for quantifying possible small changes of suspicious nevi during follow-up.

We found that the generalized entropy H^q computed using the $P(\alpha)$ spectrum is very useful for this purpose:

$$H^q = \frac{1}{1-q}\frac{1}{\ln 2}\ln \sum_i p_i^q .$$

(9.3)

From empirical tests on an extensive sample we found that the discriminative power is highest if q is chosen very small but not equal to zero. (Remember, $q = 1$ in (9.3) yields the Shannon information, while q exactly 0 gives the Hartley information.) This result implies that the diversity of existing structural elements is more important for distinguishing malignant melanomas from benign nevi than the frequency of single elements as mentioned above. This is corroborated by the finding that the solitary identification of specific patterns (dots, stripes, etc) yielded only low discrimination rates.

Classification of Melanocytic Skin Lesions: Results: The two structural measures discussed above (B, D) can be used together with the remaining

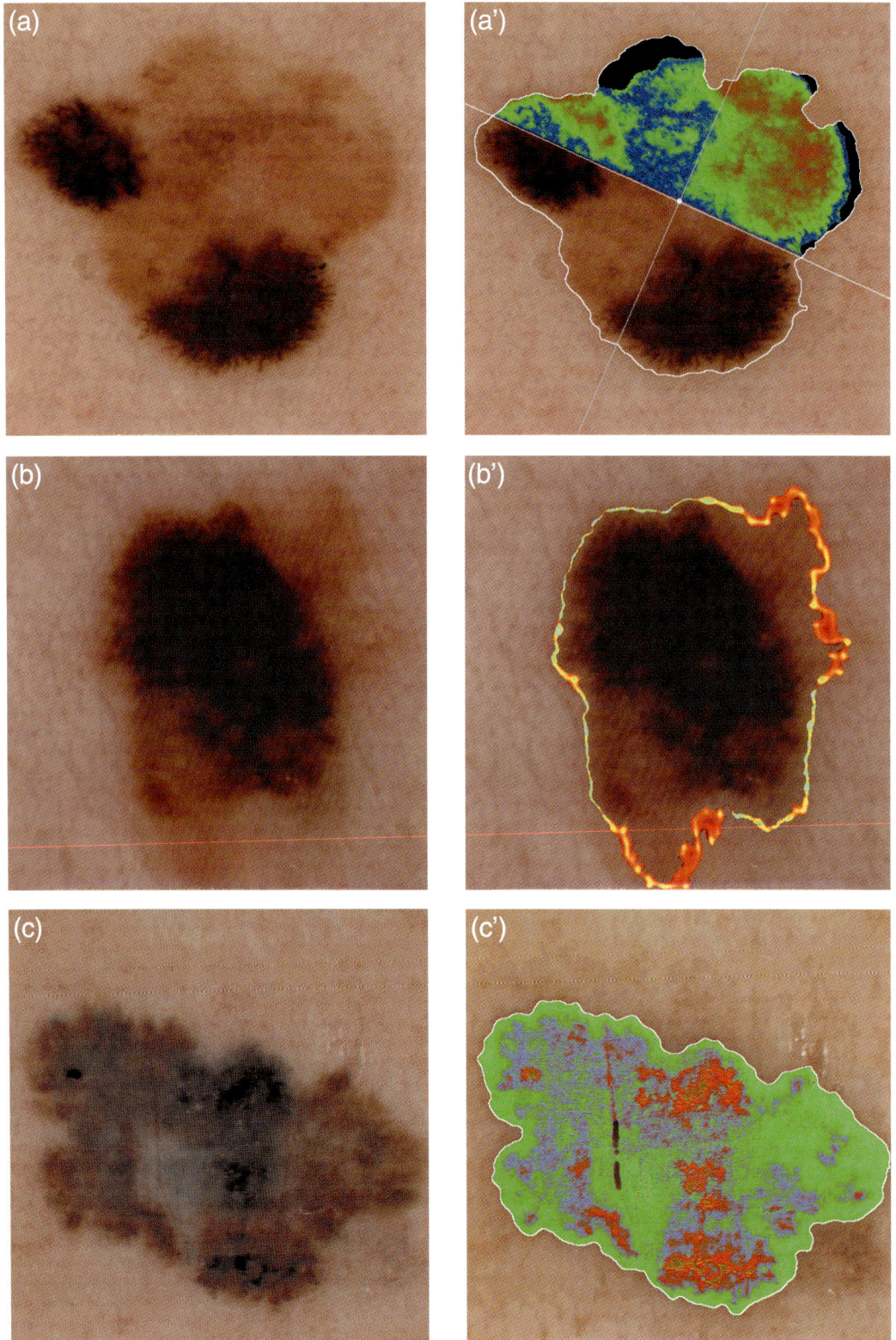

Fig. 9.7. Visualization of dermatoscopic features: the top panel shows the quantified result of color asymmetry (dermatoscopic feature) (**a**). The border characteristic (middle panel) is quantified by the scaling index method (**b**). The lower panel shows the visualization of color variety (**c**)

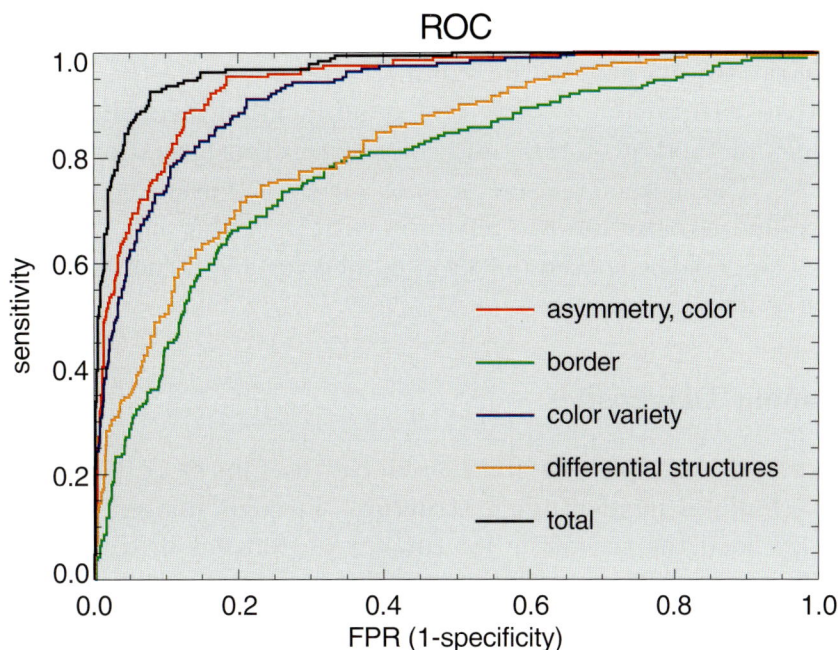

Fig. 9.8. Receiver operating characteristic (ROC) curves of the derived indicator of malignancy (printed in black) and single features. A valuable diagnostic assistance to dermatologists can be achieved by a combination of several features

parameters (A, C), and, possibly, others to identify skin cancer at an early stage of development.

To do this, the remaining complex dermatoscopic features have to be quantitatively characterized, different (reduced) measures have to be developed, their importance has to be assessed and given a 'weighting', and finally, in order to offer a computer generated suggestion for the diagnosis, these weighted measures have to be combined to a single indicator of malignancy. In this process of deriving a reliable classifier other important aspects, such as feature selection and estimation of the classification rate, have to be considered, too. The features were selected using a tenfold-crossvalidation and optimization criterion with respect to the overall classification rate on the training set. In the outcome every feature group of the dermatoscopic ABCD-rule was represented in the final classifier, albeit with different weight when compared to the traditional heuristically derived risk score.

Based on our analysis a detection efficiency in excess of 90% seems possible, thus providing powerful diagnostic support 'on site' to the medical practitioners [9.27]. Figure 9.8 shows the current standard of discrimination of malignant melanocytic lesions from benign ones. The results are based on a sample of 749 images acquired at the dermatology center of the University of Regensburg, Germany using standardized lighting conditions - providing a homogenous sample (from the point of view of data acquisition).

In the dermatoscopic ABCD-rule the term *asymmetry* (A) refers (beside the geometric aspect) also to the symmetry properties of the distribution of colors and structural components. Our quantification algorithm includes the scaling index method, geometric and color-space transformations and statistical measures. The characterisation of color variety (C) uses a specific quantized color-space and is based on the Shannon entropy. Other algorithms quantify the homogeneity of the spatial distribution of color and structural components within the lesion.

9.4.3 Tumour Diagnostics

In the previous example, we illustrated the usefulness of general pattern-analysis techniques in identifying structures in color images. More recently we have extended the technique for tumour diagnostics in three-dimensional tomographic images. The aim is to provide radiologists with better and faster methods that will allow them to determine the volume, the precise shape, and the inner structure of tumours. Radiologists need this information so that they can monitor how the tumour changes over time during treatment. The change can be either a variation in shape and volume or in the differential structure of the tumour when e.g. parts of the tumour become necrotic. Such information is invaluable, as it helps surgeons plan the surgery that a patient needs. It can also be used, for example, to monitor the success of chemotherapy or radiation treatment or to fine tune the dose and combination of drugs a patient needs [9.28, 9.29]. Currently, experts base their decisions on 3D data from X-ray or nuclear magnetic resonance tomography. However, it generally takes a trained expert over an hour to analyze these data by painstaking inspection of all slices of interest, a process which is far too long for routine use in clinics. Therefore, only a rough estimate of the shape and size of the tumours obtained by planimetric measurements in a single slide usually has to suffice for such assessment of tumours under study.

Our research aims to develop image processing tools with which it becomes possible to make an exact quantitative assessment of the shape and size of a tumour within a few minutes, and to provide experts with even more information about the tumorous tissue.

In order to calculate the exact volume of a tumour it is necessary to develop a segmentation model which divides the 3D data into several segments. From this division the radiologist can preselect the segments which constitute tumorous tissue in organs. Since the tumours often differ little from the surrounding tissue, the segmentation algorithm must be very accurate in the sense that even small changes of image features are clearly detected. A low inter- and intra-observer variability as well as high reproducibility are also mandatory.

Our algorithm, which meets the requirements mentioned above, implements elements from the well known 'simulated immersion' segmentation model [9.30] as well as region growing algorithms. Both are adapted to the three-dimensional case. For the application of this method it is necessary to calculate the gray-level gradient for each voxel. It turned out that it is particularly useful to determine these gradients with the help of scaling indices. In this case a modified version of the SIM, the so-called 'scaling vector method', which takes into account anisotropies in the scaling behavior [9.31], is applied. Therefore, the three-dimensional volume data are represented as a point distribution in a four-dimensional 'state space', which contains the three space coordinates and the gray level of each voxel. The (anisotropic) scaling index for each voxel, i, is then calculated by projecting all points within a hypercube with side length r centered around the regarded voxel onto the gray level axis. Assuming scaling behavior for the projected point distribution and calculating the logarithmic derivative as described in (9.2) leads to the anisotropic scaling index α_{gi}. With this quantity it is possible to detect even smallest gray level fluctuations with respect to each space direction, whilst at the same time filtering out any unwanted noise.

For the image segmentation we define a 'segment' as a contiguous and homogenous region. It is surrounded either by inhomogeneous regions (i.e. edges) or by other segments. In order to divide the volume under consideration into segments we start with the voxels with the lowest α_{gi}-values which characterize the homogenous regions and then proceed to the voxels with higher α_{gi}. The assignment of the voxels to the segments during the region growing process is controlled by a voxel-segment homogeneity criterion: a voxel is assigned to an already existing segment if it adjoins the segment and if its gray level differs only by a certain predefined amount from the gray level of the starting voxel of the segment. If this is not the case the voxel becomes the starting voxel of

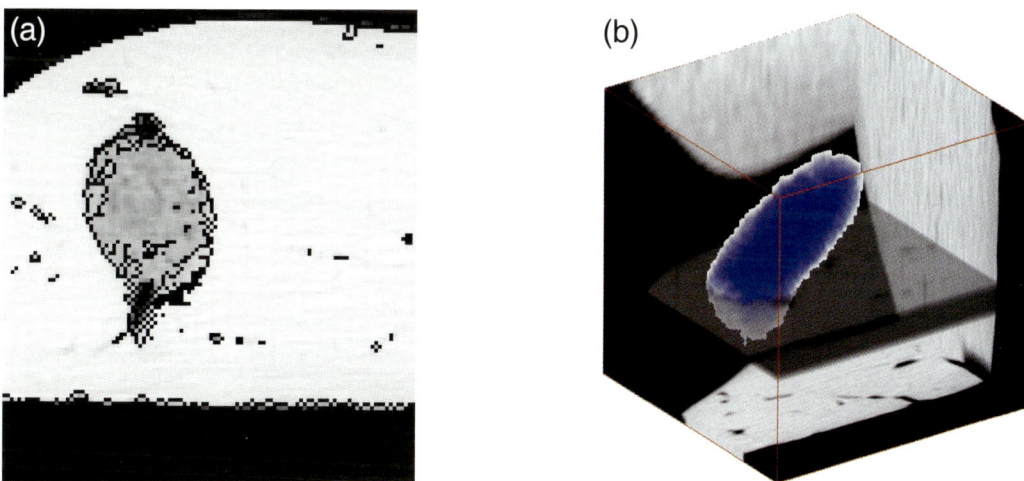

Fig. 9.9. (a) Slice of a cow's liver with implanted tissue. The black lines indicate borders of the segments. (b) 3D reconstruction of the segmented volume. The reconstructed volume amounts to 18.4 ml. The true volume of the implanted tissue is 18.1 ml

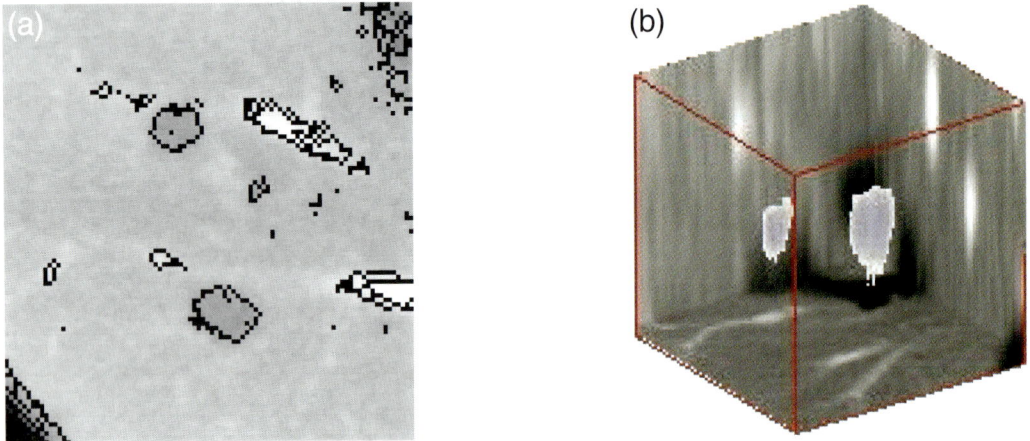

Fig. 9.10. (a) Slice of a liver with tumourous tissue. The black lines indicate borders of the segments. (b) 3D reconstruction of the segmented tumours. The reconstructed volume amounts to 0.75 ml. The volume of the tumours determined by radiologists' inspection is 0.77 ml

a new segment. In order to avoid over-segmentation adjacent segments which differ by less than a certain predefined amount in their mean gray value are then merged.

We first tested our analysis method [9.32] using computer tomography images of known samples of tissue implanted within a cow's liver, and found that we could determine the volume of the tissue implant to within 1–3% (see Fig. 9.9). In a second test we compared the segmentation results with experts' results, where a radiologist segmented tumorous regions in livers by hand. A comparison of our segmentation results with that obtained by these experts showed that our segmentation model could reproduce the radiologists' results to within 2–5% (see Fig. 9.10).

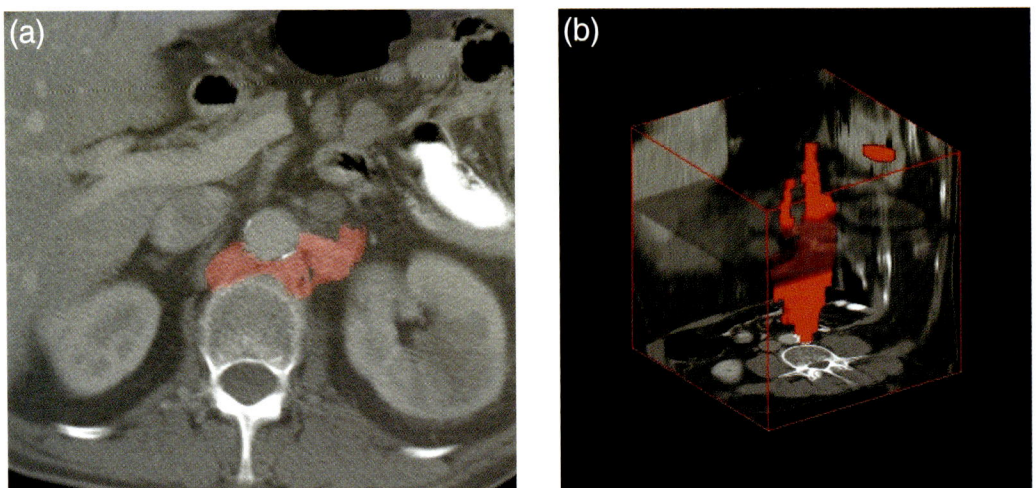

Fig. 9.11. (a) Slice of a segmented malignant lymphoma. (b) 3D reconstruction of the whole tumorous region. The segmentation model is well suited to analyze that kind of tumour, too

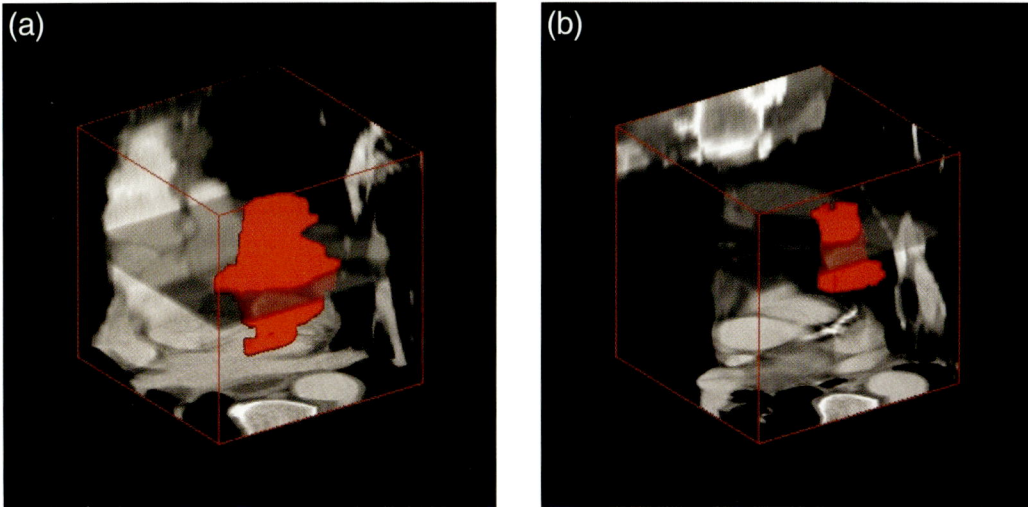

Fig. 9.12. (**a**) 3D reconstruction of a segmented tumour in the stomach before chemotherapeutical treatment. (**b**) 3D reconstruction of the same tumour after chemotherapeutical treatment. The volume decreased by more than 50%. Thus the patient responded to the treatment. This result matches well with the histomorphological result

Since then, we have optimized the analysis strategy and techniques for any kind of tomography (see Fig. 9.11). This means that our software can, in principle, accompany future advances in technology and image resolution. However, the analysis technique has to be optimized separately for different types of tumours (for example in the liver, stomach, or lungs) because of its inherent sensitivity to both the tumour and the background tissue structures.

We performed first clinical studies to assess the diagnostic value of volumetric measurements from spiral CT datasets for response evaluation in patients with adenocarcinoma of the esophago-gastric junction (AEG) and the stomach during neo-adjuvant chemotherapy (CTx). Seventy patients with advanced tumours of the AEG and the stomach, scheduled for neo-adjuvant chemotherapy, were studied. Volumetric measurements were done using the segmentation model described above. All the patients were examined pre-therapeutically and after two cycles of CTx. For the response evaluation volumetric data as obtained with our segmentation technique (see Fig. 9.12) were used. All the results were compared to endosonography (ENS) and histomorphology (HIS).

It could be shown in our study that CT with secondary reconstructions and volumetry correlates quite well with ENS and HIS results and therefore offers a reliable method to quantify response to neo-adjuvant chemotherapy.

Currently we are about to perform a similar study for tumours in the lung. We also started to analyze the differential structure of tumorous tissue in the liver based on NMR-data in order to separate benign from malignant tumours as well as metastases from malignant tumours.

9.4.4 Dynamics in Long-Term ECGs

So far we have described applications of the scaling index method to the evaluation of complex (static) patterns in images. In these cases the state space was given by the 'canonical variables' of an image, i.e. x, y, and the gray-scale, for instance. We have mentioned already that the method can be used in a similar manner for the characterization of dynamical systems. In analogy between images and time series the task is to recognize and quantify 'dynamical patterns' in time dependent signals.

We shall restrict ourselves to a single example of a univariate time series, the beat-to-beat sequence, as it can be extracted from an electrocardiogram (for some other techniques used in this field refer to Chaps. 7 and 8). The so-called RR intervals measure the time between two succeeding heartbeats, or more correctly, the timing of ventricular excitation. The more or less regular rhythm can be interrupted by diverse arrhythmias, generated in the atria or the ventricles. The observed arrhythmias are not pathological in general: to a certain degree everyone's heart shows such irregular events. A cardiologist can measure the electric signals using electrodes placed on the surface of the chest, and can follow the electrical signal propagation in great detail from the resultant electrocardiogram (ECG). Our aim was to investigate certain aspects of heart diseases, using only the easily obtained noninvasive measurements of the heart rhythm. We used our nonlinear analysis techniques to characterize the heart and found (surprisingly) that even this simple measurement contains a great deal of information.

The most important question at this point is how to construct an appropriate state space representation of a one-dimensional time series? The procedure described in the following can easily be transferred to other biosignals such as brain waves (EEGs), blood pressure, or breathing patterns.

We follow in a first step the method proposed by Packard et al. [9.33], which is widely used for the characterization of nonlinear dynamical systems. Let $\boldsymbol{x} = (x_1, x_2, ..., x_i, ..., x_n)$ be a single measured time series, sampled regularly in time or according to the natural rhythm of the system, for instance to the pacemaker of the heart (see Fig. 9.14). Then, $\boldsymbol{x}$ can generally be expanded to a vector valued time series by the following prescription:

$$\boldsymbol{x}_i := \begin{pmatrix} x_i \\ x_{i-1} \\ ... \\ x_{i-d+1} \end{pmatrix}, \forall i \geq d \,. \tag{9.4}$$

Depending on the chosen embedding dimension d each single measurement is replaced by a d-tuple of succeeding measurements and mapped into a d-dimensional Cartesian coordinate system. Assuming that the type of data $\boldsymbol{x}$ is continuous and sampled at regular time intervals (e.g. an EEG record) $\boldsymbol{X}_i$

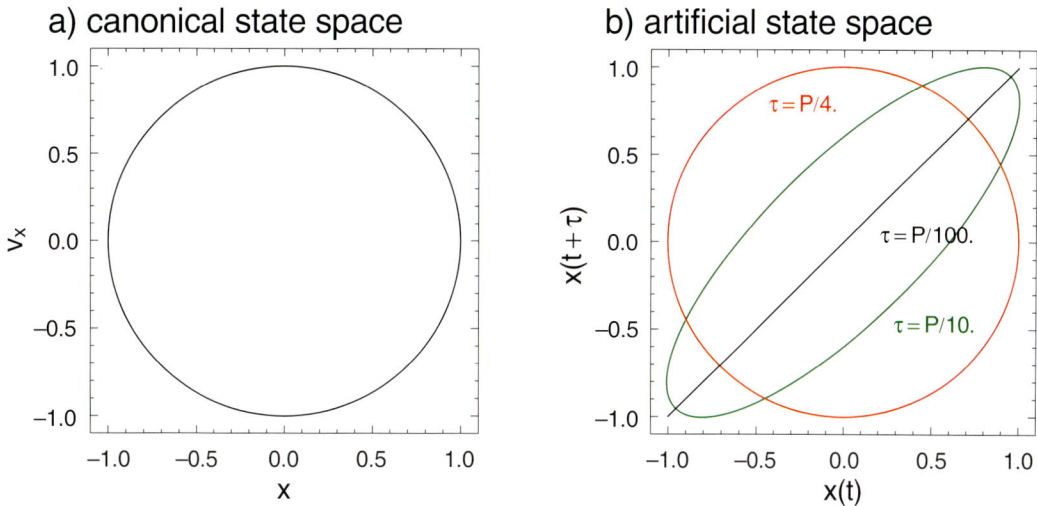

Fig. 9.13. Canonical (**a**) and artificial (**b**) state space representation of an ideal pendulum. From a theoretical point of view the structures in both representations are topologically invariant. For practical applications an appropriate choice of the delay time τ - in this case $\tau = P/4$ - can be estimated with the help of the autocorrelation function or the mutual information

is given by $\boldsymbol{X}_i = \{x(t_i), x(t_i - \tau), ..., x(t_i - (d-1)\tau)\}$, where τ is called the delay time. A mapping of the original time series in such way is called a *representation using delay coordinates*. Expression (9.4) means that a time series is drawn against itself, but time shifted.

Takens [9.34] has shown that a representation of a time dependent system in delay coordinates is in some important aspects equivalent to a representation using the canonical variables. This equivalence implies that the resulting structures in state space are topologically invariant. This theorem can be used when the canonical variables are unknown or a complete measurement of them is not possible.

For an ordinary pendulum, for instance, the known canonical variables are the position (or elongation) x of the mass and its velocity $v_x = dx/dt$. Drawing x vs. v_x yields an ellipsoidal curve in state space (see Fig. 9.13a).

If instead of v_x we use the time shifted measurement of the position $x(t-\tau)$, again an ellipse is formed in state space. Finally, if we choose for the delay time τ a quarter of the oscillation period, we can exactly reproduce the representation of the system in canonical coordinates (see red line in Fig. 9.13b)[1]. From a theoretical point of view and with respect to the proof of topological invariance, the choice of τ is irrelevant for the result, as long as $\tau \neq 0$. In practical applications, however, the right choice of τ, and particularly that of the embedding dimension d, might be crucial, because some other theoretical requirements for this artificial phase space treatment, such as the total measuring time $T \to \infty$ and measuring accuracy $\delta x/x \to 0$, cannot be fulfilled in an experiment.

---

[1] This is simply because $x \propto \sin \omega t$ and $v_x \propto dx/dt = \cos \omega t$.

(a)

(b)

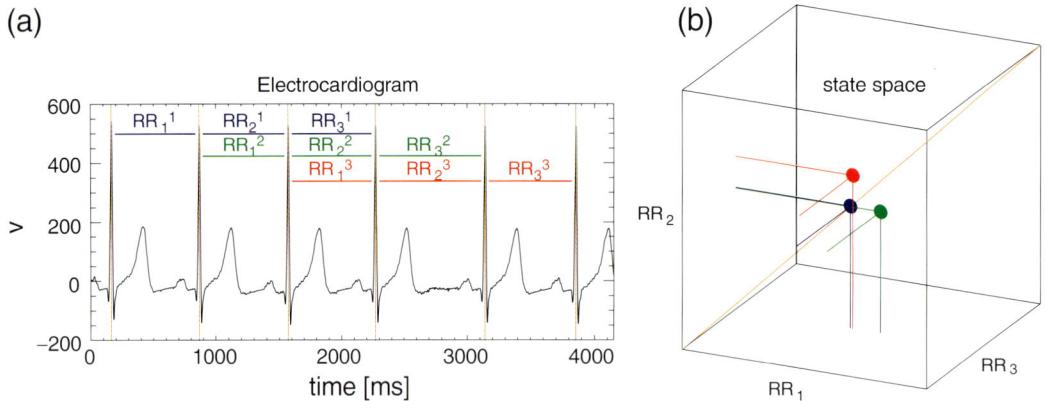

Fig. 9.14. The construction of an artificial state space representation of a single time series ECG measurement. One point in state space (**b**) is given by three coordinates extracted from the ECG-recording (**a**). Rapid heartbeats are found at the bottom left (e.g. during exercise), slow heartbeats at the top right (e.g. sleep phase)

Returning to the heart, the simplest measurement to make is the time between consecutive heartbeats. Its variability reflects the interplay between the central nervous system - which receives information from the various sensors in the body, processes them, and then translates them into the optimum heart action - and the ability of the heart to comply. If the heart is healthy, it responds rapidly. In contrast, an impaired response signifies that the heart is unhealthy. In order to identify normal and abnormal behavior patterns, the most valuable data are from an ECG measured over 24 hr. In this period, the heart experiences many different situations, both physically and psychologically, and this means that the 'system' has sampled a great many 'states'.

Ideally, our analysis should take into account all the important measurements in the ECG, together with the levels of oxygen and carbon dioxide in a patient's blood, and the blood pressure. However, it is usually impractical to measure all of these quantities, and it may even be unnecessary for some purposes, in particular in view of the fact that we do not really know the set of canonical variables describing the complex system 'heart'.

Our reduced analysis strategy involves investigating the pattern between heartbeats, adding further information as and when required [9.36, 9.37]. We have found that by analyzing the time interval between four consecutive heartbeats or 'triplets', we can accumulate a great deal of information that has powerful diagnostic and predictive value. The ECG data is plotted in a 3D state space, where each axis represents one of the consecutive beats. RR1, the time between the first two heartbeats, is plotted on the x-axis; RR2, the time between the next two, is plotted on the y-axis; and RR3, the time between the following pair, is shown on the z-axis (see Fig. 9.14).

If the heartbeat is perfectly regular over the time interval under consideration, the system will remain in an identical state and thus occupy the same

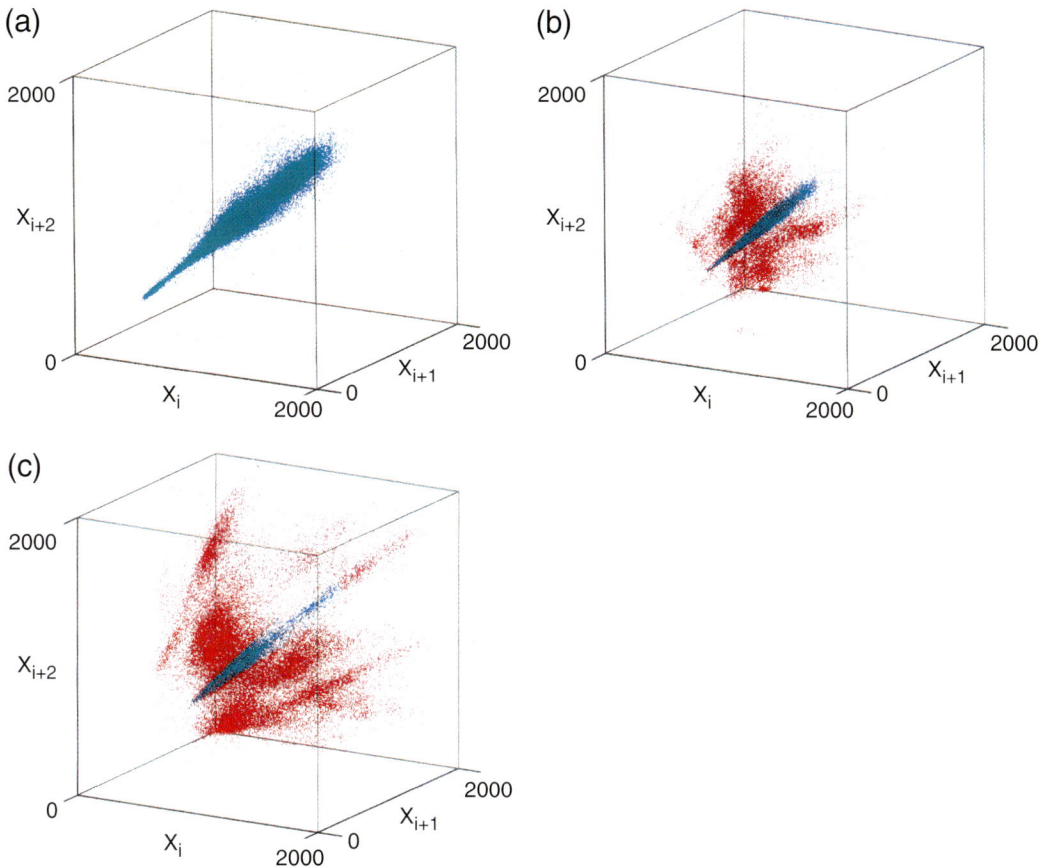

Fig. 9.15. The cloud in the upper left panel (**a**) represents the variability of the sinus-rhythm of a healthy person. The state space representations of post-infarction patients are shown in the upper right (**b**) and lower left panel (**c**). Red points indicate the presence of arrhythmias. Whereas the arrhythmias in the bottom left are characterized by well defined structures, the point distribution in the top right is more diffuse and much more irregular, indicating that the patient is at high risk with respect to sudden cardiac death

point in state space. Over a 24 hr period, however, the heart rate will change as a patient exercises or rests, for example, and the data will form a volume in state space. Our analysis therefore concentrates on the occupation density and dynamics in state space (see Fig. 9.15). For a normal, healthy person, the characteristic state space structure of the heartbeat triplets is a club shape centered along the diagonal, showing that the rapid heartbeats join smoothly with the regular and slow ones. This homogeneity in structure suggests that we are dealing with a system that can be driven dynamically from one state to the next without major disruptions (Fig. 9.15, upper left panel). For a patient who has arrhythmia, the irregularity in the heartbeat is revealed clearly in the nondiagonal regions of state space. We found that the arrhythmias appear to scale according to the healthy heart rate. In other words, if the heart is beating fast then the arrhythmias also occur fast and vice versa, thus creating coherent, off-diagonal structures. These signify that the electrical signals in the heart always travel along similar paths, for example, around a section of dead tissue in the heart wall, irrespective of the heart rate. This would appear to be an inherently stable situation, and indeed the patient (Fig. 9.15, lower left panel) was still well and in good shape 10 yr after this particular 24 hr ECG was taken. The analysis can also be used to identify patients at serious risk from sudden cardiac death. In this case, the state space appears to have much less structure (Fig. 9.15b). And while the arrhythmias are identifiable, they do not scale with heart rate in the same way as those of the previous patient. This suggests that the heartbeat irregularities do not correspond to constant and hence stable signal paths within the cardiac wall. This patient died of a sudden cardiac arrest a few weeks after the ECG was taken. Individual examples, such as those in Fig. 9.15, only illustrate the possibilities for new analysis techniques. In order to verify their usefulness in a clinical setting, we need to calibrate the technique using 'training samples', i.e. patients whose ECGs have been evaluated independently by experienced cardiologists. We then need to test the algorithms in blind studies where we do not know the clinical history of a patient. In the case of sudden cardiac death, preliminary studies have already been conducted with encouraging results (improved risk assessment up to a factor two) [9.35, 9.38].

9.5 Conclusion

We have shown that local scaling properties of different kinds of measurements, ranging from images, tomography to time-dependent biosignals, provide a powerful tool for diagnostic purposes. These local scaling properties quantify different complex patterns in the signals and characterize the local environment of each data point in the appropriate (or chosen) state space. A particular choice for quantifying the complex patterns is the 'scaling index', which is directly related to the specific information gain provided by each data point - a very important and fundamental quantity to determine. We believe that it is precisely this fundamental meaning, which makes the scaling analysis such a powerful diagnostic tool in many different applications. From our studies so far, we conclude that it would be very beneficial indeed to supplement standard (linear) analysis with such a nonlinear technique in order to characterize complex systems more thoroughly.

Acknowledgements. The authors would like to thank T. Aschenbrenner, G. Wiedenmann, V. Demmel, R. Sachs, J. Retzlaff, and P. Schücker (MPE) as well as W. Heckl, S. Thalhammer, and R. Stark (Ludwig-Maximilians-Universität, Munich), W. Stolz (University of Regensburg), and G. Schmidt, P. Gerhard, H. Helmberger, and N. Sorger (Technical University Munich). Financial support from DLR and Behrens Weise Stiftung is gratefully acknowledged.

References

9.1   P. Grassberger, Phys. Lett. A **97**(6), 227 (1983).

9.2   A.M. Fraser and H.L. Swinney, Phys. Rev. A **33**, 1134 (1986).

9.3   D. Kaplan and L. Glass, Phys. Rev. Lett. **68**(4), 427 (1992).

9.4   G. Sugihara and R. May, Nature **344**, 741 (1990).

9.5   A. Wolf, J.B. Swift, H.L. Swinney, and J.A. Vastano, Physica D **16**, 285 (1985).

9.6   B. Mandelbrot, *The Fractal Geometry of Nature* (Freeman, San Francisco, 1992).

9.7   H.-O. Peitgen, H. Jürgens, and D. Saupe, *Chaos and Fractals* (Springer, New York, 1992).

9.8   T.C. Halsey, M.H. Jensen, L.P. Kadanoff, I. Procaccia, and B.I. Shraiman, Phys. Rev. A **33**, 1141 (1986).

9.9   P. Grassberger, R. Badii, and A. Politi, J. Statist. Phys. **51**, 135 (1988).

9.10 G. Binnig, C.F. Quate, and C. Gerber, Phys. Rev. Lett. **56**(9), 930 (1986).

9.11 J.F. Jørgensen, K. Carneiro, and L.L. Masden, Nanotechnology **4**, 152 (1993).

9.12 R.J. Colton, A. Engel, J.E. Frommer, H.E. Gaub, A.A. Gewirth, R. Guckenberger, J. Rabe, W.M. Heckl, and B. Parkinson, *Procedures in Scanning Probe Microscopy* (Wiley, New York, 1998).

9.13 J. Freund, M. Edelwirth, P. Kröbel, and W.M. Heckl, Phys. Rev. B **55**(8), 5394 (1997).

9.14 J.S. Villarubia, Surf. Sci. **321**, 287 (1994).

9.15 F. Jamitzky, R.W. Stark, G. Morfill, and W.M. Heckl, J. Comput.-assist. Micros. **10**(2), 57 (1998).

9.16 F. Jamitzky, R.W. Stark, W. Bunk, S. Thalhammer, C. Räth, T. Aschenbrenner, G. Morfill, and W.M. Heckl, Ultramicroscopy **86**, 241 (2001).

9.17 A. Horsch, W. Stolz, A. Neiß, R. Abmayr, R. Pompl, A. Bernklau, W. Bunk, D. Dersch, A. Gläßl, R. Schiffner, W. Schoner, and G. Morfill, Med. Inf. Eur. 531 (1997).

9.18 W. Stolz, O. Braun-Falco, P. Bilek, M. Landthaler, and A.B. Cognetta, *Color Atlas of Dermatoscopy* (Blackwell, Oxford, 1994).

9.19 W. Stolz, M. Landthaler, T. Merkle, and O. Braun-Falco, Lancet, 864 (1989).

9.20 W. Stolz, A. Riemann, A.B. Cognetta, L. Pillet, W. Abmayr, D. Hölzl, P. Bilek, F. Nachbar, M. Landthaler, and O. Braun-Falco, Eur. J. Dermatol. **4**(7), 521 (1994).

9.21 C.M. Grin, A.W. Kopf, B. Welkovich, R.S. Bart, and M.J. Levenstein, Arch. Dermatol. **126**, 763 (1990).

9.22 B.K. Rao, A.A. Marghoob, W. Stolz, A.W. Kopf, J. Slade, Q. Wasti, S.P. Schoenbach, M. De-David, and R.S. Bart, Skin Res. Technol. **3**, 8 (1997).

9.23 W. Stolz, R. Schiffner, L. Pillet, T. Vogt, H. Harms, T. Schindewolf, M. Landthaler, and W. Abmayr, J. Am. Acad. Dermatol. **35**(2), 202 (1996).

9.24 D.J. Hand, *Construction and Assessment of Classification Rules* (Wiley, New York, 1997).

9.25 G. Plewig and P. Kaudewitz, *Empfehlungen zur Diagnostik, Therapie und Nachsorge Maligner Melanome.* Tumorzentrum München 4 (1994).

9.26 T. Schindewolf, W. Stolz, R. Albert, W. Abmayr, and H. Harms, Anal. Quant. Cytol. Histol. **15**(1), 1 (1993).

9.27 R. Pompl, W. Bunk, A. Horsch, W. Stolz, W. Abmayr, W. Brauer, A. Gläßl, and G. Morfill, in *Bildverarbeitung für die Medizin 2000*, edited by A. Horsch and T. Lehmann (Springer, Berlin, 2000), p. 234.

9.28 H. Helmberger, W. Bautz, U. Fink, U. Vogel, M. Lenz, B. Kersting-Sommerhoff, and P. Gerhardt, in *ICR 94 Singapore, Handbook* edited by L. Tan (Continental, Singapore, 1994).

9.29 H. Helmberger, W. Bautz, A. Sendler, U. Fink, and P. Gerhardt, Radiologie **35**, 587 (1995).

9.30 L. Vincent and P. Soille, IEEE Trans. Pattern. Anal. Mach. Intell. **13**, 583 (1991).

9.31 C. Räth and G. Morfill, J. Opt. Soc. Am. A **14**, 3208 (1997).

9.32 C. Räth, W. Bunk, B. Schulte, N. Sorger, C. Ganter, H. Helmberger, A. Horsch, P. Gerhard, and G. Morfill, in *Bildverarbeitung für die Medizin 2000*, edited by A. Horsch and T. Lehmann (Springer, Berlin, 2000), p. 449.

9.33 N.H. Packard, J.P. Crutchfield, J.D. Farmer, and R.S. Shaw, Phys. Rev. Lett. **45**, 712 (1980).

9.34 F. Takens, Lect. Notes Math. **898**, 366 (1980).

9.35 G. Schmidt, G. Morfill, P. Barthel, M. Hadamitzky, H. Kreuzberg, V. Demmel, R. Schneider, K. Ulm, and A. Schömig, Pacing Clin. Electrophysiol. **19**(6), 976 (1996).

9.36 G. Schmidt and G. Morfill, Pacing Clin. Electrophysiol. **17**, 2336 (1994).

9.37 G. Schmidt and G. Morfill, Pacing Clin. Electrophysiol. **17**, 1174 (1994).

9.38 G. Morfill and G. Schmidt Phys. Bl. **50**, 156 (1994).

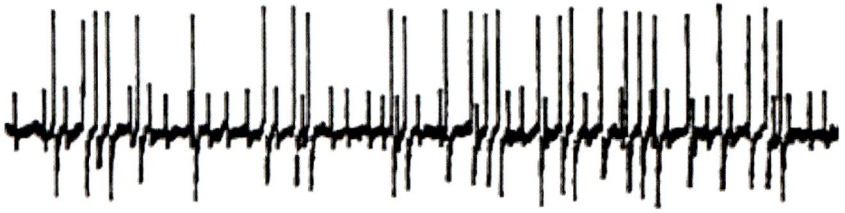

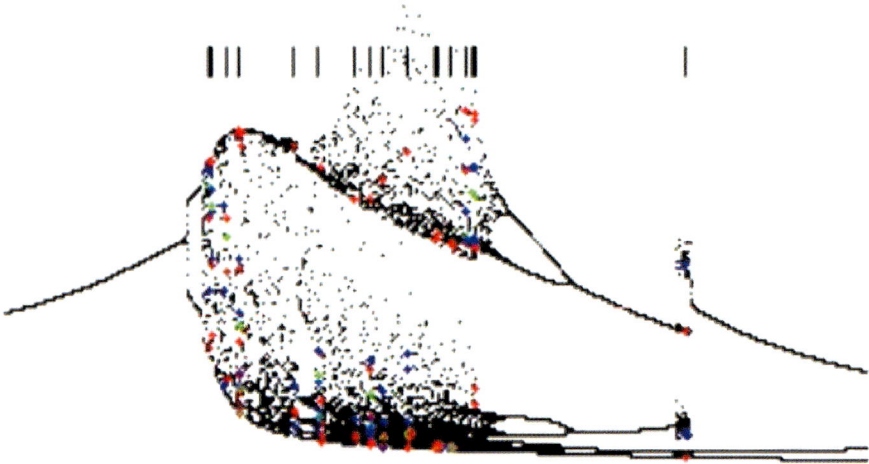

10. Unstable Periodic Orbits and Stochastic Synchronization in Sensory Biology

Frank E. Moss and Hans A. Braun

We discuss here two new methods for extracting low-dimensional dynamical objects from dynamically noisy, nonstationary datasets. The analysis and detection algorithms assess the presence or absence of such objects and return a determined statistical confidence level. The first objects we discuss are stable and unstable periodic orbits and bifurcations between these states. Second, the statistical process of synchronization between stable orbits in the presence of noise is demonstrated. Several examples based on experimental data from sensory biology and one example from the rat brain hypothalamus are presented.

10.1 Introduction

Perhaps the oldest discussion of dynamical systems involves the determination of stability or instability as expounded already in classical times, for example by Archimedes [10.1]. Systems evolving in the neighborhood of a stable point, or fixed point, exhibit a regular return to a periodic or quasiperiodic orbit. Those evolving near an unstable fixed point may remain in a larger neighborhood, called an attractor, but exhibit irregular or chaotic behavior. It is our purpose here to discuss and exhibit a relatively new method for determining

◄ **Fig. 10.0.** The paddlefish, *Polyodon spathula*, swimming against a stream of water in a viewing apparatus, called a 'swim mill'. The fish's electrosensitive rostrum, or nose, is in contact with a grill through which water is flowing. The fish is waiting for a *Daphnia*, a small plankton of about 1 mm in length that emits a weak electrical signature, to emerge through the grill. Using its electrosensory organs arrayed over its rostrum, the fish will first detect, then locate, track, and capture the *Daphnia*. This fish is a juvenile of about 20 cm total length. The two other panels show impulse recordings from electrosensory afferent nerve fibres and a bifurcation diagram of interspike intervals of a computer model

the presence of stable or unstable orbits in noisy dynamical systems. Can the noise itself be distinguished from low-dimensional dynamical processes, for example chaotic dynamics? And what about high-dimensional behavior? Here we shall define high-dimensional to mean dimensionality of four or greater, and we make no distinction between the noise, which formally can be of infinite dimension, and this high-dimensional behavior. The dynamical signatures of interest here are the projections onto a two-dimensional plane of trajectories evolving in time in three dimensions, a Poincaré section. All else we regard as the noise, which shows on the Poincaré section as a random distribution around the projection of the trajectory. We shall call these patterns 'signatures', and show that, together with a suitable statistical measure, they can distinguish low-dimensional behavioral objects, that is orbits, in noisy data. Indeed, in the case of experimental data, we often do not know the dynamical evolution equations even though we may guess at them. To find stable and unstable periodic orbits in such data may greatly help us guess.

10.1.1 Stable and Unstable Orbits, Noise, and Topology in Dynamical Systems

Random perturbations, or 'noise', of course, themselves cause an irregularity in these orbits. If the noise is large - comparable to the size of the attractor - it is difficult to distinguish stable or unstable orbits from each other and from the noise using traditional methods [10.2–10.7]. But even the noisy patterns projected onto the Poincaré section reflect simple topological properties of the orbits that will be of use. Consider Fig. 10.1. Here we show two such projections. We have labeled the axes T_{n+1} and T_n respectively, because later we shall apply this method to recordings of neuronal spike trains where they represent the time intervals between action potentials. But, in principle, the T's can represent any variable characteristic of the system. The diagonal line is called the line of periodicities, because for every point lying on the line, $T_{n+1} \equiv T_n$ which is the definition of a periodic process.

A dynamical system may have one or more characteristic periodicities represented by specific points lying on the diagonal. Here we shall consider only one characteristic period, represented by one fixed point. Now our question is how a perturbation, say an initial condition represented by a point off the diagonal, returns to the periodic fixed point. There are two ways we are concerned with here. An *unstable* periodic orbit (UPO) is represented by the topology shown in Fig. 10.1a, that is, the crossing of stable and unstable directions, or manifolds, at the unstable fixed point. These directions are represented by arrows: outward pointing representing the unstable direction and inward pointing representing the stable one. The projections of the actual trajectories will, of course, be curves, but we represent them in the figure by straight-line tangents intersecting at the fixed point. The slopes of these tangents are the

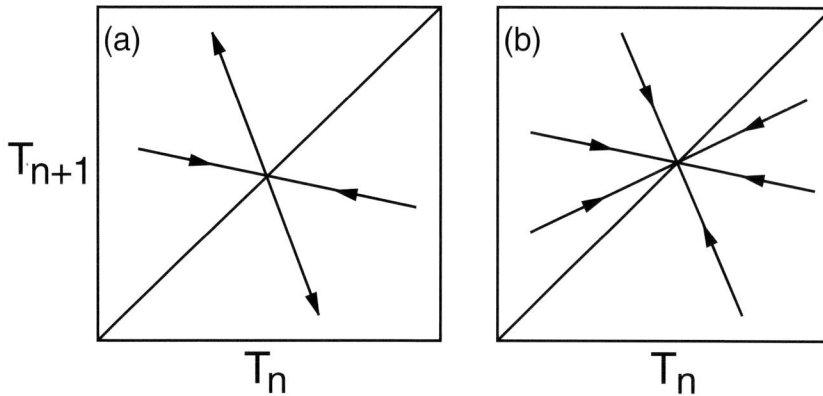

Fig. 10.1. Sections representing two-dimensional projections of three-dimensional motion: (**a**) of an unstable periodic orbit (UPO) and (**b**) of a stable periodic orbit (SPO), for example a limit cycle. Inward and outward pointing arrows respectively represent stable and unstable directions. The diagonal is the line of periodic fixed points

inverse eigenvalues of the linearized motion in the near neighborhood of the fixed point (see, for example, [10.8]). In contrast a *stable* periodic orbit (SPO) is stable to perturbations in all directions and is represented by the topology shown in Fig. 10.1b. The projections shown in Fig. 10.1 represent the signatures of period-1 orbits. In principle all periods can be represented in a dynamical system. Higher order periods are revealed by tangent lines in plots of T_{n+q} vs. T_n for an orbit of period-q. In this work we shall deal only with period-1 orbits though higher orders have been experimentally observed and scaled. Higher periods are more difficult to observe in experiments because the probability to detect them decreases exponentially with the period [10.9].

10.1.2 The Signature of an Unstable Orbit: The Encounter

We can imagine that, as it evolves in time, a recurrent trajectory in three-dimensional space passes through the Poincaré section leaving a sequence of points. After a large number of passes, a pattern is formed. An example is shown in Fig. 10.2a, which represents the time intervals between neural action potentials, or spikes, T_n, from an actual experimental recording. First return maps, often called 'scatter plots', are familiar to biologists, and efforts have been made, usually to no avail, to interpret the shape of the cloud of points, in this case a kind of fattened quarter-moon, and relate them to known biological processes. What is missing is the information contained in the *temporal sequence* of the points as they occur in the dataset, that is, as they are punched out in the Poincaré plane by the recurrences of the trajectory. Figure 10.2b shows a particular sequence: three points which converge on the fixed point (numbered 1-3) along a stable direction, followed immediately by three points which diverge (3-5, with point 3 common to both sets). Such an identified set is called an 'encounter'. The meaning of this term is that the general trajectory

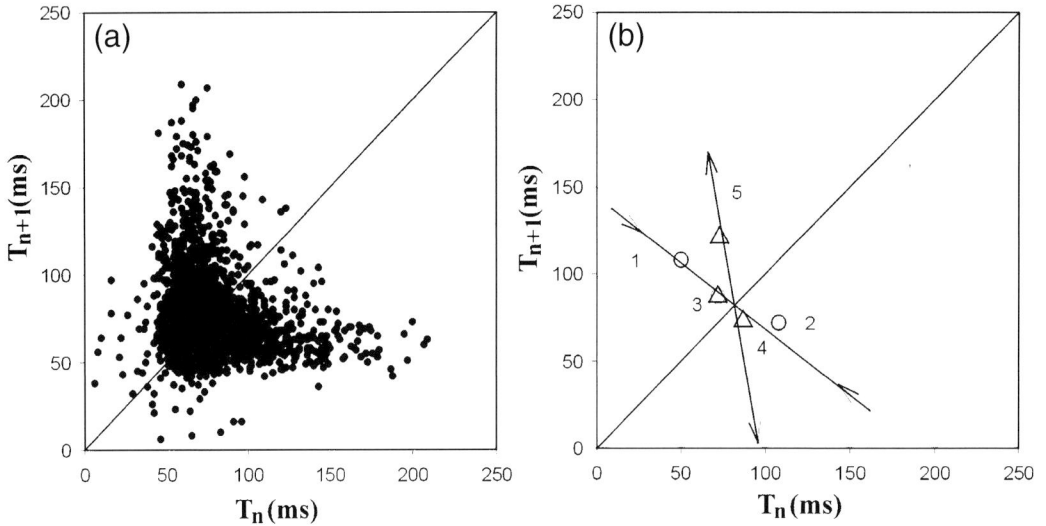

Fig. 10.2. First return maps, or Poincaré sections. (**a**) Map of the time intervals recorded from a rat facial cold receptor, a temperature sensitive sensory neuron. (**b**) A selected encounter with an UPO extracted from the data shown in (**a**). Stable and unstable directions are indicated by the arrows

of the system has encountered the period-1 UPO, and the signature of that encounter is depicted in Fig. 10.2b.

Notice also that the distances between sequential points and the fixed point decrease (in fact, in noise-free purely dynamical systems, they decrease exponentially) along the stable direction and increase (also exponentially) along the unstable direction. The multipliers in the exponents of these distances are also the eigenvalues of the linearized motion. But in our dynamically noisy systems, the exponential behavior shows up only in suitable ensemble averages. An alternate, and more sensitive, method of detecting SPOs and UPOs in noisy datasets based on a statistical measure of the eigenvalues has recently been described by Dolan [10.10].

10.1.3 Counting Encounters

It is an easy matter to program a computer to search a dataset for encounters as shown in Fig. 10.2b. Various degrees of stringency in the definition of an encounter can be applied and various numbers of points can be included in the definition. The least stringent definition might be simply to require that three points converge toward the diagonal followed by three that diverge (with one common point). But this definition is arbitrary; for example, six points without a common point could also be used. A next higher level of selectivity might be to additionally require that a straight line fit to the three converging points yields a slope (eigenvalue) that lies in a certain range, for example 0 to -1 followed by a similar slope requirement for the diverging set, for example -1 to $-\infty$. An additional criterion is that the intersection of the straight line fits of incoming and outgoing intervals must lie within a certain limiting value close to the 45° line. In one form or another all of these definitions have been used but the results of the analyses using them do not substantially or qualitatively differ. Having thus defined an encounter, the *number* of such encounters, N, in the entire dataset is counted.

As we shall discuss later, such encounters have been found, sometimes in abundant numbers, in datasets from various biological neurons. The UPOs encountered in all biological systems thus far examined have been found to have eigenvalues lying in the aforementioned ranges, namely, $0 > m \geq -1$ for the stable directions, and $-1 > m > -\infty$ for the unstable directions. These slopes, m, are consistent with the points being laid down alternately across the diagonal as shown in Fig. 10.2. Sequences can approach the fixed point, however, also from only one side of the diagonal, and these have been associated with the *precursors* of a period doubling bifurcation [10.11, 10.12].

Note that the algorithm(s) mentioned above are designed to detect the signatures of UPOs by means of the topology shown in Fig. 10.1a. As we show below, SPOs are detected using the same algorithm, whereby they are conspicuous because of the absence of encounters (for UPOs). In other words, the absence of UPOs in files that are otherwise ordered indicates the presence of SPOs. (Noting that SPOs also include quasiperiodic orbits, we can ask: what other kind of order is there?)

10.1.4 Why is a Surrogate Dataset Needed?

Noise is ubiquitous in nearly all biological and many physical systems. Moreover, the noise is *dynamical* as opposed to instrumental. Dynamical noise operates on the parameters and/or is inherent in the variables of the system. If the evolution equations that represent the system are known, then dynamical noise may be additive on one or more variables or their derivatives or it may be multiplicative, operating on a parameter which determines the operating point of the system, for example a bifurcation parameter, or both. Dynamical noise itself becomes part of the solution of the evolution equations and thus cannot be filtered out, at least by the traditional methods of linear filter theory. We use the term 'instrumental' to designate the noise that is added to the output of a system, as for example, by a noisy measuring instrument. Instrumental noise can be removed or significantly reduced by linear filters.

10.1.5 A Statistical Measure of the Number of Encounters

Because our data is inherently noisy, the conditions that define an encounter can be satisfied by chance alone. That is, false encounters will be found in files composed of random numbers or of randomized data. Such files are called 'surrogates', and a large culture has developed around the problem of choosing the 'right' surrogate for any particular application. We will not recite this discussion here, but instead cite some of the relevant literature [10.13–10.15]. For the purpose of detecting UPOs and distinguishing SPOs, we have found that surrogates made by simply shuffling the original experimental data points to random locations in the file are superior to all other methods [10.10, 10.15]. A suitable surrogate should preserve some measure of the original data. For time intervals between spikes, simple shuffling preserves the interspike interval histogram, a measure deemed important in many biological applications.

We can send the search algorithm through the surrogate file to count the number of chance encounters, N_s. It is necessary to obtain a good average of this quantity, so the data are re-randomized (re-shuffled) a number of times in order to obtain the average, $\langle N_s \rangle$ and the standard deviation, σ. Though we have found that 20 surrogate realizations are usually sufficient for reasonably good Gaussian statistics [10.10], in the applications that follow here, 100 realizations were used.

We now wish to compare the results obtained by analyzing the surrogate files with the result obtained from the original data file. This can be done in a standard way:

$$K = \frac{N - \langle N_s \rangle}{\sigma} .$$

(10.1)

Note that K is a measure of the distance between the finding in the actual data file and the mean of what would be expected by pure chance. The distance units are the standard deviation of the surrogate findings, so that K has units of *sigmas*. If the statistics are Gaussian, then $K > 2$ indicates a finding (of UPOs) in the data with better than 95% confidence, and $K > 3$ indicates a confidence level better than 99% [10.16]. Note that the confidence level is built into this measure in the following sense: if a dataset is too short to provide good statistics σ will be large, and N and $\langle N_s \rangle$ small; thus the resulting K value will not indicate statistical confidence. In the data to be displayed, we shall be segmenting out files of about 1 000 time intervals and computing the K values as indicators of the presence or absence of UPOs.

10.1.6 Discussion of UPOs

Dissipative physical systems that are chaotic exhibit a structure built of UPOs that has been called a skeleton [10.17–10.20]. The skeleton is a set of UPOs with a countable infinity of periods. It is more probable for a general trajectory to visit the lowest period orbits, and this probability scales exponentially with period [10.9]. Numerous experiments with physical systems have exhibited UPOs [10.21, 10.22]. Techniques for the control of chaos are actually based on the control of UPOs, that is to apply controlled perturbations in order to render the orbit in question stable [10.23]. Control of chaos has also been demonstrated experimentally in numerous applications including in biophysics [10.24–10.28]. A claim was made to have demonstrated the control of chaos in a hippocampal slice of rat brain with the implication that such control could be developed into an implantable therapy for the control of epilepsy [10.29]. But it was soon shown that a nonchaotic but dynamically noisy system could be controlled with the very same algorithm claimed for the control of chaos [10.30]. The cause of this confusion was the failure to account for the effects of dynamical noise that is large in biological systems and particularly so in the rat brain slice preparation.

In order that the experimental study of UPOs in dynamically noisy systems could advance, it was necessary to develop ways to count their occurrences with statistical accuracy. Such a method is the one outlined above, first developed and demonstrated in 1995 in two different physical systems (one an infinite-dimensional time delayed chaotic system) [10.31]. This statistical detection algorithm was experimentally demonstrated in sensory biology for the first time the following year in the crayfish caudal photoreceptor system [10.32, 10.33]. Since 1996, the algorithm discussed here, or small variations of it, has been used in a variety of biophysical and even medical applications [10.34–10.36] including the catfish electroreceptor organs, rat facial cold receptors and the hypothalamic regulatory neurons of rat brain slices [10.37–10.39].

Moreover, a qualitatively accurate model of the aforementioned sensory neurons incorporating a noise mediated subthreshold slow oscillator has been developed by the Marburg group [10.40–10.42]. The model also successfully describes the temperature dependence of the spike firing patterns observed with the sensory neurons. A rich structure of UPOs and SPOs has emerged. They have been categorized according to their topological connectivities [10.43]. A collision of a homoclinic orbit with a period-1 fixed point, studied with the model, has led to the discovery of an explosion of interspike time intervals at a critical temperature - an effect that has been observed experimentally in the crayfish caudal photoreceptor [10.44, 10.45].

10.1.7 Stable Orbits and Noisy Synchronization

Of course the existence of periodic orbits in a system leads one to immediately inquire into their state(s) of synchronization, a question again considered already in classical times [10.46]. Cyclic and synchronization processes are ubiquitous in biological applications [10.47, 10.48]. We shall be concerned here with *noisy* synchronization, for example as it occurs in biological prepa-

rations and in medical applications [10.49–10.51]. Dynamical systemic noise
causes the two oscillations to synchronize and drop out of synchronization alternately and randomly, a process considered originally by Stratonovich [10.52].
Early studies on bistable systems [10.53] paved the way for extensive studies
of the process in a variety of physical systems [10.54–10.57] including chaotic
synchronization [10.58–10.60].

One can consider two periodic functions. In order to assess the state of
synchronization between them, we need to measure the phase of one relative
to the other, that is the phase difference, $\Delta\varphi(t)$. This can be measured in
two ways, first by marking, for example, the zero crossings of both functions.
In the first way, one can mark the phase differences $\Delta\varphi_i$, of the zero crossings
cycle-by-cycle, that is for each cycle of one of the functions, called the reference
function. A graph of these phase differences as a function of time, or for every
cycle of the reference function, is called a *synchrogram*. In the second way,
one counts the phase differences continuously in time from some initial zero
crossing of the reference function. In this case, a phase diffusion can be defined
as we shall show further on.

In the case of biological applications one may wish to measure the state
of synchronization of a biological oscillator which is manifest as a time series of neural spikes compared to some stimulus which may be a simple si-
nusoidal function. In this case, using the first method one simply marks the
phases wherein the spikes occur relative to each successive cycle of the stimulus. These phases are then assembled into a synchrogram. In the case of the
second method, the cumulative phases of all spikes relative to some identified
time are marked on the stimulus function, for example its first zero crossing.
These are displayed graphically as a continuous function of time. The slope of
such a plot represents the mean phase diffusion. Slope zero or small signifies a
high degree of phase coherence (small diffusion) between the neural spike train
and the stimulus.

10.1.8 Statistical Measures of the Quality of Synchronization

We present here just two measures: those that emerge from the two ways of
determining the phase differences. In general, a high quality of synchronization
is shown by constant phase differences between stimulus and spike train. That
is, the set of spikes occurs at the same phases during every cycle of the stimulus.

In the first method, these phases are plotted on a synchrogram, and phase
constancy, or locking, is represented by straight lines of zero slope. In contrast,
when the phase lock is broken, the trace on the synchrogram becomes a line
with positive or negative, nonzero slope. The relative phase of the spike train is
now some function of time. If the relative velocity is constant the trace on the
synchrogram is straight and the slope is a measure of the relative phase velocity.
Synchrograms show times of phase synchronization interspersed with times of

no phase locking. In a long dataset, a statistical measure is the probability density of phase differences. Examples of synchrograms and of this probability density are shown below in Sect. 10.3.

In the second method, the quality of synchronization is measured by the mean effective phase diffusion constant:

$$D_{\text{eff}} = \frac{d}{dt}\left[\left\langle \Delta\varphi^2(t) \right\rangle - \left\langle \Delta\varphi(t) \right\rangle^2\right] = \frac{1}{\langle T_{\text{lock}} \rangle}, \qquad (10.2)$$

where the averages are over the complete dataset, and $\langle T_{\text{lock}} \rangle$ is the average time the spike phases are locked to the stimulus. The first expression on the right shows that D_{eff} is actually the slope of the graph of the mean square phase differences. We shall show examples of these measures in Sect. 10.3 below.

10.2 Unstable Periodic Orbits in Physical and Biological Systems

We wish to reiterate that the UPO detection method outlined in Sect. 10.1 contains a 'built-in' indicator of the statistical confidence level of the finding for any dataset. This makes it particularly useful for short datasets that may also come from nonstationary preparations. These characteristics make the method especially useful for analyzing data from biological or medical applications. In the following Sects. 10.2.2–10.2.4 we show some examples of the method applied to datasets from sensory neurons. The experiments whence these data were extracted have been reported in more detail in [10.37–10.39]. The three sensory neurons investigated (catfish electroreceptor, rat cold receptor, and hypothalamic neurons) have in common that they each contain an inherent subthreshold oscillator mediated by internal neuronal noise. The Marburg group discovered these oscillators many years ago in dogfish electroreceptors [10.61].

First, however, we present numerical data from a physical system - the logistic map - which demonstrates the use and effectiveness of the method for detecting period-1 UPOs buried in noise.

10.2.1 UPOs in the Logistic Map

The first demonstration is realized in the logistic map:

$$X_{n+1} = ax_n\left(1 - x_n\right) + \xi_n, \qquad (10.3)$$

where a is the bifurcation parameter and ξ_n is the dynamical noise. We examine a small portion of the bifurcation diagram that shows the famous period-3 window and some regions of chaos just before and after the window. We apply our algorithm discussed in Sect. 10.1 above and compare this to a traditional method, the Lyapunov exponent [10.2–10.7].

The results are shown in Fig. 10.3. For this map, when $a < 3.82$ and $a > 3.86$, the system is chaotic and there should be a large number of UPOs present. In the mid-section, $3.82 < a < 3.86$, approximately, the system is within the period-3 window. The dynamics is purely periodic. The bifurcation diagram without added noise ($\xi_n = 0$) is shown in Fig. 10.3a. Near the middle of the window, one can see the onset of a period doubling cascade. On each of the three branches of the periodic trace there is a cascade of bifurcations, seemingly from period-1 to period-2 to period-4, etc. (We are, however, within the period-3 window, so the first cycle is period-3, and the ensuing cascade is thus period-3 to period-6, etc.) When noise is added this cascade becomes indistinct and is indistinguishable from the chaotic regions using traditional methods. The task of our algorithm is to distinguish this periodic behavior buried within the noise, that is, not to confuse it with chaos. Moreover, we hope that our method will sharply delineate the boundaries between the two chaotic regions and the window.

Figure 10.3b shows a plot of the measured K values through the window compared to the measured Lyapunov exponents using a traditional method, for a *small* value of added noise. All measurements were obtained from stationary ($a = $ constant) datasets of 3 000 points in length. One can see that the K values easily delineate the boundaries of the window, but even at this small noise, the Lyapunov exponent is confused by the period doubling cascade. A Lyapunov exponent greater than zero is the indication of chaos. The $K = 3$ confidence level of 99% is shown by the dashed line. The algorithm clearly marks both of the boundaries at the 99% confidence level.

Figure 10.3c shows the same experiment repeated but for the largest noise possible such that the map itself does not become unstable and iterate to infinity. We note that the K values still can distinguish the boundaries sharply at the 99% confidence level, whereas the Lyapunov measure is now completely useless being greater than zero over the whole window.

We can conclude that the new algorithm based on a statistical measure of the topological signatures of low dimensional behavior is superior to traditional methods for detecting the presence of UPOs buried in the noise. Moreover, the algorithm signals the presence of SPOs by its negative value, an indicator already noticed early on [10.31].

10.2.2 UPOs and SPOs in the Catfish Electroreceptor

Bifurcations in dynamical systems are often indicators of a change in the state of stability of a system. They are controlled by one or more quantities, usually under the control of the experimenter, called bifurcation parameters. In the following dataset, taken from the catfish electroreceptor system, we show a

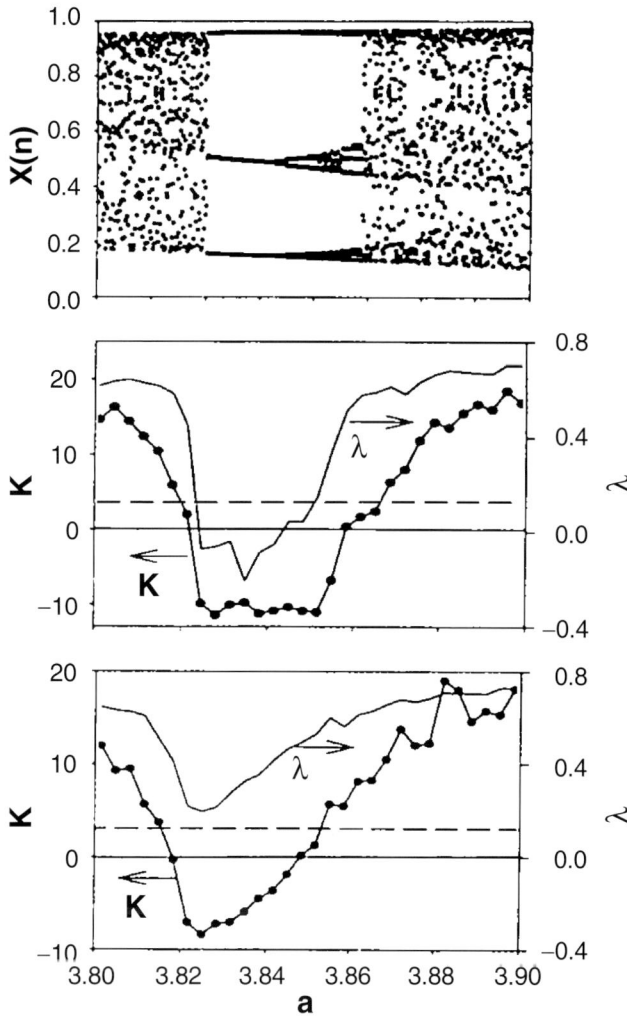

Fig. 10.3. (a) The bifurcation diagram for the logistic map in the vicinity of the period three window. (b) The Lyapunov exponent and the statistic K calculated from stationary files of 3 000 points with small noise ($\xi = 0.002$). (c) The Lyapunov exponent and the statistic K calculated from stationary files of 3 000 points with large noise ($\xi = 0.005$)

period-1 to period-2 bifurcation controlled by the temperature of the preparation (see Fig. 10.4). Thus the temperature is the bifurcation parameter. Prior to the temperature change, the distribution of time intervals is monomodal indicating a noisy period-1 behavior. A few tens of seconds after the temperature step, the bifurcation occurs with a change in the distribution to bimodal or noisy period-2 behavior. The K values for the file segments indicated by the solid bars are shown above the time interval record. We note that prior to the bifurcation, $K \cong 5.4$, a strong indication of the presence of UPOs. These persist into the period-2 region, but die away well after the bifurcation where the negative value of K indicated the presence of SPOs. Therefore, in addi-

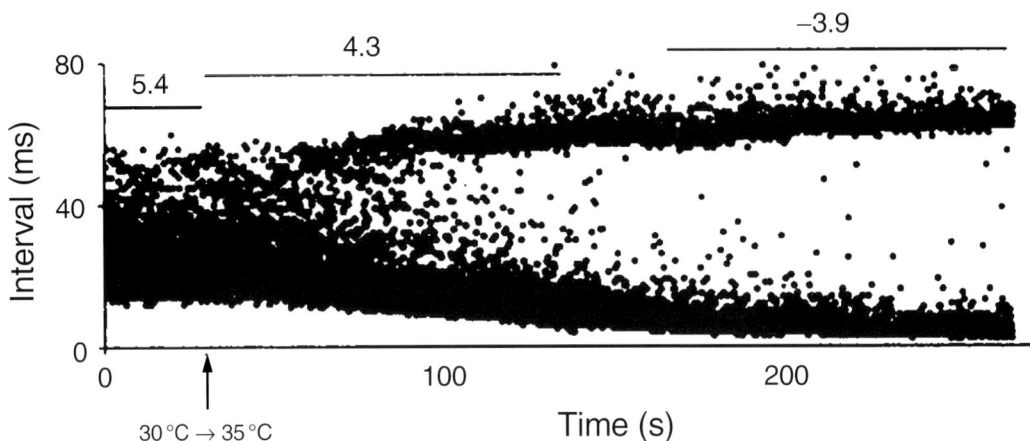

Fig. 10.4. Interspike time intervals from the catfish electroreceptor plotted as a function of time. An arrow indicates the time at which the temperature was raised from 30°C to 35°C. The statistic K is shown for three different regions

tion to the bifurcation in the order of the period, there is an accompanying transition from instability (UPOs) to stability (SPOs).

10.2.3 UPOs in the Rat Facial Cold Receptor

The rat facial cold receptor is another sensory neuron with characteristics similar to the electroreceptors, that is they are characterized by subthreshold noise-mediated slow oscillators [10.61].

We show below in Fig. 10.5 a dataset obtained by driving the preparation with temperature ramps. We see that initially and finally, in the $T = 30°C$ regions, the small K values indicate no low-dimensional dynamical objects, only noise. But after the ramp down in temperature UPOs appear in large concentrations and accompany the period-1 to period-2 bifurcation that is induced by the ramp down in temperature and persist into the constant $T = 20°C$ region. The temperature up-ramp unfolds the previous sequence of events. The UPOs vanish and are finally replaced again by noise. In Fig. 10.5, the actual encounters are indicated by enlarged symbols.

10.2.4 UPOs in a Hypothalamic Neuron

For our final example, we move to the hypothalamus of the rat brain. This part of the brain controls several autonomous functions like body temperature and fluid balance with neuronal sensitivity for diverse homeostatic parameters. Below in Fig. 10.6 we show an example of a period-1 to period-2 bifurcation and the accompanying presence of UPOs in a hypothalamic brain slice recorded during slowly increasing temperature. In this case, the dynamics are much more

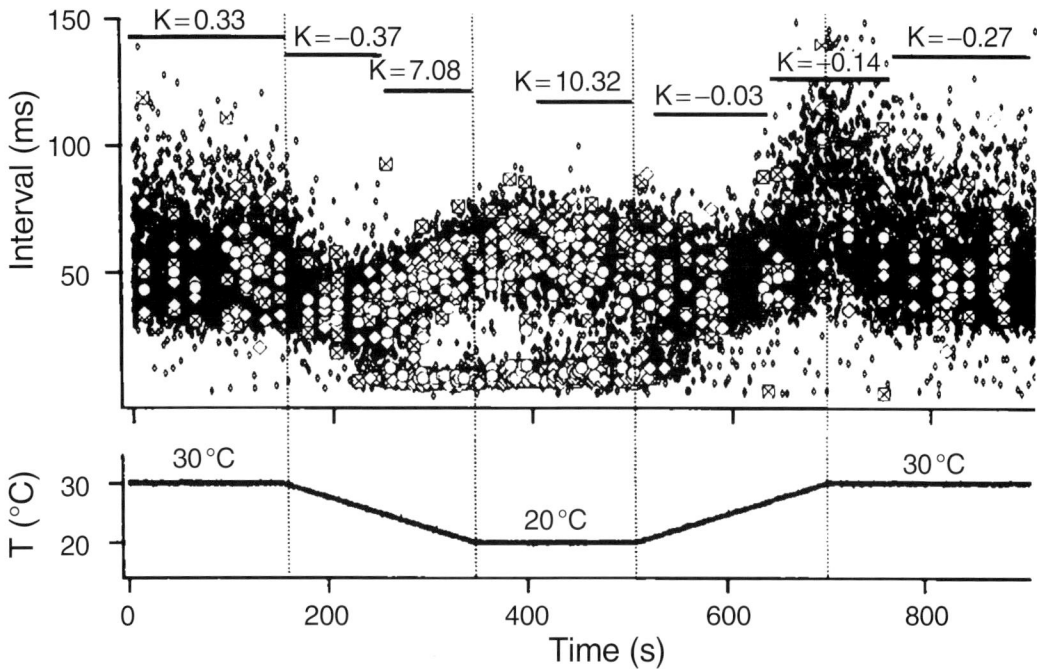

Fig. 10.5. Interspike time intervals from the rat facial cold receptor. The second panel shows the temperature as it was held constant at 30°C for 180 s, then dropped linearly to 20°C over 200 s, and held constant for 120 s, then increased linearly back to 30° over 200 s, and finally held constant for 200 s. The statistic K is shown for each region, and the individual encounters are shown with symbols. Stable points are shown with crosses, unstable with diamonds, and the common points with circles

noisy and the indication consequently arises at a smaller confidence level than in the previous examples. Here the value, $K \cong 2.9$ indicates a confidence level lying between 95% and 99%. The temperature range is much smaller, however, showing the greater sensitivity of hypothalamic neurons to slow temperature changes. We again note the familiar pattern whereby the UPOs enter as an accompaniment to the period doubling bifurcation. In this case their early appearance is also a precursor of the bifurcation [10.11, 10.12].

10.2.5 Discussion

The presence of UPOs in datasets is not necessarily an indicator of chaos. UPOs can occur in nonchaotic situations, for example, during period doubling cascades. At each bifurcation the previous (stable) periodic orbit loses stability, but persists after the bifurcation as a UPO. Our algorithm can detect these UPOs. In each of the foregoing biological examples, a period doubling bifurcation was indicated. The algorithm is detecting the appearance of the period-1 unstable orbit persisting into the period-2 regime. The algorithm did not detect similar UPOs in the period doubling cascade within the period-3 window of the logistic map, because it was set to detect period-1 UPOs only.

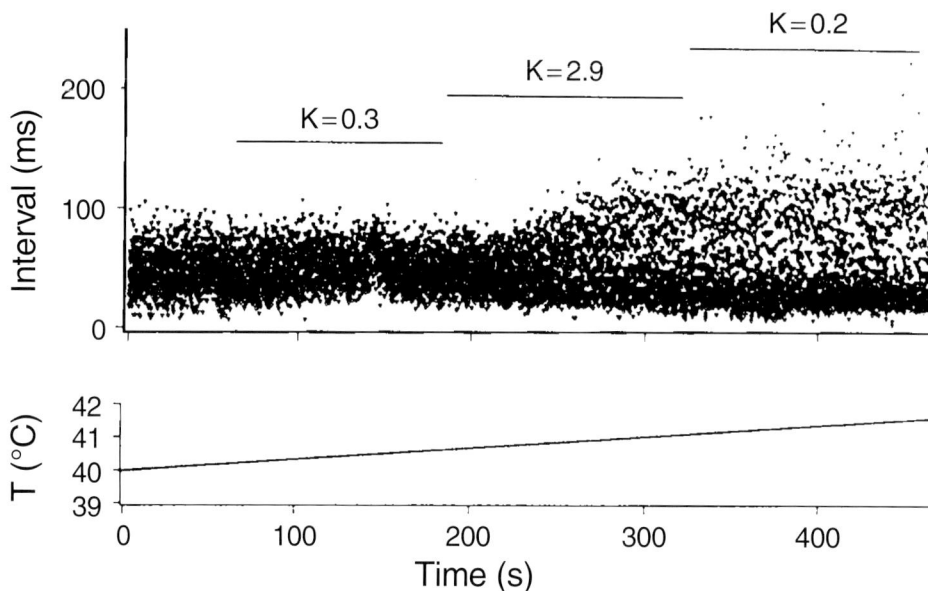

Fig. 10.6. Interspike time intervals of the rat hypothalamic neuron. The temperature was linearly swept from 40°C to 42°C, and the statistic K was calculated for three different regions

The foregoing three examples show that UPOs can be detected in datasets from sensory neurons and that they accompany period doubling bifurcations. They may also anticipate these bifurcations in some cases. The UPO analysis presented here adds a new dimension to time series analysis of biological datasets which may develop into useful indicators of biological function.

10.3 Synchronization of Stable Periodic Orbits in the Paddlefish Electroreceptor with an External Periodic Stimulus

We turn now to the paddlefish electroreceptor system. This fish is notable for a characteristic long paddle-shaped appendage, called a rostrum, preceding the animal's mouth. The rostrum is covered with thousands of electroreceptors [10.62]; see Fig. 10.7. Each electroreceptor cell is an oscillator and hence can be synchronized with an external electric field. The data shown below were obtained by applying a weak uniform electric field parallel to the surface of the rostrum. The field varied in frequency as indicated but was 'weak', that is the order of 1 to $10\,\mu$/cm. This can be compared with the behavioral sensitivity of the animal of approximately 0.5 to $1\,\mu$V/cm [10.63–10.65]. Recordings of the afferent neural spikes were obtained from a ganglion near the animal's brain. Afferent sensory neurons from receptive areas on the rostrum terminate on this ganglion. Thus we assess the state of synchronization between the spike trains from the ganglion and the external periodic signal [10.66, 10.67].

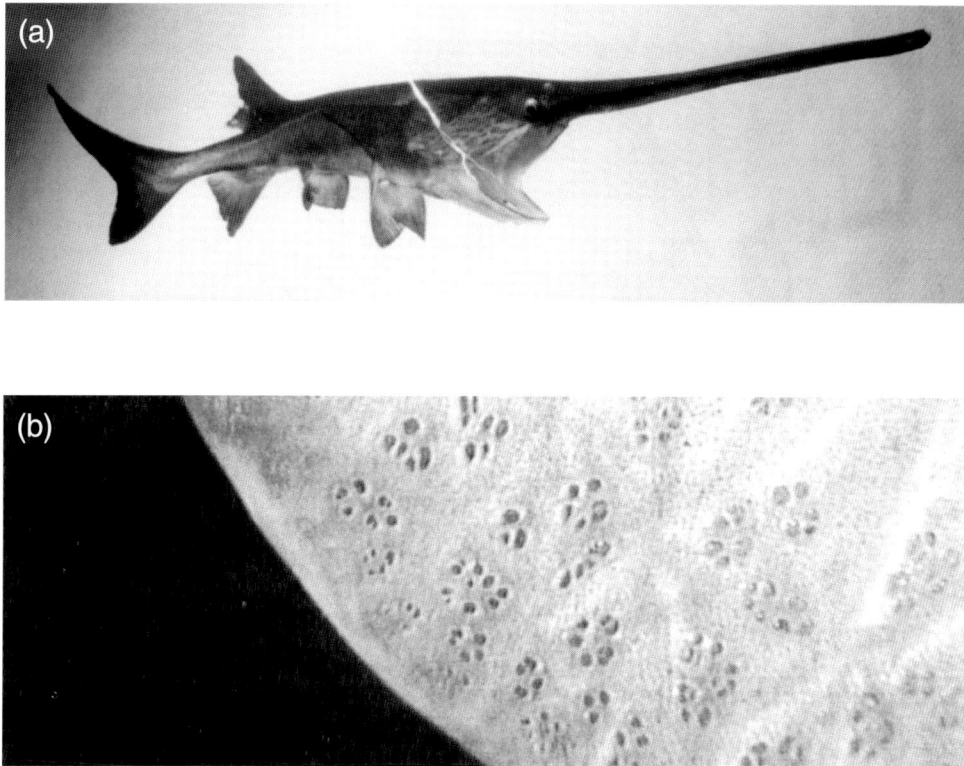

Fig. 10.7. The paddlefish (**a**) biting on an electric dipole with a 5 Hz, 10 μV periodic signal. A close-up view of the rostrum (**b**) showing the pores which comprise the receptive fields of the electroreceptor cells

10.3.1 Noisy Synchronization· The Synchrogram

As indicated in Sects. 10.1.7 and 10.1.8 above, the phase of every spike relative to the stimulus cycle can be assembled into a synchrogram as shown below in Fig. 10.8. The synchrograms indicate the best quality of synchronization for the 17 Hz stimulus frequency. This is shown by the relatively long time periods where the traces are more-or-less lines of zero slope. The upper panel for 5 Hz stimulus also shows significant synchronization, but the lower panel for 21 Hz shows virtually no synchronized activity.

10.3.2 A Statistical Measure: The Probability Density of Phases

But the synchrograms alone cannot provide a qualitative measure of the quality of synchronization. For this we assemble the probability density of phases. One can accomplish this by constructing a binning along the phase axis in Fig. 10.8, then projecting all the phase values into these bins. The result is a measured probability density of phases. The result for the data of Fig. 10.8 is shown below in Fig. 10.9.

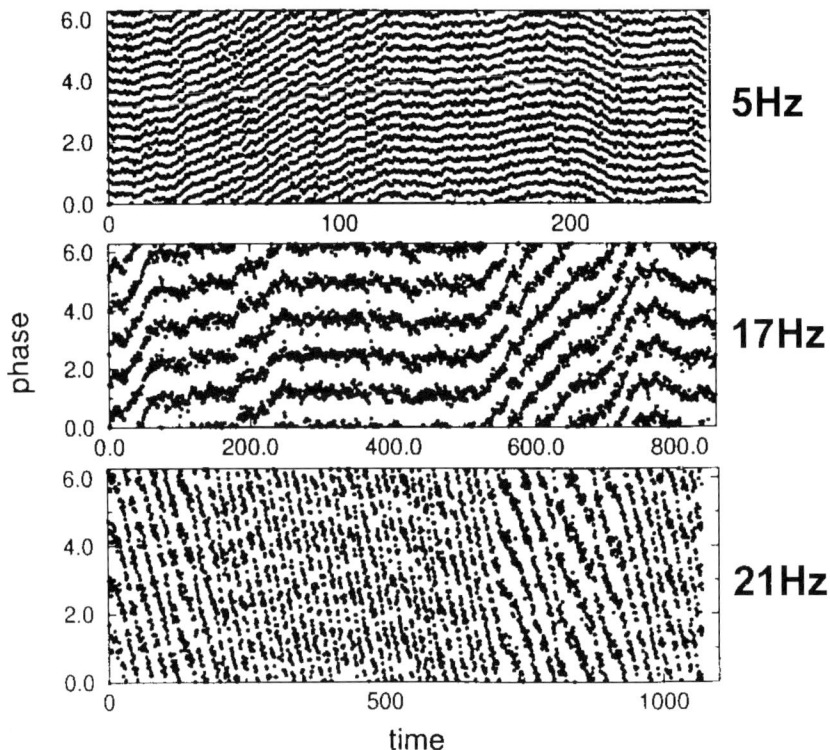

Fig. 10.8. Three synchrograms for three different stimulus frequencies as indicated. The electrore-ceptor is optimized for frequencies close to 7–12 Hz which are the characteristic frequencies emitted by its typical prey, the *Daphnia*

Measurements of synchronization such as these shown here bear on discussions relating to the encoding of stimuli by sensory neurons. Traditionally, the neuroscience community has assumed that neurons transmit information to the central nervous system via a rate coding. According to this view, all information is transmitted as the short time averaged firing rate. An alternative view - synchronization coding - is suggested by these data and analysis.

10.4 Conclusion

In this chapter we have discussed two new methods for detecting low-dimensional dynamical behavior in noisy datasets. Such datasets are common in biological and medical applications. In all cases, the dynamical noise inherent in the systems has obscured these low-dimensional objects, but the algorithms presented here are able to detect them and assess the statistical significance of their occurrences.

We have deliberately avoided speculations on the biological significance or function of these objects, leaving those questions for future study.

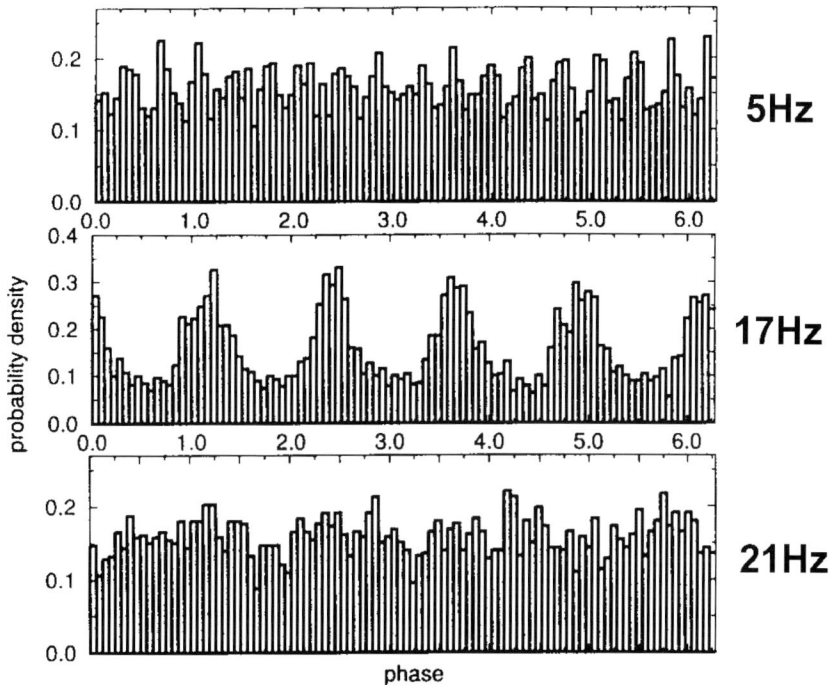

Fig. 10.9. Probability density of phases for the data shown in Fig. 10.8. The peak-to-valley relative heights are quantitative measures of the quality of synchronization. Obviously the 5 spikes/stimulus cycle shown by the density in the middle panel for 17 Hz indicate the highest quality of synchronization

Acknowledgements. We are grateful to our colleagues and collaborators who have made this work possible: in Marburg, Dr. Martin Huber, Prof. Karl-Heinz Voigt, and Mateus Dewald; and in St. Louis, Prof. Lon Wilkens, Drs. Xing Pei, Alexander Neiman, David Russell, and Kevin Dolan. This work was supported in St. Louis by the U. S. Office of Naval Research, Physics Division, and in Marburg by the Volkswagen Foundation and by the Kempkes Foundation. FM is grateful to the Alexander von Humboldt Foundation for an award that has greatly facilitated this research by making extended visits to Germany possible.

References

10.1 Archimedes, 'Planes in Equilibrium' in *Method*, currently in reconstruction; Rev: Netz, Phys. Today **53**(6), 32 (2000).

10.2 D. Ruelle, Phys. Today **47**, 24 (1994).

10.3 A. Wolf, J.B. Swift, H.L. Swinney, and J.A.Vastano, Physica D **16**, 285 (1985).

10.4 P. Grassberger and I. Procaccia, Phys. Rev. Lett. **50**, 346 (1983).

10.5 D.T. Kaplan, Physica D **73**, 38 (1994).

10.6 L.F. Olsen and W.M. Schaffer, Science **249**, 499 (1990).

10.7 G. Sugihara and R.M. May, Nature **344**, 734 (1990).

10.8 S.H. Strogatz, *Nonlinear Dynamics and Chaos* (Addison-Wesley, Reading, 1994).

10.9 X. Pei, K. Dolan, and Y.-C. Lai, Chaos **8**, 853 (1998).

10.10 K.T. Dolan, Analysis of biological and physical systems using nonlinear topological methods, (Ph.D. Dissertation, University of Missouri, Rolla, 2000).

10.11 L. Omberg, K. Dolan, A. Neiman, and F. Moss, Phys. Rev. E **61**, 4848 (2000).

10.12 M.T. Huber, J.-C. Krieg, H.A. Braun, X. Pei, A. Neiman, and F. Moss, Neurocomputing **32/33**, 823 (2000).

10.13 J. Theiler, S. Eubank, A. Longtin, B. Galdrikian, and J.D. Farmer, Physica D **58**, 77 (1992).

10.14 T. Schreiber, Phys. Rev. Lett. **80**, 2105 (1998).

10.15 K. Dolan, A. Witt, M. Spano, A. Neiman, and F. Moss, Phys. Rev. E **59**, 5235 (1999).

10.16 P.R. Bevington, *Data Reduction and Error Analysis* (McGraw-Hill, New York, 1969), p. 48.

10.17 P. Cvitanovic, Phys. Rev. Lett. **61**, 2729 (1988).

10.18 P. Cvitanovic, Physica D **51**, 138 (1991).

10.19 R. Artuso, E. Aurell, and P. Cvitanovic. Nonlinearity **3**, 361 (1990).

10.20 R. Artuso, E. Aurell, and P. Cvitanovic, Nonlinearity **3**, 325 (1990).

10.21 R. Badii, E. Brun, M. Finardi, L. Flepp, R. Holzner, J. Parisi, C. Reyl, and J. Simonet, Rev. Mod. Phys. **66**, 1389 (1994).

10.22 E.R. Hunt, Phys. Rev. Lett. **67**, 1953 (1991).

10.23 E. Ott, C. Grebogi, and J.A. Yorke, Phys. Rev. Lett. **64**, 1196 (1990).

10.24 W.L. Ditto, S.N. Rauseo, and M.L. Spano, Phys. Rev. Lett. **65**, 3211 (1990).

10.25 A. Garfinkel, M.L. Spano, W.L. Ditto, and J.N. Weiss, Science **257**, 1230 (1992).

10.26 V. Petrov, V. Gaspar, J. Masere, and K. Showalter, Nature **361**, 240 (1993).

10.27 R.W. Rollins, P. Parmananda, and P. Sherard, Phys. Rev. E **47**, R780 (1993).

10.28 R. Roy, T.W. Murphy, T.D. Maier, Z. Gillis, and E.R. Hunt, Phys. Rev. Lett. **68**, 1259 (1992).

10.29 S.J. Schiff, K. Jerger, D. Duong, T. Chang, M.L. Spano, and W.L. Ditto, Nature **370**, 615 (1994).

10.30 D.J. Christini and J.J. Collins, Phys. Rev. Lett. **75**, 2782 (1995).

10.31 D. Pierson and F. Moss, Phys. Rev. Lett. **75**, 2124 (1995).

10.32 X. Pei and F. Moss, Nature **379**, 618 (1996).

10.33 X. Pei and F. Moss, Int. J. Neural Syst. **7**, 429 (1996).

10.34 L. Menendez de la Prida, N. Stollenwerk, and J.V. Sanchez-Andres, Physica D **110**, 323 (1997).

10.35 K. Narayanan, R.B. Govindan, and M.S. Gopinathan, Phys. Rev. E **57**, 4594 (1998).

10.36 M. Le Van Quyen, J. Martinerie, C. Adam, and F.J. Varela, Phys. Rev. E **56**, 3401 (1997).

10.37 H.A. Braun, K. Schäfer, K. Voigt, R. Peters, F. Bretschneider, X. Pei, L. Wilkens, and F. Moss, J. Comp. Neurosci. **4**, 335 (1997).

10.38 H.A. Braun, M. Dewald, K. Schäfer, K. Voigt, X. Pei, K. Dolan, and F. Moss, J. Comp. Neurosci. **7**, 17 (1999).

10.39 H.A. Braun, M. Dewald, K. Voigt, M. Huber, X. Pei, and F. Moss, Neurocomputing **26/27**, 79 (1999).

10.40 H.A. Braun, M.T. Huber, M. Dewald, K. Schäfer, and K. Voigt, Int. J. Bifurcat. Chaos **8**, 881 (1998).

10.41 H. A. Braun, M. T. Huber, N. Anthes, K. Voigt, A. Neiman, X. Pei, and F. Moss, Neuro-computing **32-33**, 51 (2000).

10.42 H.A. Braun, M.T. Huber, N. Anthes, K. Voigt, A. Neiman, X. Pei, and F. Moss, Biosystems **62**, 99 (2001), [Note, for the equations and parameter values please see the URL: http://neurodyn.umsl.edu/hodgkin-huxley.html.]

10.43 R. Gilmore, X. Pei, and F. Moss, Chaos **9**, 812 (1999).

10.44 U. Feudel, A. Neiman, X. Pei, W. Wojtenek, H. Braun, M. Huber, and F. Moss, Chaos **10**, 231 (2000).

10.45 W. Braun, B. Eckhardt, H.A. Braun, and M. Huber, Phys. Rev. E **62**, 6352 (2000).

10.46 C. Huygens, *Horoloqium Oscilatorium* (Parisiis, France, 1673).

10.47 L. Glass and M.C. Mackey, *From Clocks to Chaos. The Rhythms if Life* (Princeton University Press, Princeton, 1988).

10.48 A.T. Winfree, *The Geometry of Biological Time* (Springer, New York, 1980).

10.49 C. Schäfer, M.G. Rosenblum, H. Abel, and J. Kurths, Phys. Rev. E **60**, 857 (1998).

10.50 C. Schäfer, M.G. Rosenblum, J. Kurths, and H. Abel, Nature **392**, 239 (1998).

10.51 P. Tass, M. Rosenblum, J. Weule, J. Kurths, A. Pickovsky, J. Volkmann, A. Schnitzler, and J.-H. Freund, Phys. Rev. Lett. **81**, 3291 (1998).

10.52 R.L. Stratonovich, *Topics in the Theory of Random Noise* (Gordon and Breach, New York, 1967).

10.53 A. Neiman, Phys. Rev. E **49**, 3484 (1994).

10.54 A. Neiman, L. Schimansky-Geier, A. Cornell-Bell, and F. Moss, Phys. Rev. Lett. **83** 4896 (1999).

10.55 A. Neiman, L. Schimansky-Geier, F. Moss, B. Shulgin, and J.J. Collins, Phys. Rev. E **60**, 284 (1999).

10.56 A. Neiman, A. Silchenko, V. Anishchenko, and L. Schimansky-Geier, Phys. Rev. E **58**, 7118 (1998).

10.57 B. Shulgin, A. Neiman, and V. Anishchenko, Phys. Rev. Lett. **75**, 4157 (1995).

10.58 L.M. Pecora and T.L. Carroll, Phys. Rev. Lett. **64**, 821 (1990).

10.59 A.S. Pikovsky, M.G. Rosenblum, G.V. Osipov, and J. Kurths, Physica D **104**, 219 (1996).

10.60 M.G. Rosenblum, A.S. Pikovsky, and J. Kurths, Phys. Rev. Lett. **76**, 1804 (1996).

10.61 H.A. Braun, H. Wissing, K. Schäfer, and M. Hirsch, Nature **367**, 270 (1994).

10.62 L.A. Wilkens, D.F. Russell, X. Pei, and C. Gurgens, Proc. R. Soc. B **264**, 1723 (1997).

10.63 D. Russell, L. Wilkens, and F. Moss, Nature **402**, 219 (1999).

10.64 P. Greenwood, L. Ward, D. Russell, A. Neiman, and F. Moss, Phys. Rev. Lett. **84,** 4773 (2000).

10.65 J.A. Freund, J. Kienert, L. Schimansky-Geier, B. Beisner, A. Neiman, D. Russell, T. Yakusheva, and F. Moss, Phys. Rev. E. **63**, 31910 (2001).

10.66 A. Neiman, X. Pei, D. Russell, W. Wojtenek, L. Wilkens, F. Moss, H. A. Braun, M. T. Huber, and K. Voigt, Phys. Rev. Lett. **82**, 660 (1999).

10.67 A.B. Neiman, D.F. Russell, X. Pei, W. Wojtenek, J. Twitty, E. Simonotto, B.A. Wettring, E. Wagner, L.A. Wilkens, and F. Moss, Int. J. Bifurcat. Chaos **10**, 2499 (2000).

11. Crowd Disasters and Simulation of Panic Situations

Dirk Helbing, Illés J. Farkas, and Tamás Vicsek

One of the most tragic collective behaviors is a panic stampede [11.1–11.9], as it often leads to the death of people who are either crushed or trampled down by others. While this behavior is comprehensible in life-threatening situations like fires in crowded buildings [11.10, 11.11], it is hardly understood in cases of a rush for good seats at a pop concert [11.12], or without any obvious reasons. Unfortunately, the frequency of such disasters is increasing [11.12], as growing population densities combined with easier transportation lead to greater mass events like pop concerts, sporting events, and demonstrations. Nevertheless, systematic studies of panics [11.8] are rare [11.5, 11.10, 11.12]. Moreover, there is a scarcity of quantitative theories capable of predicting the dynamics of human crowds [11.13–11.15]. Here we show that simulations of pedestrian behavior can give valuable insights into the mechanisms and preconditions of panic, jamming, and the observed 'faster-is-slower effect'. We also provide clues to practical ways of minimizing the related tragedies. Furthermore, we identify an optimal strategy for collective problem solving in crisis situations, corresponding to a suitable mixture of individualistic and herding behavior.

11.1 Introduction

With some exceptions, panics are observed in cases of scarce or dwindling resources [11.8, 11.10], which are either required for survival or anxiously desired. They are usually distinguished into escape panics ('stampedes', bank, or stock market panics) and acquisitive panics ('crazes', speculative manias) [11.5, 11.6], but in some cases this classification is questionable [11.12].

Fig. 11.0. Panicking football fans trying to escape the football stadium in Sheffield. Hardly anybody manages to pass the open door, because of the clogging effect occurring in crowds at high pressures

It is believed that panicking people are obsessed by short-term personal interests uncontrolled by social and cultural constraints [11.5, 11.10]. This is possibly a result of the reduced attention in situations of fear [11.10], which also causes alternatives like side exits to be mostly ignored [11.11]. It is, however, often attributed to social contagion [11.1–11.10, 11.12], i.e. a transition from individual to mass psychology, in which individuals transfer control over their actions to others [11.6], leading to conformity [11.16]. This 'herding behavior' (regarding the herding behavior of stock markt dealers cf. Chap. 14) is irrational, as it often leads to bad overall results like dangerous overcrowding and slower escape [11.6, 11.11, 11.12]. In this way, herding behavior increases the fatalities, or, more generally, the damage in the crisis faced.

The various socio-psychological theories for this contagion assume hypnotic effects, rapport, mutual excitation of a primordial instinct, circular reactions, social facilitation (see the summary by Brown [11.4]), or the emergence of normative support for selfish behavior [11.7]. Brown [11.4] and Coleman [11.6] add another explanation related to the prisoner's dilemma [11.17] or common goods dilemma (an example is discussed in Chap. 6) [11.18], showing that it is reasonable to make one's subsequent actions contingent upon those of others, but the socially favorable behavior of orderly walking is unstable, which normally gives rise to rushing by everyone. These thoughtful considerations are well compatible with many aspects discussed above and with the classical experiments by Mintz [11.8], which showed that jamming in escape situations depends on the reward structure ('payoff matrix').

Nevertheless and despite of the frequent reports in the media and many published investigations of crowd disasters (see Table 11.1), a quantitative understanding of the observed phenomena in panic stampedes is still lacking. Here, we add another aspect to the explanation of panics by simulating a computer model for the crowd dynamics of pedestrians. Approaches to social phenomena in the spirit of statistical physics are quite promising as they have led to a number of exciting discoveries [11.19–11.23]. We are at a point where the exact methods of physics combined with the potentials of computers start to produce relevant results on society.

During our study we managed to find answers to questions like the following:

1. Why do people get ahead more slowly when they are trying to escape fast?
2. Why does the pedestrian flow through exits become irregular in panic situations?
3. Do wide spaces along corridors and escape routes increase the efficiency of leaving?
4. Why are alternative exits overlooked or not efficiently used in escape situations?
5. Why is it possible that panics are sometimes triggered without any apparent reason?

Table 11.1. Incomplete list of major crowd disasters after Smith and Dickie [11.24], http://ourworld.compuserve.com/homepages/G_Keith_Still/disaster.htm, http://SportsIllustrated.CNN.com/soccer/world/news/2000/07/09/stadium_disasters_ap/, and other Internet sources. The number of injured people was usually a multiple of the fatalities

Date	Place	Venue	Deaths	Injured	Reason
1863	Santiago, Chile	Church	2000		
1881	Vienna, Austria	Theatre	570		
1883	Sunderland, UK	Theatre	182		
1902	Ibrox, UK	Stadium	26	517	Collapse of West Stand
1903	Chicago, USA	Theatre	602		
1943	London, UK	Underground Station	173		Stampede while air raid
1946	Bolton, UK	Stadium	33	400	Collapse of a wall
1955	Santiago, Chile	Stadium	6		Fans trying to force their way into the stadium
1961	Rio de Janeiro, Brazil	Circus	250		
1964	Lima, Peru	Stadium	318	500	Goal disallowed
1967	Kayseri, Turkey	Stadium	40		
1968	Buenos Aires, Argentina	Stadium	75	150	Fans fleeing from fire
1970	St. Laurent-du-Pont, France	Dance Hall	146		
1971	Ibrox, UK	Stadium	66	140	Collapse of barriers
1971	Salvador, Brazil	Stadium	4	1500	Fight and wild rush
1974	Cairo, Egypt	Stadium	48		Crowds break barriers
1976	Port-au-Prince, Haiti	Stadium	2		Firecracker
1979	Nigeria	Stadium	24	27	Light failure
1979	Cincinatti, USA	Stadium	11		Fans trying to force their way into the stadium
1981	Piraeus, Greece	Stadium	24		Rush of leaving fans
1981	Sheffield, UK	Stadium		38	Crowd surge
1982	Cali, Columbia	Stadium	24	250	Provocation by drunken fans
1982	Moscow, USSR	Stadium	340		Re-entering fans after last minute goal
1985	Bradford, UK	Stadium	56		Fire in wooden terrace section
1985	Mexico City, Mexico	Stadium	10	29	Fans trying to force their way into the stadium
1985	Brussels, Belgium	Stadium	38	> 400	Riots break out and wall collapses
1987	Tripoli, Libya	Stadium	2	16	Collapse of a wall
1988	Katmandu, Nepal	Stadium	93	> 100	Stampede due to hailstorm
1989	Hillsborough, Sheffield, UK	Stadium	96		Fans trying to force their way into the stadium
1990	Mecca, Saudi Arabia	Pedestrian Tunnel	1425		Overcrowding
1991	Orkney, South Africa	Stadium	> 40		Fans trying to escape fighting
1991	New York, USA	Stadium	9		Overcrowding at concert
1992	Rio de Janeiro, Brazil	Stadium		50	Part of the fence giving way
1992	Bastia, Corsica	Stadium	17	1900	
1994	Mecca, Saudi Arabia		270		Rush at 'stoning the devil' ritual

continued on next page

334   Dirk Helbing et al.

Table 11.1 (*Continued*)

Date	Place	Venue	Deaths	Injured	Reason
7/31/1996	Tembisa, South Africa	Railway Station	15	> 20	Electric cattle prods used by security guards
10/17/1996	Guatemala City, Guatemala	Stadium	80	180	Fans trying to force their way into the stadium
6/29/1997	Las Vegas, USA	Hotel	1	50	Gunshot
7/2/1997	Düsseldorf, Germany	Stadium	1	> 300	Overcrowding at concert
3/23/1998	Dhaka, Bangladesh	Multi-Storey Building	1	15	Fire stampede
4/9/1998	Mecca, Saudi Arabia	Holy Place (Meda)	107		Pilgrim stampede at 'stoning the devil' ritual
4/19/1998	Harare, Zimbabwe	Stadium	4	10	Spectators scrambled for seats
8/22/1998	Manila, Phillipines	Presidential Action Centre	2		Large crowd waiting for jobs and housing
12/1/1998	Chervonohrad, Ukraine	Cinema	4		Stampede due to in- and outcoming children
12/25/1998	Lima, Peru	Disco	9	7	Tear gas
5/31/1999	Minsk, Belarus	Subway Station	51	150	Heavy rain at rock concert
1/15/1999	Kerala, India	Hindu Shrine	> 50		Collapse of parts of the shrine
10/7/1999	Benin, Nigeria	Religious Place	14		Stampede at a Christian revivalist rally
12/5/1999	Innsbruck, Austria	Stadium	5	25	Fans re-entering the stadium?
3/3/2000	Kaloroa, Bangladesh	Examination Place	5		Stampede to enter an examination hall
3/12/2000	Mecca, Saudi Arabia	Holy Place	2	4	Pilgrims were crushed by the crowd
3/24/2000	Durban, South Africa	Disco	13	44	Tear gas
3/24/2000	Chiaquelane, Mozambique	Chiaquelane Camp	5	10	Aid chaos
4/17/2000	Lisbon, Portugal	Nightclub	7	65	Poisonous gas bombs
4/21/2000	Seville, Spain			30	Spectators panicked in a crowd during Good Friday procession
4/23/2000	Monrovia, Liberia	Stadium	3		Fans trying to force their way into the stadium
5/26/2000	Lahore, Pakistan	Circus	8	3	Guards used batons
6/5/2000	Addis Ababa, Ethiopia	Memorial Place	14		Children trying to cover from a rainstorm
6/30/2000	Roskilde, Denmark	Stadium	8	25	Failure of loudspeakers
7/9/2000	Harare, Zimbabwe	Stadium	12		Tear gas
Dec. 2000	São Januário, Brazil	Stadium		200	Oversold stadium

6. How is it possible to estimate the number of casualties in emergency situations?
7. What would be the best escape strategy from a smoky room, when the exits are not visible?

One may, therefore, speak of a breakthrough in the understanding of the special dynamic phenomena observed in crowds under emergency conditions.

11.2 Observations

After having carefully studied the related socio-psychological literature [11.4, 11.8, 11.9, 11.25], reports in the media, available video materials (see http://angel.elte.hu/~panic/), empirical investigations [11.2, 11.10, 11.11, 11.26], and engineering handbooks [11.27, 11.28], we can summarize the following characteristic features of escape panics:

1. People move or try to move considerably faster than normal [11.27].
2. Individuals start pushing, and interactions among people become physical in nature.
3. Moving and, in particular, passing of a bottleneck becomes incoordinated [11.8].
4. At exits, arching and clogging are observed [11.27].
5. Jams are building up [11.25].
6. The physical interactions in the jammed crowd add up and cause dangerous pressures up to 4450 N/m [11.11, 11.24], which can bend steel barriers or tear down brick walls.
7. Escape is further slowed down by fallen or injured people turning into 'obstacles'.
8. People show a tendency of mass behavior, i.e. to do what other people do [11.9, 11.10].
9. Alternative exits are often overlooked or not efficiently used in escape situations [11.10, 11.11].

To give a more personal impression of situations when panic strikes, we add some quotations:

1. *"They just kept pushin' forward and they would just walk right on top of you, just trample over ya like you were a piece of the ground."* (After the panic at 'The Who Concert Stampede' in Cincinatti.)
2. *"People were climbin' over people ta get in ... an' at one point I almost started hittin' 'em, because I could not believe the animal, animalistic ways of the people, you know, nobody cared."* (After the panic at 'The Who Concert Stampede'.)

3. *"Smaller people began passing out. I attempted to lift one girl up and above to be passed back ... After several tries I was unsuccessful and near exhaustion."* (After the panic at 'The Who Concert Stampede'.)
4. *"I couldn't see the floor because of the thickness of the smoke."* (After the 'Hilton Hotel Fire' in Las Vegas.)
5. *"The club had two exits, but the young people had access to only one"*, said Narend Singh, provincial minister for agriculture and environmental affairs. However, the club's owner, Rajan Naidoo, said the club had four exits, and that all were open. *"I think the children panicked and headed for the main entrance where they initially came in"*, he said. (After the 'Durban Disco Stampede'.)

11.3 Generalized Force Model of Pedestrian Motion

The above observations have encouraged us to model the collective phenomenon of escape panic in the spirit of self-driven many-particle systems. Our computer simulations of the crowd dynamics of pedestrians are based on a generalized force model [11.20, 11.29–11.40], which is particularly suited to describe the fatal build up of pressure observed during panics [11.2, 11.11, 11.12, 11.24]. We assume a mixture of socio-psychological [11.20, 11.29, 11.31] and physical forces influencing the behavior in a crowd: each of N pedestrians i of mass m_i likes to move with a certain desired speed v_i^0 in a certain direction e_i^0, and therefore tends to correspondingly adapt his or her actual velocity v_i within a certain characteristic time τ_i. Simultaneously, he or she tries to keep a velocity-dependent distance to other pedestrians j, walls W, and fire fronts F. This can be modeled by 'interaction forces' f_{ij}, f_{iW}, and f_{iF}, respectively. Additionally, we take into account fluctuations $\xi_i(t)$ of amplitude η_i. In mathematical terms, the change of velocity in time t is then given by the acceleration equation

$$m_i \frac{d\boldsymbol{v}_i}{dt} = m_i \frac{v_i^0(t)\boldsymbol{e}_i^0(t) - \boldsymbol{v}_i(t)}{\tau_i} + \sum_{j(\neq i)} \boldsymbol{f}_{ij} + \sum_W \boldsymbol{f}_{iW} + \boldsymbol{f}_{iF} + \boldsymbol{\xi}_i(t)\,, \quad (11.1)$$

while the change of position $\boldsymbol{r}_i(t)$ is given by the velocity $\boldsymbol{v}_i(t) = d\boldsymbol{r}_i/dt$. We describe the *psychological tendency* of two pedestrians i and j to stay away from each other by a repulsive interaction force $A_i \exp[(r_{ij} - d_{ij})/B_i]\,\boldsymbol{n}_{ij}$, where A_i and B_i are constants. $d_{ij} = \|\boldsymbol{r}_i - \boldsymbol{r}_j\|$ denotes the distance between the pedestrians' centers of mass, and $\boldsymbol{n}_{ij} = (n_{ij}^1, n_{ij}^2) = (\boldsymbol{r}_i - \boldsymbol{r}_j)/d_{ij}$ is the normalized vector pointing from pedestrian j to i. If their distance d_{ij} is smaller than the sum $r_{ij} = (r_i + r_j)$ of their radii r_i and r_j, the pedestrians touch each other. In this case, we assume two additional forces inspired by granular interactions [11.41, 11.42], which are essential for understanding the particular effects in panicking crowds: a *'body force'* $k(r_{ij} - d_{ij})\,\boldsymbol{n}_{ij}$ counteracting body compression and a *'sliding friction force'* $\kappa(r_{ij} - d_{ij})\,\Delta v_{ji}^t\,\boldsymbol{t}_{ij}$ impeding *relative* tangential motion, if pedestrian i comes close to j. Herein, $\boldsymbol{t}_{ij} = (-n_{ij}^2, n_{ij}^1)$ means the tangential direction and $\Delta v_{ji}^t = (\boldsymbol{v}_j - \boldsymbol{v}_i)\cdot\boldsymbol{t}_{ij}$ the tangential velocity difference, while k and κ represent large constants. In summary, we have

$$\boldsymbol{f}_{ij} = \{A_i\exp[(r_{ij} - d_{ij})/B_i] + k\Theta(r_{ij} - d_{ij})\}\,\boldsymbol{n}_{ij} + \kappa\Theta(r_{ij} - d_{ij})\Delta v_{ji}^t\,\boldsymbol{t}_{ij}\,,\tag{11.2}$$

where the function $\Theta(x)$ is zero if the pedestrians do not touch each other $(d_{ij} > r_{ij})$; otherwise it is equal to the argument x.

The interaction with the walls is treated analogously, i.e. if d_{iW} means the distance to wall W, $\boldsymbol{n}_{iW}$ denotes the direction perpendicular to it, and $\boldsymbol{t}_{iW}$ the direction tangential to it, the corresponding interaction force with the wall reads

$$\boldsymbol{f}_{iW} = \{A_i\exp[(r_i - d_{iW})/B_i] + k\Theta(r_i - d_{iW})\}\,\boldsymbol{n}_{iW}$$
$$- \kappa\Theta(r_i - d_{iW})(\boldsymbol{v}_i\cdot\boldsymbol{t}_{iW})\,\boldsymbol{t}_{iW}\,.\tag{11.3}$$

Fire fronts are treated similarly to the walls, but have a stronger psychological effect. People reached by the fire front become injured and immobile ($v_i = 0$), which replaces the physical interaction effect ($k = 0 = \kappa$).

The model parameters have been specified as follows: with a mass of $m_i = 80\,\text{kg}$, we represent an average soccer fan. The desired velocity v_i^0 can reach more than $5\,\text{m/s}$ (up to $10\,\text{m/s}$) [11.28], but the observed free velocities for leaving a room correspond to $v_i^0 \approx 0.6\,\text{m/s}$ under relaxed, $v_i^0 \approx 1\,\text{m/s}$ under normal, and $v_i^0 \lesssim 1.5\,\text{m/s}$ under nervous conditions [11.27]. A reasonable estimate for the acceleration time is $\tau_i = 0.5\,\text{s}$. With $A_i = 2\times 10^3\,\text{N}$ and $B_i = 0.08\,\text{m}$ one can reflect the distance kept at normal desired velocities [11.28] and fit the measured flows through bottlenecks [11.28], amounting to 0.73 persons per second for an effectively $1\,\text{m}$ wide door under conditions with $v_i^0 \approx 0.8\,\text{m/s}$. The parameters $k = 1.2\times 10^5\,\text{kg s}^{-2}$ and $\kappa = 2.4\times 10^5\,\text{kg m}^{-1}\text{s}^{-1}$ determine the obstruction effects in cases of physical interactions. Although, in reality, most parameters are varying individually, we chose identical values for all pedestrians to minimize the number of parameters for reasons of calibration and robustness, and to exclude irregular outflows because of parameter variations. However, to avoid model artefacts (gridlocks by exactly balanced forces in symmetrical configurations), a small amount of irregularity of almost arbitrary kind is needed. This irregularity was introduced by uniformly distributed pedestrian diameters $2r_i$ in the interval $[0.5\,\text{m}, 0.7\,\text{m}]$, approximating the distribution of shoulder widths of soccer fans.

11.4 Simulation Results

Based on the above model assumptions, we will now simulate several essential phenomena of escape panic, which are insensitive to reasonable parameter variations, but fortunately become less pronounced for broader corridors and wider exits. We will separately treat the following characteristic behaviors observed in panic situations or cases of emergency evacuation:

1. People are getting nervous, resulting in a higher level of fluctuations.
2. They are trying to escape from the source of panic, which can be reflected by a significantly higher desired velocity.
3. Individuals in complex situations, who do not know what is the right thing to do, orient at the actions of their neighbors, i.e. they tend to do what other people do. We will describe this by an additional herding interaction, but attractive interactions have probably a similar effect.

11.4.1 'Freezing by Heating' Instead of Lane Formation

For normal, relaxed situations with small fluctuation amplitudes η_i, our microsimulations of counterflows in corridors reproduce the empirically observed *formation of lanes* consisting of pedestrians with the same desired walking direction, see Fig. 11.1a [11.29–11.38, 11.43, 11.44]). If we do not assume periodic boundary conditions, these lanes are dynamically varying. Their number depends on the width of the street [11.31, 11.36], on pedestrian density, and on the noise level. Interestingly, one finds a *noise-induced ordering* [11.44, 11.45]: compared to small noise amplitudes, medium ones result in a more pronounced segregation (i.e. a smaller number of lanes), while large noise amplitudes lead to a 'freezing by heating' effect (see Fig. 11.1b).

The conventional interpretation of lane formation assumes that pedestrians tend to walk on the side which is prescribed in vehicular traffic. However, the above model can explain lane formation even without assuming a preference for *any* side [11.39, 11.44]. The most relevant point is the higher relative velocity of pedestrians walking in opposite directions. As a consequence, they have more frequent interactions until they have segregated into separate lanes. The resulting collective pattern of motion minimizes the frequency and strength of avoidance manoeuvres, if fluctuations are weak. Assuming identical desired velocities $v_i^0 = v_0$, the most stable configuration corresponds to a state with a minimization of interactions

$$-\frac{1}{N}\sum_{i\neq j}\tau \boldsymbol{f}_{ij}\cdot \boldsymbol{e}_i^0 \approx \frac{1}{N}\sum_i (v_0 - \boldsymbol{v}_i\cdot \boldsymbol{e}_i^0) = v_0(1-E)\,. \qquad (11.4)$$

(a)

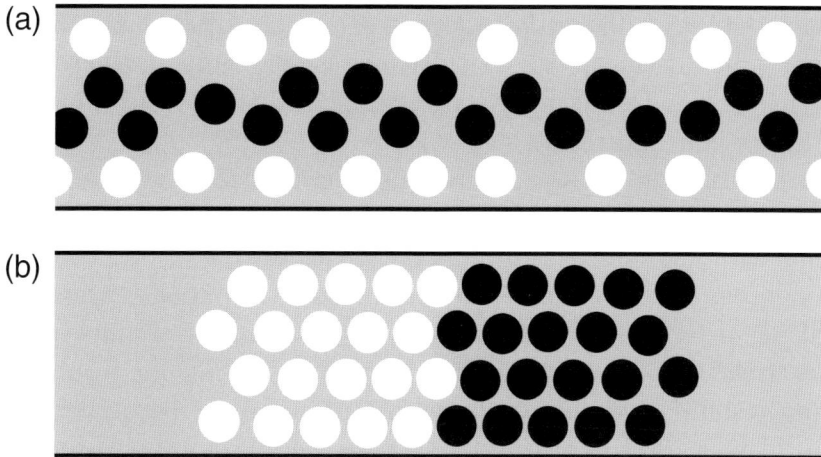

(b)

Fig. 11.1. (**a**) Formation of lanes in initially disordered pedestrian crowds with opposite walking directions and small noise amplitudes η_i (after [11.38, 11.39, 11.43]; cf. also [11.29–11.37, 11.44]). White disks represent pedestrians moving from left to right, black ones move the other way round. Lane formation does not require the periodic boundary conditions applied above; see the Java applet `http://www.helbing.org/Pedestrians/Corridor.html`. (**b**) For sufficiently high densities and large fluctuations, we observe the noise-induced formation of a crystallized, 'frozen' state (after [11.38, 11.39, 11.43])

It is related to a maximum efficiency

$$E = \frac{1}{N} \sum_i \frac{\boldsymbol{v}_i \cdot \boldsymbol{e}_i}{v_0} \tag{11.5}$$

of motion corresponding to *optimal self-organization* [11.44]. The efficiency E with $0 \leq E \leq 1$ (where $N = \sum_\alpha 1$ is the respective number of pedestrians α) describes the average fraction of the desired speed v_0 with which pedestrians actually approach their destinations. That is, lane formation 'globally' maximizes the average velocity into the respectively desired direction of motion, although the model does not even assume that pedestrians would try to optimize their behavior *locally*. This is a consequence of the symmetrical interactions among pedestrians with opposite walking directions. One can even show that a large class of driven many-particle systems, if they self-organize at all, tend to globally optimize their state [11.44].

To reflect the effect of getting nervous in panic situations, one can assume that the individual level of fluctuations is given by

$$\eta_i = (1 - n_i)\eta_0 + n_i\eta_{\max} , \tag{11.6}$$

where n_i with $0 \leq n_i \leq 1$ measures the nervousness of pedestrian i. The parameter η_0 means the normal and $\eta_{\max}$ the maximum fluctuation strength.

It turns out that, at sufficiently high pedestrian densities, lanes are destroyed by increasing the fluctuation strength (which is analogous to the temperature). However, instead of the expected transition from the 'fluid' lane state to a disordered, 'gaseous' state, a solid state is formed [11.39]. It is characterized by a blocked situation with a regular (i.e. 'crystallized' or 'frozen') structure so that we call this paradoxical transition *freezing by heating* (see Fig. 11.1b). Notably enough, the blocked state has a *higher* degree of order, although the internal energy is *increased* and the resulting state is *metastable* with respect to structural perturbations such as the exchange of oppositely moving particles. Therefore, 'freezing by heating' is just opposite to what one would expect for equilibrium systems, and different from fluctuation-driven ordering phenomena in metallic glasses and some granular systems [11.46–11.48], where fluctuations lead from a disordered *metastable* to an ordered *stable* state. A rather general model for related *noise-induced ordering* processes has been recently developed [11.45].

The preconditions for the unusual freezing-by-heating transition are the additional driving term $v_i^0 \boldsymbol{e}_i^0 / \tau_i$ and the dissipative friction $-\boldsymbol{v}_i / \tau_i$, while the sliding friction force is not required. Inhomogeneities in the channel diameter or other impurities which temporarily slow down pedestrians can further this transition at the respective places. Finally note that a transition from fluid to blocked pedestrian counter flows is also observed, when a critical particle density is exceeded [11.39, 11.49].

11.4.2 Transition to Incoordination due to Clogging

At bottlenecks such as doors, one observes jamming and, in cases of counterflows, additional oscillations of the passing direction (see Fig. 11.8). The simulated outflow from a room is well-coordinated and regular, if the desired velocities $v_i^0 = v_0$ are normal. However, for desired velocities above $1.5 \, \text{m/s}$, i.e. for people in a rush, we find an irregular succession of arch-like blockings of the exit and avalanche-like bunches of leaving pedestrians, when the arches break (see Fig. 11.2a, b). This phenomenon is compatible with the empirical observations mentioned above and comparable to intermittent clogging found in granular flows through funnels or hoppers [11.41, 11.42] (although this has been attributed to *static* friction between particles without remote interactions, and the transition to clogging has been observed for small enough openings rather than for a variation of the driving force).

11.4.3 'Faster-is-Slower Effect' due to Impatience

Since clogging is connected with delays, trying to move faster (i.e. increasing v_i^0) can cause a smaller average speed of leaving, if the friction parameter κ is large enough (see Fig. 11.2c, d). This 'faster-is-slower effect' is particularly

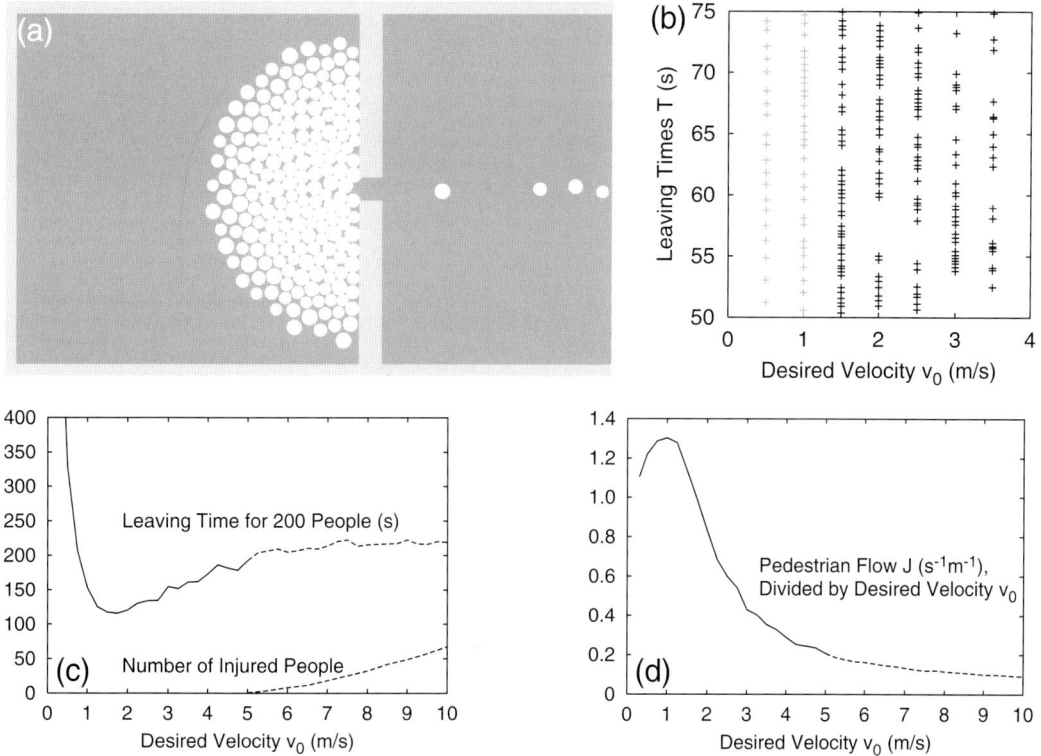

Fig. 11.2. Simulation of pedestrians moving with identical desired velocity $v_i^0 = v_0$ towards the 1 m wide exit of a room of size $15\,\text{m} \times 15\,\text{m}$ (from [11.40]). (**a**) Snapshot of the scenario. Dynamic simulations are available at `http://angel.elte.hu/~panic/`. (**b**) Illustration of leaving times of pedestrians for various desired velocities v_0. Irregular outflow due to clogging is observed for high desired velocities ($v_0 \geq 1.5\,\text{m/s}$, see dark plusses). (**c**) Under conditions of normal walking, the time for 200 pedestrians to leave the room decreases with growing v_0. (**d**) Desired velocities higher than 1.5 m/s reduce the efficiency of leaving, which becomes particularly clear when the outflow J is divided by the desired velocity. This is due to pushing, which causes additional friction effects. Moreover, above a desired velocity of about $v_0 = 5\,\text{m/s}$ (- -), people are injured and become non-moving obstacles for others, if the sum of the magnitudes of the radial forces acting on them divided by their circumference exceeds a pressure of $1\,600\,\text{N/m}$ [11.24]

tragic in the presence of fires, where fleeing people reduce their own chances of survival. The related fatalities can be estimated by the number of pedestrians reached by the fire front (see Fig. 11.3).

Since our friction term has, on average, no deceleration effect in the crowd, if the walls are sufficiently remote, the arching underlying the clogging effect requires a *combination* of several effects:

1. slowing down due to a bottleneck such as a door and

2. strong inter-personal friction, which becomes dominant when pedestrians get too close to each other.

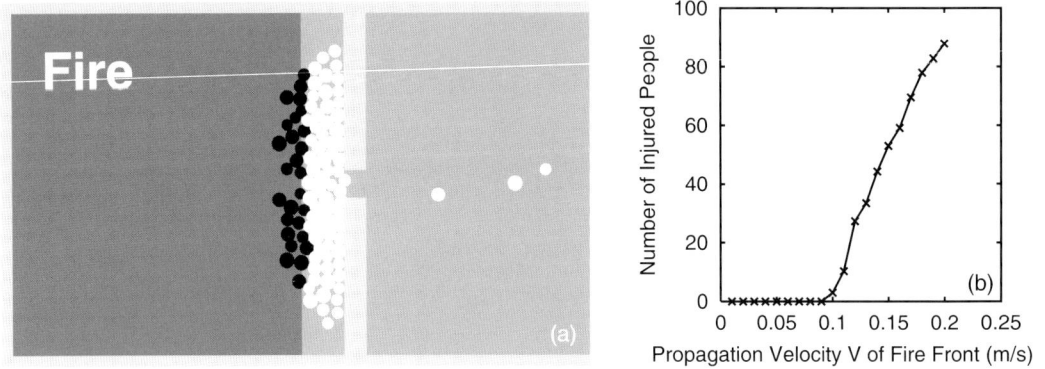

Fig. 11.3. Simulation of $N = 200$ individuals fleeing from a linear fire front, which propagates from the left to the right wall with velocity V, starting at time $t = 5\,\mathrm{s}$ (for a Java simulation applet, see http://angel.elte.hu/~panic/). (**a**) Snapshot of the scenario for a $15\,\mathrm{m} \times 15\,\mathrm{m}$ large room with one door of width $1\,\mathrm{m}$. The fire is indicated by dark gray color. Pedestrians reached by the fire front are injured and symbolised by black disks, while the white ones are still active. The psychological effect of the fire front is assumed 10 times stronger than that of a normal wall ($A_F = 10A_i$). (**b**) Number of injured persons (casualities) as a function of the propagation velocity V of the fire front, averaged over 10 simulation runs. Up to a critical propagation velocity V_{crit} (here: about $0.1\,\mathrm{m/s}$), nobody is injured. However, for higher velocities, we find a fast increase of the number of casualties with increasing V. The transition is continuous

Consequently, the danger of clogging can be minimised by avoiding bottle-necks in the construction of stadia and public buildings. Notice, however, that jamming can also occur at widenings of escape routes! This surprising result is illustrated in Fig. 11.4 and originates from disturbances due to pedestrians, who expand in the wide area because of their repulsive interactions or try to overtake each other. These squeeze into the main stream again at the end of the widening, which acts like a bottleneck and leads to jamming. *Significantly improved outflows can be reached by columns placed asymmetrically in front of the exits, which also prevent the build up of fatal pressures* (see Fig. 11.5 and the Java applets at http://angel.elte.hu/~panic/).

11.4.4 'Phantom Panics'

Sometimes, panics have occured *without* any comprehensible reasons such as a fire or another threatening event (e.g. in Moscow, 1982; Innsbruck, 1999). Due to the 'faster-is-slower effect', panics can be triggered by small pedes-trian counterflows [11.11], which cause delays to the crowd intending to leave. Consequently, stopped pedestrians in the back, who do not see the reason for the temporary slowdown, are getting impatient and pushy. In accordance with observations [11.29, 11.36], one may describe this by increasing the desired velocity, for example, by the formula

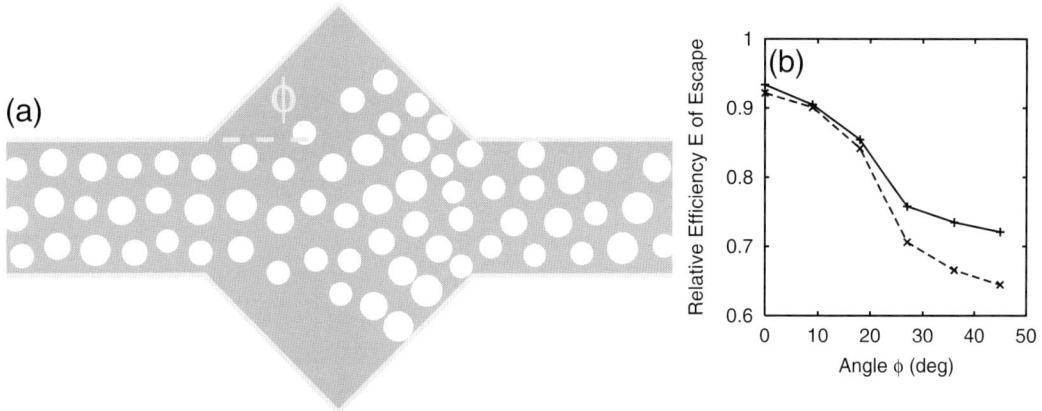

Fig. 11.4. Simulation of an escape route with a wider area (from [11.40], see also the Java applets supplied at http://angel.elte.hu/~panic/). (**a**) Illustration of the scenario with $v_i^0 = v_0 = 2\,\mathrm{m/s}$. The corridor is $3\,\mathrm{m}$ wide and $15\,\mathrm{m}$ long, the length of the triangular pieces in the middle being $2 \times 3\,\mathrm{m} = 6\,\mathrm{m}$. Pedestrians enter the simulation area on the left-hand side with an inflow of $J = 5.5\,\mathrm{s}^{-1}\mathrm{m}^{-1}$ and flee towards the right-hand side. (**b**) Efficiency of leaving as a function of the angle ϕ characterizing the width of the central zone, i.e. the difference from a linear corridor. The relative efficiency $E = \langle \boldsymbol{v}_i \cdot \boldsymbol{e}_i^0 \rangle / v_0$ measures the average velocity along the corridor compared to the desired velocity and lies between 0 and 1 (—). While it is almost one (i.e. maximal) for a linear corridor ($\phi = 0$), the efficiency drops by about 20%, if the corridor contains a widening. This becomes comprehensible, if we take into account that the widening leads to disturbances by pedestrians, who expand in the wide area due to their repulsive interactions or try to overtake each other, and squeeze into the main stream again at the end of the widening. Hence, the right half of the illustrated corridor acts like a bottleneck and leads to jamming. The drop of efficiency E is even more pronounced, (i) in the area of the widening where pedestrian flow is most irregular (- -), (ii) if the corridor is narrow, (iii) if the pedestrians have different or high desired velocities, and (iv) if the pedestrian density in the corridor is higher

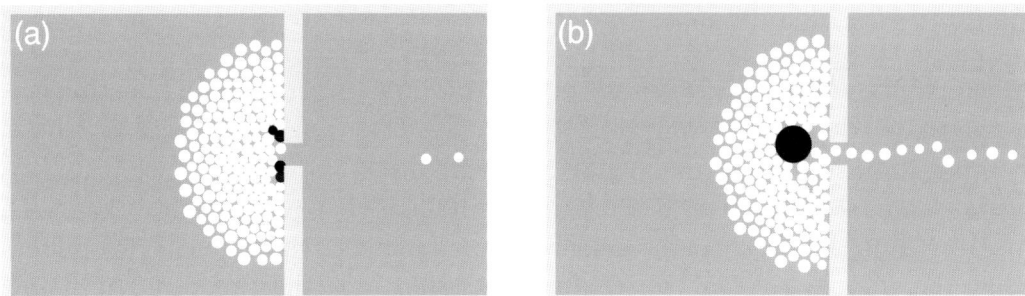

Fig. 11.5. (**a**) In panicking crowds, high pressures build up due to physical interactions. This can injure people (black disks), who turn into obstacles for other pedestrians trying to leave (see also the lower curve in Fig. 11.2c). (**b**) A column in front of the exit (large black disk) can avoid injuries by taking up pressure from behind. It can also increase the outflow by 50%. In large exit areas used by many hundred people, several randomly placed columns are needed to subdivide the crowd and the pressure. An asymmetric configuration of the columns is most efficient, as it avoids equilibria of forces which may temporarily stop the outflow

$$v_i^0(t) = [1 - n_i(t)]v_i^0(0) + n_i(t)v_i^{\max} \, . \tag{11.7}$$

Herein, $v_i^{\max}$ is the maximum desired velocity and $v_i^0(0)$ the initial one, corresponding to the expected velocity of leaving. The time-dependent parameter

$$n_i(t) = 1 - \frac{\overline{v}_i(t)}{v_i^0(0)} \tag{11.8}$$

reflects the nervousness, where $\overline{v}_i(t)$ denotes the average speed into the desired direction of motion. Altogether, long waiting times increase the desired velocity, which can produce inefficient outflow. This further increases the waiting times, and so on, so that this tragic feedback can eventually trigger so high pressures that people are crushed or falling and trampled. It is, therefore, imperative, to have sufficiently wide exits and to prevent counterflows, when big crowds want to leave [11.40].

11.4.5 Mass Behavior

Finally, we investigate a situation in which pedestrians are trying to leave a smoky room, but first have to find one of the invisible exits (see Fig. 11.6a). Each pedestrian i may either select an individual direction $\boldsymbol{e}_i$ or follow the average direction $\langle \boldsymbol{e}_j^0(t) \rangle_i$ of his neighbors j in a certain radius R_i [11.50,11.51], or try a mixture of both. We assume that both options are weighted with the nervousness n_i:

$$\boldsymbol{e}_i^0(t) = \mathcal{N}\left[(1 - n_i)\,\boldsymbol{e}_i + n_i\,\langle \boldsymbol{e}_j^0(t) \rangle_i \right] \, , \tag{11.9}$$

where $\mathcal{N}(\boldsymbol{z}) = \boldsymbol{z}/\|\boldsymbol{z}\|$ denotes normalization of a vector $\boldsymbol{z}$. As a consequence, we have individualistic behavior if n_i is low, but herding behavior if n_i is high.

Our model suggests that neither individualistic nor herding behavior performs well (see Fig. 11.6b). Pure individualistic behavior means that each pedestrian finds an exit only accidentally, while pure herding behavior implies that the complete crowd is eventually moving into the same and probably blocked direction, so that available exits are not efficiently used, in agreement with observations. According to Figs. 11.6b, c we expect optimal chances of survival for a certain mixture of individualistic and herding behavior, where individualism allows some people to detect the exits and herding guarantees that successful solutions are imitated by the others. If pedestrians follow the walls instead of 'reflecting' at them, we expect that herd following causes jamming and inefficient use of doors as well (see Fig. 11.2), while individualists moving in opposite directions obstruct each other.

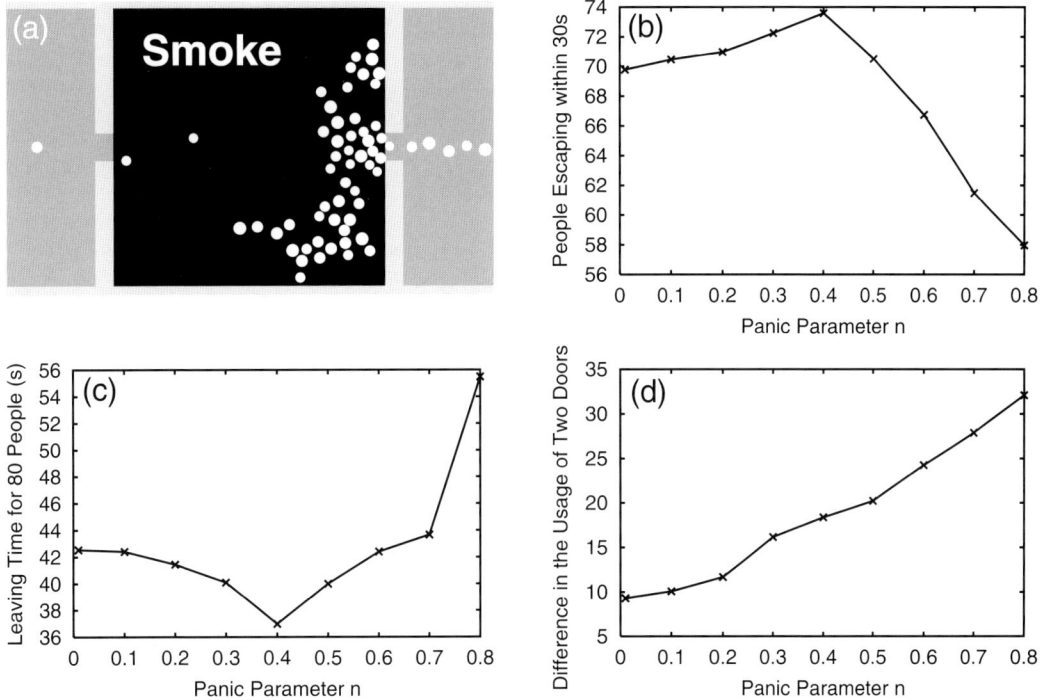

Fig. 11.6. Simulation of $N = 90$ pedestrians trying to escape a smoky room of area $A = 15\,\mathrm{m} \times 15\,\mathrm{m}$ (black) through two smoke-hidden doors of $1.5\,\mathrm{m}$ width, which have to be found with a mixture of individualistic and herding behavior (from [11.40]). Java applets are available at `http://angel.elte.hu/~panic/`. (**a**) Snapshot of the simulation with $v_i^0 = v_0 = 5\,\mathrm{m/s}$. Initially, each pedestrian selects his or her desired walking direction randomly. Afterwards, a pedestrian's walking direction is influenced by the average direction of the neighbors within a radius of, for example, $R_i = R = 5\,\mathrm{m}$. The strength of this herding effect grows with increasing nervousness parameter $n_i = n$ and increasing value of $h = \pi R^2 \rho$, where $\rho = N/A$ denotes the pedestrian density. When reaching a boundary, the direction of a pedestrian is reflected. If one of the exits is closer than $2\,\mathrm{m}$, the room is left. (**b**) Number of people who manage to escape within $30\,\mathrm{s}$ as a function of the nervousness parameter n. (**c**) Illustration of the time required by 80 individuals to leave the smoky room. If the exits are relatively narrow and the degree n of herding is small or large, leaving takes particularly long, so that only some of the people escape before being poisoned by smoke. Our results suggest that the best escape strategy is a certain compromise between following of others and an individualistic searching behavior. This fits well into experimental data on the efficiency of group problem solving [11.52–11.54], according to which groups normally perform better than individuals, but masses are inefficient in finding new solutions to complex problems. (**d**) Absolute difference $|N_1 - N_2|$ in the numbers N_1 and N_2 of persons leaving through the left exit or the right exit as a function of the degree n of herding. We find that pedestrians tend to jam up at one of the exits instead of equally using all available exits, if the nervousness is large

11.4.6 Problem Solving Behavior

In contrast to the results presented in the previous paragraphs, the discussion in the following two paragraphs will be partly speculative in nature.

The above herding model could be viewed as paradigm for problem-solving behavior in science, economics, and politics, where new solutions to complex problems have to be found in a way comparable to finding the exit of a smoky

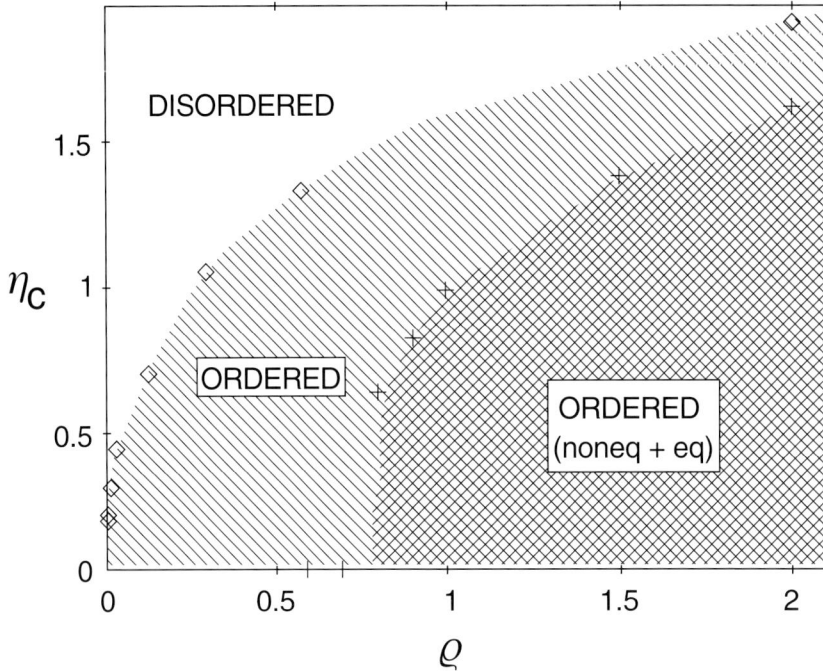

Fig. 11.7. Phase diagram of a self-propelled particle (SPP) model and the corresponding equilibrium system. The nonequilibrium SPP system with $v_i^0 = v_0 > 0$ becomes ordered in the whole region below the curve connecting the diamonds. In the static equilibrium case with $v_i^0 = v_0 = 0$, the ordered region extends only up to a finite 'percolation' density, see the beginning of the area given by the curve connecting the plus signs (after [11.51])

room. From the simulation results, we may conclude that people will not manage to cope with a sequence of challenging situations, if everyone sticks to his own idea egocentrically, but they will also fail, if everyone follows the same idea, so that the possible spectrum of solutions is not adequately explored. Therefore, the best strategy appears to be pluralism with a reasonable degree of readiness to follow good ideas of others, while a totalitarian regime would probably not survive a series of crises. This fits well into experimental data on the efficiency of group problem solving [11.52–11.54], according to which groups normally perform better than individuals, but masses are inefficient in finding new solutions to complex problems.

Considering the phase diagram in Fig. 11.7, we may also conjecture that, in the presence of a given level $\eta_i = \eta$ of fluctuations, there is a certain critical density above which people tend to show (ordered) mass behavior, and below which they behave individualistically (disordered). In fact, the tendency of mass behavior is higher in dense populations than in dilute ones.

11.4.7 Comparison with Stock Markets

Although the situation seems to be more difficult [11.55], our model might also contribute to a better understanding of other forms of panics like the ones observed at 'crashes' of stock markets [11.5, 11.10, 11.23, 11.56–11.63] (regarding this topic see also Chaps. 12-14). Such cases may again be viewed as situations with dwindling resources, in which herding behavior occurs and the collective run for a reasonable return (corresponding to the 'exit') decreases the chances of everyone to achieve it, because of competitive (analogous to 'repulsive') interactions. In other words, the reinforcement of buying and selling decisions has sometimes features of herding behavior, which is reflected in bubbles and crashes of stock markets (see, e.g. [11.61–11.63]). Remember that herding behavior is frequently found in complex situations, where individuals do not know what is the right thing to do. Everyone counts on collective intelligence then, believing that a crowd is following someone who has identified the right action. However, as has been shown for panicking pedestrians seeking for an exit, such mass behavior can have undesirable results, since alternatives are not efficiently exploited.

Another analogy between stock markets and self-driven many-particle systems are the irregular oscillations of stock prices and of pedestrian flows at bottlenecks (see Fig. 11.8). Presently it is unknown whether this analogy is just by chance, but there could be similar mechanisms at work: at the stock exchange market we have a competition of two different groups, optimistic traders ('bulls') and pessimistic ones ('bears'). The optimists count on growing stock prices and increase the price by their orders. In contrast, pessimists speculate on a decrease in the price and reduce it by their selling of stocks. Hence, traders belonging to the same group enforce each others' (buying or selling) actions, while optimists and pessimists push (the price) into opposite directions. Consequently, the situation is partly comparable with opposite pedestrian streams pushing at a bottleneck, which is reflected in a roughly similar dynamics, see Fig. 11.8b [11.38, 11.43]. The mechanism leading to alternating pedestrian flows is the following: once somebody is able to pass the narrowing, pedestrians with the same walking direction can easily follow. Hence, the number and 'pressure' of waiting and pushing pedestrians becomes less than on the other side of the narrowing where, consequently, the chance to occupy the passage grows. This leads to a deadlock situation which is followed by a change in the passing direction.

11.5 Conclusions

In summary, we have developed a continuous pedestrian model based on plausible interactions, which is, due to its simplicity, robust with respect to parameter

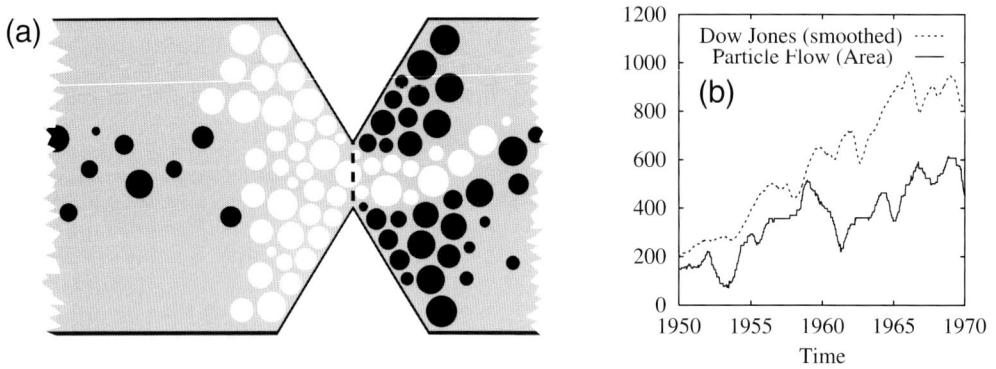

Fig. 11.8. (a) In pedestrian counterflows, one observes oscillations of the passing direction at bottlenecks (after [11.38,11.43], see also [11.30–11.37]). (b) The time-dependent difference in the area of particles that have passed the bottleneck from the left-hand side or right-hand side, respectively, looks similar to the dynamics of the Dow Jones index. The question whether this is just by chance or due to a deeper relationship of both systems, is subject to present research

variations and suitable for drawing conclusions about the possible mechanisms beyond escape panic (regarding an increase of the desired velocity, strong friction effects during physical interactions, and herding). After having calibrated the model parameters to available data on pedestrian flows, we managed to reproduce many observed phenomena including

1. the breakdown of fluid lanes ('freezing by heating'),
2. the build up of pressure,
3. clogging effects at bottlenecks,
4. jamming at widenings,
5. the 'faster-is-slower effect',
6. 'phantom panics' triggered by counterflows and impatience, and
7. inefficient use of alternative exits.

We are also able to simulate situations of dwindling resources and estimate the casualties (see Figs. 11.2c and 11.3). Therefore, the model can be used to test buildings for their suitability in emergency situations. It accounts for the considerably different dynamics both in normal and panic situations just by changing a single parameter $n_i = n$. In this way, we have proposed a consistent theoretical approach allowing a continuous switching between seemingly incompatible kinds of human behavior (individualistic, rational behavior vs. irrational herding behavior).

Moreover, the model can serve as an example linking collective behavior as a phenomenon of mass psychology (from the socio-psychological perspective) to the view of an emergent collective pattern of motion (from the perspective of physics). Our simulations suggest that the optimal behavior in escape situations is a suitable mixture of individualistic and herding behavior. This conclusion is probably transferable to many cases of problem solving in new and complex situations, where standard solutions fail. It may explain why both individualistic and herding behaviors are common in human societies. The crucial point, however, is to find the optimal mixture between them.

Finally, we point out that conclusions from findings for self-driven many-particle systems reach far into the realm of the social, economic, and psychological sciences. One of the reasons is that the competition of moving particles for limited space is analogous to the situation in various socio-economic and biological systems, where individuals or other entities compete for limited resources as well.

Acknowledgements. The authors are grateful to the Collegium Budapest-Institute for Advanced Study for the warm hospitality and the excellent scientific working conditions. D.H. thanks the German Research Foundation (DFG) for financial support by the Heisenberg scholarship He 2789/1-1. T.V. and I.F. are grateful for partial support by OTKA and FKFP. Last but not least, Tilo Grigat has helped with formatting this manuscript.

References

11.1 G. LeBon, *The Crowd* (Viking, New York, 1960).

11.2 B.D. Jacobs and P. 't Hart, in *Hazard Management and Emergency Planning*, edited by D.J. Parker and J.W. Handmer (James and James Science, London, 1992), Chap. 10.

11.3 N.J. Smelser, *Theory of Collective Behavior* (The Free Press, New York, 1963).

11.4 R. Brown, *Social Psychology* (The Free Press, New York, 1965).

11.5 D.L. Miller, *Introduction to Collective Behavior* (Wadsworth, Belmont, CA, 1985), Fig. 3.3 and Chap. 9.

11.6 J.S. Coleman, *Foundations of Social Theory* (Belkamp, Cambridge, MA, 1990), Chaps. 9 and 33.

11.7 R.H. Turner and L.M. Killian, *Collective Behavior*, 3rd edn. (Prentice Hall, Englewood Cliffs, 1987).

11.8 A. Mintz, J. Abnorm. Norm. Soc. Psychol. **46**, 150 (1951).

11.9 E. Quarantelli, Sociol. and Soc. Res. **41**, 187 (1957).

11.10 J.P. Keating, Fire J., 57+147 (May/1982).

11.11 D. Elliott and D. Smith, Ind. Env. Crisis Q. **7**(3), 205 (1993).

11.12 N.R. Johnson, Soc. Probl. **34**(4), 362 (1987).

11.13 K.H. Drager, G. Løvås, J. Wiklund, H. Soma, D. Duong, A. Violas, and V. Lanères, in *Proc.1992 Emergency Manage. Eng. Conf.* (Society for Computer Simulation, Orlando, Florida, 1992), p. 101.

11.14 M. Ebihara, A. Ohtsuki, and H. Iwaki, Microcomput. Civ. Eng. **7**, 63 (1992).

11.15 G.K. Still, Fire **84**, 40 (1993).

11.16 J.L. Bryan, Fire J., 27+86 (Nov./1985).

11.17 R. Axelrod and D. Dion, Science **242**, 1385 (1988).

11.18 N.S. Glance and B.A. Huberman, Sci. Am. **270**, 76 (1994).

11.19 W. Weidlich, Phys. Rep. **204**, 1 (1991).

11.20 D. Helbing, *Quantitative Sociodynamics. Stochastic Methods and Models of Social Interaction Processes* (Kluwer, Dordrecht, 1995).

11.21 W. Weidlich, *Sociodynamics* (Harwood, Amsterdam, 2000).

11.22 D. Helbing, H.J. Herrmann, M. Schreckenberg, and D.E. Wolf (eds.), *Traffic and Granular Flow '99: Social, Traffic, and Granular Dynamics* (Springer, Berlin, 2000).

11.23 F. Schweitzer and D. Helbing (eds.), *Economic Dynamics from the Physics Point of View*, Physica A **287**, 339 (2000).

11.24 R.A. Smith and J.F. Dickie (eds.), *Engineering for Crowd Safety* (Elsevier, Amsterdam, 1993).

11.25 H.H. Kelley, J.C. Condry Jr., A.E. Dahlke, and A.H. Hill, J. Exp. Soc. Psychol. **1**, 20 (1965).

11.26 D. Canter (ed.), *Fires and Human Behaviour* (David Fulton, London, 1990).

11.27 W.M. Predtetschenski and A.I. Milinski, *Personenströme in Gebäuden, Berechnungsmethoden für die Projektierung* (Rudolf Müller, Köln-Braunsfeld, 1971).

11.28 U. Weidmann, *Transporttechnik der Fußgänger* (Institut für Verkehrsplanung, Transporttechnik, Straßen- und Eisenbahnbau (IVT), ETH Zürich, 1993).

11.29 D. Helbing, Behav. Sci. **36**, 298 (1991).

11.30 D. Helbing, P. Molnár, and F. Schweitzer, in *Evolution of Natural Structures* (Sonderforschungsber. 230, Stuttgart, 1994), p. 229

11.31 D. Helbing and P. Molnár, Phys. Rev. E **51**, 4282 (1995).

11.32 D. Helbing, in *Traffic and Granular Flow*, edited by D.E. Wolf, M. Schreckenberg, and A. Bachem (World Scientific, Singapore, 1996), p. 87.

11.33 P. Molnár, *Modellierung und Simulation der Dynamik von Fußgängerströmen* (Shaker, Aachen, 1996).

11.34 P. Molnár, in *Social Science Microsimulation*, edited by J. Doran, N. Gilbert, U. Mueller, and K. Troitzsch (Springer, Berlin, 1996).

11.35 D. Helbing and P. Molnár, in *Self-Organization of Complex Structures: From Individual to Collective Dynamics*, edited by F. Schweitzer (Gordon and Breach, London, 1997), p. 569.

11.36 D. Helbing, *Verkehrsdynamik* (Springer, Berlin, 1997).

11.37 D. Helbing, P. Molnár, I. Farkas, and K. Bolay, Env. Planning B **28**, 361 (2001).

11.38 D. Helbing, Rev. Mod. Phys. **73**(4), 1067 (2001). submitted to Rev. Mod. Phys. (2000).

11.39 D. Helbing, I. Farkas, and T. Vicsek, Phys. Rev. Lett. **84**, 1240 (2000).

11.40 D. Helbing, I. Farkas, and T. Vicsek, Nature **407**, 487 (2000).

11.41 G.H. Ristow and H. J. Herrmann, Phys. Rev. E **50**, R5 (1994).

11.42 D.E. Wolf and P. Grassberger (eds.), *Friction, Arching, Contact Dynamics* (World Scientific, Singapore, 1997).

11.43 D. Helbing, Phys. Bl. **57**, 27 (2001).

11.44 D. Helbing and T. Vicsek, New J. Phys. **1**, 13.1 (1999).

11.45 D. Helbing and T. Platkowski, Int. J. Chaos Theory Appl. **5**, 25 (2000).

11.46 J. Gallas, H.J. Herrmann, and S. Sokołowski, Phys. Rev. Lett. **69**, 1371 (1992).

11.47 P.B. Umbanhowar, F. Melo, and H.L. Swinney, Nature **382**, 793 (1996).

11.48 A. Rosato, K.J. Strandburg, F. Prinz, and R.H. Swendsen, Phys. Rev. Lett. **58**, 1038 (1987).

11.49 M. Muramatsu, T. Irie, and T. Nagatani, Physica A **267**, 487 (1999).

11.50 T. Vicsek, A. Czirók, E. Ben-Jacob, I. Cohen, and O. Shochet, Phys. Rev. Lett. **75**, 1226 (1995).

11.51 A. Czirók, M. Vicsek, and T. Vicsek, Physica A **264**, 299 (1999).

11.52 N.H. Anderson, J. Soc. Psychol. **55**, 67 (1961).

11.53 H.H. Kelley and J.W. Thibaut, in *The Handbook of Social Psychology*, Vol. 4, edited by G. Lindzey and E. Aronson (Addison-Wesley, Reading, 1969).

11.54 P.R. Laughlin, N.L. Kerr, J.H. Davis, H.M. Halff, and K.A. Marciniak, J. Personality Soc. Psychol. **31**, 522 (1975).

11.55 L. Mann, T. Nagel, and P. Dowling, Sociometry **39**(3), 223 (1976).

11.56 J.-P. Bouchaud, A. Matacz, and M. Potters, The leverage effect in financial markets: retarded volatility and market panic (2001); [available from arXiv.org/abs/cond-mat/0101120v2].

11.57 R.N. Mantegna and E. Stanley, *Introduction to Econophysics: Correlations and Complexity in Finance* (Cambridge University Press, Cambridge, England, 1999).

11.58 J.-P. Bouchaud and M. Potters, *Theory of Financial Risk: From Statistical Physics to Risk Management* (Cambridge University Press, Cambridge, England, 2000).

11.59 H. Levy, M. Levy, and S. Solomon, *Microscopic Simulation of Financial Markets* (Academic, San Diego, 2000).

11.60 B.B. Mandelbrot, *Fractals and Scaling in Finance: Discontinuity, Concentration, Risk* (Springer, New York, 1997).

11.61 M. Youssefmir, B.A. Huberman, and T. Hogg, Comp. Econ. **12**, 97 (1998).

11.62 J.D. Farmer, Market force, ecology, and evolution, submitted to J. Econ. Behav. Org. (2000); [available from arXiv.org/abs/adap-org/9812005].

11.63 T. Lux and M. Marchesi, Nature **397**, 498 (1999).

Part IV

Nonlinear Economics

12. Investigations of Financial Markets Using Statistical Physics Methods

Rosario N. Mantegna and H. Eugene Stanley

We begin with a brief historical note concerning the growing interest of statistical physicists in the analysis and modeling of financial markets. We then briefly discuss the key concepts of arbitrage and efficient markets. We relate these concepts to apparently 'universal' aspects observed in the empirical analysis of stock price dynamics in financial markets. In particular, we consider (i) the empirical behavior of the probability density function for the return of an economic time series to where it started and (ii) the content of economic information in a financial time series.

12.1 Introduction

The quantitative modeling of financial markets started in 1900 with the pioneering work of the French mathematician Bachelier [12.1]. Since the 1950s, the analysis and modeling of financial markets have become an important research area of economics and financial mathematics [12.2]. The researches pursued have been very successful, and nowadays a robust theoretical framework char-

---

◄ **Fig. 12.0.** Color representation quantifying the stability in time of the eigenvectors of the correlation matrix that deviate from random-matrix bounds. Two partially overlapping time periods A and B, of four months each, were analyzed, January 1994 – April 1994 and March 1994 – June 1994. Each of the 225 squares has a rainbow color proportional to the scalar product ('overlap') of the largest 15 eigenvectors of the correlation matrix in period A with those of the same 15 eigenvectors from period B. Perfect stability in time would imply that this pixel representation of the overlaps has ones (the red end of the rainbow spectrum) in the diagonal and zeros (violet) in the off-diagonal. The eigenvectors are shown in inverse rank order (from smallest to largest), and we note that the pixels near the upper right corner have colors near the red end of the spectrum, corresponding to the fact that the largest 6–8 eigenvectors are relatively stable; in particular, the largest 3–4 eigenvectors are stable for very long periods of time. The remainder of the pixels are distributed toward the violet end of the spectrum, corresponding to the fact that the overlaps are not statistically significant, and corroborating the finding that their corresponding eigenvalues are random. This figure is kindly contributed by P. Gopikrishnan and V. Plerou

acterizes these disciplines [12.3–12.6]. In parallel to these studies, starting from the 1990s a group of physicists became interested in the analysis and modeling of financial markets by using tools and paradigms of their own discipline (for an overview, consider, for example, [12.7–12.10]). The interest of physicists in such systems is directly related to the fact that, during the years, predictability has assumed a meaning in physics, which is quite different from the one originally associated with the predictability of, for example, a Newtonian linear system. The degree of predictability of physics systems is nowadays known to be essentially limited in nonlinear and complex systems. This makes the physical prediction less strong, but on the other hand the area of research covered by physical investigations and of its application may increase [12.11].

In addition to the above observations, there are a series of reasons explaining why this discipline emerged during the last decade. We will try to discuss some of them hereafter.

Since the 1970s, a series of significant changes has taken place in the world of finance. One key year was 1973, when currencies began to be traded in financial markets and their values determined by the foreign exchange market, a financial market active 24 hr a day all over the world. During that same year, Black and Scholes [12.12] published the first paper that presented a rational option-pricing formula.

Since that time, the volume of foreign exchange trading has been growing at an impressive rate. The transaction volume in 1995 was 80 times what it was in 1973. An even more impressive growth has taken place in the field of derivative products. The notional amount of financial derivative market contracts issued in 1999 was 81 trillion US dollars. Contracts were negotiated in the over-the-counter market (i.e. directly between firms or financial institutions), and in specialized exchanges that deal only in derivative contracts. Today, financial markets facilitate the trading of huge amounts of money, assets, and goods in a competitive global environment.

A second revolution began in the 1980s when electronic trading was adapted to the foreign exchange market. The electronic storing of data relating to financial contracts - or to bid and ask quotes issued by traders - was put in place at about the same time that electronic trading became widespread. One result is that today a huge amount of electronically stored financial data is readily available. These data are characterized by the property of being high-frequency data - the average time delay between two records can be as short as a few seconds. Between the available databases it may be worth mentioning the Olsen and Associates database comprising all the bid and ask quotes of the foreign exchange market collected from information vendors, including Reuters, Knight-Ridder, and Telerate since 1986 and the Trade and Quote (TAQ) database, which comprises all the trades and quotes related to all the securities listed in the New York Stock Exchange, Nasdaq National Market System, and SmallCap issues. The TAQ database is available monthly on CD-

Rom from the New York Stock Exchange. The enormous expansion of financial markets requires strong investments in money and human resources to achieve reliable quantification and minimization of risk for the financial institutions involved.

12.2 Econophysics

The research approach of physicists to financial modeling aims to be complementary to the ones of financial mathematicians and economists. The main goals are (i) to contribute to a better understanding and modeling of financial markets and (ii) to promote the use of physical concepts and expertise in the multidisciplinary approach to risk management.

This research area is often addressed as *econophysics*. The word econophysics describes the present attempts of a number of physicists to model financial and economic systems using paradigms and tools borrowed from theoretical and statistical physics.

Financial markets exhibit several of the properties that characterize complex systems. They are open systems in which many subunits interact nonlinearly in the presence of feedback. In financial markets, the governing rules are rather stable and the time evolution of the system is continuously monitored. It is now possible to develop models and to test their accuracy and predictive power using available data, since large databases exist even for high-frequency data.

A research community has begun to emerge in econophysics starting from the 1990s. New interdisciplinary journals have been published, conferences have been organized, and a set of potentially tractable scientific problems has been provisionally identified. The research activity of this group of physicists is complementary to the most traditional approaches of finance and mathematical finance. One characteristic difference is the emphasis that physicists put on the empirical analysis of economic data. Another is the background of theory and method in the field of statistical physics developed over the past 30 yr that physicists bring to the subject. The concepts of scaling, universality, disordered frustrated systems, and self-organized systems might be helpful in the analysis and modeling of financial and economic systems. One argument that is sometimes raised at this point is that an empirical analysis performed on financial or economic data is not equivalent to the usual experimental investigation that takes place in physical sciences. In other words, it is impossible to perform large-scale experiments in economics and finance that could falsify any given theory.

We note that this limitation is not specific to economic and financial systems, but also affects such well developed areas of physics as astrophysics, atmospheric physics, and geophysics. Hence, in analogy to activity in these more established areas, we find that we are able to test and falsify any theories associated with the currently available sets of financial and economic data provided in the form of recorded files of financial and economic activity.

Among the important areas of physics research dealing with financial and economic systems, one concerns the complete statistical characterization of the stochastic process of price changes of a financial asset. Several studies have been performed that focus on different aspects of the analyzed stochastic process, e.g. the shape of the distribution of price changes [12.9, 12.13–12.17], the temporal memory [12.18–12.21], and the higher-order statistical properties [12.22–12.24]. This is still an active area, and attempts are ongoing to develop the most satisfactory stochastic model describing all the features encountered in empirical analyses. One important accomplishment in this area is an almost complete consensus concerning the finiteness of the second moment of price changes. This has been a longstanding problem in finance, and its resolution has come about because of the renewed interest in the empirical study of financial systems.

A second area concerns the development of a theoretical model that is able to encompass all the essential features of real financial markets. Several models have been proposed [12.25–12.39], and some of the main properties of the stochastic dynamics of stock price are reproduced by these models as, for example, the leptokurtic non-Gaussian shape of the distribution of price differences. Parallel attempts in the modeling of financial markets by taking into account some results observed in the empirical analyses have been also developed by economists [12.40–12.42, 12.87].

One of the more active areas in finance is the pricing of derivative instruments. In the simplest case, an asset is described by a stochastic process and a derivative security (or contingent claim) is evaluated on the basis of the type of security and the value and statistical properties of the underlying asset. This problem presents at least two different aspects: (i) 'fundamental' aspects, which are related to the nature of the random process of the asset, and (ii) 'applied' or 'technical' aspects, which are related to the solution of the option-pricing problem under the assumption that the underlying asset performs the proposed random process.

In this area the investigations which are considering the problem of the rational pricing of a derivative product when some of the canonical assumptions of the Black and Scholes model are relaxed [12.9, 12.43, 12.44]. Other autors focus on aspects of portfolio selection and its dynamical optimization [12.45–12.49]. A further area of research considers analogies and differences between price dynamics in a financial market and such processes as turbulence [12.15, 12.19, 12.50] and the dynamics of ecological systems [12.16, 12.51].

Another common theme encountered in econophysics concerns the time correlation of financial series. The detection of the presence of a higher-order correlation in price changes has motivated a reconsideration of some beliefs of what is termed 'technical analysis' [12.52].

In addition to the studies that analyze and model financial systems, there are studies of the income distribution of firms and studies of the statistical properties of their growth rates [12.53–12.56]. The statistical properties of the economic performances of complex organizations such as universities or entire countries have also been investigated [12.57].

This brief presentation of some of the current efforts in this emerging discipline has only illustrative purposes and cannot be exhaustive. For a more complete overview, consider, for example, the proceedings of conferences dedicated to these topics [12.7, 12.8].

From the above overview we see that econophysics started as an emerging discipline during the 1990s. However before the starting up of the discipline a series of physicists and mathematicians have investigated on an individual basis financial and economic problems. Some of these pioneering approaches are discussed in the next section.

12.3 An Historical Note

The interest of the physics community in financial and economic systems has roots that date back to 1942, when a paper from Majorana on the essential analogy between statistical laws in physics and in the social sciences was published [12.58]. In his contribution, he wrote that "......It is important that the principles of quantum mechanics have led to recognize (...) the statistical char acter of the fundamental laws of elementary processes. This conclusion makes essential the analogy between physics and social sciences, between which there is an identity of values and method". This unorthodox point of view was considered of marginal interest until recently. Indeed, prior to the 1990s, very few professional physicists did any research associated with social or economic systems. The exceptions included Kadanoff [12.59], Montroll and Badger [12.60], and a group of physical scientists at the Santa Fe Institute [12.61].

In this chapter we briefly discuss the application to financial markets of such concepts as power-law distributions, correlations, scaling, unpredictable time series, and random processes. During the past 30 yr, physicists have achieved important results in the field of phase transitions, statistical mechanics, nonlinear dynamics, and disordered systems. In these fields, power-laws, scaling, and unpredictable (stochastic or deterministic) time series are present and the current interpretation of the underlying physics is often obtained using these concepts.

With this background in mind, it may surprise scholars trained in the natural sciences to learn that the first use of a power-law distribution - and the first mathematical formalization of a random walk - took place in the social sciences. In 1897 the Italian social economist Pareto investigated the statistical character of the wealth of individuals in a stable economy by modeling them using the distribution

$$y \sim x^{-\nu}, \tag{12.1}$$

where y is the number of people having income x or greater than x and ν is an exponent that Pareto estimated to be 1.5 [12.62]. Pareto noticed that his result was quite general and applicable to nations 'as different as those of England, of Ireland, of Germany, of the Italian cities, and even of Peru'.

It should be fully appreciated that the concept of a power-law distribution is counterintuitive, because it lacks any characteristic scale. This property prevented the use of power-law distributions in the natural sciences until the recent emergence of new paradigms (i) in probability theory, thanks to the work of Lévy [12.63] and thanks to the application of power-law distributions to several problems pursued by Mandelbrot [12.64]; and (ii) in the study of phase transitions, which introduced the concepts of scaling for thermodynamic functions and correlation functions [12.65].

Another concept ubiquitous in the natural sciences is the random walk. The first theoretical description of a random walk in the natural sciences was performed in 1905 by Einstein [12.66] in his famous paper dealing with the determination of the Avogadro number. In subsequent years, the mathematics of the random walk was made more rigorous by Wiener [12.67], and now the random walk concept has spread across almost all research areas in the natural sciences.

The first formalization of a random walk was not in a publication by Einstein, but in the doctoral thesis by Bachelier [12.1]. Bachelier presented his thesis to the faculty of sciences at the Academy of Paris on 29 March 1900, for the degree of *Docteur en Sciences Mathématiques*. The first page of his thesis is shown in Fig. 12.1. His advisor was Poincaré, one of the greatest mathematicians of his time. The thesis, entitled *Théorie de la Spéculation*, is surprising in several respects. It deals with the pricing of options in speculative markets, an activity that today is extremely important in financial markets where derivative securities - those whose value depends on the values of other more basic underlying variables - are regularly traded on many different exchanges. To complete this task, Bachelier determined the probability of price changes by writing down what is now called the Chapman-Kolmogorov equation and recognizing that what is now called a Wiener process satisfies the diffusion equation (this point was re-discovered by Einstein in his 1905 paper on Brownian motion). Retrospectively analyzed, Bachelier's thesis lacks rigor in some of its mathematical and economic points. Specifically, the determination of a Gaussian distribution for the price changes was - mathematically speaking - not sufficiently motivated. On the economic side, Bachelier investigated price changes, whereas economists are mainly dealing with changes in

THÉORIE

DE

LA SPÉCULATION,

Par M. L. BACHELIER.

INTRODUCTION.

Les influences qui déterminent les mouvements de la Bourse sont innombrables, des événements passés, actuels ou même escomptables, ne présentant souvent aucun rapport apparent avec ses variations, se répercutent sur son cours.

A côté des causes en quelque sorte naturelles des variations, interviennent aussi des causes factices : la Bourse agit sur elle-même et le mouvement actuel est fonction, non seulement des mouvements antérieurs, mais aussi de la position de place.

La détermination de ces mouvements se subordonne à un nombre infini de facteurs : il est dès lors impossible d'en espérer la prévision mathématique. Les opinions contradictoires relatives à ces variations se partagent si bien qu'au même instant les acheteurs croient à la hausse et les vendeurs à la baisse.

Le Calcul des probabilités ne pourra sans doute jamais s'appliquer aux mouvements de la cote et la dynamique de la Bourse ne sera jamais une science exacte.

Mais il est possible d'étudier mathématiquement l'état statique du marché à un instant donné, c'est-à-dire d'établir la loi de probabilité des variations de cours qu'admet à cet instant le marché. Si le marché, en effet, ne prévoit pas les mouvements, il les considère comme étant

Fig. 12.1. First page of the PhD thesis of Luis Bachelier. This work was written in 1900, almost 50 yr before the concept of random walk was currently used to model asset price dynamics in a financial market and 73 yr before the first publication of a rational option price procedure

the logarithm of price. However, these limitations do not diminish the value of Bachelier's pioneering work.

To put Bachelier's work into perspective, the Black and Scholes option-pricing model - considered the milestone in option-pricing theory - was published in 1973, almost three-quarters of a century after the publication of his thesis. Moreover, theorists and practitioners are aware that the Black and Scholes model needs correction in its application, meaning that the problem of which stochastic process describes the changes in the logarithm of prices in a financial market is still an open one.

The problem of the distribution of price changes has been considered by several authors since the 1950s, which was the period when mathematicians began to show interest in the modeling of stock market prices. Bachelier's original proposal of Gaussian distributed price changes was soon replaced by a model in which stock prices are log-normal distributed, i.e. stock prices are performing a geometric Brownian motion. In a geometric Brownian motion, the differences of the logarithms of prices are Gaussian distributed. This model is known to provide only a first approximation of what is observed in real data. For this reason, a number of alternative models have been proposed with the aim of explaining

(i) the empirical evidence that the tails of measured distributions are fatter than expected for a geometric Brownian motion; and
(ii) the time fluctuations of the second moment of price changes.

Among the alternative models proposed, 'the most revolutionary development in the theory of speculative prices since Bachelier's initial work' [12.2], is Mandelbrot's hypothesis that price changes follow a Lévy stable distribution [12.68]. Lévy stable processes are stochastic processes obeying a generalized central limit theorem. By obeying a generalized form of the central limit theorem, they have a number of interesting properties. They are stable (as are the more common Gaussian processes) - i.e. the sum of two independent stochastic processes x_1 and x_2 characterized by the same Lévy distribution of index α is itself a stochastic process characterized by a Lévy distribution of the same index. The shape of the distribution is maintained (is stable) by summing up independent and identically distributed Lévy stable random variables.

Lévy stable processes define a basin of attraction in the functional space of probability density functions. The sum of independent and identically distributed stochastic processes $S_n \equiv \sum_{i=1}^{n} x_i$ characterized by a probability density function with power-law tails,

$$P(x) \sim x^{-(1+\alpha)}, \tag{12.2}$$

will converge, in probability, to a Lévy stable stochastic process of index α when n tends to infinity [12.69].

This property tells us that the distribution of a Lévy stable process is a power-law distribution for large values of the stochastic variable x. The fact that power-law distributions lack a typical scale is reflected in Lévy stable processes by the property that the variance of Lévy stable processes is infinite for $\alpha < 2$. Stochastic processes with infinite variance, although well defined mathematically, are extremely difficult to use and, moreover, raise fundamental questions when applied to real systems. For example, in physical systems the second moment is often related to the system temperature, so infinite variances imply an infinite (or undefined) temperature. In financial systems, an infinite variance would complicate the important task of risk estimation.

12.4 Key Concepts

In the model of financial markets a pair of concepts are crucial to understand the current theories modeling them. In this section we will discuss two relevant concepts namely the concept of absence of arbitrage and the concept of efficient market. Financial markets are systems in which a large number of traders interact with one another and react to external information in order to determine the best price for a given asset. The goods might be as different as animals, ore, equities, currencies, or bonds - or derivative products issued on those underlying financial goods. Some markets are localized in specific cities (e.g. New York, Tokyo, and London) while others (such as the foreign exchange market) are delocalized and accessible all over the world.

When one inspects a time series of the time evolution of the price, volume, and number of transactions of a financial product, one recognizes that the time evolution is unpredictable. At first sight, one might sense a curious paradox. An important time series, such as the price of a financial good, is essentially indistinguishable from a stochastic process. There are deep reasons for this kind of behavior, and in this chapter we will examine some of these.

12.4.1 Arbitrage

A key concept for the understanding of markets is the concept of arbitrage - the simultaneous purchase and sale of the same or equivalent security in order to profit from price discrepancies.

The presence of traders looking for arbitrage conditions contributes to a market's ability to evolve the most rational price for a good. To see this, suppose that one has discovered an arbitrage opportunity. One will exploit it and, if one succeeds in making a profit, one will repeat the same action. This action will increase the demand of a given good at a given place or time and will simultaneously increase the supply of the same good at another time or place. The modification of the demand and supply levels forces the price of the considered good to attain a more rational value in a given place or time.

To summarize: (i) new arbitrage opportunities continually appear and are discovered in the markets but (ii) as soon as an arbitrage opportunity begins to be exploited, the system moves in a direction that gradually eliminates the arbitrage opportunity.

12.4.2 Efficient Market Hypothesis

Markets are complex systems that incorporate information about a given asset in the time series of its price. The most accepted paradigm among scholars in finance is that the market is highly efficient in the determination of the most rational price of the traded asset. The efficient market hypothesis (EMH) was originally formulated in the 1960s [12.70]. A market is said to be efficient if all the available information is instantly processed when it reaches the market and it is immediately reflected in a new value of prices of the assets traded.

The theoretical motivation for the efficient market hypothesis has its roots in the pioneering work of Bachelier [12.1], who proposed that the price of assets in a speculative market be described as a stochastic process. This work remained almost unknown until the 1950s, when empirical results [12.2] about the serial correlation of the rate of return showed that correlations on a short time scale are negligible and that the approximate behavior of return time series is indeed similar to uncorrelated random walks.

The EMH was formulated explicitly in 1965 by Samuelson [12.71], who showed mathematically that properly anticipated prices fluctuate randomly. Using the hypothesis of rational behavior and market efficiency, he was able to demonstrate how Y_{t+1}, the expected value of the price of a given asset at time $t+1$, is related to the previous values of prices $Y_0, Y_1, \ldots, Y_t$ through the relation

$$E\{Y_{t+1}|Y_0, Y_1, \ldots, Y_t\} = Y_t. \tag{12.3}$$

Stochastic processes obeying the conditional probability given in (12.3) are called martingales [12.72]. The notion of a martingale is, intuitively, a probabilistic model of a 'fair' game. In gambler's terms, the game is fair when gains and losses cancel, and the gambler's expected future wealth coincides with the gambler's present assets. The fair game conclusion about the price changes observed in a financial market is equivalent to the statement that there is no way of making a profit on an asset by simply using the recorded history of its price fluctuations. The conclusion of this 'weak form' of the EMH is then that price changes are unpredictable from the historical time series of those changes.

Since the 1960s, a great number of empirical investigations have been devoted to testing the efficient market hypothesis [12.73]. In the great majority of the empirical studies, the time correlation between price changes has been found to be negligibly small, supporting the efficient market hypothesis. However, it was shown in the 1980s that by using the information present in additional time series such as earnings/price ratios, dividend yields, and term-structure variables, it is possible to make predictions of the rate of return of a given asset on a long time scale, much longer than a month. Thus empirical observations have challenged the stricter form of the efficient market hypothesis.

Thus empirical observations and theoretical considerations show that price changes are difficult if not impossible to predict if one starts from the time series of price changes. In its strict form, an efficient market is an idealized system. In actual markets, residual inefficiencies are always present. Searching out and exploiting arbitrage opportunities is one way of eliminating market inefficiencies.

12.5 Idealized Systems in Physics and Finance

The efficient market is an idealized system. Real markets are only approximately efficient. This fact will probably not sound too unfamiliar to physicists because they are well acquainted with the study of idealized systems. Indeed, the use of idealized systems in scientific investigation has been instrumental in the development of physics as a discipline. Where would physics be without idealizations such as frictionless motion, reversible transformations in thermodynamics, and infinite systems in the critical state? Physicists use these abstractions in order to develop theories and to design experiments. At the same time, physicists always remember that idealized systems only approximate real systems, and that the behavior of real systems will always deviate from that of idealized systems. A similar approach can be taken in the study of financial systems. We can assume realistic 'ideal' conditions, e.g. the existence of a perfectly efficient market, and within this ideal framework develop theories and perform empirical tests. The validity of the results will depend on the validity of the assumptions made.

The concept of the efficient market is useful in any attempt to model financial markets. After accepting this paradigm, an important step is to fully characterize the statistical properties of the random processes observed in financial markets.

12.6 Empirical Analysis

Econophysics is an interdisciplinary research area, with a growing number of practitioners. We shall briefly describe the spirit and substance of some recent work that focuses on universal aspects observed in the empirical analysis of different (in location and time period) financial markets.

12.6.1 Statistical Properties of Price Dynamics

The knowledge of the statistical properties of price dynamics in financial markets is fundamental. It is necessary for any theoretical modeling aiming to obtain a rational price for a derivative product issue on it [12.74] and it is the starting point of any valuation of the risk associated with a financial position [12.75]. Moreover, it is needed in any effort aiming to model the system. In spite of this importance, the modeling of such variable is not yet conclusive.

Several models exist which are showing partial successes and unavoidable limitations. In this research, the approach of physicists maintains the specificity of their discipline namely to develop and modify models by taking into account results of empirical analysis.

Several models have been proposed and we will not review them here. Here we wish only to focus on the aspects which are *universally* observed in various stock price and index price dynamics.

12.6.2 Short- and Long-Range Correlations

In any financial market - either well established and highly active as the New York stock exchange, *emerging* as the Budapest stock exchange, or *regional* as the Milan stock exchange - the autocorrelation function of returns is a monotonic decreasing function with a very short correlation time. High frequency data analyses have shown that correlation times can be as short as a few minutes in highly traded stocks or indices [12.19, 12.76]. A fast decaying autocorrelation function is also observed in the empirical analysis of data recorded transaction by transaction. By using as time index the number of transactions occurred from a selected origin, a time memory as short as a few transactions has been detected in the dynamics of most traded stocks of the Budapest 'emerging' financial market [12.77].

The short-range memory between returns is directly related to the necessity of absence of continuous arbitrage opportunities in efficient financial markets. In other words, if correlation were present between returns this would allow devising trading strategies that would provide a net gain continuously and without risk. The continuous search for and exploitation of arbitrage opportunities from traders focused on this kind of activity drastically diminish the redundancy in the time series of price changes. Another mechanism reducing the redundancy of stock price time series is related to the presence of so-called 'noise traders'. With their action, noise traders add into the time series of stock price information, which is unrelated to the economic information decreasing the degree of redundancy of the price changes time series.

It is worth pointing out that not all the economic information present in stock price time series disappears due to these mechanisms. Indeed the redundancy that needs to be eliminated concerns only price change and not any nonlinear functions of it [12.78].

The absence of time correlation between returns does not mean that returns are identically distributed over time. In fact different authors have observed that nonlinear functions of return such as the absolute value or the square are correlated over a time scale much longer than a trading day. Moreover the functional form of this correlation seems to be power-law up to at least 20 trading days approximately [12.20, 12.21, 12.24, 12.76, 12.79–12.81].

A final observation concerns the degree of stationary behavior of the stock price dynamics. Empirical analysis shows that returns are not strictly sense stationary stochastic processes. Indeed the volatility (standard deviation of returns) is itself a stochastic process. Although a general proof is still lacking, empirical analyses performed on financial data of different financial markets suggest that the stochastic process is locally nonstationary but asymptotically stationary. By asymptotically stationary we mean that the probability density function (PDF) of returns measured over a wide time interval exists and it is uniquely defined. A paradigmatic example of simple stochastic processes which are locally nonstationary but asymptotically stationary is provided by ARCH [12.82] and GARCH [12.83] processes.

12.6.3 The Distribution of Returns

The PDF of returns shows some 'universal' aspects. By 'universal' aspects we mean that they are observed in different financial markets at different periods of time provided that a sufficiently long time period is used in the empirical analysis. The first of these 'universal' or stylized facts is the leptokurtic nature of the PDF. A leptokurtic PDF characterizes a stochastic process having small changes and very large changes more frequent than in the case of Gaussian distributed changes. Leptokurtic PDFs have been observed in stocks and indices time series by analyzing both high-frequency and daily data. An example is provided in Fig. 12.2. Thanks to the recent availability of transaction-by-transaction data, empirical analyses on a transaction time scale have also been performed. One of these studies performed by analyzing stock price in the Budapest stock exchange show that return PDF are leptokurtic down to a 'transaction' time scale [12.77]. See a direct example in Fig. 12.3.

The origin of the observed leptokurtosis is still debated. There are several models trying to explain it. Just to cite (rather arbitrarily) a few of them: (i) a model of Lévy stable stochastic process [12.68]; (ii) a model assuming that the non-Gaussian behavior occurs as a result of the uneven activity during market hours [12.84]; (iii) a model where a geometric diffusive behavior is superimposed to Poissonian jumps [12.85]; (iv) a quasi-stable stochastic process with finite variance [12.86]; and (v) a stochastic process with rare events described by a power-law exponent not falling into the Lévy regime [12.17, 12.41, 12.87]. The above processes are characterized by finite or infinite moments. In the attempt to find the stochastic process that best describes stock price dynamics, it is then important to try to preliminary conclude about the finiteness or infiniteness of the second moment.

The above answer is not simply obtained [12.88] and careful empirical analyses must be performed to reach a reliable conclusion. It is our opinion that an impressive amount of empirical evidence has been recently found supporting the conclusion that the second moment of the return PDF is fi-

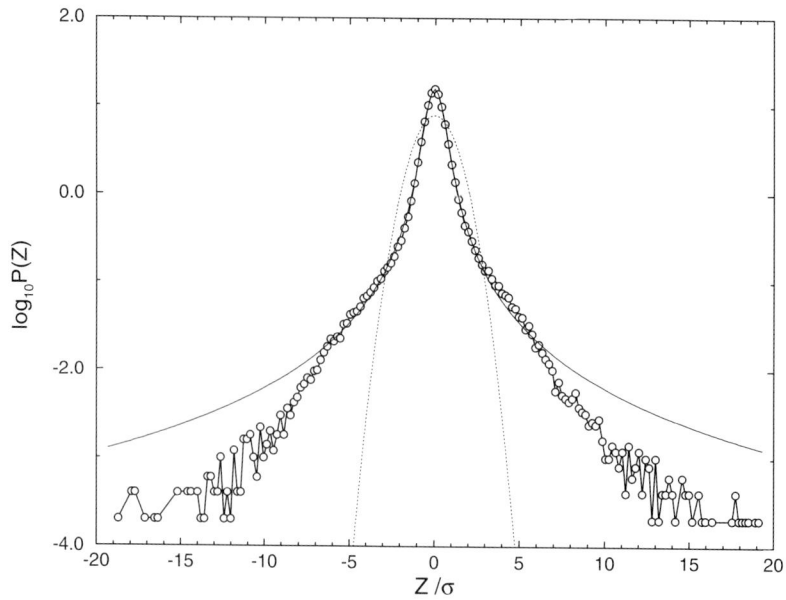

Fig. 12.2. Logarithm of the probability density function of the S&P 500 index high-frequency changes computed at a $\Delta t = 1$ min time horizon. The probability density function is compared with the symmetrical Lévy stable distribution of index $\alpha = 1.40$ and scale factor $\gamma = 0.00375$ (solid line). The dotted line is the Gaussian distribution with standard deviation σ equal to the experimental value 0.0508. The variations of price are normalized to this value. Deviations from the Lévy stable profile are observed for $|z|/\sigma \gtrsim 6$. From [12.14]

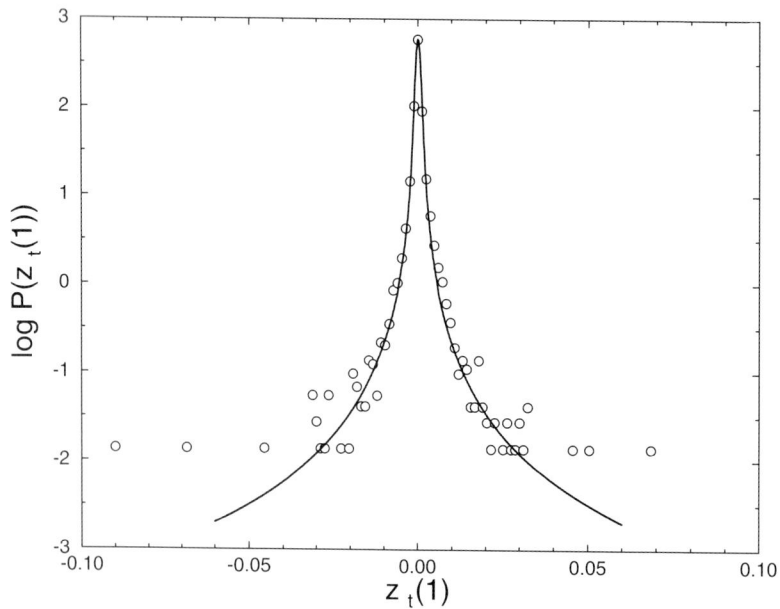

Fig. 12.3. Semilogarithmic plot of the probability density function of tick by tick log price changes measured for the MOL Company (the time period is the second quarter of 1998). The solid line is a symmetrical Lévy distribution of index $\alpha = 1.60$ and scale factor $\gamma = 0.000005$. From [12.77]

nite [12.14, 12.17, 12.41, 12.87, 12.89, 12.90]. This conclusion has a deep consequence on the stability of the return PDF. The finiteness of the second moment and the independence of successive returns imply that the central limit theorem asymptotically applies. Hence the form expected for the return PDF must be Gaussian for very long time horizons. We then have two regions - at short time horizons we observe leptokurtic distributions whereas at long time horizons we expect a Gaussian distribution. A complete characterization of the stochastic process needs an investigation performed at different time horizons. During this kind of analysis, non-Gaussian scaling and its breakdown has been detected [12.14, 12.19].

12.7 Collective Dynamics

In the previous sections we saw that 'universal' facts suggest that the stock price change dynamics in financial markets is well described by an unpredictable time series. However, this does not imply that the stochastic dynamics of stock price time series is a random walk with independent identically distributed increments. Indeed the stochastic process is much more complex than a customary random walk.

One key question in the analysis and modeling of a financial market concerns the independence of the price time series of different stocks traded simultaneously in the same market. The presence of cross-correlations between pairs of stocks has been known for a long time and it is one of the basic assumptions of the theory of the selection of the most efficient portfolio of stocks [12.91]. Recently, physicists have also started to investigate empirically and theoretically the presence of such cross-correlations.

It has been found that a meaningful economic taxonomy may be obtained by starting from the information stored in the time series of stock price only. A simple example is shown in Fig. 12.4. This has been achieved by assuming that a metric distance can be defined between the synchronous time evolution of a set of stocks traded in a financial market and under the essential *Ansatz* that the subdominant ultrametric associated with the selected metric distance is controlled by the most important economic information stored in the time evolution dynamics [12.92].

Another approach to detect collective movement of stock price fluctuations involves the comparison of the statistics of the measured equal-time cross-correlation matrix C against a 'null hypothesis' of a 'random' cross-correlation matrix, constructed from mutually uncorrelated time series [12.93, 12.94]. The comparison is performed in the diagonal basis, and it is found that $\approx 98\%$ of the eigenvalues of C are consistent [12.93, 12.94] with that of a random cross-correlation matrix (see Fig. 12.5). There are also deviations [12.93, 12.94] for $\approx 2\%$ of the eigenvalues at both edges of the eigenvalue spectrum, which

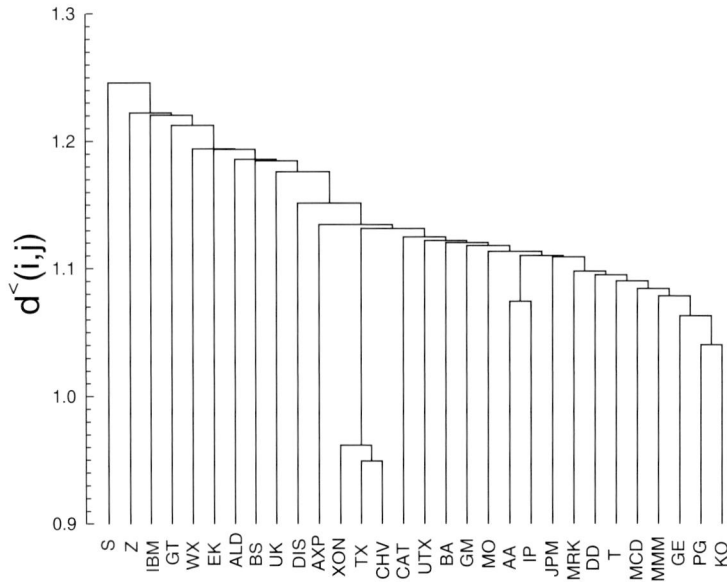

Fig. 12.4. Hierarchical tree associated to the stock portfolio of 30 stocks used to compute the Dow Jones Industrial Average index in 1998. The 30 stocks are labeled by their tick symbols. The hierarchical tree is obtained with the correlation based clustering procedure of [12.92]. In the hierarchical tree, several groups of stocks are detected. They are homogeneous with respect to the economic activities of the companies: (i) oil companies (Exxon (XON), Texaco (TX), and Chevron (CHV)); (ii) raw material companies (Alcoa (AA) and International paper (IP)); and (iii) companies working in the sectors of consumer nondurable products (Procter and Gamble (PG)) and food and drinks (Coca Cola (KO)). The distance between each stock and the others is the ultrametric distance $d^<(i,j)$ computed starting from the correlation coefficient matrix. From [12.92]

are found to correspond mainly to conventionally identified sectors of business activity [12.96].

The observation of the presence of a certain degree of statistical synchrony in the stock price dynamics suggests the following conclusion. Consideration of the time evolution of only a single stock price could be insufficient to reach a complete modeling of all essential aspects of a financial market.

12.8 Conclusion

This chapter briefly discusses the goals and scopes of econophysics, the motivations and precursors of physicists involved in the analysis and modeling of financial markets, and some of the stylized 'universal' facts that are observed in financial markets and are considered robust by several researchers working in the field. Starting from these results one can devise studies trying to enrich and expand this knowledge to provide theoreticians and computer scientists with the empirical facts that need to be explained by their models progressively proposed. The ultimate goal is to contribute to the search for the best model describing a financial market, one of the most intriguing *complex systems*.

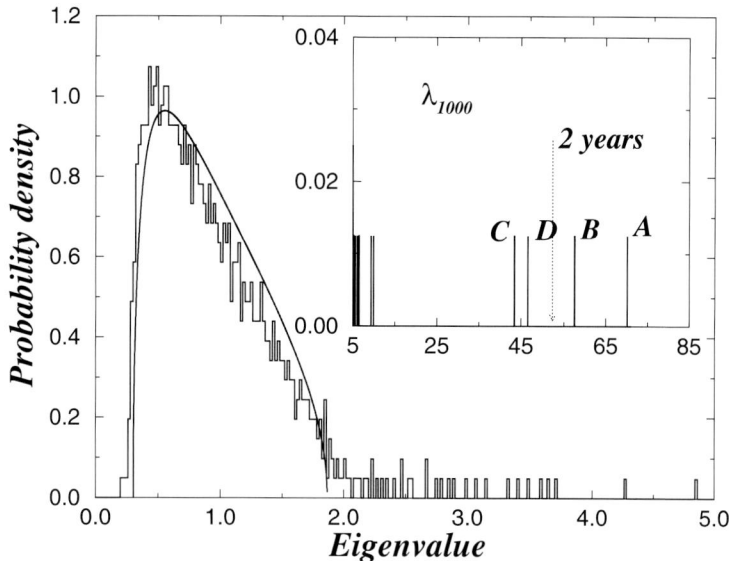

Fig. 12.5. The probability density of the eigenvalues of the equal-time cross-correlation matrix C constructed from price fluctuations of 1 000 largest stocks in the TAQ database for the 2 yr period 1994–1995. Recent analytical results [12.95] for cross-correlation matrices constructed from mutually uncorrelated time series predict a distribution of eigenvalues within a finite range depending on the ratio R of the length of the time series to the dimension of the matrix (solid curve). In our case $R = 6.448$ corresponding to eigenvalues distributed in the interval $0.37 \leq \lambda_k \leq 1.94$ [12.93–12.95]. However, the largest eigenvalue for the 2 yr period (inset) is approximately 30 times larger than the maximum eigenvalue predicted for uncorrelated time series. The inset also shows the largest eigenvalue for the cross-correlation matrix for four half-yr periods - denoted A, B, C, D. The arrow in the inset corresponds to the largest eigenvalue for the entire 2 yr period, $\lambda_{1\,000} \approx 50$. The distribution of eigenvector components for the large eigenvalues, well outside the bulk, shows significant deviations from the Gaussian prediction of RMT, which suggests 'collective' behavior or correlations between different companies. The largest eigenvalue corresponds to the correlations within the entire market [12.93, 12.94]

Acknowledgements. We thank our many collaborators for joining us in work reported herein. These include L.A.N. Amaral, G. Bonanno, P. Cizeau, P. Gopikrishnan, F. Lillo, Y. Liu, M. Meyer, Z. Palagyi, V. Plerou, and B. Rosenow. Finally, we thank INFM, MURST, and NSF for financial support.

References

12.1 L. Bachelier, Ann. Sci. l'Ecole Norm. Supér. **III-17**, 21 (1900).

12.2 P.H. Cootner (ed.), *The Random Character of Stock Market Prices* (MIT, Cambridge, 1964).

12.3 J.E. Ingersoll Jr., *Theory of Financial Decision Making* (Rowman and Littlefield, Savage, 1987).

12.4 R.C. Merton, *Continuous-Time Finance* (Blackwell, Cambridge, 1990).

12.5 I. Karatzas and S.E. Shreve, *Brownian Motion and Stochastic Calculus, 2nd edn.* (Springer, Berlin, 1991).

12.6 J.Y. Campbell, A.W. Lo, and A.C. MacKinlay, *The Econometrics of Financial Markets*, (Princeton University Press, Princeton, 1997).

12.7 R.N. Mantegna (ed.), *Proceedings of the International Workshop on Econophysics and Statistical Finance*, Physica A [special issue] **269** (1999).

12.8 J.-P. Bouchaud, K. Lauritsen, and P. Alstrom (eds.), *Proceedings of the International Conference on Applications in Finacial Analysis*, Int. J. Theor. Appl. Finance [special issue] **3**(3), 309 (2000).

12.9 J.-P. Bouchaud and M. Potters, *Theory of Financial Risk* (Cambridge University Press, Cambridge, 2000).

12.10 R.N. Mantegna and H.E. Stanley, *An Introduction to Econophysics: Correlations and Complexity in Finance* (Cambridge University Press, Cambridge, 2000).

12.11 G. Parisi, Physica A **263**, 557 (1999).

12.12 F. Black and M. Scholes, J. Polit. Econ. **81**, 637 (1973).

12.13 R.N. Mantegna, Physica A **179**, 232 (1991).

12.14 R.N. Mantegna and H.E. Stanley, Nature **376**, 46 (1995).

12.15 S. Ghashghaie, W. Breymann, J. Peinke, P. Talkner, and Y. Dodge, Nature **381**, 767 (1996).

12.16 M. Potters, R. Cont, and J.-P. Bouchaud, Europhys. Lett. **41**, 239 (1998).

12.17 P. Gopikrishnan, M. Meyer, L.A.N. Amaral, and H.E. Stanley, Eur. Phys. J. B **3**, 139 (1998).

12.18 W. Li, Int. J. Bifurcat. Chaos **1**, 583 (1991).

12.19 R.N. Mantegna and H.E. Stanley, Nature **383**, 587 (1996).

12.20 Y. Liu, P. Cizeau, M. Meyer, C.-K. Peng, and H.E. Stanley, Physica A **245**, 437 (1997).

12.21 R. Cont, M. Potters, and J.-P. Bouchaud, in *Scale Invariance and Beyond*, edited by B. Dubrulle, F. Graner and D. Sornette (Springer, Berlin, 1997).

12.22 A. Arneodo, J.F. Muzy, and D. Sornette, Eur. Phys. J. B **2**, 277 (1998).

12.23 U. A. Müller, M.M. Dacorogna, R.B. Olsen, O.V. Pictet, M. Schwarz, and C. Morgenegg, J. Banking Finance **14**, 1189 (1995).

12.24 P. Cizeau, Y. Liu, M. Meyer, C.-K. Peng, and H.E. Stanley, Physica A **245**, 441 (1997).

12.25 J.-P. Bouchaud and R. Cont, Eur. Phys. J. B **6**, 543 (1998).

12.26 H. Takayasu, H. Miura, T. Hirabayashi, and K. Hamada, Physica A **184**, 127 (1992).

12.27 P. Bak, K. Chen, J. Scheinkman, and M. Woodford, Ric. Econ. **47**, 3 (1993).

12.28 D. Challet and Y.-C. Zhang, Physica A **256**, 514 (1998).

12.29 M. Lévy and S. Solomon, Int. J. Mod. Phys. C **7**, 595 (1996).

12.30 G. Caldarelli, M. Marsili, and Y.-C. Zhang, Europhys. Lett. **40**, 479 (1997).

12.31 M. Lévy, H. Lévy, and S. Solomon, J. Phys. I France **5**, 1087 (1995).

12.32 B.B. Mandelbrot, *Fractals and Scaling in Finance* (Springer, New York, 1997).

12.33 S. Maslov and Y.-C. Zhang, Physica **262**, 232 (1999).

12.34 P. Bak, M. Paczuski, and M. Shubik, Physica A **246**, 430 (1997).

12.35 A.H. Sato and H. Takayasu, Physica A **250**, 231 (1998).

12.36 D. Sornette and A. Johansen, Physica A **261**, 581 (1998).

12.37 D. Stauffer, Ann. Phys.-Berlin **7**, 529 (1998).

12.38 D. Stauffer and T.J.P. Penna, Physica A **256**, 284 (1998).

12.39 H. Takayasu, A.H. Sato, and M. Takayasu, Phys. Rev. Lett. **79**, 966 (1997).

12.40 T. Lux, J. Econ. Dyn. Control **22**, 1 (1997).

12.41 T. Lux, J. Econ. Behav. Organ. **33**, 143 (1998).

12.42 T. Lux and M. Marchesi, Nature **397**, 498 (1999).

12.43 J.-P. Bouchaud and D. Sornette, J. Phys. I France **4**, 863 (1994).

12.44 E. Aurell and S.I. Simdyankin, Int. J. Theor. Appl. Finance **1**, 1 (1998).

12.45 R. Baviera, M. Pasquini, M. Serva, and A. Vulpiani, Int. J. Theor. Appl. Finance **1**, 473 (1998).

12.46 S. Galluccio and Y.-C. Zhang, Phys. Rev. E **54**, R4516 (1996).

12.47 S. Galluccio, J.-P. Bouchaud, and M. Potters, Physica A **259**, 449 (1998).

12.48 M. Marsili, S. Maslov, and Y.-C. Zhang, Physica A **253**, 403 (1998).

12.49 D. Sornette, Physica A **256**, 251 (1998).

12.50 R.N. Mantegna and H.E. Stanley, Physica A **239**, 255 (1997).

12.51 J.D. Farmer, Market force, ecology, and evolution; [available from `xxx.lanl.gov/abs/adap-org/9812005`].

12.52 N. Vandewalle and M. Ausloos, Physica A **246**, 454 (1997).

12.53 M.H.R. Stanley, L.A.N. Amaral, S.V. Buldyrev, S. Havlin, H. Leschhorn, P. Maass, M.A. Salinger, and H.E. Stanley, Nature **379**, 804 (1996).

12.54 L.A.N. Amaral, S.V. Buldyrev, S. Havlin, H. Leschhorn, P. Maass, M.A. Salinger, H.E. Stanley, and M.H.R. Stanley, J. Phys. I France **7**, 621 (1997).

12.55 L.A.N. Amaral, S.V. Buldyrev, S. Havlin, M.A. Salinger, and H.E. Stanley, Phys. Rev. Lett. **80**, 1385 (1998).

12.56 H. Takayasu and K. Okuyama, Fractals **6**, 67 (1998).

12.57 Y. Lee, L.A.N. Amaral, D. Canning, M. Meyer, and H.E. Stanley, Phys. Rev. Lett. **81**, 3275 (1998).

12.58 E. Majorana, Scientia **36**, 58 (1942).

12.59 L.P. Kadanoff, Simulation **16**, 261 (1971).

12.60 E.W. Montroll and W.W. Badger, *Introduction to Quantitative Aspects of Social Phenomena* (Gordon and Breach, New York, 1974).

12.61 P.W. Anderson, J.K. Arrow, and D. Pines (eds.), *The Economy as an Evolving Complex System* (Addison-Wesley, Redwood City, 1988).

12.62 V. Pareto, *Cours d'Economie Politique*, (F. Rouge, Lausanne and Paris, 1897).

12.63 P. Lévy, *Calcul des Probabilités* (Gauthier-Villars, Paris, 1925).

12.64 B.B. Mandelbrot, *The Fractal Geometry of Nature* (Freeman, San Francisco, 1982).

12.65 H.E. Stanley, *Introduction to Phase Transitions and Critical Phenomena* (Oxford University Press, Oxford, 1971).

12.66 A. Einstein, Ann. Phys. **17**, 549 (1905).

12.67 N. Wiener, J. Math. Phys. **2**, 131 (1923).

12.68 B.B. Mandelbrot, J. Bus. **36**, 394 (1963).

12.69 B.V. Gnedenko and A.N. Kolmogorov, *Limit Distributions for Sums of Independent Random Variables* (Addison-Wesley, Cambridge, 1954).

12.70 E.F. Fama, J. Finance **25**, 383 (1970).

12.71 P.A. Samuelson, Industrial Management Rev. **6**, 41 (1965).

12.72 J.L. Doob, *Stochastic Processes* (Wiley, New York, 1953).

12.73 E.F. Fama, J. Finance **46**, 1575 (1991).

12.74 J.C. Hull, *Options, Futures, and Other Derivatives, 3rd edn.* (Prentice-Hall, Upper Saddle River, 1997).

12.75 D. Duffie and J. Pan, J. Deriv. [spring issue], 7 (1997).

12.76 Y. Liu, P. Gopikrishnan, P. Cizeau, M. Meyer, C.-K. Peng, and H.E. Stanley, Phys. Rev. E **60**, 1390 (1999).

12.77 Z. Palágyi and R.N. Mantegna, Physica A **269**, 132 (1999).

12.78 R. Baviera, M. Pasquini, M. Serva, D. Vergni, and A. Vulpiani, Efficiency in foreign exchange markets (1999); [available from `xxx.lanl.gov/abs/cond-mat/9901225`].

12.79 M.M. Dacorogna, U.A. Müller, R.J. Nagler, R.B. Olsen, and O.V. Pictet, J. Int. Money Finance **12**, 413 (1993).

12.80 M. Pasquini and M. Serva, Physica A **269**, 140 (1999).

12.81 M. Raberto, E. Scalas, G. Cuniberti, and M. Riani, Physica A **269**, 148 (1999).

12.82 R.F. Engle, Econometrica **50**, 987 (1982).

12.83 T. Bollerslev, J. Economet. **31**, 307 (1986).

12.84 P.K. Clark, Econometrica **41**, 135 (1973).

12.85 R.C. Merton, J. Financ. Econ. **3**, 125 (1976).

12.86 R.N. Mantegna and H.E. Stanley, Phys. Rev. Lett. **73**, 2946 (1994).

12.87 T. Lux, Appl. Financ. Econ. **6**, 463 (1996).

12.88 A.L. Tucker, J. Bus. Econ. Statist. **10**, 73 (1992).

12.89 P. Gopikrishnan, V. Plerou, L.A.N. Amaral, M. Meyer, and H.E. Stanley, Phys. Rev. E **60**, 5305 (1999).

12.90 V. Plerou, P. Gopikrishnan, L.A.N. Amaral, M. Meyer, and H.E. Stanley, Phys. Rev. E **60**, 6519 (1999).

12.91 H. Markowitz, *Portfolio Selection: Efficient Diversification of Investment* (Wiley, New York, 1959).

12.92 R.N. Mantegna, Eur. Phys. J. B **11**, 193 (1999).

12.93 L. Laloux, P. Cizeau, J.-P. Bouchaud, and M. Potters, Phys. Rev. Lett. **83**, 1468 (1999).

12.94 V. Plerou, P. Gopikrishnan, B. Rosenow, L.A.N. Amaral, and H.E. Stanley, Phys. Rev. Lett. **83**, 1471 (1999).

12.95 A.M. Sengupta and P.P. Mitra, Phys. Rev. E **60**, 3389 (1999).

12.96 P. Gopikrishnan, B. Rosenow, V. Plerou, and H.E. Stanley, Phys. Rev. E. Rapid Communications **64**, 035106-1 (2001).

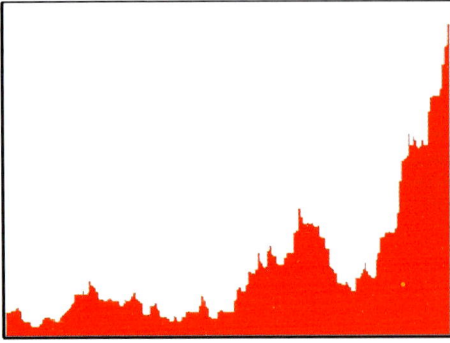

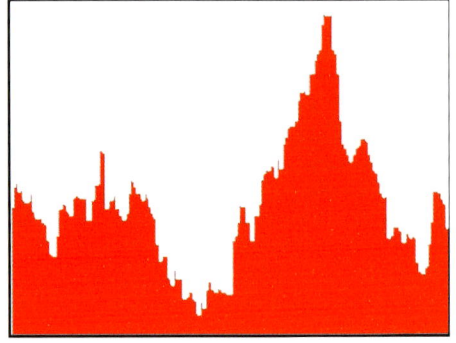

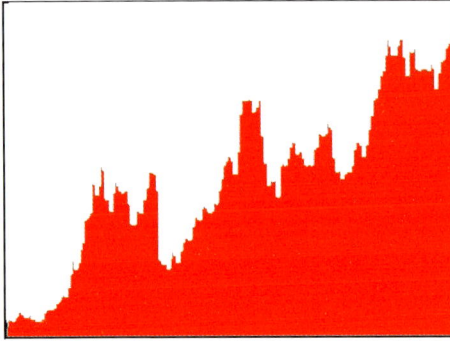

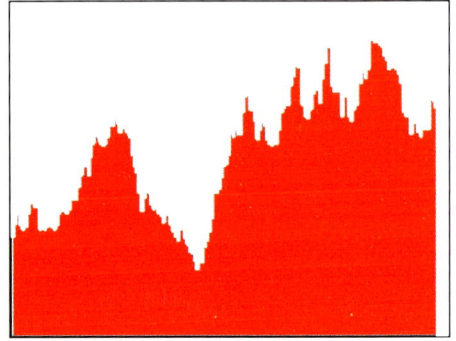

13. Market Fluctuations I: Scaling, Multiscaling, and Their Possible Origins

Thomas Lux and Marcel Ausloos

In this chapter, we provide a survey of research on scaling phenomena in financial data pursued by physicists and compare their methodology and results with the approach of economists dealing with the same topic. We also try to put this work into perspective by discussing in how far it is reconcilable with traditional models in finance (the efficient market hypothesis) or whether it leads to a new viewpoint on market interactions.

13.1 Introduction

In the following we present a selective review of research on the statistical properties of financial data with an emphasis on those features that have also been the focus of recent analyses by physicists. We highlight that many of the observations known as 'stylized facts' (economics terminology) of financial markets can be interpreted as 'scaling laws' in the sense of statistical physics. Although the latter term has been virtually unknown in the pertinent economics literature until recently, some ubiquitous statistical findings like the 'fat tails' of the distribution of returns and the phenomenon of volatility clustering can, in fact, be cast into the framework of scaling theory. Reviewing work on these empirical characteristics by both economists and physicists, we compare their respective methodology and results. Although we find that, in

---

◀ **Fig. 13.0.** Real and synthetic financial 'time series': two of the four exhibits are daily records of well-known financial indices. The two remaining plots are generated from numerical simulations of two recently proposed models. One of them is a realization of a new type of stochastic process: the multifractal model proposed in [13.114]. In this approach, a multifractal cascade serves as a time transformation which expands and contracts the timescale of a homogeneous incremental Brownian motion. The last graph stems from simulations of the behavioral model proposed in [13.109, 13.110]. Here the price dynamics results from a microscopic model of a large ensemble of interacting agents. The main text identifies the model-generated records and empirical data

the end, many of the results obtained by both groups are in good *qualitative* agreement, comparison of the outcome of different test procedures also suggests that one should be cautious in attaching excessive credibility to numerical parameter estimates. We also discuss the implications of scaling in finance for the theoretical modeling of the price formation process. On the one hand, the empirical scaling laws are reconcilable with traditional models of financial markets: since, in principle, we could trace back scaling of prices and returns to similar behavior of the flow of new information, the finding of scaling laws per se does not serve as evidence against perfectly efficient and unbiased information processing (the efficient market hypothesis (EMH)). On the other hand, the experience that scaling laws could often be explained by the working of systems of many interacting units in statistical physics suggests a different approach: from this perspective, some of the salient features of the data might be explained by the interactions of a large ensemble of heterogeneous market participants (the interacting agent view). Some recent models provide first steps into this direction.

13.2 Scaling in the Probability Distribution of Returns

13.2.1 Early Work and Theoretical Background

Research on the statistical properties of financial time series has started 101 yr ago, when Louis Bachelier wrote a Ph.D. Thesis with the title 'Théorie de la Spéculation' (reprinted in [13.34], for the first page see Chap. 12, p. 359). In fact, in proposing the normal distribution as a model of price variations, Bachelier seemed to have been the first scientist who formulated a testable hypothesis on the statistical behavior of financial data.

Although it turned out later that the normal distribution can usually be overwhelmingly rejected by explicit statistical tests, due to its familiarity it is still often used in theoretical work as well as by market practitioners. Theoretically, the appeal of the Gaussian stems, of course, in finance like in other fields, from its stability-under-addition property: i.e. the Gaussian distribution is the limit distribution for sums of independent and identically distributed (i.i.d.) random variables (the central limit theorem). This property, in fact, comes into play quite naturally when dealing with financial prices: the most easily available data, price changes on a daily frequency, can be considered as the sums of a multitude of smaller increments on the intra-daily level. If it is reasonable to assume that these high-frequency variations are i.i.d. random variables, then the central limit law should apply to price changes of a lower frequency and it, then, seems natural to postulate that their distribution approaches a Gaussian shape. But why should we assume that single price changes are governed by random motion? Shouldn't there be systematic economic factors behind price variations in financial markets? Thus, before looking at empirical results, we will shortly outline the *economic* reasoning behind the i.i.d. assumption for price changes.

For a long time, the only available theoretical background to the statistical behavior of asset prices has been the so-called *efficient market hypothesis* (EMH) (cf. [13.48]). It states that, at any point in time, asset prices should reflect the discounted expected stream of earnings from holding the underlying asset. Thereby, the expectations are to be understood in a mathematical sense as the expected future values of stochastic quantities, which are computed using all presently known information about future circumstances.

Although mutatis mutandis the efficient market hypothesis can be applied to other financial markets as well (for example, foreign exchange, precious metal, or future markets), we will concentrate on share markets for ease of exposition. In this case, future earning streams are most easily identified as dividend payments. The price predicted by the EMH should, therefore, be given by:

$$p_t = \sum_{i=1}^{\infty} \delta^i E\Big[d_{t+i}|I_t\Big].$$
(13.1)

In (13.1) p_t denotes the price at time t, d_{t+i} denotes the dividend paid in a future period $t + i$, and $\delta < 1$ is a discount factor reflecting the fact that the expected payment of one dollar in the future has less value today than immediate availability of the same amount. $E[\cdot]$, finally, is the mathematical expectation operator which is computed conditional on the current information set I_t.

We can also express the asset pricing equation in a somewhat different manner if we do not assume that the investor will hold the asset forever, but rather considers to sell it at time $t+1$ in order to consume his receipts or make another investment. Under this perspective, future earnings are composed of the dividend plus the price at $t + 1$ which are both uncertain in t. Hence, the price should equal:

$$p_t = \delta E\Big[\, p_{t+1} + d_{t+1}|I_t \,\Big].$$
(13.2)

Iterating (13.2) and repeatedly replacing prices by future dividends, it is, in fact, easy to show that both variants are identical. However, there is a slight difference between (13.1) and (13.2): in order to make sure that the sum on the right hand side of (13.1) converges, we have to assume that the mathematical 'transversality condition' $\lim_{i \to \infty} \delta^i E_t[p_{t+i}] \to 0$ holds. Although this condition may appear innocuous, a large body of literature has developed about this and similar conditions in applications in finance [13.51] as well as in other branches of economics [13.72]. In a finance context, its violation allows for price paths which do not conform to the efficient market hypothesis and give rise to so-called *rationally expected speculative bubbles*[1].

Let us assume that the transversality condition holds and prices can be expressed by (13.1). Then, *price changes* come about by arrival of some *new* items of information about future dividends (i.e. an increase of the information set) which changes the rationally computed expectations of future prices and dividends. Of course, such information continuously hits the market: one can, for example, think of such elementary items like weather or climatic events affecting output in certain industries, political changes (like introduction of new taxes on emissions, changes of monetary policy) as well as any kind of firm-specific events (e.g. all types of news about management policies). According to the EMH, the entirety of all these factors would explain the history of price changes: *the 'news arrival process' is crucial for and, in fact, governs the characteristics of the distribution of price increments.* As 'news' should be events that are independent of the preceding ones, the 'independence' part of the i.i.d. assumption can be immediately justified by the above considerations. If we are prepared to accept the second part as well (identical distribution), we end up with the central limit law for low-frequency price changes as the aggregate of high-frequency ones. Note that under the EMH, the distribution of price changes is, then, a mere reflection of the distribution of arriving news (which, however, is an aggregate of diverse individual events that cannot be quantified). Scaling laws for price changes would, therefore, have to be explained by similar scaling laws for the 'news arrival process'.

13.2.2 Mandelbrot's Stable Distribution Hypothesis and Alternative Models

During the 1960s and 1970s, the above picture has undergone some changes: first, it has been found that price increments themselves are less suitable for statistical analysis than *relative* changes. One, therefore, routinely transforms the raw time series of prices into returns: $r_t = \frac{p_t - p_{t-1}}{p_{t-1}} \approx \ln(p_t) - \ln(p_{t-1})$ for statistical analysis. A second and more important modification was brought about by Mandelbrot's and Fama's finding that daily returns are decisively

---

[1] The most elementary example is a price path containing a bubble term B_t which develops as $B_{t+1} = B_t/\delta$. It is easy to check that this would be in harmony with (13.2), cf. [13.17]. Nevertheless, the price would not be an efficient one just because it incorporates the bubble term B_t. As the literature on rational bubbles has only weak links to the statistical literature we dispense with a more detailed treatment of this research here. In [13.111] it is shown that scaling laws derived from theoretical models of rational bubbles are quite different from empirical ones. In particular, instead of the ubiquitous empirical finding of approximately cubic scaling in the tail regions of the unconditional distribution (cf. Sects. 13.2.3 and 13.2.4 below), rational bubbles give rise to exponents smaller than 1.

nonnormal (cf. [13.47, 13.112]). In fact, standard tests for normality usually reject their null hypothesis at extremely high levels of significance.

Giving up normality, but maintaining stability under aggregation, Mandelbrot [13.112] proposed the Lévy stable distributions as an alternative model. Under this new hypothesis on the distribution of returns, relative price changes would have no finite second and higher moments and would, therefore, obey the *generalized* central limit law under time aggregation. Besides their desirable stability property, the Lévy distributions are also in harmony with the usual visual appearance of the empirical distribution for returns: they possess more probability mass in their tails and center than the Gaussian (i.e. they are *leptokurtotic*). Unfortunately, the stable Lévy distributions lack an analytical closed-form solution and can only be described by their characteristic function:

$$
\log E(\mathrm{e}^{\mathrm{i}xt}) =
$$
$$
\begin{cases}
\mathrm{i}\delta t - |ct|^{\alpha_s}\left[1 - \mathrm{i}\beta\,\mathrm{sign}(t)\tan(\pi\,\alpha_s/2)\right] & \text{if}\quad \alpha_s \neq 0, \\
\mathrm{i}\delta t - |ct|\left[1 + \mathrm{i}\beta(2/\pi)\,\mathrm{sign}(t)\log|t|\right] & \text{if}\quad \alpha_s = 1,
\end{cases}
\tag{13.3}
$$

with parameters α_s, β, c, and δ determining the shape, skewness, width, and location. The distribution function and density of x can only be obtained numerically by evaluating the inverse Fourier transform of the characteristic function (13.3).

Furthermore, at the time of publication of Mandelbrot's and Fama's papers, almost no statistical method was known for estimating the parameters and testing goodness-of fit of the Lévy distributions. Nevertheless, the argument of convergence of aggregate returns towards one of the stable laws appeared so convincing, that a number of researchers accepted indication of $\alpha_s < 2$ as evidence in favor of Lévy stable distributions without further testing of fit. Examples in the literature include [13.35, 13.138, 13.139, 13.151]. At the same time, Mandelbrot's hypothesis was questioned by others on the basis of cleverly designed tests for stability-under-addition itself. These tests often produced nonstationary results for the characteristic exponent α_s at different levels of time aggregation, which is in contradiction to the assumed stability of the distributions (cf. [13.55, 13.70, 13.78, 13.153]). Similarly, analysis of the behavior of higher moments produced evidence against the Lévy model in that the divergence proceeded more slowly than expected [13.89][2].

For more than 25 yr, these two conflicting strands of research coexisted in the literature and the issue of the appropriate distributional assumptions remained basically undecided. However, from a practical point of view, a number

---

[2] From his recent publications, it seems that Mandelbrot has also converged to acceptance of finite second moments, cf. [13.50].

of alternative, more easily tractable statistical models have also been proposed, although they lack a strong theoretical foundation. From the wealth of papers that appeared over the decades, only a few are mentioned here emphasizing the broad spectrum of distributions that have been proposed:

- mixtures of normal distributions with different means or variances [13.87, 13.152].
- subordinated processes, most notably Clark's model [13.30] of a log-normal subordinator for the variance of a Gaussian incremental distribution. Interestingly, Clark used volume data as a measure of trading intensity in order to get a fit of the log-normal.
- compound diffusion-jump processes with the jump component reflecting 'extraordinary' events [13.54].
- the Student's t-distribution as a fat tailed distribution with finite higher moments [13.18].
- Tuckey's '$g \times h$ distributions' (transformations of the normal that allow for skewness and leptokurtosis) [13.121].
- hyperbolic distributions [13.43].

13.2.3 The Contribution of the Statistical Theory of Extremes

This huge variety of distributional forms makes it hard to find universal laws in the statistical behavior of financial prices. Starting in the early 1990s, a new type of analysis has appeared in empirical finance, that, in fact, highlights typical features by abstracting from a specific distributional shape. This strand of literature took up a proposal by DuMouchel [13.42] to concentrate on the behavior of the tails instead of trying to fit the entire distribution. The theoretical background to this research program is provided by statistical extreme value theory (laid out, for example, in the following recent textbooks: [13.13, 13.133]). One of the basic results of the theory of extremes is typology of the limiting distributions for maxima (or minima) from i.i.d. random variables with continuous distributions. Namely, denoting by $M = \max(x_1, x_2, \dots, x_n)$ the maximum of a sample of observations $\{x_i\}$, it can be shown that after appropriate change of scale and location the limiting distribution of M belongs to one of only three classes of distribution functions. Expressed more formally, the limiting distribution of the normalized maximum, $P[a_n M + b_n \leq x]$, with a_n and b_n denoting normalizing constants, converges to one of the following types of *generalized extreme value* (GEV) distributions:

$$G_{1,\alpha}(x) = \begin{cases} 0 & x \leq 0\,, \\ \exp\left(-x^{-\alpha}\right) & x > 0\,, \end{cases} \tag{13.4}$$

$$G_{2,\alpha}(x) = \begin{cases} \exp\left(-(-x)^{\alpha}\right) & x \leq 0\,, \\ 1 & x > 0\,, \end{cases} \tag{13.5}$$

$$G_{3}(x) = \exp\left(-e^{-x}\right) \qquad x \in \mathbb{R}\,. \tag{13.6}$$

As an alternative to (13.4) to (13.6) one also encounters the so-called von Mises representation of the GEV distributions which provides a unified framework for all three cases:

$$G_{\gamma}(x) = \exp\left(-(1+\gamma x/\sigma)^{-1/\gamma}\right)\,. \tag{13.7}$$

In this formalization, $\sigma > 0$ is a scale parameter and the three elementary types of extremal behavior are characterized by $\gamma > 0$, $\gamma < 0$, and the limit in the case $\gamma \to 0$. For type 1 (2), the shape parameters of both representations (1) and (2) are related to each other by the following identities: $\gamma = 1/\alpha$ ($\gamma = -1/\alpha$).

From this classification of *extrema*, a similar classification of the behavior in the distribution's outer parts can be derived. Namely, denoting the probabilities $\text{Prob}[X_i \leq x] \equiv W$, it follows directly from the classification of extremes in (13.4)–(13.6) that if the maximum of a distribution follows a GEV of type i ($i = 1, 2, 3$) then the upper tail of the distribution is close to

$$W_{1,\alpha} = 1 - x^{-\alpha}, \qquad x \geq 1\,, \tag{13.8}$$
$$W_{2,\alpha} = 1 - (-x)^{\alpha}, \qquad -1 \leq x \leq 0\,, \tag{13.9}$$
$$W_{3} = 1 - \exp(-x), \qquad x \geq 0\,. \tag{13.10}$$

i.e. a so-called generalized Pareto distribution (GPD) having the same tail shape parameter α (cf. [13.133], Chap. 5).

As with the GEVs, one can also integrate the three laws (13.8)–(13.10) into one unifying representation:

$$W_{\gamma} = 1 - (1+\gamma x/\sigma)^{-1/\gamma}\,, \tag{13.11}$$

The three elementary types of tail behavior can be described as hyperbolic decline (first case), distributions with finite endpoints (case 2), and exponential decline (case 3). They are characterized by $\gamma > 0$, $\gamma < 0$, and the limit in the $\gamma \to 0$ case. For type 1 (2), the shape parameters of both representations (13.8)–(13.10) and (13.11) are again related to each other by the following identities: $\gamma = 1/\alpha$ ($\gamma = -1/\alpha$).

A variety of estimators has been developed for the 'tail index' α or γ. Estimation of this quantity for the tail shape alone is favorable over an overall fit of the empirical distribution for a number of reasons: first, it gives an indication of the *type* of behavior in the outer parts and, therefore, may allow us to exclude a number of candidate processes from the outset (e.g. an estimate of γ significantly different from zero would simultaneously imply rejection of the normal, mixtures of normals as well as diffusion-jump processes, as they all lack the required behavior of large returns). Second, in financial applications, one is often more interested in extreme realizations (crash risks!) than in the exact shape of minor fluctuations. As in any fit of the overall distribution, the extremes only constitute a small part of the sample, their 'influence' on the final estimation result is almost negligible. Furthermore, the choice of a particular overall model already restricts the results that can be found for the tail behavior (for example, choice of the Lévy distributions always implies that tails are assumed to be extremely heavy and restricts the tail index to the region $\alpha < 2$).

The majority of studies in the literature applied maximum likelihood estimators for the tail index α or γ. Because of its simplicity the 'Hill tail index estimator' [13.74] has become the standard work tool in most studies of tail behavior of economic data. The Hill estimate $\hat{\gamma}_H$ is obtained by maximization of the likelihood of the relevant tail function conditional on the chosen size of the 'tail'. It is computed as:

$$\hat{\gamma}_H = \left(\hat{\alpha}_H\right)^{-1} = \frac{1}{k}\sum_{i=1}^{k}\left[\log x_{(n-i+1)} - \log x_{(n-k)}\right]. \tag{13.12}$$

In (13.12), sample elements are put in descending order: $x_{(n)} \geq x_{(n-1)} \geq \ldots \geq x_{(n-k)} \geq \ldots \geq x_{(1)}$ with k the number of observations located in the 'tail'.

The results from a large body of research in this vein turned out to yield astonishingly uniform behavior for data from different markets in that *one usually finds a tail index (exponent of the cumulative distribution) in the range 2.5 to about 4*. Examples include [13.81, 13.86, 13.99, 13.100, 13.102]. Surveys of this literature can be found in [13.126, 13.162].

Later on, refined statistical methods with data-dependent choice of the tail region (i.e. the number k of observations that follow the approximate tail distribution) confirmed these findings [13.37, 13.107, 13.108]. Similarly, analysis of the tails of high-frequency, intra-daily data also led to results in accordance with those for daily data. This squares well with the theoretically expected invariance of tail behavior at different levels of time aggregation [13.1, 13.124]. It is worth emphasizing that this invariance holds for processes with $\alpha > 2$ *despite* convergence of the overall distribution to the Gaussian. Therefore, the laws governing the tails will become less and less 'visible' under aggregation and will only show up in the extreme outer parts of extremely large datasets. As the financial data of the formerly prevalent daily frequency are only available in limited numbers, the increased availability of high-frequency data appears particularly valuable. In fact, with intra-daily records, one not only enjoys a tremendous increase of the sheer number of data points to be used, but also has available data at an 'earlier' stage of time aggregation, so that one could hope for an extended range of validity of the pertinent tail laws.

13.2.4 Recent Contributions by Physicists

Physicists would, of course, interpret the findings of large returns (r_t) following a relationship like $\text{Prob}(r_t > x) \sim x^{-\alpha}$ implied by the tails obeying (13.8) as the typical appearance of a power-law. However, one should note the difference between the statistical arguments outlined above and the concept of scaling laws in physics: while scaling laws should ideally extend over a large range of observations (typically several decades), the arguments from extreme value theory only imply that the most extreme part of the distribution will asymptotically converge to a limiting power-law. For a particular dataset, the region where this power-law can be 'observed' may be rather small and the analysis of a larger portion of the data with similar tools may already give different results.

The developments in empirical finance sketched above are mirrored to some extent in the recent work by physicists on distributional aspects of financial data. To our knowledge, the first contribution in this vein is an article by Mantegna [13.116] who estimates the parameters of the stable laws (cf. (13.3)) for various sectoral indices of the Milan stock exchange. His results are comparable to those obtained by economists in similar studies. In later work, Mantegna and Stanley [13.117] as well as Cont et al. [13.33] remarked that Lévy-type behavior does not extend to the most extreme parts of the distribution and replaced the original Lévy hypothesis by that of a *truncated* Lévy process whose tails fall off exponentially after some threshold value (cf. also Chap. 12). A few years later, a paper from the same group [13.63] arrived at the conclusion that tails are rather governed by a cubic power-law, that is, their tail index estimate is close to three. Thus, it seems that researchers from economics and physics have completely converged on this issue.

As an interesting side aspect of the above discussion, it appears that the *central* parts of the empirical distributions of returns are closer to the shape of the stable laws than the extremal region. This can be seen in Mantegna and Stanley [13.117] who have shown that the estimated Lévy law provides a good fit to high-frequency returns of the S&P 500 index up to some high

Table 13.1. Scaling of extreme returns in various financial time series. Asymptotic 95% confidence intervals of the Hill estimator [13.74] are given in parentheses. The time intervals and number of observations are: DAX: 10/59–12/98 ($n = 9\,818$), NYCI (New York Stock Exchange Composite Index): 01/66–12/98 ($n = 8\,308$), USD-DEM: 01/74–12/98 ($n = 6\,140$), Gold price: 01/78–12/98 ($n = 5\,140$). Note that all estimates point to a scaling exponent around three. Furthermore, the confidence intervals of the Hill estimates from the 5% tail as well as from the usually smaller 'optimal' tail allow rejection of the stable law hypothesis $\alpha < 2$. 'Optimal' tail sizes are computed with the stopping algorithm of [13.39]

Data	Regression estimate	Hill 5%	Hill estimate based on optimal tail size
DAX	2.92	3.01 (2.74, 3.27)	3.04 (2.80, 3.29)
NYCI	2.31	3.58 (3.24, 3.93)	3.42 (2.34, 4.41)
USD-DEM	3.35	3.69 (3.28, 4.10)	3.60 (2.31, 4.88)
Gold	2.67	2.51 (2.20, 2.81)	3.78 (2.69, 4.87)

threshold value (see Chap. 12, Fig. 12.2). Similarly, Groenendijk et al. [13.67] highlight the fact that certain estimation methods for the parameters of stable distributions based on the interquartile range give quite uniform results across assets. Again, the likely reason is that the implicit restriction to the inner parts of the distribution with some methods makes empirical records look closer to the stable laws than investigation of the tails. Note also that similar conclusions are confirmed by tail index estimation itself. If we were to consider a tail size of 30% and more appropriate for maximum likelihood estimation of α, we would end up with a tail index in the range 1 to 2, that is in the basin of attraction of the stable laws. However, data-driven choice of the tail fraction to be used in this type of computation tells us that we have to go further out into the tail in order to get a reliable estimate of α - with the usual finding that α is around 3–4 as outlined above[3].

Figure 13.1 and Table 13.1 provide illustrations of the typical scaling behavior of the tails of the unconditional distribution of financial returns. In Fig. 13.1 we exhibit results obtained with both daily and intra-daily recordings of the German share price index DAX. Figure 13.1a shows that the cumulative distributions of both time series exhibit clear deviations from the exponential decline of a normal distribution and are better characterized by hyperbolic power-law

[3] At the end, the Lévy type appearance of the distribution's center may have a quite mundane explanation: on the highest frequency, the peak at the center is simply due to a very large number of observations exactly equal to zero (no trade). Furthermore, at this frequency, the distribution is discontinuous because the minimum tick size is finite in most markets. Temporal aggregation, then, amounts to smoothing over this initially kinked peak at zero which may produce the artificial impression of a smooth shape (personal communication by P. Gopikrishnan).

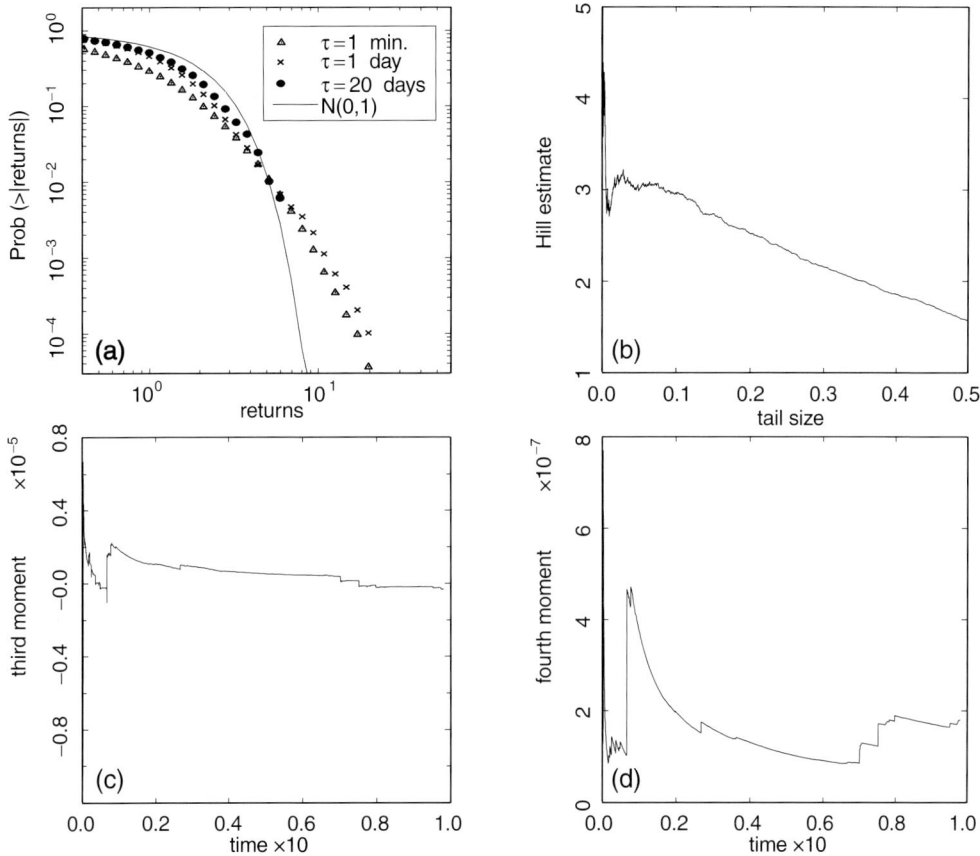

Fig. 13.1. Scaling of extreme returns from a typical financial time series (returns of the German share price index DAX). (**a**) The complement of the cumulative distribution function at different levels of time aggregation. The data under consideration are daily observations over the years 1959 to 1998 and minute-to-minute returns during opening hours from 1988 to 1995. For both daily and intra-daily frequencies, we observe approximate power-law behavior of the tails. Performing a least squares fit of the most extreme 20% of the data, we find exponents 2.92 (daily data) and 3.06 (minute-to-minute returns). At lower frequencies (e.g. returns over 20 days) a crossover to the normal distribution (with mean zero and unit variance) is observed. In order to enhance comparability, returns have been rescaled by dividing by the sample standard deviation at all levels of time aggregation. (**b**) Exhibits the results of maximum likelihood estimation of the tail index for daily DAX data. When varying the number of data considered to make up the 'tail', a plateau with $\alpha \approx 3.1$ is found and a monotonic decline of the estimate thereafter. Statistical methods for selecting the optimal tail size usually give numbers in the vicinity of the plateau. The bottom panel shows the time development of both the third moment (**c**), and fourth moment (**d**). Apparent convergence of the third moment and seeming nonconvergence of the fourth moment square well with our estimates of the tail index

behavior. Since one usually finds only slight differences between the left and right tails, we have merged both sides by using absolute values of returns. Log-log regression estimation of the shape of the distribution in the extremal region yields very similar results with estimates around three for both time series. However, at a higher level of time aggregation (exemplified by returns over 20 days), the visibility of power-law scaling breaks down and the distribution approaches the normal. Maximum likelihood estimates using the Hill technique [13.74] are exhibited in Fig. 13.1b. Here we do not restrict ourselves to only one estimate, but show the typical variation of the tail indices with increasing number of tail elements, k. Upon inspection, one detects a plateau of the estimate with $\hat{\alpha}_H \approx 3.1$ for tail sizes between about 5 and 10% of the data, and monotonic decline thereafter. Usually, refined methods for endogenous selection of k tend to choose estimates around this almost stable region. In this case, the outcome is in almost perfect agreement with that obtained by log regression. The validity of the results is further confirmed by inspection of the development of the third and fourth sample moments (computed with daily data): while the third moment appears to converge with sample size, for the fourth moment convergence appears doubtful and it seems likely that this quantity suffers from nonstationarity (Fig. 13.1c-13.1d).

Table 13.1 shows numerical results obtained with various important financial time series: besides the DAX, we also consider the New York Stock Exchange Composite Index (NYCI), the USD/DEM exchange rate, and the price of gold[4]. Although we find some variation across assets, the picture confirms the overall impression from the literature with estimates hovering between 2.5 and 4. Comparing regression estimates with two sets of Hill estimates computed from both the 5 percent tails and an endogenous selection method for k [13.39], we see some fluctuations, but hardly any systematic differences. It is also interesting to note that explicit hypothesis testing based on the 95% confidence intervals exhibited for the Hill estimates allows us to reject the Lévy hypothesis $\alpha < 2$ in all cases.

13.3 Temporal Dependence

13.3.1 Dependence in Raw Returns: Theoretical Background and Empirical Evidence

As pointed out above, the efficient market hypothesis implies that price changes are caused by newly arriving information which should itself follow some sort of stochastic process. Denoting price increments by ε_t and maintaining the assumption that increments are i.i.d. stochastic variables with mean zero, the price itself would, then, follow a random walk:

$$p_{t+1} = p_t + \varepsilon_t \ . \tag{13.13}$$

This implies that the best forecast of tomorrow's price is the price observed today:

$$E\left[p_{t+1}|I_t\right] = p_t \ . \tag{13.14}$$

---

[4] The bottom panels of Fig. fig:lux show parts of the DAX (left-hand side) and USD/DEM time series (right-hand side).

Strictly speaking, however, (13.1) and (13.2) do not imply that prices themselves should follow a random walk. Rather, one can show that this property holds for a closely related quantity, the value of an initial investment whose returns are continuously reinvested in the same asset (cf. [13.91]). Nevertheless, in empirical research, the efficient market hypothesis has often been identified with the random walk character of prices or their logs: $\ln(p_{t+1}) = \ln(p_t) + \varepsilon_t$. The latter form, of course, is in harmony with returns rather than price increments following an i.i.d. stochastic process.

As certain statistical features of empirical data (to be dealt with in detail below) are not consistent with the random walk model, it has later on mostly been replaced in the financial literature by the wider concept of a *martingale process*. This concept implies that prices (or their logs) are still assumed to follow a process of the type of (13.13) but with the only restriction imposed on increments that $E[\varepsilon_t] = 0$, $\forall t$. This allows for nonidentical distributions of the increments as well as for many forms of dependence in higher moments and is, nevertheless, in harmony with the EMH since price changes remain unpredictable under the martingale hypothesis (cf. [13.27]).

A wealth of research has tried to find evidence in favor of or against the hypothesis in (13.13). *Supporting evidence* comes form so-called unit-root tests: if one embeds (13.13) into a time series model of the more general type $p_{t+1} = \rho\, p_t + \varepsilon_t$ one can perform a test of the hypothesis $\rho = 1$ (unit-root test). This is equivalent to a standard t-test of a hypothesized coefficient in a regression, but since the value $\rho = 1$ is at the boundary between the stationary and nonstationary cases, the distribution of the test statistic is nonstandard (cf. [13.40]). Results of this and a number of related tests are quite uniform in that one is usually *unable to reject* the above null hypothesis for financial prices (or their logs).

On the other hand, i.i.d.-ness of (relative) price changes would imply absence of any significant autocorrelation in the returns series. In a strict sense, this implication of the random walk model is often violated in that one finds significant correlation over one or two lags for data on a daily level (cf. [13.27]), and over a couple of minutes for intra-daily data (cf. Chap. 7 in [13.119]). However, the economic interpretation is that these slightly significant correlations can be attributed to market frictions and market imperfections especially with thinly traded assets and, therefore, may not really be considered as a rejection of the underlying economic idea of the EMH[5]. More refined tests try to show, for example, that prices systematically tend to overreact to news [13.38] or that prices are excessively volatile compared with fundamental factors [13.137]. These as well as related issues are still in the center of current debates. From

---

[5] For indices, some small positive correlation can even come about in an artificial way from the aggregation of stocks itself.

the literature available so far, evidence seems to be in favor of the overreaction and excess volatility hypotheses (cf. [13.52, 13.163]).

Another possible avenue for attacking the unit-root character of prices is to demonstrate prevalence of hidden nonlinear structure. Again, this has been a very active area of research over the last decade, so that we cannot give an exhaustive overview of relevant work here. Introduction of concepts from chaos theory started with Scheinkman and LeBaron [13.135] and Frank and Stengos [13.53]. While the first vintage of investigations seemed somewhat supportive of chaotic attractors, these results have been questioned by later authors (cf. [13.60, 13.134]). After about one decade of research in this vein, it seems that the search for a low-dimensional attractor in financial data has practically been given up and that the majority view among economists is that price dynamics *are more complicated* than dynamics arising from standard models of low-dimensional chaotic behavior (for example, the Lorenz attractor).

On the other hand, some newly developed explicit tests of the i.i.d. hypothesis (with the so-called Brock-Dechert-Scheinkman (BDS) test the most popular variant among economists) routinely *reject* i.i.d.-ness of returns (cf. [13.24]). The results of this literature (reviewed in [13.11]) clearly speak against (log) prices following a random walk. However, as the source of rejection of the null hypothesis could be temporal dependence in variances and higher moments, these results can not necessarily be interpreted as evidence against the more general martingale hypotheses.

Still different, but somewhat related attacks on the pure stochastic nature of returns are based on the successes of certain trading strategies. Note that even with the martingale model, no useful information about the *direction* of future price changes can be inferred from the past price history. Hence, not only all those strategies that are known under the heading of 'chartism' but also refined time series techniques like neural networks are bound to remain useless and should not have any advantage over the simple 'buy-and-hold' device. Put the other way round, demonstration of excess returns earned through chartist techniques would point towards violation of the martingale assumption. Again, a large number of relevant studies exist on this issue and we shall confine ourselves to mentioning only a few typical examples: Brock et al. [13.25] show that popular chartist techniques like moving average and resistance lines would have earned small excess returns when used over very long horizons, while Caginalp and Laurent [13.26] demonstrate the astonishing predictive capability of 'candlestick' techniques[6]. It has often been noticed that tests for the profitability of trading rules (as well as tests of various 'anomalies' of price records) do suffer from a data mining bias which emerges from hundreds of practitioners

---

[6] [13.7, 13.157, 13.158] discuss relationships between moving average techniques and certain concepts from statistical physics.

and academic researchers looking for conspicuous properties and performing hypothesis tests on one and the same dataset. As is well known from introductory statistics, such 'data snooping' should be avoided, as seemingly significant results will eventually emerge by chance with repeated tests. In fact, the recent development of statistical methods to account for data snooping biases allowed a reconsideration of the Brock et al. [13.25] analysis of trading rules with the result that in an out-of-sample test excess returns now turned out to be insignificant [13.145]. The statistical test design is a bootstrap algorithm from the whole universe of similar trading rules (a total of 7 846 trading rules). In a similar vein, Sullivan et al. [13.146] show that the seemingly ubiquitous 'calendar effects' (abnormal returns related to the day of the week, week of the month etc.) do not survive a correction for data mining. Hence, both earlier results on the profitability of trading rules and the appearance of 'anomalies' have to be interpreted with care.

Interestingly, besides the trading methods developed by market practitioners, recent literature has also investigated more refined mathematical methods for pattern recognition. As an example, we mention Neely et al. [13.125] who show that exploitable patterns can be detected in foreign exchange rates by genetic programming (for an introduction to genetic programming cf. [13.88]).

As a summary of our tour d'horizon on the search for dependencies in raw returns, it turns out form a large body of diverse approaches that on a first view and with standard statistical techniques prices look like realizations of a martingale process and, therefore, seem essentially unpredictable. At a closer look and with more refined statistical instruments, however, one can detect a number of slight deviations from pure randomness, so that one might conclude prices are only *approximately efficient* (cf. [13.49])[7] The deviations from i.i.d.-ness that are statistically obvious appear, however, to be caused mainly by temporal structure in volatility (second moments) and are hardly noticeable in first moments.

13.3.2 Dependence in Squared and Absolute Returns

Our claim of the almost complete absence of dependence in raw returns is illustrated in the lower left hand panel of Fig. 13.2: autocorrelations of the DAX returns, in fact, do exceed the 95% bounds for the first two lags but for lags larger than two days the autocorrelations are not significantly different from zero any more. However, the picture changes when considering simple transformations of the original data. As also shown in Fig. 13.2 (middle and

---

[7] A similar view is also expressed in Zhang [13.164]. However, it is not clear whether his supporting evidence for inefficiencies, i.e. sign patterns of returns, is statistically significant and, if so, whether significance would survive on condition that a correction for data mining is exerted. In any case, it should be noted that these patterns have been known for a long time in the economics literature, e.g. [13.77]. The paper by Caginalp and Laurent [13.26] provides a more rigorous statistical analysis of sign patterns.

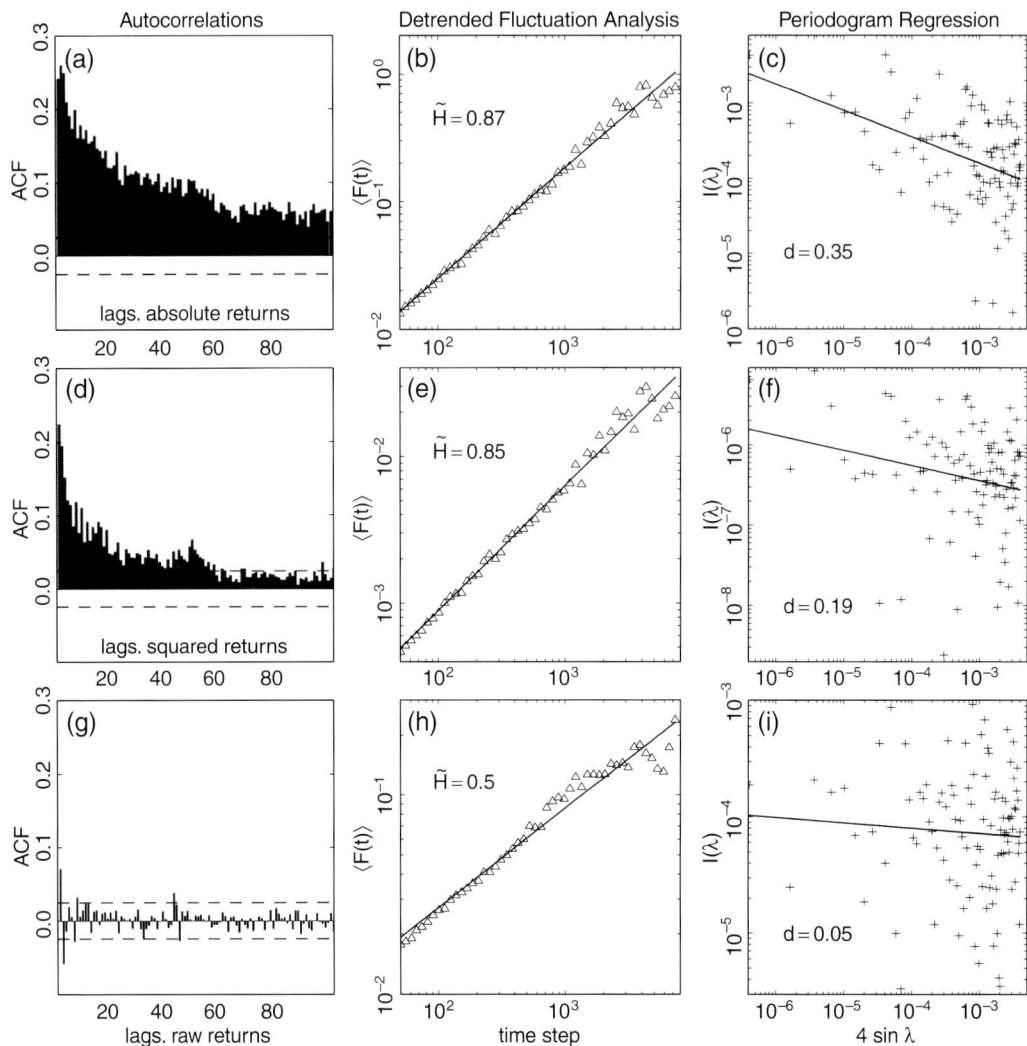

Fig. 13.2. Analysis of temporal dependence in raw (**g**)–(**i**), squared (**d**)–(**f**), and absolute returns (**a**)–(**c**). Data are again daily observations of the DAX returns (1959–1998). For each time series the autocorrelation function over a hundred lags is shown on the left. In the middle part, we exhibit the results of detrended fluctuation analysis (DFA, see (13.18) and Sect. 5.4.2) which can be compared to the estimates of the fractional differencing parameters obtained through periodogram regression, cf. [13.58] (right). The DFA exponent $\tilde{H}$ is identical to the Hurst exponent H. Note that for a time series with finite variance, the two estimates, $\tilde{H}$ and d, are related by the identity: $d = (\tilde{H} - 0.5)$. We see that results obtained by both methods are in satisfactory agreement for both raw and squared returns, while there is some divergence concerning the degree of long-term dependence in absolute returns (cf. Table 13.1)

upper left hand panels) for *squares* and *absolute values* of returns temporal independence is strongly rejected. On the contrary, we find significant positive autocorrelation coefficients over an extended time horizon. As absolute and squared returns only preserve the scale and neglect the direction of the increments, we can interpret both transformations as simple measures of the scale

Table 13.2. Temporal scaling in various financial time series. Asymptotic 95% confidence intervals of the Geweke/Porter-Hudak (GPH) estimates [13.58] of the parameter of fractional differencing are given in parentheses. For information about the data, see Table 13.1. Note: based on the GPH confidence intervals, we can never reject absence of long memory in raw returns (although the DFA occasionally gave scaling exponents above 0.6). For squared and absolute returns, on the contrary, we always accept the long-memory property

Method	Data	Raw returns	Squared returns	Absolute returns
H from DFA	DAX	0.50	0.85	0.87
	NYCI	0.50	0.92	0.93
	USD-DEM	0.61	0.87	0.92
	Gold	0.64	1.06	1.07
d from GPH	DAX	0.05 $(-0.09,\, 0.18)$	0.19 $(0.05,\, 0.33)$	0.35 $(0.21,\, 0.49)$
	NYCI	0.07 $(-0.07,\, 0.22)$	0.15 $(0.01,\, 0.29)$	0.41 $(0.26,\, 0.55)$
	US-DEM	0.07 $(-0.09,\, 0.23)$	0.24 $(0.08,\, 0.40)$	0.29 $(0.13,\, 0.45)$
	Gold	-0.02 $(-0.19,\, 0.14)$	0.62 $(0.45,\, 0.78)$	0.57 $(0.40,\, 0.73)$

of fluctuations, i.e. *volatility*. The above findings, then, imply that *volatility is strongly correlated over time*. Hence, expected volatility in the next periods is the higher the more volatile today's market is. Visually, we observe typical alternations between turbulent and tranquil episodes in the data (volatility clustering).

Economists have developed a class of time series models which captures this ubiquitous feature of financial data. This so-called GARCH (generalized autoregressive conditional heteroscedasticity) model assumes that returns are drawn from a normal distribution whose variance follows an autoregressive process including dependence on past squared increments [13.19, 13.44].

In its general form, with an arbitrary number of lags included, this model reads:

$$r_t = h_t \varepsilon_t, \qquad \text{with } \varepsilon_t \sim N(0,1),$$
$$h_t = \alpha_0 + \sum_{i=1}^{p} \alpha_i r_{t-i}^2 + \sum_{j=1}^{q} \beta_j h_{t-j}, \ \alpha_0 > 0, \ \alpha_i, \beta_j \geq 0. \tag{13.15}$$

Using statistical information criteria in order to discriminate between different variants of the GARCH(p,q) model, it mostly turns out that the GARCH(1,1) model is not outperformed by more complicated versions (cf. [13.15, 13.20]). It, therefore, turns out to be sufficient in most cases to include only last period's squared return (r_{t-1}^2) and last period's variance (h_{t-1}) in the difference equation. Empirical estimation yields results that are also relatively constant across markets and time horizons. Interestingly, the sum of the coefficient $\alpha_1 + \beta_1$

comes usually close to one, that is, it approaches the threshold where the process becomes nonstationary[8]. Refinements of GARCH models have been a very active research field over the last two decades. The surveys by Bollerslev et al. [13.20] and Bera and Higgins [13.15] give an overview of wealth of the variations of the ARCH theme alternative specifications of the dependence structure or alternative distributions of the increments (e.g. Student's t-test, stable Lévy increments).

In the light of Sect. 13.2, it is interesting to notice that despite its normal incremental distribution the GARCH model leads to power-law tails [13.68] which is in harmony with empirical evidence. However, it has also often been found that the GARCH variance dynamics is unable to capture all of the deviations from Gaussian behavior in the data, as the residuals from an estimated GARCH model are often still non-Gaussian and leptokurtotic (cf. [13.126]). Furthermore, as has been shown by [13.140], the theoretical tail indices calculated for the estimated theoretical GARCH processes are often not identical to those directly estimated from the data.

An even more important disadvantage of the GARCH class is that the decay of the autocorrelations is exponential, i.e. it covers only short-term dependence[9]. Typical plots of the empirical ACF in Fig. 13.2 are, however, quite suggestive of hyperbolic decline rather than exponential decline.

13.3.3 Long-Term Dependence

Hyperbolic decline of the autocorrelation function is a defining property of stochastic processes exhibiting long memory. Well-known examples of long-memory processes are *fractional Brownian motion* (fBm, cf. [13.115]) which can be generated as a moving average from standard Brownian increments $dB(t)$:

$$x(t) = \int_{-\infty}^{t} (t - s)^{H-1/2} \, dB(s) \,, \tag{13.16}$$

or *fractionally integrated autoregressive moving average* (ARFIMA) models [13.64]:

$$\Phi(\mathfrak{B})(1 - \mathfrak{B})^d x(t) = \Theta(\mathfrak{B})\varepsilon(t) \,, \tag{13.17}$$

with $\mathfrak{B}$ the backward shift operator defined by $\mathfrak{B}x(t) = x(t - 1)$, $\Phi(\mathfrak{B})$ and $\Theta(B)$ the autoregressive and moving-average polynomials and $\varepsilon(t)$ a noise term following the normal distribution. It has been shown that both the decay of the

---

[8] GARCH models are usually estimated by constraint maximum likelihood which constrains the sum of the coefficients to be smaller than 1.

[9] [13.120] also show that empirical autocorrelation functions are usually outside the 95% confidence bands obtained with GARCH processes. Note, however, that some recent developments allow for long-term dependence in an ARCH framework, cf. the fractionally integrated GARCH model (FIGARCH) [13.8].

autocorrelation function and the spectral density function behave identically for both processes with a relationship $d = H - 0.5$ of the crucial long memory parameters H (known as the *Hurst exponent,* cf. [13.115] and d (the *degree of fractional differencing,* cf. [13.58]). As the temporal development of many quantities of interest can be characterized solely by the exponents H or d, long memory processes are also often encountered under the label of self-similar stochastic processes. Positive autocorrelation (persistence) is obtained with $H > 1/2$ or $d > 0$, while the case $H < 0.5$ ($d < 0$) gives rise to antipersistent (self-avoiding) processes.

Convincing empirical evidence for long memory properties in squared and absolute returns has been reported by a number of papers both from economists and physicists over the last couple of years. In economics, this property seems to have been recognized first by Ding et al. [13.41]. Their findings have been confirmed in a number of other studies recently [13.36, 13.103, 13.122]. The consensus now is that this feature appears in virtually all financial prices [13.98]. More or less independently, physicists also reported findings of long-term dependence in financial data, e.g. [13.57, 13.96, 13.154, 13.161].

Here again, it is puzzling to see that while the results are almost identical, the physicists' research has been carried out without knowledge of the relevant literature in economics. One of the reasons for the apparent lack of contact and knowledge of parallel studies besides different cultural background and the lack of common meetings (until recently) may be the different preferences concerning statistical methods and the different attitudes towards drawing inferences from statistical results. While most economists prefer methods that allow explicit hypothesis testing, many physicists apparently have a preference for graphical tools of inference. In testing for long-term dependence, the most popular techniques among physicists are the rescaled range (R/S) method (see Chap. 8) or the more robust detrended fluctuation analysis (DFA) (see Chaps. 5, and 7-8 for examples) proposed by [13.128]. The recent papers by economists, on the other hand, mostly use the Geweke/Porter-Hudak [13.58] variant of periodogram regression (for which asymptotic results on the distribution of the test statistic are available) and Lo's [13.97] method which transforms the R/S methodology into an explicit test of the null-hypothesis of short vs. long-term dependence.

Figure 13.2 illustrates the typical outcome of some of the most popular methods (the detrended fluctuation analysis and Geweke/Porter-Hudak methods) for the German share price index DAX. DFA works as follows: consider the cumulative sum of a time series: $y_t = \sum_{\tau=1}^{t} x_\tau$ and divide the resulting sequence $\{y_t\}$ into T/l nonoverlapping boxes of length l. Computing the local trend within each box as the result of a least-squares regression the average standard deviation of the detrended walk over the T/l blocks, $\overline{F(l)}$, is expected to scale with l as:

$$\overline{F(l)} \sim l^H ,\tag{13.18}$$

where H again is the Hurst exponent. Equation (13.18) allows us to estimate H from a linear regression in a log-log plot. The DFA introduced here is referred to as DFA1 in Chap. 5, where also the other variants of the detrended fluctuation analysis are being discussed in detail. In Sect. 5.4.2, the fluctuation exponent was denoted by α. For long-range correlated sequences, the autocorrelation function $C(l)$ decays by a power-law according to (5.1), with an exponent γ that is related to H by $H = 1 - \gamma/2$.

The second method, in contrast, estimates d in (13.17) from a linear regression of the log-periodogram $\ln\{I(\lambda_j)\}$ on transformations of small Fourier frequencies $\ln\{4\sin^2(\lambda_j/2)\}$ which for a fractionally differenced ARMA process should obey:

$$E[\ln\{I(\lambda_j)\}] = c - d\ln\{4\sin^2(\lambda_j/2)\} .\tag{13.19}$$

Economists usually consider it an advantage that knowledge on the asymptotic distribution of the estimator $\hat{d}$ from (13.19) is available while for the R/S technique despite its long usage no such results could be derived so far.

Evidence in favor of long memory in both squared and absolute DAX returns in Fig. 13.2 is confirmed by similar outcomes for other assets as reported in Table 13.2. Together with the large body of available literature on this issue, they confirm that long-term dependence in squared and absolute returns are an ubiquitous feature of financial data. We therefore have another *scaling law* here: *hyperbolic decline of the autocorrelation function of these measures of volatility.*

As also shown in Fig. 13.2, raw returns themselves do not need to be characterized by long-term dependence. Given the absence of significant autocorrelation in the raw data, this result may not come as a surprise (although one could imagine appropriate mixtures of, say, negative short-term dependence with positive long-term dependence that could obscure the underlying dependence structure leading to apparently insignificant correlations). Nevertheless, a lot of research has been devoted to testing for long-term dependence in raw returns, so that a word on this issue is in order here. Quite some time ago, a number of papers in finance have applied rescaled range analysis to raw returns and have reported positive evidence for a self-similarity parameter $H > 0.5$ [13.21, 13.66, 13.83]. However, later reconsiderations of this issue with other techniques (e.g. [13.61, 13.95, 13.97]) could not find convincing evidence in favor of such behavior[10]. From this literature, a certain majority opinion

---

[10] The recent paper by Taqqu and Teverovsky [13.149], on the other hand, reconsiders the S&P 500 data used by Lo [13.97] and reports some support in favor of small degrees of long-term dependence.

seems to have emerged among economists that the long memory property is confined to power transformations of returns but appears to be absent in the raw data. This is also in harmony with economic intuition as even slight deviations from $H = 0.5$ for returns could be easily exploited by appropriate trading strategies [13.6, 13.75] and would, of course, strongly violate the martingale hypothesis[11].

The conclusion of early findings of $H > 0.5$ being due to the upward bias of the R/S technique for processes with self-similarity parameter around 0.5 is also confirmed by recent Monte Carlo studies of the performance of various estimators. [13.129, 13.148, 13.150] all found the R/S technique to have larger mean squared error than alternative techniques, so that H estimates in the range $[0.4, 0.6]$ as reported in earlier studies could arise quite easily from sampling fluctuations. Particular doubts on the relevance of some Hurst exponent estimates above 0.5 are raised by Pilgram and Kaplan [13.129] who report that they found the 'typical value' $H = 0.75$ for simulated time series of synthetic $1/f^\alpha$ noises with true H in the range 0.25 to 0.75!

On the other hand, *some* (negative) deviations from $H = 0.5$ that have recently been reported may be easily explained from an economic point of view by regulations imposed on the market by monetary authorities. As an interesting example, Vandewalle and Ausloos [13.154, 13.155] report estimates of H for a variety of exchange rates (their results are obtained from DFA analysis of multifractality, see below). Many of them, in particular, those of members of the European Monetary System (EMS) of the 1980s and 1990s, turn out to yield $H < 0.5$. However, within the EMS, the obligations of central banks to keep fluctuations within the band creates the necessity of interventions if the margin is approached. As essentially the exchange rate band acts as a reflecting boundary (either explicitly or through the expectations of agents), the managed floating of European currencies contains an element of antipersistence, and $H < 0.5$ could be seen as the immediate statistical consequence of this mechanism.

13.4 Multiscaling, Multifractality, and Turbulence in Financial Markets

Absence of long memory in raw returns and prevalence of long-term dependence in squared and absolute returns can be put into a broader context by considering a *continuum of power transformations* of the raw data. The first paper to look at power transformations other than $q = 1$ and $q = 2$ (abso-

---

[11] Volatility clustering and long-term dependence in second moments are, however, still consistent with the martingale model and the EMH. Given these empirical regularities, the EMH, however, would imply that news come in clusters and with long-term dependence in their scale.

lute and squared returns), was [13.41]. For a long series of the daily S&P 500 index variations (ranging from 1929 to 1991) they showed that the strongest correlation is obtained around $q = 1$ (absolute returns). This finding has later on been confirmed for other financial data as well (e.g. [13.122]). For physicists, variation of the scaling exponent for various powers of the original data is reminiscent of *multifractal* or *multi-affine* behavior[12], a feature also encountered in data from turbulent flows. In fact, the first attempt at recovering traces of multifractality also dates back to the early 1990s [13.160]. A few years later, this topic has been taken up by several other groups (see for example [13.14, 13.59, 13.80, 13.136, 13.155, 13.156]).

Let us review and illustrate the typical physical approach for determining multifractal structure within a dataset following the lines of [13.80, 13.155, 13.156]. In order to establish deviations from monofractal behavior, one usually investigates the scaling of the moments of a signal $x(t)$, the time series of log prices in our case. One considers the cumulated sum $y(t)$ of the signal, which is also called the 'landscape' of the signal, and calculates the height-height correlation (or structure) function $c_q(\tau)$:

$$c_q(\tau) = E\left[|y(t + \tau) - y(t)|^q\right] . \tag{13.20}$$

For arbitrary q, a *generalized* Hurst exponent can be defined through the relation:

$$c_q(\tau) \sim \tau^{qH_q} . \tag{13.21}$$

If a linear dependence is obtained, i.e. H_q turns out to be constant, then the signal is monofractal or, in other words, the data follows a simple scaling law for all values of q. This would hold for fractional Brownian motion. However, with a *nonlinear* development of H_q, the data generating process is more complicated and will be said to be of a multifractal nature. The behavior of the underlying process can, then, only be characterized fully by the whole spectrum of its local Hurst exponents at various powers q. Since some curvature of the H_q spectrum is a necessary condition, one may also extract information from the local slopes. For example, Vandewalle and Ausloos [13.155] report the estimate of $C_1 = -\frac{dH_q}{dq}|_{q=1}$ (known as the *intermittency* or *sparseness* parameter) for various exchange rates.

Examples of empirical scaling exponents of financial data (i.e. the development of the exponent qH_q in (13.21)) are exhibited in Fig. 13.3. This plot nicely confirms that some curvature in the scaling exponent can be found for various time series from different types of financial markets. The deviations from the linear development expected under uni-fractal Brownian motion suggest that

---

[12] Cf. [13.45, 13.46, 13.71], for introductory sources to the multifractal formalism, and [13.10], for more details on the structure function approach introduced below.

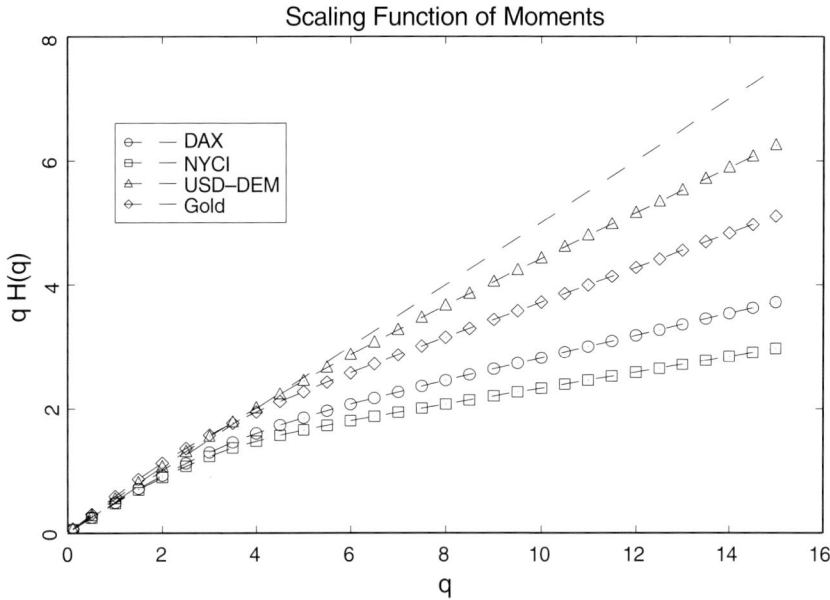

Fig. 13.3. Scaling exponents of moments for four financial time series: the share price indices DAX and NYCI, the Dollar-Deutschmark exchange rate, and the price of gold. The broken line gives the expected scaling $qH_q = q/2$ under Brownian motion. For all time series under consideration, one observes curvature in the scaling exponent and, therefore, deviations from monofractal behavior

the data under consideration are characterized by multifractal behavior. Since this result is shared by all the studies mentioned above, the multifractal nature of financial returns may be added as a new stylized fact, which extends and generalizes prior insights on the temporal characteristics of both returns and volatility.

Disturbingly, *none* of the time series models that are currently used in the applied financial literature is capable of systematically reproducing this property of empirical data. However, whether *spurious* multifractal behavior could originate from simpler models seems to be an open question. While [13.155,13.156] report negative results from Monte Carlo simulations of ARCH models, Baviera et al. [13.12] found transient behavior of ARCH models resembling that of a nonlinear H_q spectrum. Furthermore, Berthelsen et al. [13.16] have already demonstrated that short time series from random walks may appear to exhibit multifractality and [13.22] give another example of a monofractal process that shows apparent multiscaling. Much future research is, therefore, needed to single out which types of processes could conform to the seemingly universal finding of a nonlinear H_q spectrum in financial returns.

If multiscaling does not turn out to be a spurious characteristic of 'simpler' time series models, one of the major contributions of physicists to finance may consist in their development of asset pricing models with genuine multifractal behavior. First important steps in this direction seem to have been made with the application of combinatorial models of multifractal cascades to financial data (cf. [13.50, 13.113, 13.114, 13.136]). Interestingly, Mandelbrot et al. [13.114] provide a model in which a multifractal cascade acts like a *time transformation*, a concept that also appeared in various econometric models of asset returns. First experiments indicate that these models (originating from the physical literature on turbulence) seem to be able to outperform GARCH type specifications in terms of the fit of the unconditional distribution, although, essentially, they are one-parameter families of stochastic processes (cf. [13.106]). The upper right-hand panel of Fig. 13.0 provides an illustration of an artificial price series generated from a stochastic process with a multifractal time transformation. An alternative approach building upon physical models of turbulence is due to Moffat [13.123], who developed a dynamical (behavioral) theory of capital markets based on a continuous time description similar to that for hydrodynamic flow. He determined the onset of turbulent behavior and showed that intermittency exists for the time series spectra of volatility distributions in this model. A certain drawback of the multifractal apparatus has been the combinatorial nature of the models available from statistical physics literature which does not enable one to formulate forecasts based on an estimated model. However, recent developments of iteratively soluble stochastic cascade models seem to be able to overcome this problem allowing for short-term forecasts of volatility based on the multifractal properties of past data [13.23, 13.56].

It appears noteworthy that many physicists seem to have been fascinated (and motivated to approach financial data) by the similarity between certain empirical characteristics in finance and turbulence. These similarities seem to have been noticed first by Vassilicos [13.159]. Further contributions are [13.59, 13.73, 13.76, 13.118]. In fact, a visual inspection of time series plots of both financial returns and velocity differences in turbulent flows shows perplexing similarities (cf. [13.118]). Similarities between both types of data include the phenomenon of volatility clustering and the leptokurtotic shape of the unconditional distribution. On the surface we find, in fact, two of the most elementary 'stylized facts' of empirical finance to be shared by turbulent dynamics! However, although far-reaching analogies have been drawn on the basis of these similarities [13.59], it soon turned out that there are also some important differences: (i) while price changes are almost uncorrelated (over time spans longer than a few minutes), velocity changes *do* exhibit significant antipersistence as is apparent from their typical diffusion exponent $\sim 1/3$ [13.118], (ii) the approach to the Gaussian shape under time aggregation appears to occur much faster with turbulence data than with financial returns. The latter feature (which might be due to a higher degree of long-term dependence in volatility in financial data) led Holdom to conclude: 'The turbulence in the financial markets is stronger in this sense than the turbulence in the fluid' [13.76]. Hence, although there are indeed interesting (and stimulating) similarities, the every day notion of 'turbulence' in financial markets does not extend to all quantitative details of the physical phenomenon. Whether the mathematical routes to multifractal or multi-affine behavior are similar in turbulence and finance, therefore, remains an open question.

This ends our discussion of empirical scaling laws in financial data and we now turn to attempts towards an explanation of these statistical characteristics.

13.5 Explanations of Financial Scaling Laws

13.5.1 Methodological Background

As already mentioned several times in the preceding sections, the main tenets of the *efficient market hypothesis* can also be interpreted as hypotheses on the origin of the statistical characteristics of prices and returns: if prices follow a martingale process and their increments reflect forthcoming information on future earning prospects in an immediate and unbiased manner, then the distribution of returns must be a mere reflection of the distribution of news. Hence, the news arrival process would have to share the features of fat tails and volatility clustering and should even exhibit multifractal features. It is worth emphasizing that under this hypothesis the origin of scaling laws in financial time series is to be found in exogenous forces that cover a wide variety of influences (e.g. climatic and political factors). Only part of the complex bundle of news factors results from economic variables in a strict sense (e.g. macroeconomic factors) and an explanation of the scaling laws would, therefore, lead us outside the realm of economics. Furthermore, this implication of the EMH cannot be tested as the news arrival process is not observable itself.

Although one should surely not deny the paramount importance of news for the price formation in financial markets, one may argue that a perfect one-to-one relationship between news and returns appears too farfetched. News are incorporated into prices through the interplay of demand and supply: in most cases new information will be available to some market participants prior to others. If the news is favorable, the better-informed traders will buy additional units of the asset in order to gain from their informational advantage. This will drive up prices and thereby incorporate the knowledge of the insiders into prices. Mutatis mutandis, the same should happen in the case of adverse information hitting the market. However, although we might suspect that the 'invisible hand' of the market mechanism may operate relatively efficiently and smoothly most of the time, we also have evidence of various deviations from purely information-driven price formation. Much of the scattered empirical work on this topic has been mentioned in Sect. 13.3.1 above.

From a common sense aspect, one may argue that one usually observes a quite diverse number of behavioral variants among traders with informational trading constituting only part of the overall picture. With a large fraction of agents following chartist practices, large companies pursuing portfolio insurance and synthetic hedging programs, and some traders arguably being guided by herd instincts, we have a broad range of trading motives in real-life markets. Since large fractions of demand and supply originate from these diverse backgrounds, they will also exert an influence on the price formation process. The question, then, is whether all these factors (agents) only add some slight amount of noise to the efficient formation of prices, or whether, on the contrary, they are crucial to the resulting market outcome. The latter view would, of course, imply that markets are subject to some kind of endogenous dynamics originating from the *interaction* of individual traders. The perplexing similarity of the statistical characteristics of very different markets could, then, be explained by the similarity of the behavior of traders. The idea that the financial scaling laws have their origin in the trading process with its interaction of a large ensemble of heterogeneous traders has been denoted the *interacting agent hypothesis* (cf. [13.109]). Its implementation requires to construct structural behavioral models of financial markets whose stochastic characteristics match the stylized facts found in real-life data.

Unfortunately, standard modeling practices in economics have rather tried to avoid heterogeneity and interaction of agents as far as possible. Instead, one often restricted attention to the thorough theoretical analysis of the decisions of one (or a few) *representative* agents. Since, in recent years, the consciousness of the importance of *heterogeneity* has been increasing, a couple of papers have appeared that implement various types of microscopic behavioral models of financial markets.

For physicists, the explanation of scaling laws through the collective behavior of a multitude of elementary units (traders in our case) is a rather familiar approach. It is, therefore, not astonishing that quite a number of multi-agent models have been developed by physicists, partly in collaboration with economists and partly without contact to the economics community. Interest in modeling markets arose from the recognition of the universal character of scaling laws in finance and their resemblance to scaling laws found in various physical multi-unit systems. The resulting research program is described in [13.141] as follows:

"Statistical physicists have determined that physical systems which consist of a large number of interacting particles obey universal laws that are independent of the microscopic details. This progress was mainly due to the development of scaling theory. Since economic systems also consist of a large number of interacting units, it is plausible that scaling theory can be applied to economics."

Quite at the same time, a number of authors in economics proposed a statistical approach for modeling micro-economic heterogeneity among agents [13.2, 13.3, 13.85, 13.101, 13.132]. As stressed by the last author, such an approach bears implications very different from the formerly dominant representative agent methodology: while analysis and results are the same at both the micro and macro level with representative agents, the statistical approach may lead to emergent properties on the macro level that result from the *interaction* of microscopic units and cannot be inferred from the observable characteristics on the micro level themselves. Interestingly, one of the arguments in [13.104, 13.132] is that a multi-agent approach may allow an endogenous explanation of time-varying variances.

13.5.2 Variants of Microscopic Models

In economics, the first microscopic models[13] have not been developed so much to explain the stylized facts as emerging properties of an artificial market. Rather, their focus has been on learning of agents from experience and whether learning leads agents to converge with their plans to some theoretical equilibrium of the model. Typical examples are the papers [13.4, 13.29] and the Santa Fe stock market model [13.90, 13.127]. All of these examples use either genetic algorithms (GAs) or genetic programming (GP) to formalize the information processing of individual traders. Arifovic [13.4] builds upon a well-known static model of a foreign exchange market. She encodes agents' choice variables (consumption and holding of foreign currency) by binary numbers and applies the usual GA operations of selection, crossover, and mutation to these simple 'strategies'. Over time, successful behavior is on average maintained, whereas unsuccessful agents will adapt the 'strategies' of the more lucky ones. Mutation also allows to try new variants which are not in the pool of alternatives

---

[13] Even prior to the studies mentioned in the main text, Stigler [13.144] applied elementary Monte Carlo models in a simulation of a financial market. The early microstructure literature covers some more examples of microsimulations (e.g. [13.31, 13.69]). Kim and Markowitz [13.84] simulate markets with an ensemble of investors pursuing different versions of portfolio insurance strategies. They investigate whether this type of activity leads to excessive volatility, a question which has been raised after the 1987 stock market crash.

at time zero. One of the main conclusions of this model is that the time series from the simulations share the random walk character of empirical data and also allow to explain some features that emerged in experimental markets (laboratory markets with a small number of human participants interacting via computer).

Chen and Yeh's work [13.28, 13.29], on the other hand, has agents forming a theory on the relevant variables via GPs. That is, a given input time series (for example, the price record) is fitted by varying functional forms which are developed out of a tree-like structure from some elementary operations $(+, -, /, \times, \cos, \sin, \text{etc})$. The emerging population of genetic programs is subject to similar processes of selection, crossover, and mutation like the simpler genetic algorithm agents of Arifovic [13.4]. In early papers, e.g. [13.28], the main focus is on the convergence of the market to its theoretical equilibrium. In the most recent extension of their model, Chen and Yeh [13.29] distinguish between two subsystems: traders who use GPs in order to predict prices and volatility and business schools who produce and supply these GPs in some kind of academic competition. Among other things, they show that often the relative price changes from the simulations appear to be uncorrelated. Hence, the martingale property of prices seems to emerge without traders actually *believing* in such behavior.

Finally, we mention the Santa Fe stock market model which has been elaborated by a number of authors (e.g. [13.82, 13.90, 13.127]) over more than a decade. The structure of this model is even more complex in that GAs are combined with classifier systems. The latter part contains diverse trading signals which include both fundamental factors (e.g. relationship between price and dividends) and chartist factors (e.g. moving averages of various length). From the activated classifiers (whose if-part conforms with the environment), one is chosen according to a random draw with past success governing the probabilities of activation. The prescribed trading leads to a change of individual wealth and also allows to update the strength of the classifier. Again, classifiers are subject to the usual GA operations after a certain number of trading periods. In the simulation experiments, both the macroscopic development of prices and volume in this artificial market as well as the rate of activation of various classifiers are observed. The first important finding was a bifurcation in the model dynamics when changing the frequency of revaluation of the classifiers (via the GA mechanism): with more frequent revaluation, a dominance of chartist devices emerges while with less frequent application of GAs, the traders learn to prefer fundamentalist techniques. In a recent paper [13.90] the capability of the Santa Fe model to match some of the stylized facts of financial markets (volatility clustering, leptokurtosis) is emphasized. As only some rough statistics are provided, it is, however, not entirely clear so far how closely this popular model is able to mimic and explain the stylized facts.

Much simpler interaction models (from the economist's point of view) have been proposed by physicists: [13.32, 13.79, 13.142, 13.143] all construct models based on lattice structures well known in percolation theory and Ising type models. Denoting the states of each agent (knot) as 'buying', 'selling' and 'inactive', dynamics arise from cluster formation and activation processes that are propagated through the system. Since such models are well-known to produce power-laws at least in the vicinity of a critical point (the percolation threshold), it is not astonishing that modeling market interaction in this way leads to similar results. Recent papers have developed this approach further by introducing features to produce volatility clustering as well as power-laws close to the numerical values obtained with empirical data. As this branch of literature is *authoritatively* surveyed in the companion paper by Sornette, Stauffer, and Takayasu (see Chap. 14), we dispense with a detailed treatment here. Another approach with a rather direct inspiration by physical models is to be found in a pioneering paper by Takayasu et al. [13.147]. Here, the behavior of traders is formalized using threshold values for the decisions to buy or sell the asset. Agents who have just carried out a transaction, adjust their threshold values before participating in further trading rounds. In Takayasu et al. [13.147] thresholds are increased (decreased) if the agent had bought (sold) in his last transaction. Simulated time series show that these threshold dynamics can generate some realistic features like crashes in prices. A somewhat similar model is developed in Bak et al. [13.9], where a new threshold is chosen randomly from those of the remaining agents. Bak et al. show theoretically that their model leads to a self-similarity parameter $H = 0.25$ for prices, which, of course, is in contrast to empirical results. In an extended version time dependence of price changes is built into the model through some kind of volatility feedback (thresholds are now also updated by an amount proportional to past price movements). Not entirely surprising, this leads to measurable long-term dependence in price changes with an estimated $H = 0.65$. Although this appears to conform to some results of the older empirical literature, it should be kept in mind that these findings have been disputed in more recent research and, thus, the long-term dependence in price levels resulting from the volatility feedback, seems also to be at odds with empirical findings (cf. Sect. 13.3.3).

In personal communication, we found very different reactions from scientists from the economics and physics communities to these models. While, on the one hand, many physicists would consider the first type of models unnecessarily complicated, economists, on the other hand, would bemoan the almost entire absence of key economic concepts like utility or profit maximization, as well as the neglect of the issue of expectation formation, in the physicists' models. Nevertheless, these approaches are often considered as very useful and necessary steps towards incorporating new modeling techniques and new perspectives on interaction in markets. However, in order to take into account the peculiarities of economic systems (namely, in order to account for the nonmechanical nature of individual decisions even within a multi-agent framework), it seems indispensable to attempt a synthesis of both approaches.

The models by Levy, Levy, and Solomon [13.92–13.94] and Lux and Marchesi (variants of which have been published in from both economics and physics journals) may constitute some modest steps in this direction. [13.92–13.94] allow agents to split their wealth between a safe bond with constant interest rate and a risky asset with stochastic returns and derive their decisions from a standard utility-of-wealth function. Prices are determined from a market clearing condition (so that demand equals supply in any period). With explicit wealth maximization, expectation formation of agents about future prices has to be included in the model. This is formalized in a relatively simple manner by assuming that agents have different memory spans. Each agent, then, computes the expectation of next period's price as the average of past prices observed over the length of his time horizon. For many choices of parameters, the model produces very spectacular (but not entirely realistic) crashes and booms of the market price [13.92]. Furthermore, varying combinations of time horizons among traders lead to diverse and sometimes surprising results for the time development of the share of wealth owned by different groups [13.93]. The existing papers are silent on the scaling laws for prices and returns, but Levy and Solomon [13.94] show that with the (crucial) addition of 'social security' paid out to investors whose wealth falls below some threshold value in the course of events, realistic power-laws (in accordance with Pareto's famous law) for the distribution of wealth among agents can be obtained.

The simulation framework of Lux and Marchesi [13.109, 13.110] builds upon earlier theoretical work [13.101, 13.104, 13.105] in which the features of simple multi-agent models are analyzed via mean field approximation and master equation techniques. The model contains two groups of traders: fundamentalists who buy (sell) when the price is believed to be below (above) the fundamental value, and chartists, who follow trends and are subject to some sort of herd behavior. The latter group is divided into two subgroups: optimistic individuals who believe in an increase of the price in the near future and pessimistic ones who believe that the price will fall. Furthermore, agents are not bound to remain within one group for all time, but are allowed to change behavior by switching between these three behavioral variants. Switches between the optimistic and pessimistic chartist group are governed by the joint influence of the observed (temporary) price trend and the observed majority opinion among their fellow traders: rising prices together with an optimistic majority will induce former pessimists to join this majority and vice versa. On the other hand, with conflicting signals (for example, rising prices together with a pessimistic majority), the incentive to change behavior will be considerably weaker. Switches between the chartist and fundamentalist groups are governed by the difference of profits earned by traders from both groups. Of course, if one strategy has an advantage, traders from the other group will be attracted to this alternative. Given the distribution of traders over the three groups at any instant in time, their demand and supply decisions follow from their particular strategy and their opinion about the future development of the market. If imbalances between demand and supply result, prices are adjusted in the usual manner. With changes in market prices, trends, majority opinions, and profitability of strategies may also undergo changes, which leads to revisions of agents' strategies and a new distribution over groups.

As Lux and Marchesi [13.109] are mainly interested in the emergence of scaling laws from the market interaction, they perform the following 'experiment': changes of the fundamental value of the asset are introduced as external input. However, this 'news arrival process' is assumed to lack all the typical characteristics of real-life data (fat tails and heteroscedasticity), but is instead modeled as a white noise process. According to the efficient market hypothesis, returns should reflect this 'innocuous' distribution of the news, whereas endogenous emergence of scaling laws would lend some support to the interacting agent hypothesis. In fact, it turns out that scaling laws with realistic exponents emerge for a wide range of model parameters. Hence, as illustrated in a typical simulation in Fig. 13.4, it appears that the market interactions of agents *transform* exogenous noise (news) into fat tailed returns with clustered volatility[14]. The upper left-hand panel of Fig. 13.0 shows a typical price path from a simulation of the model which also is hard to distinguish from empirical records. The paper also provides an avenue towards reconciliation of the seemingly adverse views of the market efficiency and the interacting agent approaches in that prices are shown to closely follow the motion of the fundamental factors. In fact, a number of standard tests for the implications of the efficient market hypothesis (e.g. unit root tests) proved unable to recover any 'inefficiency' in the price series from this artificial market. Hence, the emergence of scaling laws does not come along with deviations from efficient price formation that are easily recognized.

13.6 Conclusion

In this chapter, we have surveyed research on scaling phenomena in financial data pursued by physicists and have compared their methodology and results with the approach of economists dealing with the same topic. We have also tried to put this work into perspective by discussing how far it is reconcilable

---

[14] This even happens when we assume a stationary fundamental value, cf. [13.110].

'artificial' financial market

DAX, 1979 – 1993

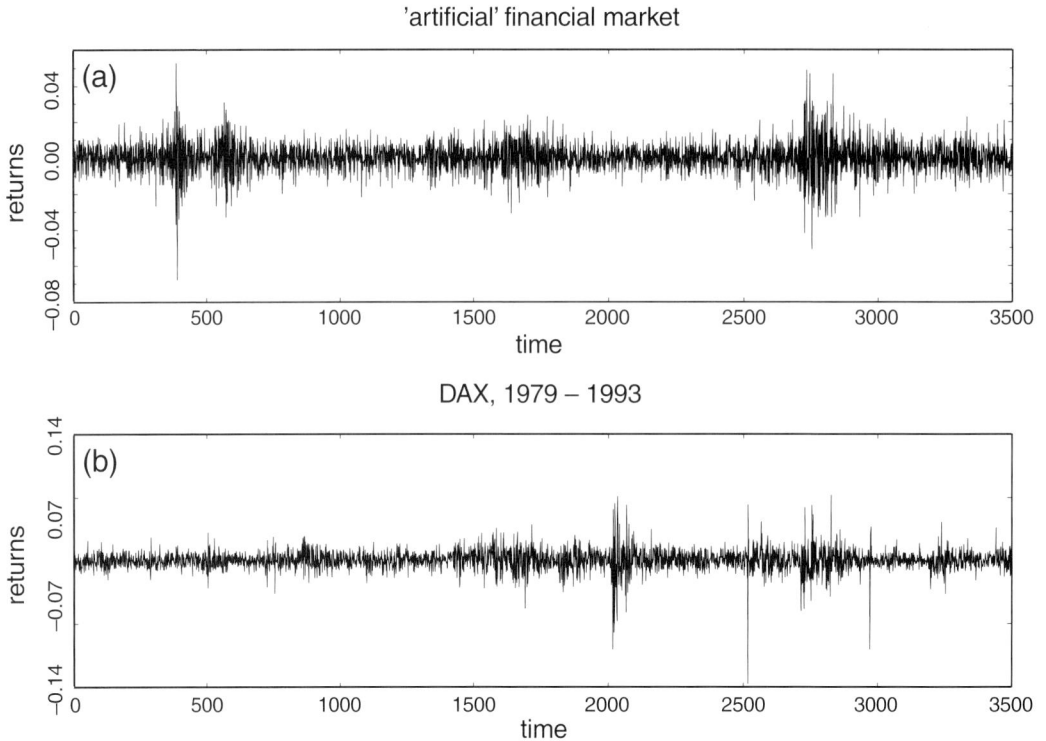

Fig. 13.4. A typical time series of returns from the model of Lux/Marchesi [13.109] (**a**) compared with the empirical record of daily returns of the German share price index DAX (**b**). Both time series share the clustered volatility phenomenon and a higher concentration of large returns than would occur under a normal distribution. As shown in [13.109] for the artificial financial market, both the tail index estimates and the estimates of the statistics for long-term dependence are in good agreement with empirical results

with traditional models in finance (the efficient market hypothesis) or whether it leads to a new viewpoint on market interactions.

As it turns out, many of the scaling phenomena highlighted in this literature (i.e. the cubic power-law of the empirical distribution of returns or the power-law in temporal dependence of squared and absolute returns) have been in no way alien to economics, but under the headings of 'fat tails' and 'volatility clustering' count as ubiquitous stylized facts of financial data. In most cases, there are no material differences in the empirical results reported by researchers from economics and finance-aficionados from physics (the convergence of initially different outcomes has been pointed out above). The key difference is rather that both groups often tend to express one and the same phenomenon in a somewhat different language. Right here the danger emerges, that parallel research remains unnoticed and earlier investigations by researchers from the other tradition are repeated. The history of the Lévy distribution hypothesis and its abandonment provides an example of this possibility.

However, more recently, the focus of analysis has been broadened and new work on additional scaling laws has been carried out. Interesting examples include the analysis of scaling in the distribution of intra-daily transaction numbers [13.130] or scaling in waiting times in tick-by-tick data [13.131] which unambiguously add new insights to our body of knowledge on empirical characteristics of financial data. Given the currently burgeoning interest in high-frequency data in finance [13.62], the increased availability of large datasets at intra-daily frequencies, and the experience of physicists to deal with such large datasets, in fact, suggest that a lot remains to be detected by such data analytical exercises.

From the conceptual point of view, the major innovative contributions by physicists are, however, not so much in the purely empirical work, but in suggesting new approaches in the modeling of the underlying phenomena: first, the introduction of the multifractal framework allows a unifying treatment of the different scaling laws in various powers of returns[15] and may provide us with a clue to a new vintage of empirical asset pricing models able to account for these phenomena. Second, from the theoretical side, the main contribution lies in the recognition of statistical physics, that scaling laws like the ones found in financial markets can often be explained by interactions in multi-particle systems (cf. Chap. 11). For a long time, such a perspective has indeed been alien to economics with its paradigm of the representative agent, but it seems to be gaining ground. The recent attempts at designing multi-agent models of financial markets that were sketched in the last part of our survey may pave the way to a new branch of descriptive models for financial markets.

Acknowledgements. We are grateful to Armin Bunde and Hans Joachim Schellnhuber, the organizers of the Heraeus Workshop on 'Facets of Universality: Climate, Biodynamics and Stock Markets' at Giessen University, June 1999, for inviting us to this exciting interdisciplinary event and for asking us to jointly contribute a chapter on financial markets to this volume. We are also grateful to Armin Haas and Dietrich Stauffer for careful reading and very detailed remarks on earlier versions. As it seems to us, this joint undertaking greatly improved our understanding of the theoretical and methodological background of the sometimes different approaches by economists and physicists. It should, therefore, be obvious to the reader that this review has also benefited immensely from our discussions with scientists from both fields over the last couple of months. In fact, the number of people involved via discussions is too long to thank all of them personally.

---

[15] With respect to these features Clive Granger wrote: "I feel that when a satisfactory theory for this area is found, it may unlock a rush of new results having real practical importance." [13.65]

References

13.1   A. Abhyankar, L.S. Copeland, and W. Wong, Appl. Econ. Lett. **2**, 288 (1995).

13.2   M. Aoki, J. Econ. Dyn. Control **18**, 865 (1994).

13.3   M. Aoki, *New Approaches to Macroeconomic Modeling: Evolutionary Stochastic Dynamics, Multiple Equilibria, and Externalities as Field Effects* (Cambridge, University Press, Cambridge, 1996).

13.4   J. Arifovic, J. Polit. Econ. **104**, 510 (1996).

13.5   M. Ausloos, Physica A **284**, 385 (2000).

13.6   M. Ausloos and N. Vandewalle, *Using DFA for Profit Making in the FX Market* (mimeo, University of Liège, 2000); [in press] Int. J. Theoret. Appl. Finance (2000).

13.7   M. Ausloos, N. Vandewalle, P. Boveroux, A. Minguet, and K. Ivanova, Physica A **274**, 229 (1999).

13.8   R. Baillie, T. Bollerslev, and H. Mikkelsen, J. Economet. **74**, 3 (1996).

13.9   P. Bak, M. Paczuski, and M. Shubik, Physica A **246**, 430 (1997).

13.10 A.-L. Barabasi and T. Vicsek, Phys. Rev. A **44**, 2730 (1991).

13.11 W. Barnett and A. Serletis, J. Econ. Dyn. Control **24**, 703 (2000).

13.12 R. Baviera, L. Biferale, R.N. Mantegna, and A. Vulpiani, Transient multiaffine behaviors in ARCH and GARCH processes, paper presented at Int. Workshop Econophys. Statist. Finance, Palermo, September 28 - 30, 1998.

13.13 J. Beirlant, J.L. Teugels, and P. Vynckier, *Practical Analysis of Extreme Values* (Leuven University Press, Leuven, 1996).

13.14 A. Bershadskii, Eur. Phys. J. B **11**, 361 (1999).

13.15 A.K. Bera and M.L. Higgins, J. Econ. Surv. **7**, 305 (1993).

13.16 C.L. Berthelsen, J.A. Glazier, and S. Raghavachari, Phys. Rev. E **49**, 1860 (1994).

13.17 O.J. Blanchard and M.W. Watson, in *Economic and Financial Structure: Bubbles, Bursts, and Shocks*, edited by P. Wachtel (Lexington Books, Lexington, 1982), p. 295.

13.18 R.C. Blattberg and N.J. Gonedes, J. Bus. **47**, 244 (1974).

13.19 T. Bollerslev, J. Economet. **31**, 307 (1986).

13.20 T. Bollerslev, R.Y. Chou and K.F. Kroner, J. Economet. **52**, 5 (1992).

13.21 G.G. Booth, F.R. Kaen, and P.E. Koveos, J. Monetary Econ. **10**, 407 (1982).

13.22 J.-P. Bouchaud, M. Potters, and M. Meyer, Eur. Phys. J. B **13**, 595 (2000).

13.23 W. Breymann, S. Ghashghaie, and P. Talkner, Int. J. Theor. Appl. Finance **3**, 357 (2000).

13.24 W. Brock, D. Hsieh, and B. LeBaron, *Nonlinear Dynamics, Chaos, and Instability: Statistical Theory and Economic Inference* (MIT, Cambridge, 1991).

13.25 W. Brock, J. Lakonishok, and B. LeBaron, J. Finance **47**, 1731 (1992).

13.26 G. Caginalp and H. Laurent, Appl. Math. Finance **5**, 181 (1998).

13.27 J. Campbell, A. Lo, and A. MacKinlay, *The Econometrics of Financial Markets* (Princeton, University Press, Princeton, 1997).

13.28 S.-H. Chen and C.-H. Yeh, in *Evolutionary Programming VI*, edited by P.J. Angeline (Springer, Berlin, 1997), p. 137.

13.29 S.-H. Chen and C.-H. Yeh, On the emergent properties of artificial stock markets: the efficient market hypothesis and the rational expectations hypothesis, J. Econ. Beh. Organ. [in press] (2001).

13.30 P.K. Clark, Econometrica **41**, 135 (1973).

13.31 K.J. Cohen, S.F. Maier, R.A. Schwartz, and D.K. Whitcomb, *The Microstructure of Securities Markets* (Prentice-Hall, Englewood Cliffs, 1986).

13.32 R. Cont and J.-P. Bouchaud, Macroecon. Dyn. **4**, 170 (2000).

13.33 R. Cont, M. Potters, and J.-P. Bouchaud, in *Scale Invariance and Beyond*, edited by B. Dubrulle, F. Graner, and D. Sornette (Springer, Berlin, 1997).

13.34 P.H. Cootner (ed.), *The Random Character of Stock Market Prices* (MIT, Cambridge, 1964).

13.35 R.W. Cornew, D.E. Town, and L.D. Crowson, J. Futures Markets **4**, 531 (1984).

13.36 N. Crato and P.J.F. de Lima, Econ. Lett. **45**, 281 (1994).

13.37 J. Danielsson and C.G. de Vries, J. Empir. Finance **4**, 241 (1997).

13.38 W. DeBondt and R.H. Thaler, J. Finance **40**, 793 (1985).

13.39 H. Drees and E. Kaufman, Stoch. Proc. Appl. **75**, 149 (1998).

13.40 D.A. Dickey, W.R. Bell, and R.B. Miller, Am. Statist. **40**, 12 (1986).

13.41 Z. Ding, C.W.J. Granger, and R.F. Engle, J. Empir. Finance **1**, 83 (1993).

13.42 W. DuMouchel, Ann. Statist. **11**, 1019 (1983).

13.43 E. Eberlein and U. Keller, Bernoulli **1**, 281 (1995).

13.44 R.F. Engle, Econometrica **50**, 987 (1982).

13.45 C.J.G. Evertz and B. Mandelbrot, in *Chaos and Fractals: New Frontiers of Science*, edited by H.-O. Peitgen, H. Jürgens, and D. Saupe (Springer, Berlin, 1992), p. 921.

13.46 K. Falconer, *Fractal Geometry: Mathematical Foundations and Applications* (Wiley, New York, 1990).

13.47 E. Fama, J. Bus. **35**, 420 (1963).

13.48 E. Fama, J. Finance **25**, 383 (1970).

13.49 D. Farmer and A. Lo, *Frontiers of Finance: Evolution and Efficient Markets* (mimeo, Santa Fe Institute, 1999).

13.50 A. Fisher, L. Calvet, and B. Mandelbrot, *Multifractality of Deutschemark/US Dollar Exchange Rates* (mimeo, Cowles Foundation for Research in Economics, 1997).

13.51 R. Flood and P. Garber, *Speculative Bubbles, Speculative Attacks, and Policy Switching.* (MIT, Cambridge, 1994).

13.52 W.P. Forbes, J. Econ. Surv. **10**, 123 (1996).

13.53 M. Frank and T. Stengos, Rev. Econ. Stud. **56**, 553 (1989).

13.54 B.M. Friedman and D.I. Laibson, Brookings Papers Econ. Activity **2**, 137 (1989).

13.55 D. Friedman and S. Vandersteel, J. Int. Econ. **13**, 171 (1982).

13.56 R. Friedrich, J. Peinke, and C. Renner, Phys. Rev. Lett. **84**, 5224 (2000).

13.57 S. Galluccio, G. Caldarelli, M. Marsili, and Y.-C. Zhang, Physica A **245**, 423 (1997).

13.58 J. Geweke and S. Porter-Hudak, J. Time Ser. Anal. **4**, 221 (1983).

13.59 S. Ghashghaie et al., Nature **381**, 767 (1996).

13.60 C.G. Gilmore, J. Econ. Beh. Organ. **22**, 209 (1993).

13.61 W.N. Goetzman, J. Bus. **66**, 249 (1993).

13.62 C. Goodhart and M.O'Hara, J. Empir. Finance **4**, 73 (1997).

13.63 P. Gopikrishnan, M. Meyer, L.A.N. Amaral, and H.E. Stanley, Eur. Phys. J. B **3**, 139 (1998).

13.64 C. Granger, J. Economet. **14**, 227 (1980).

13.65 C. Granger, J. Bus. Econ. Statist. **16**, 268 (1998).

13.66 M.T. Greene and B.D. Fielitz, J. Financ. Econ. **4**, 399 (1977).

13.67 P.A. Groenendijk, A. Lucas and C.G. de Vries, J. Empir. Finance **2**, 253 (1995).

13.68 L. de Haan, S.I. Resnick, H. Rootzén, and C.G. de Vries, Stoch. Proc. Appl. **32**, 213 (1989).

13.69 N. Hakansson, A. Deja, and J. Kale, J. Finance **40**, 1 (1985).

13.70 J.A. Hall, B.W. Brorsen, and S.H. Irwin, J. Financ. Quant. Anal. **24**, 105 (1989).

13.71 T. Halsey, M. Jensen, L. Kadanoff, I. Procaccia, and B. Shraiman, Phys. Rev. A **33**, 1141 (1986).

13.72 J. Hamilton and C. Whiteman, J. Monetary Econ. **16**, 353 (1985).

13.73 A. Hilgers and C. Beck, Int. J. Bifurcat. Chaos **7**, 1855 (1997).

13.74 B.M. Hill, Ann. Statist. **3**, 1163 (1975).

13.75 S. Hodges, *Arbitrage in Fractional Brownian Motion Market* (mimeo, University of Warwick, 1995).

13.76 B. Holdom, Physica A **254**, 569 (1998).

13.77 D. Hood, P. Andreassen, and S. Schachter, J. Econ. Behav. Organ. **6**, 331 (1985).

13.78 D.-A. Hsu, R.B. Miller, and D.W. Wichern, J. Am. Statist. Assoc. **69**, 1008 (1974).

13.79 G. Iori, *A Microsimulation of Traders Activity in the Stock Market: The Role of Heterogeneity, Agents' Interaction and Trade Frictions* J. Econ. Behav. Organ. [in press] (2002).

13.80 K. Ivanova and M. Ausloos, Eur. Phys. J. B **8**, 665 (1999).

13.81 D.W. Jansen and C.G. de Vries, Rev. Econ. Statist. **73**, 18 (1991).

13.82 S. Joshi, J. Parker, and M. Bedau, *Financial Markets can be at Suboptimal Equilibria*, Comput. Econ. [in press] (2001).

13.83 F.R. Kaen and R.E. Rosenman, Am. Econ. Rev. **76**, 212 (1986).

13.84 G. Kim and H. Markowitz, J. Portfolio Manage. **16**, 45 (1989).

13.85 A. Kirman, Q. J. Econ. **108**, 137 (1993).

13.86 K.G. Koedijk, M.M.A. Schafgans, and C.G. de Vries, J. Int. Econ. **29**, 93 (1990).

13.87 S.J. Kon, J. Finance **39**, 147 (1984).

13.88 J. Koza, *Genetic Programming: On the Programming of Computers by Means of Natural Selection* (MIT, Cambridge, 1992).

13.89 A.H.-L. Lau, H.-S. Lau, and J.R. Wingender, J. Bus. Econ. Statist. **8**, 217 (1990).

13.90 B. LeBaron, W.B. Arthur, and R. Palmer, J. Econ. Dyn. Control **23**, 1487 (1999).

13.91 S.F. Leroy, J. Econ. Lit. **27**, 1583 (1989).

13.92 M. Levy, H. Levy, and S. Solomon, Econ. Lett. **45**, 103 (1994).

13.93 M. Levy, H. Levy, and S. Solomon, J. de Physique I France **5**, 1087 (1995).

13.94 M. Levy and S. Solomon, Int. J. Mod. Phys. C **7**, 65 (1996).

13.95 W. Li, Int. J. Bifurcat. Chaos **1**, 583 (1991).

13.96 Y. Liu, P. Cizeau, M. Meyer, C.-K. Peng, and H.E. Stanley, Physica A **245**, 437 (1997).

13.97 A.W. Lo, Econometrica **59**, 1279 (1991).

13.98 I.N. Lobato and N.E. Savin, J. Bus. Econ. Statist. **16**, 261 (1998).

13.99 F.M. Longin, J. Bus. **69**, 383 (1996).

13.100 M. Loretan and P.C.B. Phillips, J. Empir. Finance **1**, 211 (1994).

13.101 T. Lux, Econ. J. **105**, 881 (1995).

13.102 T. Lux, Appl. Financ. Econ. **6**, 463 (1996).

13.103 T. Lux, Appl. Econ. Lett. **3**, 701 (1996).

13.104 T. Lux, J. Econ. Dyn. Control **22**, 1 (1997).

13.105 T. Lux, J. Econ. Behav. Organ. **33**, 143 (1998).

13.106 T. Lux, *Multi-Fractal Processes as a Model for Financial Returns: A First Assessment* (mimeo, University of Bonn, 1999); Quant. Finance, [in press] (2002)

13.107 T. Lux, Appl. Financ. Econ. **11**, 299 (2001).

13.108 T. Lux, Empir. Econ. **25**, 641 (2000).

13.109 T. Lux and M. Marchesi, Nature **397**, 498 (1999).

13.110 T. Lux and M. Marchesi, Int. J. Theor. Appl. Finance **3**, 675 (2000).

13.111 T. Lux and D. Sornette, *On Rational Speculative Bubbles and Fat Tails* (mimeo, University of Bonn and CNRS, 1999).

13.112 B. Mandelbrot, J. Bus. **35**, 394 (1963).

13.113 B. Mandelbrot, Sci. Am., February, 50 (1999).

13.114 B. Mandelbrot, A. Fisher, and L. Calvet, *A Multifractal Model of Asset Returns* (mimeo, Cowles Foundation for Research in Economics, 1997).

13.115 B. Mandelbrot and J. Wallis, Water Res. Res. **5**, 228 (1969).

13.116 R.N. Mantegna, Physica A **179**, 232 (1991).

13.117 R.N. Mantegna and H.E. Stanley, Nature **376**, 46 (1995).

13.118 R.N. Mantegna and H.E. Stanley, Nature **383**, 587 (1996).

13.119 R.N. Mantegna and H.E. Stanley, *An Introduction to Econophysics. Correlation and Complexity in Finance* (Cambridge University Press, Cambridge, 2000).

13.120 T. Mikosch and C. Starica, *Limit Theory for the Sample Autocorrelations of a GARCH(1,1) Process* (mimeo, Chalmers University of Technology, 1999).

13.121 T.C. Mills, The Statistician **44**, 323 (1995).

13.122 T.C. Mills, Appl. Financ. Econ. **7**, 599 (1997).

13.123 J.W. Moffat, Physica A **264**, 532 (1999).

13.124 U.A. Müller, M.M. Dacarogna, and O. Pictet, in *A Practical Guide to Heavy Tails*, edited by R. Adler, R. Feldman, and M.S. Taqqu (Birkhäuser, Basel, 1998), p. 55.

13.125 C. Neely, P. Weller, and R. Dittmar, J. Financ. Quant. Anal. **32**, 405 (1997).

13.126 A. Pagan, J. Empir. Finance **3**, 15 (1996).

13.127 R.G. Palmer, W.B. Arthur, J.H. Holland, B. LeBaron, and P. Tayler, Physica D **75**, 264 (1994).

13.128 C.-K. Peng, S.V. Buldyrev, S. Havlin, M. Simons, H.E. Stanley, and A.L. Goldberger, Phys. Rev. E **49**, 1685 (1994).

13.129 B. Pilgram and D.T. Kaplan, Physica D **114**, 108 (1998).

13.130 V. Plerou, P. Gopikrishnan, L. Amaral, X. Gabaix, and H. Stanley, *Diffusion and Economic Fluctuations* (mimeo, Boston University, 2000).

13.131 M. Raberto, E. Scalas, R. Gorenflo, and F. Mainardi, *Scaling of Waiting-Time Distribution in Tick-by-Tick Financial Data* (mimeo, University of Genoa, 2000).

13.132 J.B. Ramsey, J. Economic Behavior and Organization **30**, 275 (1996).

13.133 R.-D. Reiss and M. Thomas, *Statistical Analysis of Extreme Values with Applications in Finance, Hydrology and Other Fields* (Birkhäuser, Basel, 1997).

13.134 D. Ruelle, Proc. R. Soc. Lond. **427A**, 241 (1990).

13.135 J.A. Scheinkman and B. LeBaron, J. Bus. **62**, 311 (1989).

13.136 F. Schmitt, D. Schertzer, and S. Lovejoy, Appl. Stoch. Models Data Anal. **15**, 29 (1999).

13.137 R.J. Shiller, Am. Econ. Rev. **71**, 421 (1981).

13.138 M.A. Simkowitz and W.L. Beedles, J. Am. Statist. Assoc. **75**, 306 (1980).

13.139 J.C. So, Rev. Econ. Statist. **69**, 100 (1987).

13.140 C. Starica, *The Tales the Tails of GARCH(1,1) Processes Tell* (mimeo, Chalmers University, 1999).

13.141 M. Stanley et al., Fractals **4**, 415 (1996).

13.142 D. Stauffer and T. Penna, Physica A **256**, 284 (1998).

13.143 D. Stauffer, P. de Oliveira, and A. Bernardes, J. Theor. Appl. Finance **2**, 83 (1999).

13.144 G. Stigler, J. Bus. **37**, 117 (1964).

13.145 R. Sullivan, A. Timmerman, and H. White, J. Finance **54**, 1647 (1999).

13.146 R. Sullivan, A. Timmerman, and H. White, *Dangers of Data-Driven Inference: The Case of Calendar Effects in Stock Returns* (mimeo, University of California at San Diego, 1999).

13.147 H. Takayasu, H. Mura, T. Hirabayashi, and K. Hamada, Physica A **184**, 127 (1992).

13.148 M.S. Taqqu and V. Teverovsky, in *A Practical Guide to Heavy Tails*, edited by R. Adler, R. Feldman, and M.S. Taqqu (Birkhäuser, Basel, 1998), p. 177.

13.149 M.S. Taqqu and V. Teverovsky, Finance Stoch. **3**, 1 (1999).

13.150 M.S. Taqqu, V. Teverovsky, and W. Willinger, Fractals **3**, 785 (1995).

13.151 J. Teichmoeller, J. Am. Statist. Assoc. **66**, 282 (1971).

13.152 A.L. Tucker and L. Bond, Rev. Econ. Statist. **70**, 638 (1988).

13.153 D.E. Upton and D.S. Shannon, J. Finance **34**, 131 (1979).

13.154 N. Vandewalle and M. Ausloos, Physica A **246**, 454 (1997).

13.155 N. Vandewalle and M. Ausloos, J. Mod. Phys. C **9**, 711 (1998).

13.156 N. Vandewalle and M. Ausloos, Eur. Phys. J. B **4**, 257 (1998).

13.157 N. Vandewalle and M. Ausloos, Phys. Rev. E **58**, 6832 (1998).

13.158 N. Vandewalle, M. Ausloos, and P. Boveroux, Physica A **269**, 170 (1999).

13.159 J.C. Vassilicos, Nature **374**, 408 (1995).

13.160 J.C. Vassilicos, A. Demos, and F. Tata, in *Applications of Fractals and Chaos*, edited by A.J. Crilly, R.A. Earnshaw, and H. Jones (Springer, Berlin, 1993), p. 249.

13.161 R.F. Voss, in *Fractal Geometry and Computer Graphics*, edited by J.L. Encarnacao et al. (Springer, Berlin, 1992), p. 45.

13.162 C.G. de Vries, in *The Handbook of International Macroeconomics*, edited by F. van der Ploeg (Blackwell, Oxford, 1994), p. 348.

13.163 K.D. West, J. Finance **43**, 639 (1988).

13.164 Y.-C. Zhang, Physica A **269**, 30 (1999).

14. Market Fluctuations II: Multiplicative and Percolation Models, Size Effects, and Predictions

Didier Sornette, Dietrich Stauffer, and Hideki Takayasu

We present a set of models of the main stylized facts of market price fluctuations. These models comprise dynamical evolution with threshold dynamics and a Langevin price equation with multiplicative noise, percolation models to describe the interaction between traders, and hierarchical cascade models to unravel the possible correlation across time scales, including the log-periodic signatures associated with financial crashes. The main empirical knowledge is summarized and some key empirical tests are presented.

14.1 Stylized Facts of Financial Time Series

The attraction of physicists to finance and to the study of stock markets is grounded on several factors.

- Physics and finance are both fundamentally based on the theory of random walks (and their generalizations to higher dimensions) and on the collective behavior of large numbers of correlated variables. Finance thus offers another fascinating playground for the application of concepts and methods developed in the natural sciences which have traditionally focused their attention on a description and understanding of the surrounding inanimate world at all possible scales.
- Stock markets offer may be one of the simplest real life experimental systems of coevolving competing learning agents and can thus be thought of as a proxy for studying biological evolution [14.1–14.4].
- It is tempting to believe that the technical abilities developed in the physical sciences could help to 'beat the market': predicting a complex time series

---

Fig. 14.0. Rigorous market analysis by rational econophysicists like one of the present authors (from [14.100] and The Economist, October 1997)

like the market price evolution shown in Fig. 14.1 is an exciting intellectual challenge as well as potentially rewarding financially.

In this short review, we present a series of models of stock markets that each provide a particular window of understanding. The different models do not play the same role. In its broadest sense, recall that a model (usually formulated using the language of mathematics) is a mathematical representation of a condition, process, concept, etc, in which the variables are defined to represent inputs, outputs, and intrinsic states and equations or inequalities are used to describe interactions of the variables and constraints on the problem. In theoretical physics, models take a narrower meaning, such as in the Ising, Potts, ..., percolation models. In economy and finance, the term 'model' is usually used in the broadest sense. Here, we will use both types of models: the microscopic threshold models (regarding other types threshold models cf. Chap. 6) of the interplay between demand and supply discussed in Sect. 14.2 and the percolation models of Sect. 14.3 fall in the second category. The cascade of correlations across scales [14.5, 14.6] briefly summarized in Sect. 14.4 belongs to this first class of models. The log-periodic signatures preceding crashes also discussed in Sect. 14.4 rely on both types of models. This diversity of models reflects our burgeoning understanding of this field which has not yet fully matured.

The first most striking observation of a stock market is that price variations seem to fluctuate randomly, leading to a price trajectory as a function of time which looks superficially similar to a random walk with Markovian increments, as shown in the upper left panel (a) of Fig. 14.1. This view was first expressed by Bachelier in his 1900 thesis [14.7] (see, Chap. 12, p. 359) and later formalized rigorously by Samuelson [14.8]. This fundamental thesis in finance is called the efficient market hypothesis and states in a nutshell that price variations are essentially random as a result of the incessant activity of traders who attempt to profit from small price differences (so-called arbitrage opportunities); the mechanism is that their investment strategies produce feedbacks on the prices that become random as a consequence. One important domain of research consists in determining the detailed mechanisms by which this feedback operates dynamically and statistically. Correlatively, the search for deviations from this efficient market state may lead to significant understanding of the way the markets function.

At first glance, the concept that price variations are uncorrelated is confirmed by looking at the two-point correlation function of the price increments. For liquid markets such as the Standard and Poor's (S&P 500), it is found significantly different from zero with statistical confidence only for very short times of the order of a few minutes, as shown in panel (a') of Fig. 14.1. On the other hand, the correlations of the amplitude (absolute value) of the variations, called the volatility, are very long-ranged. This can be visualized qualitatively by looking at panel (b) of Fig. 14.1 which constructs a random walk by successive addition of the logarithm of a measure of the amplitude of

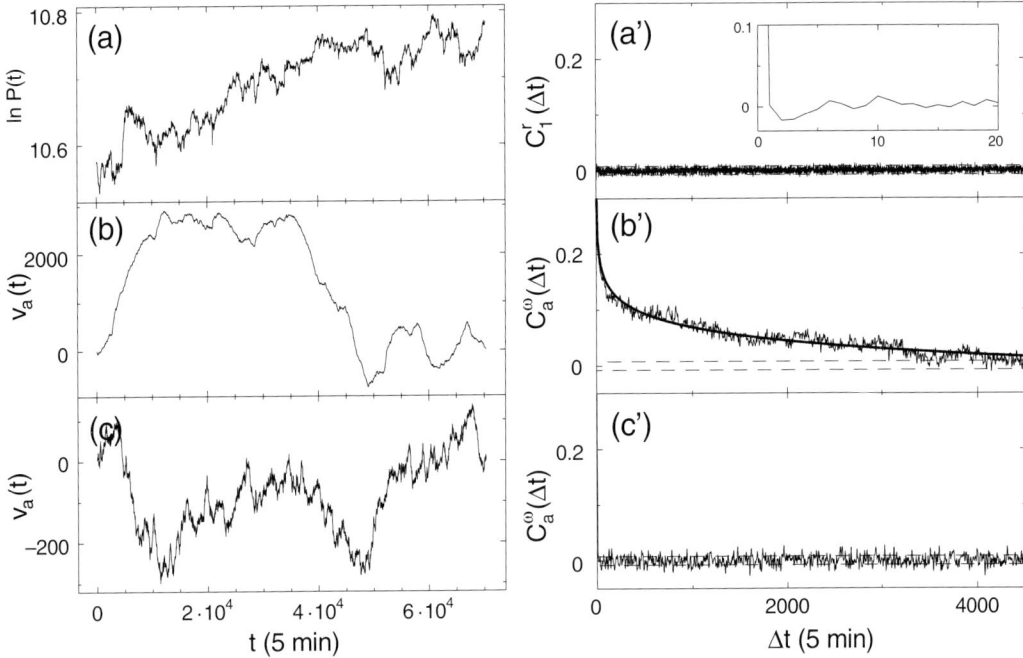

Fig. 14.1. (**a**) Time evolution of $\ln P(t)$, where $P(t)$ is the S&P 500 US index, sampled with a time resolution $\delta t = 5$ min in the period October 1991-February 1995. The data have been preprocessed in order to remove 'parasitic' daily oscillatory effects: if m_i and σ_i are respectively the mean and the r.m.s. of the signal within the ith 5 min interval of a day, the value of the signal $x(i)$ has been replaced by $m + \sigma(x(i) - m_i)/\sigma_i$. (**b**) The corresponding 'centered log-volatility walk', $v_a(t) = \sum_{i=0}^{t} \tilde{\omega}_a(i)$, as computed with the derivative of the Haar function as analyzing wavelet for a scale $a = 4$ ($\simeq 20$ min). (**c**) $v_a(t)$ computed after having randomly shuffled the increments of the signal in (**a**). (**a'**) The 5 min ($a = 1$) return correlation coefficient $C_1^r(\Delta t)$ vs. Δt. (**b'**) The correlation coefficient $C_a^\omega(\Delta t)$ of the log-volatility of the S&P 500 at scale $a = 4$ ($\simeq 20$ min); the solid line corresponds to a fit of the data using (14.23) with $\lambda^2 \simeq 0.015$ and $T \simeq 3$ months. (**c'**) same as in (**b'**) but for the randomly shuffled S&P 500 signal. In (**a'**)-(**c'**) the dashed lines delimit the 95% confidence interval. Taken from [14.5, 14.6]

the price variation obtained through a wavelet transform (see [14.5, 14.6] for details, and Chap. 2). One observes long periods of persistences which can be compared with the random walk in panel (c) obtained by first reshuffling the price variations and performing the same analysis as for panel (b). Panels (b') and (c') show the correlation functions of the volatilities corresponding respectively to panels (b) and (c). For the real S&P 500 time series, one observes an extremely slow approximately power-law decay, which provides a measure of the well-documented clustering or persistence of volatility [14.9–14.14].

In addition to the almost complete absence of correlation of price increments and the long-range correlation of volatilities, the last striking stylized fact is the 'fat tail' nature of the distribution of price variations or of returns. The qualification 'fat tail' is used to stress that the distributions of price variations decay usually more slowly than a Gaussian, which is taken as the reference that would be valid under the random walk hypothesis. Exponentially truncated Lévy laws [14.15–14.18] with exponent around $\alpha \approx 1.5$

(see definition in (14.15)) for the 6 yr period 1984–1989 and power-laws with exponents $\alpha \approx 3$ [14.19–14.21], superposition of Gaussians motivated by an analogy with turbulence [14.5, 14.6, 14.22], or stretched exponentials [14.23] have been proposed to describe the empirical distribution of price returns in organized markets.

In addition to these three well-documented stylized facts, many other studies have been performed to test for the possible existence of dependence between successive price variations at many different time scales that go beyond the method of correlation functions. There is indeed increasing evidences that even the most competitive markets are not completely free from correlations (i.e. are not strictly 'efficient') [14.4]. In particular, a set of studies in the academic finance literature have reported anomalous earnings which support technical analysis strategies [14.24–14.27] (see [14.28] for a different view). A recent study of 60 technical indicators on 878 stocks over a 12 yr period [14.29] finds that the trading signals from technical indicators do on average contain information that may be of value in trading, even if they generally underperform (without taking due consideration of risk-adjustments of the returns) a buy-and-hold strategy in a rising market by being relatively rarely invested.

In this review, we are going to present several models that propose to explain some of these empirical observations. Microscopic models of unbalance between supply and demand discussed in the next section provide an understanding for the possible existence of different market phases, such as random, bubble-like, and cyclic as well as a simple mechanism for 'fat tails' based on multiplicative noise. As we will summarize in Sect. 14.3, the percolation models give probably the simplest possible mechanisms for the observation of 'fat tails' and long-range volatility clustering while being compatible with the absence of correlations of price variations. They also teach us that finite-size effects are probably important, i.e. the number N of traders that count on the market is perhaps no more than a few hundreds to a few thousands. This is because most of the complexity observed in many microscopic models disappears when the limit of large N is taken. If correct, this suggests that a suitable modeling of the stock market belongs to the most difficult intermediate asymptotics between a few degrees of freedom and the thermodynamic limit. As a consequence, it should take into account effects of discreteness. This might be one of the ingredients at the basis of the observation of the remarkable log-periodic signatures preceding large crashes discussed in Sect. 14.4. The accumulated evidence now comprises more than 35 crashes and to our knowledge, no major financial crash preceded by an extended bubble has occurred in the past two decades without exhibiting log-periodic signature. The exception is the East-European stock markets which seem to be following a completely different logic than their larger Western counterparts and their indices do not resemble those of the other markets. In particular, we find that they follow neither power-law accelerations nor log-periodic patterns though large crashes certainly occur.

14.2 Fluctuations of Demand and Supply in Open Markets

14.2.1 Optimization of Supply Faced to an Uncertain Demand

Contrary to common sense in economics, demand and supply do not balance in reality. You can find that all shelves are always full of commodities in any department store in developed countries implying that supply is in excess. On the other hand, people are sometimes making queues in front of a popular bakery shop and fresh baked croissants are sold out immediately, which clearly shows that demand is in excess. Such excess-supply or excess-demand states can be shown to be maximal profit strategies if we take into account the fluctuations of demand as follows.

Let us define the variables needed to describe the bakery's strategy:

1. x, the selling price of a croissant;
2. y, its production cost;
3. s, the production number of croissants per a day;
4. n, the number of croissants requested by customers per day;
5. d, the demand which is the averaged value of n.

We assume that n fluctuates in the interval $[d-\delta, d+\delta]$ uniformly for simplicity, and we also assume that the remainders will be thrown away at the end of each day. The problem is what is the best s which maximizes the total profit. Obviously if $s = d + \delta$ then the bakery does not miss any customer's request and the gross sale is maximal. However, there is a possibility that it will have many unsold croissants when n is small and in that case the production cost of the remainders may cause a big loss. At the other extreme, if $s = d - \delta$, the bakery sells all its croissants and has no loss but on the other hand misses good selling opportunities. Therefore, there should be an optimal value of s between these two extremes that maximizes the expectation of the total profit. Let us denote the expectation of the total profit by $L(s)$, then we have the following evaluation,

$$
\begin{aligned}
L(s) &= \langle x \min(n, s) - ys \rangle \\
&= (x - y)(d - \delta y/x) - \frac{x}{4\delta}\{s - d - \delta(1 - 2y/x)\}^2 .
\end{aligned}
\tag{14.1}
$$

The maximal value of L is given by the following value of s;

$$
s^* = d + \delta(1 - 2y/x) .
\tag{14.2}
$$

From this equation, it is clear that in the case where the sale price is not very high, here $x < 2y$, the optimal producing s^* is smaller than the average demand d. This corresponds to the excess-demand state which the popular bakery shop follows. On the other hand, if the sale price x is higher than twice the production price y, then the best strategy is to keep the excess-supply state just like all department stores actually do ($s^* > d$). It should be noted that the balanced state of $s^* = d$ is the best strategy only when $x = 2y$. This is the reason why almost all commodities in our daily life are out of the balance of demand and supply. The coefficient 2 is of course modified if we assume a different probability density for the fluctuation of demand n. The key point in this discussion is the fluctuation of demand that is inevitable in any free economy society, and the best strategy taking such effect into account proves that the balance of demand and supply should almost always be broken to earn largest income on average.

A similar result is obtained if the bakery follows a different strategy, i.e. strives to minimize its probability of loss: the probability of losing is the same as the probability that $x \min(n, s) - ys$ be negative. In the interesting regime where $x > y$, this probability is the same as the probability for the total sale xn to be less than the total production cost ys. This leads to a probability to lose equal to

$$\text{Prob}_{\text{loss}} = \frac{y}{2} x \delta \left[s - s^{**} \right] , \tag{14.3}$$

where

$$s^{**} = (d - \delta)\frac{x}{y} . \tag{14.4}$$

We see that the production s^{**} that gives no loss with certainty is larger than the average demand d only if the sale price x is larger than $(d/(d - \delta))\, y$.

14.2.2 Consequence for the Bid-Ask Spread in Liquid Markets

In an open market such as stock markets or foreign currency exchange markets, the situation is very different because there are speculative dealers who try to earn money by changing their position from a seller to a buyer or vice versa rather frequently. By this effect, the demand and supply cannot be regarded as independent functions and furthermore we need to introduce a dynamic model to describe the pricing process correctly. Willing to buy from a market maker for instance, you will buy a stock at the 'ask' price $p_{\text{ask}} \equiv x$ and resell it at the 'bid' price $p_{\text{bid}} \equiv y$. The spread $\delta p_{\text{spread}} = p_{\text{ask}} - p_{\text{bid}} = x - y$ is usually small. Indeed, the relevant situation for a liquid market is that the 'ask' price $p_{\text{ask}} \equiv x$ is only slightly larger than the 'bid' price $p_{\text{bid}} \equiv y$:

$$x = y(1 + \epsilon) , \quad \text{with } \epsilon \ll 1 . \tag{14.5}$$

Expanding (14.2) for small ϵ gives

$$s^* - \left(\frac{d - \delta)}{\delta}\right) \propto \frac{\delta p_{\text{spread}}}{p_{\text{bid}}} ,\qquad (14.6)$$

i.e. the relative over-supply with respect to the minimum possible value $d - \delta$ is essentially equal to the relative spread. The implication of this result (14.6) is the following: reading (14.6) from right to left, we find that a market maker will be tempted to increase the spread between bid and ask if he has difficulty in getting rid of excess inventory, but this will be a smaller effect, the larger are the fluctuations of the demands, i.e. the possibility of selling in future occasions.

14.2.3 Microscopic Model of Market with Threshold Dynamics

We assume that every dealer in an open market has two prices in mind, the selling and buying prices. For each dealer, the buying price is always lower than the selling price, and the difference of these prices may represent his greediness. A dealer's action is rather simple, namely, if the market price is higher than the selling price in mind he will sell, and if the market price is lower than the buying price in mind he will buy. Let us assume the simplest case that there are only two dealers, A and B, and let their prices in mind be $p_{\text{b}}(A)$, $p_{\text{s}}(A)$, $p_{\text{b}}(B)$, and $p_{\text{s}}(B)$, where the subscripts b and s represent the buying and selling prices and the capital letters specify the dealers, A and B. When these prices are changed continuously, a trade occurs suddenly when either of the following two conditions is realized [14.30]:

$$p_{\text{b}}(A) \geq p_{\text{s}}(B) \quad \text{or} \quad p_{\text{b}}(B) \geq p_{\text{s}}(A). \qquad (14.7)$$

Note that the occurrence of a trade is characterized by a nonlinear function such as a step function.

As the greediness of the dealers always requires $p_{\text{b}}(A) < p_{\text{s}}(A)$ and $p_{\text{b}}(B) < p_{\text{s}}(B)$, there is no possibility of realizing the two conditions of (14.7) simultaneously, namely, the transaction is microscopically one-sided or irreversible. After the trade, these dealers renew their prices in their mind so that the trade condition does not hold any more.

Due to the nonlinear and irreversible nature of trades, dynamic models of dealers generally behave chaotically even if the dynamics is deterministic. There is a nonlinear effect that enhances any microscopic difference, but the estimated maximum Lyapunov exponent is 0 implying that the system is at the edge of chaos [14.30].

There are two extreme cases in this type of deterministic dealer models: one is the large asset limit and the other is the small asset limit.

- In the case of large asset limit, dealers are assumed to have an infinite amount of asset and all dealers can keep their positions, namely, a buyer can be always a buyer and a seller can be always a seller. In this limit, it is shown that there is a kind of phase transition behavior between excess-demand and excess-supply states as a function of the number ratio of buyers to sellers. In the excess-demand state, there are more buyers than sellers and the prices fluctuate with a linear upgrade trend [14.32]. In the excess-supply phase, the situation is just opposite. At the critical point, that is realized when the numbers of buyers and sellers are the same, there is no trend and the power spectrum of the price fluctuations follows an inverse square law implying that the fluctuations are quite similar to the Brownian motion.

- In the small asset limit, each dealer changes position alternatively between a buyer and a seller, namely, after the dealer bought a stock, he tries to sell the stock. As all dealers change their positions alternatively, the numbers of demand and supply automatically balance and the system always shows critical behavior, namely, the price fluctuations are similar to the Brownian motion even though the dynamics is deterministic [14.44]. This result indicates that the existence of speculative dealers who frequently change their positions is essential for the market to follow a random walk scale-free behavior. Note that this kind of stationary self-organized criticality must be distinguished from the critical behavior describing large crashes as described in Sect. 14.4. The two phenomena are not mutually exclusive as shown for instance in [14.31].

As dealers in any open market are sensitive to the market price changes, it is important to introduce a response effect in the dealer model to explain the fat tail distribution of price changes as reported e.g. by Mantegna and Stanley [14.15, 14.16] (see also Chap. 12). When dealers change their buying and selling prices in mind based on their own strategy independent of market price changes, the resulting price change distribution does not have long tails of a power-law. However, by adding the term that uniformly shifts all the dealers prices in mind proportional to the latest market price change, the distribution of market price changes become a power-law in general [14.44].

14.2.4 Derivation of Langevin Market Dynamics with Multiplicative Noise

The reason for the fat tail distributions can be theoretically explained by introducing a Langevin type stochastic equation with multiplicative noise:

$$\Delta P(t + \Delta t) = B(t)\Delta P(t) + F(t). \tag{14.8}$$

Here, $P(t)$ represents the market price at time step t and $\Delta P(t) \equiv P(t) - P(t - \Delta t)$ is the price change where Δt is the unit time interval. The effect of dealers' response on the market price change is given by $B(t)$, which is regarded as a random variable. The random additive term $F(t)$ is due to the chaotic behavior inherent in the dealer model.

In the low asset limit, it can be shown that the market price changes of the deterministic dealer model are nicely approximated by the multiplicative stochastic process described by (14.8) [14.44]. We now present a more direct derivation of (14.8) by considering the dealers' dynamics in a macroscopic way [14.45]. Let $p_b(j,t)$ and $p_s(j,t)$ be the jth dealer's buying and selling prices at time t, then the total balance of demand and supply in the market is described by the following function called the cumulative demand, $I(P,t)$;

$$I(P,t) = \sum_j \Theta(p_b(j,t) - P) - \Theta(P - p_s(j,t)),\qquad(14.9)$$

where $\Theta(x)$ is the step function which is 0 for $x < 0$ and is 1 for $x > 0$. When P is such that $I(P,t) > 0$, the number of buyers is larger than that of sellers at the price. Therefore, the balanced price at time t, $P^*(t)$, is given by the equation $I(P^*(t),t) = 0$. It is a natural assumption for an open market that the price change in a unit time is proportional to $I(P(t),t)$ when the market price is $P(t)$; therefore, we have the following equation:

$$P(t + \Delta t) - P(t) \propto I(p(t),t).\qquad(14.10)$$

As the buying and selling prices are not announced openly, no one knows the value of $P^*(t)$. Traders can only estimate it from the past market price data $\{P(t - \Delta t), P(t - 2\Delta t), ...\}$. Taking into account the effect that each dealer thinks in a different way, we can write down the time evolution equation of $P^*(t)$ as follows:

$$P^*(t + \Delta t) = P^*(t) + F(t) + W(P(t), P(t - \Delta t), ...)\,.\qquad(14.11)$$

Here, $F(t)$ represents a random variable showing the statistical fluctuation of dealers' expectation, and $W(P(t), P(t - \Delta t), ...)$ is the averaged dealers' response function. Considering the simplest nontrivial case, we have the following set of linear equations:

$$P(t + \Delta t) = P(t) + A(t)(P^*(t) - P(t))\,,\qquad(14.12)$$

$$P^*(t + \Delta t) = P^*(t) + F(t) + B(t)(P(t) - P(t - \Delta t)).\qquad(14.13)$$

Here, $A(t)$ is given by the inverse of the slope of $I(P(t),t)$ at $P = P^*(t)$ which is proportional to the inverse of the price elasticity coefficient in economics, and $B(t)$ shows the dealers' mean response to the latest market price change, and both of these coefficients can be random variables. If we can assume that $P(t)$ and $P^*(t)$ are always very close, the set of (14.12) and (14.13) becomes identical to (14.8). Namely, if the market price always follows the motion of the balanced price and if the dealers' responses to the latest price change averaged over all the dealers fluctuates randomly for different times, then the market price fluctuation is well-approximated by the Langevin type equation (14.8).

It is well known that such a stochastic process (14.8) generally produces large fluctuations following power-law distributions when $B(t)$ takes larger values than unity with finite probability [14.33–14.43]. A important condition to get a power-law distribution is that the multiplicative noise $B(t)$ must sometimes take values larger than one, corresponding to intermittent amplifications. This is not enough: the presence of the additive term $F(t)$ (which can be constant or stochastic) is needed to ensure a 'reinjection' to finite values, susceptible to the intermittent amplifications. It was thus shown [14.40–14.42] that (14.8) is only one among many convergent ($\langle \ln B(t) \rangle < 0$) multiplicative processes with repulsion from the origin (due to the $F(t)$ term in (14.8)) of the form

$$B(t+1) = e^{H(x(t),\{b(t),f(t),...\})} \, B(t) \; , \qquad (14.14)$$

such that $H \to 0$ for large $x(t)$ (leading to a pure multiplicative process for large $x(t)$) and $H \to \infty$ for $x(t) \to 0$ (repulsion from the origin). H must obey some additional constraint such a monotonicity which ensures that no measure is concentrated over a finite interval. All these processes share the same power-law probability density function (PDF)

$$P(x) = Cx^{-1-\alpha} \qquad (14.15)$$

for large x with α the solution of

$$\langle B(t)^\alpha \rangle = 1 \; . \qquad (14.16)$$

The fundamental reason for the existence of the power-law PDF (14.15) is that $\ln x(t)$ undergoes a random walk with drift to the left and which is repelled from $-\infty$. A simple Boltzmann argument [14.40–14.42] gives an exponential stationary concentration profile, leading to the power-law PDF in the $x(t)$ variable.

These results were proved for the process (14.8) by Kesten [14.34] using renewal theory and were then revisited by several authors in the differing contexts of ARCH processes in econometry [14.35] and 1D random-field Ising models [14.36] using Mellin transforms, and more recently using extremal properties of the $G-harmonic$ functions on noncompact groups [14.37–14.39] and the Wiener-Hopf technique [14.40–14.42].

In the case that $B(t)$ depends on $\Delta P(t)$, especially when it does not take a large value if the magnitude of price change exceeds a threshold value, exponential cutoffs appear in the tails of distribution of price changes resulting in a more realistic distribution [14.44].

There are cases where the behaviors of the set of equations, (14.12) and (14.13), deviate from that of (14.8). For example, in the special case that $B(t)$ is larger than 1 and $A(t)$ is smaller than 1 for a certain time interval then both $P(t)$ and $P^*(t)$ grows nearly exponentially and the difference of these values also grow exponentially. This case corresponds to the phenomenon called a bubble [14.46]. We can also find an oscillatory behavior of market price when $A(t) > 1$. Namely, the set of price equations derived theoretically can show typical behaviors of a second order difference equation for different parameter combinations, as also proposed in [14.3, 14.47]. Real data analysis based on this formulation is now under intensive study.

14.3 Percolation Models

14.3.1 Basic Percolation Model of Market Price Dynamics in Two to Infinite Dimensions

Besides the Levy-Levy-Solomon model [14.48–14.50], the Cont-Bouchaud model [14.51] seems to be the one investigated by the largest number of different authors. It uses the well-known percolation model and applies its cluster concept to groups of investors acting together. This percolation model thus, similar to the random-field Ising markets [14.52], applies physics knowledge collected over decades, instead of inventing new models for market fluctuations.

In percolation theory [14.53–14.55], every site of a large lattice is occupied randomly with probability p and empty with probability $1 - p$; a *cluster* is a group of neighboring occupied sites. For p above some percolation threshold p_c, an infinite cluster appears spanning the lattice from one side to the opposite side. The average number $n_s(p)$ of clusters containing s sites each varies for large s right at the percolation threshold as a power-law:

$$n_s \propto s^{-\tau}, \tag{14.17}$$

with an exponent τ increasing from about 2.05 in two dimensions to $5/2$ in six and more dimensions. Close to p_c a scaling law for large s holds:

$$n_s = s^{-\tau} f((p - p_c)s^\sigma), \tag{14.18}$$

with $\sigma \simeq 0.5$. For $p < p_c$, the cluster numbers decay asymptotically with a simple exponential, while above p_c they follow a stretched exponential with $\log(n_s) \propto -s^{1-1/d}$ in d dimensions.

Quite similar results are obtained if we switch from this site percolation problem to bond percolation, where all sites are occupied but the bonds between nearest neighbors are occupied with probability p; then clusters are groups of sites connected by occupied bonds.

For dimensionality $d > 6$, the critical exponents like τ are those of the Bethe-lattice or mean-field approximation, invented by Flory in 1941 [14.56], where no closed loops are possible and for which analytic solutions are possible: $\tau = 5/2$, $\sigma = 1/2$, $f =$ Gaussian. In three dimensions, most percolation results are only estimated numerically. Infinite-range bond percolation is also called random graph theory; then every site can be connected with all other sites, each with probability p. This infinite-range bond percolation limit was selected by Cont and Bouchaud [14.51] in order to give exact solutions, while the later simulations concentrated on two- or three-dimensional site percolation, with nearest neighbors only forming the clusters.

For market applications, the occupied sites are identified with investors, and the percolation clusters are groups of investors acting together. Thus at each iteration, every cluster has three choices: all investors belonging to the cluster buy (probability a); all of them sell (also probability a); and none of them acts at this time step (probability $1 - 2a$). Thus the activity a measures the time with which we identify one iteration: if this time step is one second, a will be very low since very few investors act every second; if the time step is one year, a will be closer to its maximum value $1/2$. All investors trade the same amount, and have an infinite supply of money and stocks to spend. Summation over all active clusters gives the difference between demand and supply and drives the price $P(t)$:

$$R(t) = [P(t + 1) - P(t)]/P(t) \propto \sum_{\text{buy}} n_s s - \sum_{\text{sell}} n_s s . \qquad (14.19)$$

In this way, the Cont-Bouchaud model has for a given lattice very few free parameters: the occupation probability p and the activity a. Moreover, algorithms to find the clusters in a randomly occupied lattice have been known for decades [14.53–14.55], and thus a computer simulation is quite simple if one has already a working (Hoshen-Kopelman) algorithm to find clusters.

Without any simulation [14.51], one can predict the results for very small a. If for a lattice of $N = L^d$ sites, we have a of order $1/N$, then typically no cluster, or only one, is active during one iteration. The price change then is either zero or $\pm$ the size s of the cluster. The distribution of absolute returns $|R|$ thus is identical to the cluster size distribution n_s, apart from a large contribution at $R = 0$. In particular, right at the percolation threshold $p = p_c$ we have a distribution $\pi(|R|)$ of returns obeying a power-law

$$\pi(|R|) \propto 1/R^\tau \qquad (14.20)$$

for not too small $|R|$, similar to Mandelbrot's Lévy-stable Pareto distributions [14.57–14.59]. The probability to have a jump of at least $|R|$ then decays asymptotically as $1/|R|^\alpha$ with $\alpha = \tau - 1$ between 1 and $3/2$. The volatility or variance of the return distribution is thus infinite at the percolation threshold, apart from finite-size and finite-time corrections; the same holds for skewness and kurtosis.

For larger activities, but still $a \ll 1$, scaling holds [14.60]: if we normalize height and width of the return distribution to unity, the curves for various activities a at $p = p_c$ overlap, and thus still give the above power-law. This scaling is no longer valid for large $a \simeq 1/3$ where the curves become more like a Gaussian.

This model thus reproduces some stylized facts of real markets, when inflation effects are subtracted: (i) the average return $\langle R \rangle$ is zero. (ii) There is no correlation between two successive returns or two successive volatilities, since all active clusters decide randomly and without memory whether to buy or to sell, and since the occupied sites are distributed randomly. (iii) At the percolation threshold, a simple asymptotic power-law holds for small activities (short times) and becomes more Gaussian for large a (long times).

This latter crossover to Gaussians, seen also in some analyses of real markets [14.20, 14.21, 14.61], is not seen if we replace the percolation model by a Lévy walk for the price changes [14.62], where the return is a sum of steps distributed with the same power-law exponent τ as the above percolation clusters. In this simplification, the power-law remains valid also for large a without a crossover to Gaussians. Note that in percolation, as opposed to Lévy walks, the clusters are correlated by the sum rule $\sum_s n_s s = pN$. See [14.63] for a theory.

14.3.2 Improvements of the Percolation Model

Clustering by Diffusion: Another advantage of this percolation model compared with Lévy walks is volatility clustering. While $\langle R(t)R(t+1) \rangle$ in real markets decays rapidly to zero (but see [14.4] for different information), the autocorrelations of the absolute returns [14.64] $\langle |R(t)R(t+1)| \rangle$ decay slowly as shown in panel (b′) of Fig. 14.1 (see also e.g. Fig. 14.2 of [14.65]: a turbulent day on the stock market is often followed by another turbulent day, though the sign of change for the next day is less predictable. We simulate [14.66] this volatility clustering by letting the investors diffuse slowly on the lattice; thus in the above picture, a small fraction of the investors move to another neighborhood of the city where they get different advice from a different expert. Now the autocorrelation functions decay smoothly, with unexplained size effects [14.66].

Feedback from the Last Price: So far the model assumes the investors or their advisors to be complete idiots: they decide randomly whether to buy or to sell, without regard to any economic facts. Such an assumption is acceptable for the author from Cologne since the local stock market is in Dusseldorf, not Cologne. However, the discussions of log-periodic oscillations earlier in this review made clear that not everything should be regarded as random. The simplest way to include some economic reason is the assumption that prudent investors prefer to sell if the price is high and to buy if it is low. Thus the probabilities to buy or to sell are no longer a but are changed by an amount proportional to the difference between the actual price and the initial price; the latter one is regarded as the fundamental or just price. Surprisingly, simulations [14.67–14.71] show that the distribution $\pi(R)$ is barely changed; as expected the price itself is now stabilized to values close to the fundamental price. Little changes if we allow the fundamental price to undergo Gaussian fluctuations as in [14.64].

The distribution of the wealths of the investors can be investigated only if one gives each investor a finite initial capital, and adds to it the profits and subtracts the losses made by the random decisions to buy and sell. Bankrupt investors are removed from the market. Simulations [14.72] give reasonable return distributions, but in disagreement with reality [14.73] no clear power-laws with universal exponents.

How to get the Correct Empirical Exponent $\alpha \approx 3$? The above power-law $\pi \propto 1/R^\tau$ with $2 < \tau \leq 2.5$ may have been sufficient some time ago [14.57–14.59, 14.74] for which Lévy stable distributions requiring $\tau = \alpha + 1 < 3$ could be qualified, but today's more accurate statistics shows fat tails decaying faster with $\tau > 3$, though slower than a Gaussian. There may be such power-laws multiplied with an exponential function, also called truncated Lévy distributions [14.15, 14.16, 14.18], or stretched exponentials [14.23], or most likely power-laws with an exponent near $\alpha = \tau - 1 = 3$ [14.19–14.21].

Several ways were invented to correct this exponent and get $\alpha \simeq 3$. One may work with p slightly above p_c, where an effective power-law with $\alpha = 3$ can be seen over many orders of magnitude [14.66]. (In this case, as is traditional for percolation studies, one omits the contribution from the infinite cluster.) Or one integrates over all p between zero and the percolation threshold, thus avoiding the question how investors work at $p = p_c$ without ever having read a percolation book [14.53–14.55]; now $\alpha = \tau - 1 + \sigma \simeq 1.5$ to 2 [14.75]. Much better agreement with the desired $\alpha \simeq 3$ is obtained if we follow Zhang [14.4] and take the price change R not linear in the difference between supply and demand, as assumed above, but proportional to the square root of this difference. Then $\alpha = 2(\tau + \sigma - 1)$ is about 2.9 in two dimensions, just as desired. Numerically [14.75], this power-law could be observed over five orders of magnitude, similar to reality [14.20, 14.21]. Changing the activity a proportionally to the last known price change breaks the up-down symmetry for price movements; now sharp peaks in the price, with high activity, are followed by calmer periods with low prices and low activity [14.76].

Figure 14.2 shows price change vs. time, both in arbitrary units, for 0.001 as the lower limit for the activity in the model of [14.76]. Clearly, we see sharp peaks but not equally sharp holes (the downward trend also indicates the survival probability of the first author if the Nikkei index fails to obey the prediction of Fig. 14.6 below). Figure 14.3 shows the desired slow decay of the autocorrelation function for the volatility of this market model, and for the same simulation Fig. 14.4 gives the histogram of price changes.

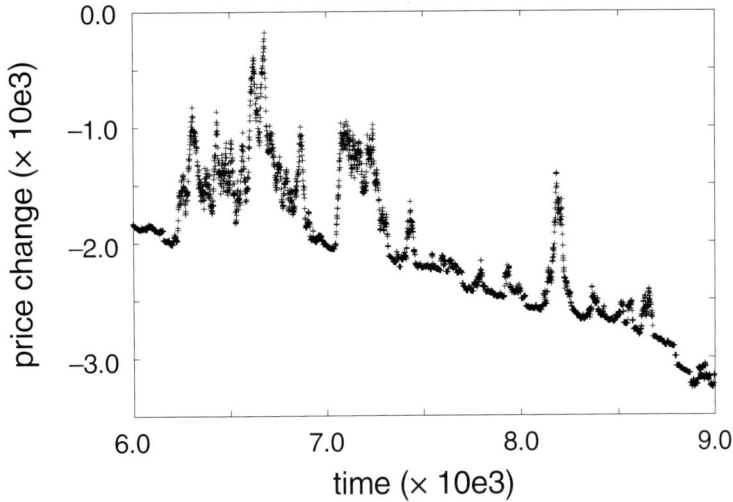

Fig. 14.2. Single run of price variations vs. time over 3 000 iterations (in arbitrary units), where the activity is between 0.001 and 0.5 at percolation threshold. Such a simulation takes less than a minute on a workstation. The units for the price change and the time are arbitrary [14.76]. The choice of parameters exaggerates on purpose the asymmetry between flat valleys and sharp peaks

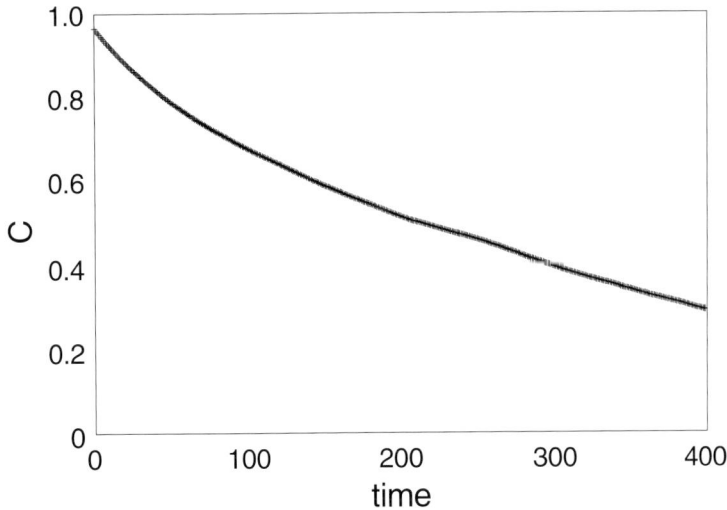

Fig. 14.3. Autocorrelations for the volatility for $0 < p < p_c$ in a square lattice of size 201 by 201 averaged over 4 800 simulations similar to Fig. 14.2, requiring 10^2 hr simulation time on a Cray-T3E

In the opposite direction, Focardi et al. [14.77] assume the price change to be *quadratic* in the difference between supply and demand when the market gets into a crash. Using also other modifications of the infinite-range Cont-Bouchaud model, their simulations show exponentially growing prices followed, at irregular time intervals, by rapid crashes.

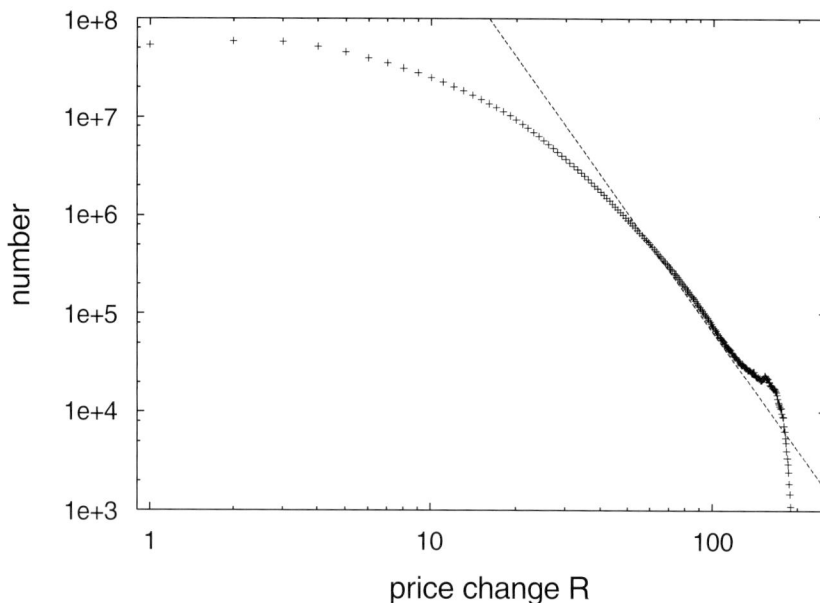

Fig. 14.4. Histogram for positive price changes on double-logarithmic scales. The straight line has slope $-(1+\alpha) = -4$ [14.76] and is shown for comparison. Negative price changes show the same behavior

Log-Periodicity and Finite-Size Effects: None of these models has the ingredients which seem needed for log-periodic oscillations before or after crashes, see Sect. 14.4.2. Percolation can give such oscillations if we let particles diffuse on the occupied sites of the infinite cluster for $p > p_c$, and if there is one preferred and fixed direction for this diffusion ('bias') [14.78]. Now the r.m.s. displacement of the diffusors varies approximately as a power t^k of time, and the effective exponent $k(t)$ approaches unity with oscillations $\propto \sin(\lambda \ln(t))$. However, here the percolation clusters remain fixed while some additional probing particle diffuses through the disordered medium; in the above algorithm to produce volatility clustering, the investors themselves diffuse and there is no additional probing particle. Only in a very different way did Pandey and Stauffer [14.68–14.71] such oscillations (approximately) in the Cont-Bouchaud percolation model for markets.

The Cont-Bouchaud percolation model is particularly suited to look at effects of finite lattice sizes, since size effects at such critical points have been studied for decades. In most of the other microscopic models [14.79], the 'thermodynamic limit' $N \to \infty$ means that the fluctuations die out or become nearly periodic. Real markets, according to these models, are dominated by the 10^2 most important players and not by millions of small investors. Also for the present Cont-Bouchaud model, the behavior becomes unrealistic (Gaussian $\pi(R)$) in this limit if $p < p_c$. Right at $p = p_c$, however, the lattice is no longer self-averaging, and the simulated return distributions keep the same shape for $N = 10^3$ to 10^6.

Of course, the extreme tails are always dominated by size effects: no investor can own more than 100% of the market, and no cluster can contain more than the $N = L^d$ lattice sites of the model. However, this trivial limit is relevant mainly above p_c; at the percolation threshold, the largest cluster is a fractal and contains on average $\propto L^D$ sites, where the fractal dimension $D = d/(\tau - 1)$ is smaller than d. Investigations of the distribution of sizes for the largest critical cluster have only begun [14.80–14.82]. Further details are reviewed in [14.84, 14.85].

14.4 Critical Crashes

14.4.1 Multiplicative Cascades on the Stock Market

The analogy between finance and hydrodynamic turbulence developed by Ghashghaie et al. [14.22] implicitly assumes that price fluctuations can be described by a *multiplicative cascade* along which the return r at a given time scale $a < T$, is given by:

$$r_a(t) \equiv \ln P(t + a) - \ln P(t) = \sigma_a(t)u(t) , \qquad (14.21)$$

where $u(t)$ is some scale independent random variable, T is some coarse 'integral' time scale and $\sigma_a(t)$ is a positive quantity that can be multiplicatively decomposed, for any decreasing sequence of scales $\{a_i\}_{i=0,...,n}$ with $a_0 = T$ and $a_n = a$, as [14.22]

$$\sigma_a = \prod_{i=0}^{n-1} W_{a_{i+1},a_i} \sigma_T . \qquad (14.22)$$

Equation (14.21) together with (14.22) writes that the logarithm of the price is a multiplicative process. But, this is different from a standard multiplicative processes due to the tree-like structure of the correlations that are added by the hierarchical construction of the multiplicands. We use $w_a(t) \equiv \ln \sigma_a(t)$ as a natural variable.

If one supposes that W_{a_{i+1},a_i} depends only on the scale ratio a_{i+1}/a_i and are i.i.d. variables with log-normal distribution of mean $-H \ln 2$ and variance $\lambda^2 \ln 2$, one can show [14.5, 14.6] that the correlation function of the volatility field $w_a(t)$ averaged over a period of length T is given by

$$C_a^\omega(\Delta t) = \lambda^2 \left(\log_2 \frac{T}{\Delta t} - 2 + 2\frac{\Delta t}{T} \right) + \lambda_T^2 , \qquad (14.23)$$

for $a \leq \Delta t \leq T$ (λ_T^2 is the variance of w_T). For $\lambda^2 \simeq 0.015$ that can be obtained independently from the fit of the PDFs, (14.23) provides a very good fit of the data (Fig. 14.1b') for the slow decay of the correlation coefficient with only one adjustable parameter $T \simeq 3$ months. Let us note that $C_a^\omega(\Delta t)$ of Fig. 14.1 can be equally well fitted by a power-law $\Delta t^{-\alpha}$ with $\alpha \approx 0.2$. In view of the small value of α, this is undistinguishable from a logarithmic decay. Moreover, (14.23) predicts that the correlation function $C_a^\omega(\Delta t)$ should not depend of the scale a provided $\Delta t > a$ in agreement with data [14.5, 14.6].

Another very informative quantity is the cross-correlation function of the volatility measured at different time scales:

$$C_{a_1,a_2}^\omega(\Delta t) \equiv \mathrm{var}(\omega_{a_1})^{-1}\mathrm{var}(\omega_{a_2})^{-1}\overline{\tilde{\omega}_{a_1}(t)\tilde{\omega}_{a_2}(t+\Delta t)} \ . \qquad (14.24)$$

It is found that $C_{a_1,a_2}^\omega(\Delta t) > C_{a_1,a_2}^\omega(-\Delta t)$ if $a_1 > a_2$ and $\Delta t > 0$. From the near-Gaussian properties of $\omega_a(t)$, the mean mutual information of the variables $\omega_a(t+\Delta t)$ and $\omega_{a+\Delta a}(t)$ reads:

$$I_a(\Delta t, \Delta a) = -0.5\log_2\left(1 - (C_{a,a+\Delta a}^\omega(\Delta t))^2\right) \ . \qquad (14.25)$$

Since the process is causal, this quantity can be interpreted as the information contained in $\omega_{a+\Delta a}(t)$ that propagates to $\omega_a(t+\Delta t)$. The remarkable observation [14.5, 14.6] is the appearance of a nonsymmetric propagation cone of information showing that the volatility at large scales influences in the future the volatility at shorter scales. This clearly demonstrates the pertinence of the notion of a cascade in market dynamics.

14.4.2 Log-Periodicity for 'Foreshocks'

A hierarchical cascade process as just described implies the existence of a discrete scale invariance if the branching ratio and scale factor along the tree are not fluctuating too much [14.83]. This possibility is actually born out by the data under the frame of log-periodic oscillations.

As alluded to in the section on percolation models, log-periodicity refers is this context to the accelerating oscillations that have been documented in stock market prices prior and also sometimes following major crashes. The formula typifying this behavior is the time to failure equation

$$I(t) = p_c + B(t_c - t)^m\left[1 + C\cos\left(2\pi\frac{\log(t_c - t)}{\log \lambda} + \Psi\right)\right] \ , \qquad (14.26)$$

where I is the price when the crash is a correction for a bubble developing above some fundamental value (it is the logarithm of the price if the crash drop is proportional to the total price), t_c is the critical time at which the crash is the most probable, m is a critical exponent, and Ψ is a phase in the cosine that can be get rid of by a change of time units. λ is the preferred scale factor of the accelerating oscillations giving the ratio between the successive shrinking periods. This expression (14.26) reflects a discrete scale invariance

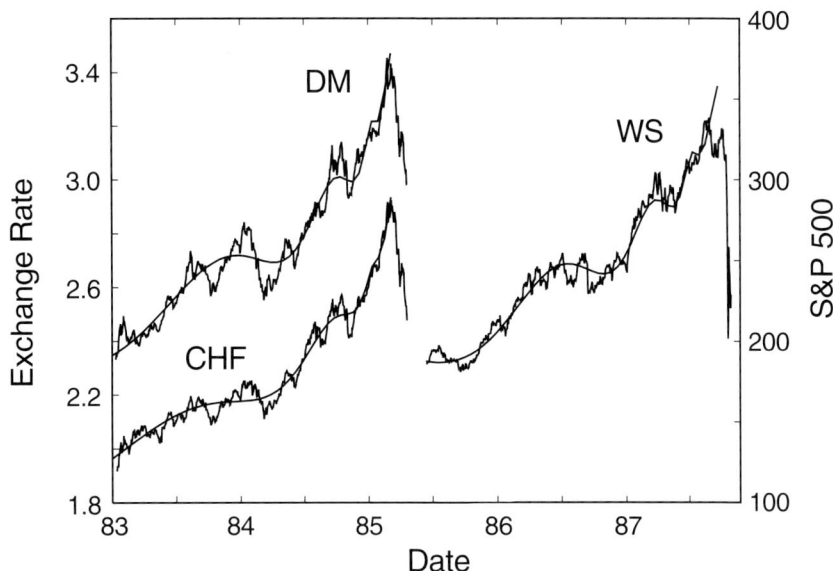

Fig. 14.5. The S&P 500 US index prior to the October 1987 crash on Wall Street and the US$ against German mark (DEM) and Swiss franc (CHF) prior to the collapse mid-85. The fit to the S&P 500 is (14.26) with $p_c \approx 412$, $B \approx -165$, $BC \approx 12.2$, $m \approx 0.33$, $t_c \approx 87.74$, $\Psi \approx 2.0$, $\lambda \approx 2.3$. The fits to the DEM and CHF currencies against the US dollar give $p_c \approx 3.88$, $B \approx -1.2$, $BC \approx 0.08$, $m \approx 0.28$, $t_c \approx 85.20$, $\Psi \approx -1.2$, $\lambda \approx 2.8$ and $p_c \approx 3.1$, $B \approx -0.86$, $BC \approx 0.05$, $m \approx 0.36$, $t_c \approx 85.19$, $\Psi \approx -0.59$, $\lambda \approx 3.3$, respectively. Reproduced from [14.86]

of the price around the critical time, i.e. the price exhibits self-similarity only under magnifications around t_c that are integer powers of λ. Figure 14.5 shows three cases illustrating the behavior of market prices prior to large crashes [14.86].

Since our initial proposition of the existence of log-periodicity preceding stock market crashes [14.87–14.89], several works have extended the empirical investigation [14.90–14.96].

A recent compilation of many crashes [14.86, 14.97–14.99] provides increasing evidence of the relevance of log-periodicity and of the application of the concept of criticality to financial crashes. The events that have been found to qualify comprise

- the Oct. 1929, the Oct. 1987, the Hong Kong Oct. 1987, and the Aug. 1998 crashes, which are global market events,
- the 1985 foreign exchange event on the US dollar,
- the correction of the US dollar against the Canadian dollar and the Japanese Yen starting in Aug. 1998,
- the bubble on the Russian market and its ensuing collapse in 1997–1998,
- twentytwo significant bubbles followed by large crashes or by severe corrections in the Argentinian, Brazilian, Chilean, Mexican, Peruvian, Venezuelan, Hong Kong, Indonesian, Korean, Malaysian, Philippine, and Thai stock markets [14.99].

In all these cases, it has been found that log-periodic power-laws adequately describe speculative bubbles on the western as well as on the emerging markets with very few exceptions.

The underlying mechanism which has been proposed [14.86, 14.97, 14.98] is that a bubble develops by a slow build-up of long-range time correlations reflecting those between traders leading eventually to a collapse of the stock market in one critical instant. This build-up manifests itself as an over-all power-law acceleration in the price decorated by 'log-periodic' precursors. This mechanism can be analyzed in an expectation model of bubbles and crashes which is essentially controlled by a crash hazard rate becoming critical due to a collective imitative/herding behavior of traders (cf. Chap. 11) [14.86, 14.97, 14.98]. A remarkable *universality* is found for all events, with approximately the same value of the fundamental scaling ratio λ characterizing the log-periodic signatures.

To test for the statistical significance of these analyses, extensive statistical tests have been performed [14.98, 14.100] to show that the reported 'log-periodic' structures essentially never occurred in $\approx 10^5$ yr of synthetic trading following a 'classical' time series model, the GARCH(1,1) model with Student-t statistics (which has a power-law tail with exponent $\alpha = 4$), often used as a benchmark in academic circles as well as by practitioners. Thus, the null hypothesis that log-periodicity could result simply from random fluctuations is strongly rejected.

14.4.3 Log-Periodicity for 'Aftershocks'

Log-periodic oscillations decorating an overall acceleration of the market have their symmetric counterparts after crashes. It has been found [14.101, 14.102] that imitation between traders and their herding behavior not only lead to speculative bubbles with accelerating over-valuations of financial markets possibly followed by crashes, but also to 'antibubbles' with decelerating market devaluations following all-time highs. The mechanism underlying this scenario assumes that the demand decreases slowly with barriers that progressively quench in, leading to a power-law decay of the market price decorated by decelerating log-periodic oscillations. This mechanism is actually very similar to that operating in the random walk of a Brownian particle diffusing in a random lattice above percolation in a biased field [14.78].

The strongest signal has been found on the Japanese Nikkei stock index from 1990 to present and on the Gold future prices after 1980, both after their all-time highs. Figure 14.6 shows the Nikkei index representing the Japanese market from the beginning of 1990 to present. The data from 1 Jan. 1990 to 31 Dec. 1998 has been fitted (the ticked line) by an extension of (14.26) using the next order terms in the expansion of a renormalization group equation [14.101, 14.102]. This fit has been used to issue a forecast in early January 1999

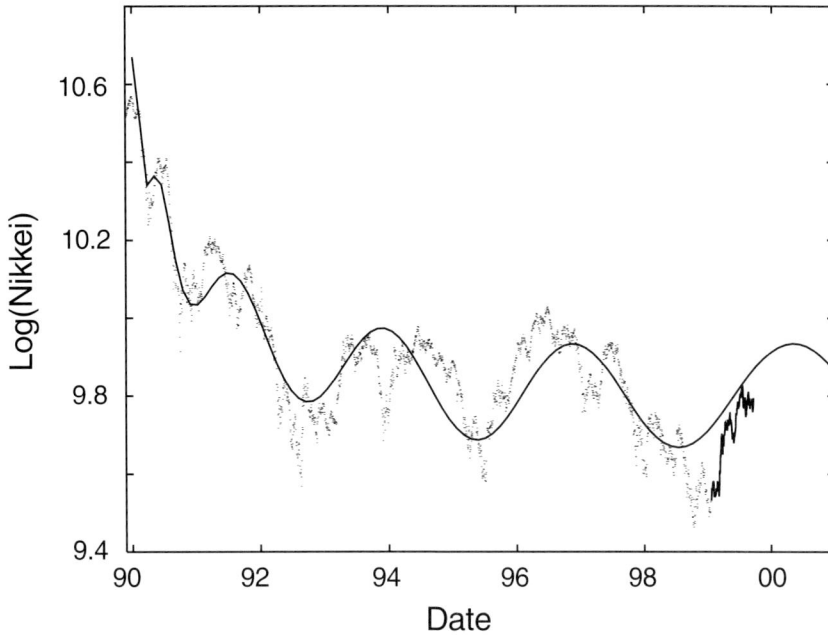

Fig. 14.6. In [14.101, 14.102], the Nikkei was fitted from 1 Jan. 1990 to 31 Dec. 1998 with an extended log-periodic formula and its extrapolation predicted that the Japanese stock market should increase as the year 2000 was approached. In this figure, the value of the Nikkei is represented as the solid line after the last point used in the analysis (31 Dec. 1998) until 21 Sep. 1999 and can be compared with our prediction (the ticked line). The dots after Dec. 1989 until 31 Dec. 1998 represent the data used in the prediction. This figure is as in [14.101, 14.102] except for the solid line starting 3rd Jan. 1999 which represents the realized Nikkei prices since the prediction was issued. For further arguments and later times see [14.103]

for the recovery of the Nikkei in 1999 [14.101, 14.102]. The forecast, performed at a time when the Nikkei was at its lowest, has correctly captured the change of regime and the overall upward trend since the beginning of this year. This prediction has first been released in January 1999 on the Los Alamos server at http://xxx.lanl.gov/abs/cond-mat/9901268. The detailed publication for IJMPC [14.101, 14.102] was mentioned already with its prediction in a wide-circulation journal which appeared in May 1999 [14.104]. One of the authors would not survive a Nikkei drop since another author relied on the Nikkei prediction and invested in Japan.

A set of secondary western stock market indices (London, Sydney, Auckland, Paris, Madrid, Milan, Zurich) as well as the Hong Kong stock market have also been shown to exhibit well-correlated log-periodic power-law antibubbles over a period 6–15 months triggered by a rash of crises on emerging markets in the early 1994 [14.99]. As the US market declined by no more than 10% during the beginning of that period and quickly recovered, this suggests that these smaller stock western markets can 'phase lock' (in a weak sense) not only because of the over-arching influence of Wall Street but also independently of the current trends on Wall Street due to other influences.

14.5 Conclusion

This review has attempted to present results that may advance our understanding of the working of stock markets. First, we proved that the demand and supply should be deviating from the balanced point in general cases when there are fluctuations in demand. In the case of an open market in which prices can change instantly following the unbalance of demand and supply, the speculative actions of dealers can be modeled numerically by models with threshold dynamics. The resulting market price fluctuations are characterized by a fat tail distribution when the dealers' response to latest price change is positive. We have also shown that a simple Langevin equation with multiplicative noise accounts for the threshold-type dynamics of traders and rationalize the 'fat tail' nature of distribution of returns. By solving the set of macroscopic market price equations, we have shown that there are three types in price changes: (1) stationary fluctuations, (2) bubble behavior, and (3) oscillatory phase. We believe that price fluctuations in open markets should be better understood by considering such a dynamical effect that has been neglected in the ordinary approach of financial technology. We have also presented models inspired by percolation that are probably the simplest microscopic models capturing the effect of imitation/clustering of traders in groups of various sizes. Improvements of the model provide reasonable agreement with the empirical value of the exponent α of the distribution of price variations. Clustering of volatility can also emerge rather naturally by the feedback effect of the price on the activity of the traders. The initial main weakness of the model, namely the fact that the connectivity had to be tuned to its critical value, has also been cured by allowing it to become a dynamical variable. The review ends up by summarizing the evidence for critical behavior associated with the formation of speculative bubbles in large stock markets and their associated log-periodicity, corresponding to accelerated oscillations up to the time of crashes. Whether this will allow us to prevent future crashes remains to be seen. Further microscopic models are reviewed by Samanidou et al. [14.85].

The overall picture that emerges is quite interesting: a mixture of more or less stationary self-similar statistical time series, may be self-organized critical, with cascades of correlations across time scales and once in a while a (truncated) divergence reflecting probably the crowd effect between traders culminating in a critical point with rather specific log-periodic signatures. According to the different models that we have presented, a crash has probably an endogenous origin and is constructed progressively by the market as a whole. In this sense, this could be termed a systemic instability. Further study might clarify what could be the regulations and information that should be released to stabilize the market and prevent these systemic instabilities.

Acknowledgements. We thank R.B. Pandey for a critical reading of the manuscript. DS (fat, old, drunk, overcited) thanks Naeem Jan for hospitality at St. Francis Xavier University where his part of the review was written up. DS (thin, young, sober, undercited) is grateful to A. Johansen for preparing Figs. 14.5 and 14.6 and for a very stimulating and enjoyable collaboration over many years.

References

14.1 R.G. Palmer, W.B. Arthur, J.H. Holland, B. LeBaron, and P. Tayler, Physica D **75**, 264 (1994).
14.2 G. Caldarelli, M. Marsili, and Y.-C. Zhang, Europhys. Lett. **40**, 479 (1997).
14.3 J.D. Farmer, Market force, ecology and evolution (2000); [available from `xxx.lanl.gov/abs/adap-org/9812005`].
14.4 Y.-C. Zhang, Physica A **269**, 30 (1999).
14.5 A. Arneodo, J.-F. Muzy, and D. Sornette, Eur. Phys. J. B **2**, 277 (1998).
14.6 J.-F. Muzy, D. Sornette, J. Delour, and A. Arneodo, Quantitative Finance **1**, 131 (2001).
14.7 L. Bachelier, *Theory of Speculation* (Gauthier-Villars, Paris, 1900).
14.8 P.A. Samuelson, Ind. Manage. Rev. **6**, 41 (1965).
14.9 C.W.J. Granger, *Forecasting in Business and Economics* (Academic, Boston, 1989).
14.10 C.W.J. Granger and T. Terasvirta, *Modelling Nonlinear Economic Relationships* (Oxford University Press, Oxford, 1993).
14.11 J.M. Poterba, M. James, L.M. Summers, and H. Lawrence, Am. Econ. Rev. **76**, 1142 (1986).
14.12 B. Arshanapalli and W. Nelson, J. Bus. Econ. Stud. **3**, 43 (1997).
14.13 Z. Ding and C.W.J. Granger, J. Economet. **73**, 185 (1996).
14.14 G. Koutmos, U. Lee, and P. Theodossiou, J. Econ. Bus. **46**, 101 (1994).
14.15 R.N. Mantegna and H.E. Stanley, Nature **376**, 46 (1995).
14.16 R.N. Mantegna and H.E. Stanley, *An Introduction to Econophysics* (Cambridge University Press, New York, 2000).
14.17 A. Arneodo, J.-P. Bouchaud, R. Cont, J.-F. Muzy, M. Potters, and D. Sornette, Comment on Turbulent cascades in foreign exchange markets (1996) (Comment on [14.22]; [available from `xxx.lanl.gov/abs/cond-mat/9607120`]).
14.18 J.P. Bouchaud and M. Potters, *Theory of Financial Risks* (Cambridge University Press, Cambridge, 2000).
14.19 T. Lux, Appl. Financ. Econ. **6**, 463 (1996) [and economic literature cited therein].
14.20 P. Gopikrishnan, M. Meyer, L.A. Nunes Amaral, and H.E. Stanley, Eur. Phys. J. B **3**, 139 (1998)
14.21 P. Gopikrishnan, M. Meyer, L.A. Nunes Amaral, and H.E. Stanley, Phys. Rev. E **60**, 5305 (1999).
14.22 S. Ghashghaie, W. Breymann, J. Peinke, P. Talkner, and Y. Dodge, Nature **381**, 767 (1996).
14.23 J. Laherrère and D. Sornette, Eur. Phys. J. B **2**, 525 (1998).
14.24 O.M. Joy, J. Portfolio Manage. **49**, 12 (1986).
14.25 N. Jagadeesh, J. Finance **45**, 881 (1990).
14.26 B.N. Lehmann, Q. J. Econ. **105**(1), 1 (1990).
14.27 W. Brock, J. Lakonishok, and B. LeBaron, J. Finance **47**, 1731 (1992).
14.28 J.A. Murphy, J. Futures Markets, Summer, 175 (1986).
14.29 R.J. Bauer and J.R. Dahlquist, *Technical Market Indicators, Analysis and Performance* (Wiley, New York, 1999).
14.30 H. Takayasu, H. Miura, T. Hirabayashi, and K. Hamada, Physica A **184**, 127 (1992).
14.31 Y. Huang, H. Saleur, C.G. Sammis, and D. Sornette, Europhys. Lett. **41**, 43 (1998).
14.32 T. Hirabayashi, H. Takayasu, H. Miura, and K. Hamada, Fractals **1**, 29 (1993).
14.33 D.G. Champenowne, Econ. J. **63**, 318 (1953).

14.34 H. Kesten, Acta Math. **131**, 207 (1973).

14.35 L. de Haan, S.I. Resnick, H. Rootzén, and C.G. de Vries, Stoch. Proc. Appl. **32**, 213 (1989).

14.36 C. de Calan, J.-M. Luck, T.M. Nieuwenhuizen, and D. Petritis, J. Phys. A **18**, 501 (1985).

14.37 S. Solomon and M. Levy, Int. J. Mod. Phys. C **7**, 745 (1996).

14.38 O. Biham, O. Malcai, M. Levy, and S. Solomon, Phys. Rev. E **58**, 1352 (1998).

14.39 O. Malcai, O. Biham, and S. Solomon, Phys. Rev. E **60**, 1299 (1999).

14.40 D. Sornette and R. Cont, J. Phys. I France **7**, 431 (1997).

14.41 D. Sornette, Physica A **250**, 295 (1998).

14.42 D. Sornette, Phys. Rev. E **57**, 4811 (1998).

14.43 H. Takayasu, A.-H. Sato, and M. Takayasu, Phys. Rev. Lett. C **79**, 966 (1997).

14.44 A.-H. Sato and H. Takayasu, Physica A **250**, 231 (1998).

14.45 H. Takayasu and M. Takayasu, Physica A **269**, 24 (1999).

14.46 O.J. Blanchard, Econ. Lett. **3**, 387 (1979).

14.47 J.-P. Bouchaud and R. Cont, Eur. Phys. J. B **6**, 543 (1998).

14.48 M. Levy, H. Levy, and S. Solomon, Econ. Lett. **94**, 103 (1994).

14.49 M. Levy, H. Levy, and S. Solomon, J. Physi. I France **5**, 1087 (1995).

14.50 M. Levy, H. Levy, and S. Solomon, *Microscopic Simulation of Financial Markets* (Academic, New York, 2000).

14.51 R. Cont and J.P. Bouchaud, Macroecon. Dyn. **4**, 170 (2000).

14.52 G. Iori, Int. J. Mod. Phys. C **10**, 1149 (1999).

14.53 D. Stauffer and A. Aharony, *Introduction to Percolation Theory* (Taylor and Francis, London, 1994).

14.54 A. Bunde and S. Havlin, *Fractals and Disordered Systems* (Springer, Berlin, 1996).

14.55 M. Sahimi, *Applications of Percolation Theory* (Taylor and Francis, London, 1994).

14.56 P. Flory, J. Am. Chem. Soc. **63**, 3083, 3091, and 3096 (1941).

14.57 B.B. Mandelbrot, J. Bus. **36**, 349 (1963).

14.58 B.B. Mandelbrot, J. Bus. **39**, 242 (1966).

14.59 B.B. Mandelbrot, J. Bus. **30**, 393 (1967).

14.60 D. Stauffer and T.J.P. Penna, Physica A **256**, 284 (1998).

14.61 L. Kullmann, J. Töyli, J. Kertész, A. Kanto, and K. Kaski, Physica A **269**, 98 (1999).

14.62 D. Chowdhury and D. Stauffer, Eur. Phys. J. B **8**, 477 (1999).

14.63 L. Kullmann and K. Kertész, Int. J. Mod. Phys. C **12**(8), 1211 (2001).

14.64 T. Lux and M. Marchesi, Nature **297**, 498 (1999).

14.65 H.E. Stanley et al., Physica A **269**, 156 (1999).

14.66 D. Stauffer, P.M.C. de Oliveira, and A.T. Bernardes, Int. J. Theor. Appl. Finance **2**, 83 (1999).

14.67 I. Chang and D. Stauffer, Physica A **264**, 294 (1999).

14.68 R.B. Pandey and D. Stauffer, Int. J. Theor. Appl. Finance **3**, 479 (2000).

14.69 I. Chang, D. Stauffer, and R.B. Pandey, Asymmetrics, correlations and fat tails in percolation market model (2001); [available from `xxx.lanl.gov/abs/cond-mat/0108345`].

14.70 D. Sornette and K. Ide, Theory of self-similar oscillatory finite-time singularities in finance, population and rupture (2001); [available from `xxx.lanl.gov/abs/cond-mat/0106054`].

14.71 A. Proykova, L. Russenova, and D. Stauffer, Nucleation of market stocks in Sornette-Ide model (2001); [available from `xxx.lanl.gov/abs/cond-mat/0110124`].

14.72 J. Liebreich, Int. J. Mod. Phys. C **10**, 1317 (1999).

14.73 M. Levy and S. Solomon, Physica A **242**, 90 (1997).

14.74 R.N. Mantegna, Physica A **179**, 232 (1991).

14.75 D. Stauffer and D. Sornette, Physica A **271**, 496 (1999).

14.76 D. Stauffer and N. Jan, Physica A **277**, 215 (2000).

14.77 S. Focardi, S. Cincotti, and M. Marchesi, WEHIA e-print (1999); [available from `dibe.unige.it/wehia`].

14.78 D. Stauffer and D. Sornette, Physica A **252**, 271 (1998).

14.79 D. Stauffer, WEHIA e-print (1999); [available from `dibe.unige.it/wehia`].

14.80 P. Sen, Int. J. Mod. Phys. C **10**, 747 (1999).

14.81 M.I. Zeifman and D. Ingman, J. Appl. Phys. **88**, 76 (2000).

14.82 D. Stauffer, Int. J. Mod. Phys. C **11**, 519 (2000).

14.83 D. Sornette, Phys. Rep. **297**, 239 (1998).

14.84 D. Stauffer, Adv. Complex Syste. **4**, 19 (2001).

14.85 E. Samanidou, E. Zschischang, D. Stauffer, and T. Lux, in *Microscopic Models for Economic Dynamics*, edited by F. Schweitzer (Springer, Berlin, 2002).

14.86 A. Johansen and D. Sornette, Risk **12**, 91 (1999).

14.87 D. Sornette, A. Johansen, and J.-P. Bouchaud, J. Phys. I France **6**, 167 (1996).

14.88 D. Sornette and A. Johansen, Physica A **245**, 411 (1997).

14.89 D. Sornette and A. Johansen, Physica A **261**, 581 (1998).

14.90 J.A. Feigenbaum and P.G.O. Freund, Int. J. Mod. Phys. B **10**, 3737 (1996).

14.91 J.A. Feigenbaum and P.G.O. Freund, Int. J. Mod. Phys. B **12**, 57 (1998).

14.92 S. Gluzman and V.I. Yukalov, Mod. Phys. Lett. B **12**, 75 (1998).

14.93 N. Vandewalle, Ph. Boveroux, A. Minguet, and M. Ausloos, Physica A **255**, 201 (1998).

14.94 N. Vandewalle, M. Ausloos, Ph. Boveroux, and A. Minguet, Eur. Phys. J. B **4**, 139 (1998).

14.95 A. Johansen and D. Sornette, Eur. Phys. J. B **9**, 167 (1999).

14.96 S. Drozdz, F. Ruf, J. Speth, and M. Wojcik, Eur. Phys. J. B **10**, 589 (1999).

14.97 A. Johansen, D. Sornette, and O. Ledoit, J. Risk **1**, 5 (1999).

14.98 A. Johansen, O. Ledoit, and D. Sornette, Int. J. Theor. Appl. Finance **3**, 219 (2000).

14.99 A. Johansen and D. Sornette, Log-periodic power-law bubbles in Latin-American and Asian markets and correlated anti-bubbles in Western stock markets: an empirical study; [available from `xxx.lanl.gov/abs/cond-mat/990728`].

14.100 A. Johansen, Discrete scale invariance and other cooperative phenomena in spatially extended systems with threshold dynamics (Ph.D. Thesis, Niels Bohr Institute, 1997); [available from `www.nbi.dk/~johansen/pub.html`].

14.101 A. Johansen and D. Sornette, Int. J. Mod. Phys. **10**, 563 (1999).

14.102 A. Johansen and D. Sornette, Int. J. Mod. Phys. **11**, 359 (2000).

14.103 D. Sornette and A. Johansen, Quantitative Finance **1**, 452 (2001).

14.104 D. Stauffer, Phys. Bl. **55**, 49 (1999).

Glossary

AEG	adenocarcinoma of the esophago-gastric junction
AFM	atomic force microscopy
AOGCM	atmosphere ocean general circulation model
ARCH	autoregressive conditional heteroscedasticy
ASA	autocorrelation spectrum analysis
BDS	Brock Dechert Scheinkman test
CFC	chlorofluorocarbons
CHF	congestive heart failure
CSIRO	Australia's Commonwealth Scientific and Industrial Research Organization
CSIRO-Mk2	CSIRO Atmospheric Research Mark 2 Climate Model
CT	computer tomography
CTx	chemotherapy
DAX	Deutscher Aktienindex
DFA	detrended fluctuation analysis
DNA	deoxyribonucleic acid
DOF	degree of freedom
ECG	electrocardiogram
ECHAM4	European Community Hamburg Model #4
ECMWF	European Center for Medium-Range Weather Forecasting
EEG	electroencephalogram
EMBL	European Molecular Biology Laboratory
EMC	effective measure of complexity
EMG	electromyogram
EMH	efficient market hypothesis
EMS	European Monetary System
ENS	endosonography
ENSO	El Niño/Southern Oscillation
EOF	empirical orthogonal function

EOG	electrooculogram
fBm	fractional Brownian motion
FIGARCH	fractionally integrated GARCH
FX	foreign exchange
GA	genetic algorithm
GARCH	generalized autoregressive conditional heteroscedasticy
GARP	Global Atmospheric Research Program
GCM	general circulation model
GEV	generalized extreme value
GFDL	Geophysical Fluid Dynamics Laboratory
GFDL-R15-a	GFDL climate model (rhomboidal resolution of 15 waves)
GHG	greenhouse gases
GP	genetic programming
GPD	generalized Pareto distribution
HADCM3	Hadley Center Climate Model #3
HARCH	heterogeneous ARCH
HIS	histomorphology
HRV	heart rate variability
i.i.d.	independent and identically distributed
IGBP	International Geosphere Biosphere Program
IPCC	International Panel of Climate Change
ISCCP	International Satellite Cloud Climatology Project
MESA	maximum entropy spectral analysis
MRM	multiple regression model
MSL	mean sea level
NADW	North Atlantic deep water formation
NAO	North Atlantic Oscillation
NCAR	National Center for Atmospheric Research
NMR	nuclear magnetic resonance
NNM	neural network model
NYCI	New York Stock Exchange Composite Index
OPYC3	Ocean Isopycnal Model #3
PC	principal components
PDF	probability density function
PLC	power-law correlation
PNA	Pacific/North America teleconnection pattern
QBO	quasi-biennial oscillation
REM	rapid eye movement
r.m.s.	root mean square
RNA	ribonucleic acid

ROC	receiver operating characteristic
RR	blood pressure (Riva-Rocci)
S&P 500	Standard and Poor's 500 Index
Sc	Stratocumulus
SF	structure function
SIM	scaling-index method
SOI	southern oscillation index
SPO	stable periodic orbit
SSA	singular spectrum analysis
SU	sulfate aerosol
TAQ	Trade and Quote
THC	thermohaline circulation
UPO	unstable periodic orbit
VF	ventricular fibrillation
WCRP	World Climate Research Program
WT	wavelet transform
WTMM	wavelet transform modulus maxima
WTMMM	WTMM maxima

Subject Index

Entries in this index are generally sorted with page number as they appear in the text. Page numbers that are marked in **bold face** indicate that the entry appears in a title or subheading. Page numbers that are marked in *italics* indicate that the entry appears in the caption of a table or figure.

Symbols

$\mathcal{M}$-cascade 44

$\mathcal{W}$-cascade
- 1D log-normal 73
- 2D random 73
- log-normal 45, *80*, 84, 86
- log-normal random . *42*, *43*, 80, *88*, 92
- random 39, 41, 46, **85**, 86

$\mathcal{W}$-cascade
- log-normal random *70*, *93*
- log-normal random 2D *71*, *72*

1/f power spectrum 225

2D WTMM 69, **75**, 76, 77, *80*

2D wavelet transform modulus maxima (2D WTMM) **65**

A

ABCD rule 289, 295, 296, 299

absolute returns **387**, 404

adenine 47

adenocarcinoma of the esophagogastric junction (AEG) 303

advective feedback 205

AEG 303

AFM **290**

AFM image 290, **291**, *292*

aftershock **430**

albedo 77

algorithm
- bootstrap 387
- box-counting 30
- genetic (GA) 399
- Hoshen Kopelman 422
- Lempel Ziv compression 11

analysis
- autocorrelation 285
- autocorrelation spectrum (ASA) 109, *110*
- correlation **181**, 225
- detection and attribution . *163*, 167
- detrended fluctuation (DFA) ...49, 173, 174, **178**, 222, **227**, 231, 247, 249, 263, **264**, *265*, 274, *389*, 391
- dispersion **262**, **274**
- DNA walk 47
- eigenvector 118
- Eulerian 105
- fluctuation **177**
- Fourier 225, 285

– fractal 47, 277
– image 289
– Lagrangian 105
– maximum entropy spectral (MESA) 109
– maximum entropy spectrum (MESA) *110*
– monofractal **227**
– multifractal 27, 31, **32**, 35, 37, **59**, 67, 76, **181**, **183**, *184*, **240**, 243, 247
– multiple regression 230
– optimal fingerprint 159
– persistence **171**
– power spectrum 46, **109**, 224, 225, 227, 272
– rescaled range (R/S) 178, 262, **263**, 264, 279, **279**, 391
– risk 286
– singular spectrum (SSA) . 109, *110*
– spectral 75, 76, 109, 225, 272, 273, 290
– wavelet . 53, 84, 109, 114, 237, *238*
– Zipf 46
angina pectoris 231
anisotropic dilation 62
anomaly
– climate 114, 134, 145
– sea-ice 151
– temperature *135*
anthropogenic signal 174
antibubble 430, 431
anticorrelation 226, 228, *233*, 234–236
– long-range .**228**, 237, 239, 247, 274
– power-law 225
AOGCM *136*, 146, 148, *155*, 158, *158*, 185
arbitrage 361, **361**, 364
ARCH 14, 89, 93, 365, 390, 420
Archaeglobus fulgidus *54*, 56
arrhythmia 304, 307, *307*
ASA 109
asset limit 417–419

asset price 375
atomic force microscopy (AFM) .289, 290
attractor 311, 312
– chaotic 386
– Lorenz 386
– low-dimensional 386
– strange 30
autocorrelation 175, 387
– long-range 266

B

bacterial genome 56
bakery's strategy 415
baroreceptor reflex 267
BAU 154, *156*, *161*
behavior
– chaotic 418
– herding 332, 344, *345*, 347, 430
– individualistic 344, 348
– mass 335, **344**
– multi-affine 397
– pedestrian 331
– power-law .30, 35, 36, 58, 107, 183, 239, 382, 383
– problem solving **345**
– scaling 266, 274
– selfish 332
Bernoulli experiment 117
bid and ask quote 354
bid-ask spread **416**
bifurcation ..141, 156, 157, 164, 166, 167, 203, 206, *207*, 209, 311, 315, 319, 320, 322–324, 400
bimanual rhythm 19
bimodality 114
bioeconomic system 195, 196
biosphere 144
bistability 205
bistable system 318
blood pressure 225, 267, 304, 306
blue whale hunting 200, *201*

bottleneck ...335, 337, 340, 341, *343*, 347, *348*
brain hypothalamus 311
brainstem 267, 270
bubble 414, 420, 428–430
– rational 376
– speculative 376, 430, 432
business-as-usual (BAU) ...154, *154*, *155*, 185, *193*, 209
buy-and-hold strategy 414

C

candlestick technique 386
Canny's multiscale edge detector . 60
cardiac risk 18
carrying capacity 198
cascade 320
– deterministic 44
– fractal 239
– multifractal 396
– multiplicative 41, **427**
– period doubling 320, 323
– process 85
– random 30, 44
– self-affine 237, 254
– self-similar 43, 237
– volatility 94
central limit theorem .. 360, 367, 374
CFC 153
change
– climate 106, 107, 113, 127, 129, 132, 134, 142, 146, 153, 154, 159, 161, *161*, 165–167, 185, 206, 208, 211, 212
– global 137, 153, 194
– land use 153
– price ..12, 356, 357, 359, 360, 362, 364, 375, 376, 384, 385, 400, 401, 418, 419, 423, 424, *426*
chaotic system 142, 165
chartism 386
chemotherapy (CTx) 303
Cheyne Stokes syndrome 228

chromosome 293
circulation
– ocean 144, 152, 155
– thermohaline (THC) 147, **193**, 194, 204, **204**, *209*
circulation type 125
climate classification
– effective 123
– genetic 123
climate means **113**
climate modeling 75
climate predictability ..143, **144**, 166
climate trend **129**
climate zone **113**, **122**, 123, 124, **131**
clogging 335, **340**, *341*, 342, 348
cloud effect 133
cloud physics 176
cloud-radiation interaction 75, 77
cognitive psychology **18**
complexity **3**, **16**
– algorithmic 7, 21, *22*
– approximative 7
– effective measure (EMC) 8
compositional patchiness **47**
computer tomography (CT) 289
conditional probability 362
confidence level 12, 14, 109, 111, *112*, 316, 319, 320, 323
congestive heart failure *221*, 222, 228, *229*, 231, *233*, 244, *245*, 246, *247*, 250
constraint
– cultural 331
– qualitative *200*
– social 331
continental drift 107
control
– cardiac 247
– nervous 267
– neuroautonomic 235, 236, *247*
– parasympathetic 228
– sympathetic 228
control system
– cardiovascular 268

– respiratory 268

convective feedback 205

conveyor belt 118, 155, 167, 204

coronary artery disease 231

correlation

– cross- 92, 135, 367, *369*

– higher-order 357

– long-range 4,
    10, 11, 46, 49, 50, 52, **53**, 174–178,
    180, 182, 183, 186, 226, 227, 229,
    229, 231, 232, 237, 248, 249, 260–
    262, 266, 274, **364**, 413, 430

– Pearson 111

– power-law (PLC) 53, 175, 177, 237,
    266

– short-range 262, 274, 276, *276*, **364**

– spurious 174

– temperature *112*

crash ... 380, 401, 402, 412, 418, 425,
    426, 428, 429, *429*, 430

– critical **427**

– financial 414, 429

– stock market 399, 429

critical point 200, **200**, 401, 418, 426

crowd disaster 332, *333*

crowd dynamics 332, 336

cryosphere 144

CT 303

CTx 303

cycle

– carbon 153

– hydrological 75

cyclogenesis 121

cyclolysis 121

cyclone density 121, *122*

cyclone path **118**, 121, 131

cytosine 47

D

Dansgaard-Oeschger event 207

Daphnia 311

DAX ... 382, *382, 383*, 384, 387, *388*,
    391, 392, *395, 404*

degree of freedom (DOF) .. 111, 113,
    122, 124, 125, *126*, 414

demand and supply ... 361, 397, 398,
    403, 412, **415**, 416, 418, 422, 432

dermatoscope 295

detection and attribution problem
    153, **159**

DFA 176, 178, *179*, 181, *182, 184*, 186,
    187, 188, 228, 230, 234, 237, 243,
    264–266, *275*, 279, 388, 391, 393

diagram

– bifurcation *311*, 319, 320, *321*

– state transition .199, 201, 202, *202*

differential game 197

diffusion limited aggregation 240

digraph 199

dilemma

– common goods 332

– prisoner's 332

dimension

– fractal 7, 21, 28,
    59, 121, 241, 246, *247, 248*, 260–262,
    278, *278*, 286, 287, 289, 296, 427

– generalized fractal 30

– Hausdorff 30, 31, 34, 66

distribution

– Lévy 91, 380

– bimodal 321

– Dirac 34, 37, 39

– frequency 288

– function 377

– Gaussian 52,
    56, 90, 259, 261, 266, 279, 358, *366*,
    367, 374, 378, 380, 413

– generalized Pareto (GDP) 379

– Gibbs 9

– hyperbolic 378

– leptokurtic 367

– Lévy stable 360, *366*, 377, 424

– multifractal 27, 31

– non-Gaussian 356

– normal 374, 382, 390

– Pareto 358, 422

– power-law .260, 263, 266, 280, 357, 358, 360, 420
– probability ..45, 116, 149, 164, 182
– return 422, 424, 426, 432
– unconditional 396
– volatility 396
distribution of returns 373, **374**, 377, 382, 397, 404
DNA 50, 52, 56, 57, 59, 178, 180, 289–292
DOF 123, 126, *126*
double pendulum 284
Dow Jones 12, *348*, *368*
drought *171*
dyadic grid 44, *86*
dyadic tree 39, 85
dynamical optimization 356
dynamics
– atmospheric 75, 83, 113
– cardiac 225, 228, 231, 234, 236, *238*, 239, 244, 249, 289
– collective **367**
– disaster 194
– heartbeat ..17, 219, **228**, 232, 237, 240, 243, 244, 246–248
– price .356, *359*, **363**, 364, 365, 367, 368
– symbolic 19, 21
– threshold 204, 401, 411, 432
Dzerdzeevskii catalogue 126

E
ECG ...229, 230, 268, *268*, **304**, 306, *306*, 308
economic loss 203
economic taxonomy 367
econophysics 355, **355**, 357, 368
eddy 106, 118
edge detection **60**
EEG 268, *268*, 269, 273, 304
efficient portfolio 367
eigenfrequency 151
eigenmode 151, 171

El Niño 142, *145*, 146, *146*, *147*, 156, 164, 165, 176
El Niño/Southern Oscillation (ENSO) 107, 145, 174
electrocardiogram (ECG) ..224, 268, 289, 304
EMC 21
EMG 268, *268*, 269, 273
EMH 362, 375, 376, 385, 393, 397
emission corridor *211*, 212
emission scenario *212*
encounter **313**, **314**, *314*, **316**
endosonography (EUS) 303
energy dissipation 286
ENSO .118, 125, 135, **144**, 145, 147, 149, 153, 166, 171
entropy **3**, **12**, 30, 286
– Boltzmann Gibbs Shannon 7
– conditional 3, 4, **4**, 10, *14*
– generalized 297
– Kolmogorov Sinai 3, 5, 13
– Renyi 17
– Shannon 7–9, 17, 300
– Shannon *n*-gram 5
Eo climate 124, *125*
EOF *110*, 111, 113, 114, *115*
EOG 269, 273
equation
– Chapman-Kolmogorov 358
– Langevin 411, 418, 419, 432
– master 402
– qualitative differential (QDE) 197, 199
escape panic 338
escape strategy *345*
Escherichia coli *27*, *54*, *55*, 57
estimator
– Hill 380, *382*
– likelihood 380, 383
– maximum likelihood 382, *383*
European climate **105**
European Monetary System (EMS) 393

exponent
- Hölder ...29, 31, *33*, 34–36, 39, 41, 46, 62, *64*, 66–68, 74, 79, 83
- Hurst 28, 29, 40, *51*, 68, *219*, 241, *242*, 243, 244, 246, 248, *250*, *252*, *253*, *260*, 261, 263, 264, 391, 393, 394
- Lyapunov 286, 320, *321*, 417
- roughness 59

F

Fano factor 262
faster-is-slower effect ..331, **340**, 342, 348
fat tail ...90, 373, 378, 397, 403, 404, 413, 414, 418, 424, 432
fBm 40, 41, *43*, 44, 51, 68, 69, 79
financial asset 356
fingerprint 184, 293
finite-size effect 29, 59, 414, 426
fire front 337, 341, *342*
fluctuation
- climate 142, 164
- heartbeat .225, 226, 234–237, **237**, 239, 247, 254, 274
- market 92, **373**, 421
- stationary 432
- weather 143, 166
flux correction 185, 186
food security *193*, 195
forcing
- anthropogenic .134, 160, 162, *163*, 164, 172, 174
- ENSO *135*
- external ..105, 129, **133**, 156, 159, 164, 174
- greenhouse 155
- radiative ..133–135, *136*, 155, 157, 158, 161
- solar 134, 161, 166
- stochastic .150, *150*, 152, *152*, 164, 166
- white noise 150

foreshock **428**
formatio reticularis 267
fractal branching 49
Framingham Heart Study 230
freezing-by-heating ...**338**, *339*, 340, 348
freshwater flux *207*
function
- autocorrelation 6, 46, 109, 220, 241, **262**, *285*, *305*, 364, 390, 392, 423, 424
- correlation ..11, 40, 58, 68, 85, 87, *88*, 177, 414, 427
- cost 196
- cross-correlation 428
- demand 197
- empirical orthogonal (EOF) ..109, 160, 173
- fractal 28, 29, 35, 68
- Gaussian *33*, 37, 49, 60, 244
- Haar *413*
- Heaviside 287
- multi-affine 29, 44, 241
- multifractal 29, 44, 73
- periodic 318
- probability density (PDF) .30, 51, *55*, 66, 90, *260*, **261**, **277**, 365, 420
- random 74
- recruitment 197
- self-affine 28
- self-similar 68
- space-scale correlation **84, 85**
- structure (SF) 31, **38**

G

GA 399
GARCH ...14, 90, 93, 365, 389, 390, 396, 430
Gaussian wavelets 180
GCM 206, *207*
generalized autoregressive conditional heteroscedasticity (GARCH) ..389
generalized extreme value (GEV) 378

genetic programming (GP) 399
GEV 378, 379
GHG 135, 136, *136*, *154*, *161*
global warming ..146, **153**, *160*, 167,
    172, 174, 187, 194, 208, *210*, 212
GP 400
GPD 379
greenhouse gas (GHG) 134, 142, 153,
    162, 172, 174, 176, 185
Grosswetterlage .111, **113**, **122**, **124**,
    127, 128, 131, **131**, *132*, 173, 182
Grosswetterlagen catalogue .125, 126
guanine 47
guardrail 212
Gulf Stream 167, *193*, 204

H
Haar wavelets 180
heart disease 277
heart failure .235, *235*, 236, 248, *248*
heartbeat 17, 264, *265*, 267, 304, 306,
    307
Heinrich event 208
higher-order statistical properties 356
histomorphology (HIS) 303
homeostasis 221, 227
homeostatic constancy 240
hypothalamic neuron .319, **322**, *324*
hypothesis
– Lévy distribution 404
– efficient market (EMH) .**361**, 362,
    374, 376, 384, 397, 403, 404, 412
– interacting agent 398, 403
– Lévy 381, 384
– martingale 385, 386
– null 367, 385, 386, 430
– stable distribution **376**
– volatility 386
hysteresis *207*, 208, *209*

I
i.i.d. ..86, *86*, 375, 376, 378, 384, 385
ice age 206
image

– Landsat ..77, *78*, 79, *80*, 81, *82*, 83
– satellite 27, 75, **75**, 83, 94
impatience **340**
impulse response approach 158
incoordination **340**
independent and identically dis-
    tributed (i.i.d.) 360, 367,
    374
information
– Hartley 297
– Kullback 8
– mutual 3–6, 10, 12, 46, 92, *93*, *305*,
    428
– Shannon 21, 297
interglacial 206
interval
– confidence *382*, 389, *413*
– heartbeat .**224**, 226, *232*, *239*, 243,
    249, 259, *270*, 271, 274
– interbeat ..*221*, 222, 225, *225*, 226,
    226–229, 231, *232*, 240, 248, *248*,
    250, *265*, 267, 271, 274, 280
– interspike 316
– RR 17, 304
investment decision 197
isotropic dilation 62, 68

J
jamming *343*
jetstream 118

K
Köppen scheme 124
knowledge
– vague 194
kurtosis 422

L
La Nina *145*
Lamb circulation pattern 132
landmark 198
Landsat **76**, 77, **82**, 84
lane formation **338**, *339*
latent energy 106

lattice anisotropy 74
law
– central limit 377
– Fick's diffusion 178
– Lévy 381, 413
– scaling 373, 374, 381, 402, 421
– tail 381
– universal persistence 175
leaving time *341*
Leonardo da Vinci *219*
leptokurtic 390, 396
leptokurtosis 365, 378, 400
likelihood domain *212*
linear damped oscillator 150
linear regression 69, *80*
linear response relation **156**
liquid water content (LWC) 75
liquid water path (LWP) 76
local order 3
log-periodicity ... 426, **428**, **430**, 432
logistic map 8, **319**, *321*, 323

M
malaria 146, *147*
malignant lymphoma *302*
malignant melanoma . 294, **294**, *294*,
    295, *296*
management 195, 376
marine fisheries **193**
market participant 195
market price .197, 402, 412, 417–419,
    421, 429
market risk 94
mass psychology 332, 348
maximum likelihood technique 40
mean sea level (MSL) 130, *130*
mean-field approximation 421
measure
– entropy 289
– multifractal 44
MESA 109
methane 156
method

– box-counting 31
– scaling index (SIM) **286**, 287, 295,
    296, *296*, 297, 300
– scaling vector 301
Mexican hat *27, 55*, 60, *61*
Milankovitch cycles 166, 206
Mises representation 379
mitigation policy 156
Modane wind tunnel 83
model
– ARCH 395
– asset pricing 395
– atmosphere ocean general circula-
    tion (AOGCM) 134, 144,
    176
– behavioral 398
– bioeconomic 200, *200*
– Black and Scholes 356, 359
– box *207*
– carbon cycle 158, *158*
– cascade 44, 91, 411
– Clark 378
– CLIMBER-2 .... *210*
– Cont Bouchaud . 421, 422, 425, 426
– deterministic dealer 419
– Einstein Uhlenbeck 151
– GARCH 389
– general circulation (GCM) 77, 175,
    207
– generalized force **336**
– herding 345
– hybrid 148
– impulse-response 158
– intermediate complexity 175
– Ising 401, 412, 420
– Kolmogorov log-normal 42
– Levy Levy Solomon 421
– Lévy 377
– market 424
– microscopic **399**
– Monte Carlo 399
– multi-agent 398
– multifractal cascade 44

– multiple regression (MRM) ...135, *135, 136*
– multiplicative cascade 90, 91
– neural network 135, *135*
– pedestrian 347
– percolation ... **411**, 412, 414, **421**, 423, 428
– qualitative **196**, *198*, 200
– Santa Fe stock market 400
– self-propelled particle *346*
– stochastic cascade 396
– stochastic climate **147**
– stochastic forcing 152
model drift 185
moment 262
– first 116, 129, 271, 387
– fourth *383*, 384
– higher 149, 377
– second 116, 131, 261, 271, 356, 360, 365, 377, 387
– third 250, *383*, 384
monofractal **39**, *42, 70*, 181, 222, 241, *242*, 244, 246, 248, 394
monofractal scaling 50, 52
monofractality 51, 243
motion
– Brownian .40, 68, 91, 94, 150, 151, 358, 418
– fractional Brownian (fBm) .39, *42, 233, 242*, 244, 261, 390, 394
– geometric Brownian 360
MRM 135, 136, *136*
MSL 130
multicriteria evolution 203
multifractal *42, 70*, 76, 181, 222, 246, 260, 394
multifractal measures 32
multifractality 74, 184, *219*, 240, **241**, **244**, 245, 246, **247**, 248, 250, *250*, 251, **393**, 394, 395
multiple equilibrium 205
multivariate statistics 160
musical string 12

myocardial infarction 230

N

NADW 172, 204, *210*
NAO ...114, *115*, 120, 124, *125*, 127, 130
New York Stock Exchange Composite Index (NYCI) 384
Nikkei 424, 430, 431, *431*
NMR 303
noise ... 148, 155, 159, 160, 166, 222, 234, *292*, 312, 315, 318, 338, *339*
– 1/f .22, 40, 228, 241, 254, 274, 393
– background 290
– Brownian 228, 234
– dynamical 315, 319, 326
– multiplicative ..414, 418, **418**, 420
– neuronal 319
– pseudo-random 41
– random 234
– red 107, 109, 174
– white 69, *87, 93*, 107, 109, 150, 174, 226
nonequilibrium phenomena 221
nonstationarity ...114, 174, 176, 180, 220, 225–227, 234, **241**, 384
nonstationary 7, 14, 311, 365
North Atlantic deep water formation (NADW) 172, *205*
North Atlantic Drift 204
nuclear magnetic resonance (NMR) 300
nucleotide *48*, 49, 50, 53
NYCI *382, 395*

O

oceanic heat transport 204
open market **415**, 419
optimization paradigm 196
optimization strategy 196
orbit
– quasiperiodic 315
– stable periodic (SPO) ..**311, 312**, *313*, **324**

– unstable periodic (UPO) **311, 312**, *313*, **319**

oscillation

– North-Atlantic (NAO) ...109, 114, 129, 132, 142, 146, 165, 166, 174
– quasi-biennial (QBO) 107, 109
– Southern 145

overcapitalization 200, 203

overexploitation *193*

overturning 206

oxygen saturation 268, 271

ozone 153

P

paddlefish *311*

palaeoclimatology 106

panic situation 339

panic stampede 332

panicking crowds 337, *343*

panicking pedestrians 347

parasympathetic activity 272

pattern recognition **290**

PC 114

PDF 39, 67, 69, *72*, 74, 81, 82, *82*, 90, 266, **271**, *271*, 274, *278*, 279, 365, 367, 427

– power-law 420

pedestrian crowds *339*

pedestrian flow 332

pedestrian motion **336**

percolation cluster 422, 423, 426

persistence 263, 264, 391, 413

– long-range 182
– short-term 173
– volatility 413
– weather 171, *171*, 173

persistence characteristics 173

phantom panic **342**

pharynx collapse 270

Pinatubo eruption 134, *136*

plasmid **291**, *292*

Plasmodium falciparum *147*

Plasmodium vivax *147*

PLC 56–58

Pleistocene Holocene transition ..129

plethysmography 268

Poincaré section ...94, 312, 313, *314*

Poissonian jumps 365

Polyodon spathula *311*

portfolio 89, 398

portfolio selection 356

precautionary management 195

precautionary policy **211**

prediction limit 142

prediction problem 142

price elasticity 197, 419

price return 414

price variation 412

principal components (PC) . 114, 173

probability of loss 416

process

– additive **88**
– ARMA 392
– Fibonacci multiplicative 95
– Gaussian 40, 360
– hierarchical cascade 428
– Lévy stable 360, 365
– Lévy 381
– macrophysical 143
– Markov 4
– microphysical 143
– multifractal 37
– multifractal cascade 27
– multiplicative .45, **88**, 90, 420, 427
– multiplicative cascade 31
– random 356, 357, 363
– self-affine 68
– self-similar 41
– stochastic 28, 356, 359, 365
– Wiener 358

production cost 415, 416

profit

– discounted 196
– total 415

purine 47

pyrimidine 47

Q

qualitative phase plot *201*
qualitative trajectory 198, 199
qualitative variable *200*
quantity space 198, *200*

R

r.m.s. . . . *27*, 29, *54*, 90, *119*, *413*, 426
R/S 178, 263, 265, 391–393
radiance field 83
radiance image 81
radiative transfer 76, 77, 176
receiver operating characteristic .297, *299*
receptor
– catfish electro- .317, 319, **320**, *322*
– paddlefish electro- **324**
– rat facial cold ..*314*, 317, **322**, *323*
regime
– circulation 125–127
– climate 111, 124, 172
– general weather 173
– Lévy 365
– meridional 127
– meso-scale 108
– totalitarian 346
regulation
– heart rate 267, **267**, 276
– heartbeat 274
– neuroautonomic 220
REM sleep ..237, 269, *270*, 271, 272, *274*, *275*, 276, *276*
– non- 271, 274
resource exploitation **195**
respiratory rhythm 273
return 414, 423, 427
Reynolds number 75, 82
Richardson experiment 121
risk assessment 286
risk estimation 360
risk management 355
RNA 57
root-mean square (r.m.s.) 28, 152

S

Rossby deformation radius 119
Rossby wave 111
rough surface **59**, **62**, **68**, *70, 71*

S

S&P 500 *87, 88*, 91–93, 381, 392, 394, 412, 413, *413*, *429*
Saccharomyces cerevisiae . *27*, 53, *54*, *55*
safety standard 212
Sc 77, *78*, 79, *80*, 82, 83
scale invariance . . . 44, 46, 53, 59, 62, 107, *108*, 226, 237
seesaw
– bipolar 208
– Iceland low and Azores high . . 120
self-affine fractal 59, 62
self-driven many-particle system 336, 349
self-organization 339
self-organized criticality 418
self-similarity . . .52, 53, 237, 392, 429
sensory biology **311**
sequence
– DNA 10, 11, 27, 29, 46, **46**, 47, 51, **53**, *54*, 227, 291, 293
– genomic 47
– human DNA **50**
– nucleotide 28
– protein 11
SF 39
shift of climate zones *105*, **132**
signal-to-noise ratio 160, 167, 172
significance level 160, 161, *161*
SIM 301
singularity .29, 32, *33*, **34**, 35, 62, 63, *64*, **127**, 128, *129*, *224*, 241, 244
skewness 377, 378, 422
skin cancer 283, 286, 289, 295
sleep apnea ..239, *239*, 270, 271, *271*, 272, *273*, *275*, *276*
– obstructive *221*, 269, 276
sleep physiology **268**

social contagion 332

solar electromagnetic radiation ... 20

solar radio emission 21

spectrum
– multifractal . 37, 43, *71*, 78, 79, *79*, 83, 244, *247*, 249, 251
– power ... 11, 29, 41, 107, 121, 223, *225, 226*, 243, 247, 248, 273, *285*
– singularity ... 27, 30, 31, 34–36, 39, 46, 66, 67, 74, 79, 80, *80*, 83, 240
– variance 149, 151

spike
– neural 318, 324
– neuronal 312
– solar **20**

SPO 313, 320, 322

stability limit *209*

statistic
– Eulerian 118
– Gaussian 14, 316

statistical theory of extremes ... **378**

stimulation
– parasympathetic 224, 236, 267
– sympathetic 224, 236, 267

stock exchange 364, 365, 381

stock market 13, 14, 27, **84**, **89**, 92, 360, 411, 412, 414, 416, 423, **427**, 428, 429, *431*, 432

stock price .. 356, 360, 364, 365, 367, 368

storm track **113**, 114, 118, **118**, *119*, 121, **131**

stratocumulus (Sc) 76, **76, 77**

stylized fact 365, 373, 395, 396, 398–400, 404, **411**, 414, 423

subsidies 203

sudden cardiac death .. **17**, *221*, 228, 229, 289, *307*, 308

sunspot numbers (SRN) 134

sustainability science 194

sympathetic activity 272

synchrogram 318, *326*

synchronization *327*

– noisy **317, 325**

T

T4 bacteriophage *54*, 57

tail index 380–382, *383*

technique
– box-counting ..27, 35, 37, 240, 243
– Wiener Hopf 420

tele-medicine 295

temperature fluctuations 180

temperature variation 109

test
– Brock Dechert Scheinkman (BDS) 386
– detection and attribution 160
– Student's t- *232*, 235, 390

THC 205, 206, **206**, 208

theorem of Mangasarian 197

theory
– Black and Scholes option pricing 89, 354, 359
– chaos 141
– information 286
– kinetic gas 143
– percolation 421
– random graph 422
– spin-glass 37
– wavelet 243

Thompson rule 132

threshold
– critical 194, 209
– percolation 401, 421, 422, 424, *425*, 427

thymine 47

time
– interspike *322–324*
– recurrence 117
– residence ..106, 127, 131, *133*, 134, 137
– sampling 117

tomographic image 300

topological invariance 305

trade and quote 354

transform
- Fourier ..11, 69, 109, 248, 280, 377
- Legendre 30, 36, 39, 41, 45, 66, 69, 79, 246, *247*
- Mellin 420
- wavelet (WT) 31, 32, **32**, 239, 243, 244, 249, *249*, 413
- wavelet 2D 61
transinformation 5, 6, 286, 300
transversality condition 375, 376
trend elimination 178
tumour diagnostic 283, **300**
turbulence 30, 42, 44, 46, 82, **82**, 83, 90, 91, 94, 95, 240, 396, 397, 414, 427

U
uncertain demand **415**
uncertainty 5, 10, 13, *13*, *15*, 22, 75, 81, 155, 166, 167, 195, 197, *198*, 199, 204
UPO ...312, 313, *314*, 315, **316**, 319, **319**, 320, 322, 323
urban warming 183

V
variability
- atmospheric 175
- climate 105–107, *108*, 111, 114, 131, 142, 146, 149, 150, *150*, 153, 155, 159, 160, 164, 166, 172
- ENSO 132
- heart rate ..17, 224, 230, 231, 272, *273*, 274, 277, 283
- interbeat 222
- natural 164, 174
- spatial climate *115*
- weather 149, 150, *150*, 164, 172
ventricular contractions **277**
ventricular fibrillation *221*, 277
ventricular tachyarrhythmia 230
ventricular tachycardia **277**
virus

- Adeno 57
- Epstein Barr *54*, 57
- Hepadna 57
- Herpes 57
- HIV2BEN 10
- Melanoplus sanguinipes *54*, 57
- Papova 57
- retro- 57
viscous fingering 240
volatility *88*, 89–91, 365, 387, 389, 392, 395, 401, 412, 413, *413*, 414, 422–424, 427, 428
volatility clustering ...373, 389, 396, 397, 400, 404, 414, 423, 426, 432
volcanism 134

W
walk
- DNA *27*, 29, 35, 46, 47, *48*, 49–51, 53, *54*, *55*, 56–58, 94
- Lévy 423
- random 28, 40, 47, 50, 177, 234, 248, 357, 358, *359*, 362, 367, 384–386, 411–413, 418, 420
wavelet techniques (WT) ..174, **179**, 222
wavelet transform modulus maxima (WTMM) ..27, **32**, **34**, 35, **35**, **59**, **62**, 181, 239, 244
wavelet-based multifractal formalism **35**
weak prognosis 194
wealth maximization 402
weather **142**
weather forecasting 141, 142, 144
WT 34, 36, 38, 39, *48*, 49, 57, 58, 63, 66, 69, *70*, 74, 79, 81, 84, 85, 176
WTMM 36, 38, 39, 41, *43*, 44, 45, 47, 51, *51*, 52, 57, 60, 63, *64*, 66, 74, 84
WTMM maxima (WTMMM) 63
WTMMM .63, *64*, 66, 69, *70*, *72*, 78, 79, 81, *82*